TABLE OF CONTENTS

Sr. No.	Title	Page No.
1	**Advanced Chemical Concentration Control for Fabrication of Devices Using SiC** J. Boecker, I. Kashkoush, and D. Waugh **Paper ID: P201-201907301401**	1-3
2	**The Synthesis of Large-size Silicon Carbide Powder for Crystal Growth** Fang Jiao, Dianpeng Cui, Mulong Yang, Zhouli Wu and Boyu Dong **Paper ID: P201-201909091152**	4-6
3	**Investigation of Defect Levels of Al/Ti 4H-SiC Schottky structures by Deep Level Transient Spectroscopy** Yawei He, Guoguo Yan, Xingfang Liu, Zhanwei Shen, Wanshun Zhao, Lei Wang, Feng Zhang, Guosheng Sun, Yiping Zeng **Paper ID: P201-201909171928**	7-10
4	**Polytype Transformation in 4H-SiC single crystals grown on on-axis Seeds** Xianglong Yang, Yan Peng, Xiufang Chen, Xuejian Xie, Jinying Yu, Xiaobo Hu, Xiangang Xu **Paper ID: P201-201909270936**	11-13
5	**Application of In-situ Pre-epi Clean Process for Next Generation Semiconductor Devices** Darian Waugh, Gim Chen, Jennifer Boecker and Ismail Kashkoush **Paper ID: P202-201907301402**	14-17
6	**An Improved Composite JTE Termination Technique for Ultrahigh Voltage 4H-SiC Power Devices** Rui Hu, Xiaochuan Deng, XiaoJie Xu, Xuan Li, Juntao Li , Zhiqiang Li, Yourun Zhang, and Bo Zhang **Paper ID: P202-201909140022**	18-21
7	**Electro-thermal Analysis of 1.2kV-100A SiC JBS Diodes Under Current Overload** Wei Zhong, Yidan Tang, Chengzhan Li, Hong Chen, Yourun Zhang, Yun Bai, Xinyu Liu **Paper ID: P202-201909141826**	22-25
8	**Simulation Study on Current Collapse Effect of E GaN HEMT** Guo Weiling, Ma Qijing, Du Shuai, Lin Tianyu, Zhu Yanxu **Paper ID: P202-201909151329**	26-28
9	**A New SiC Split-gate MOSFET Structure with Protruded P-base and the Mesa above JFET for Improving HF-FOM** Kunlin Li, Yourun Zhang, Wei Zhong, Xiaochuan Deng, Xiao Yang, Hang Chen, Bo Zhang **Paper ID: P202-201909151518**	29-32
10	**6.5kV Silicon Carbide Discontinuous Trenched Junction Barrier Schottky Diode** Zhiyu Chen, Xuan Li, Xiaochuan Deng, Juntao Li, Zhiqiang Li, Yourun Zhang, and Bo Zhang **Paper ID: P202-201909151530**	33-36
11	**Design and Characteristics of an Etching Field Limiting Ring for 10kV SiC Power Device** Yi Wen, Xiaojie Xu, Hao Zhu, Xuan Li, Xiaochuan Deng, Fei Yang, Juntao Li and Bo Zhang **Paper ID: P202-201909211821**	37-41

Sr. No.	Title	Page No.
12	**Improvement of Clamped Inductive Turn-Off Ruggedness of Trench IGBT at Overcurrent Condition with Optimized Split Gate Structure** Jiang Lu, Jiawei Liu, Xiaoli Tian, Hong Chen, Fei Liang, Yun Bai **Paper ID: P202-201909230956**	42-45
13	**Optimized Design of 4H-SiC VDMOSFET for Low ON-resistance** Defu Yin, Zhiming Wu, Xian Zou, Yongqiang Sun, Yaping Wu, Weiping Wang, Xu Li, and Junyong Kang **Paper ID: P202-201909301547**	46-49
14	**Design and Fabrication of 3300V 100mΩ 4H-SiC MOSFET with Stepped p-body Structure** Weijiang Ni, Xiaoliang Wang, Miaoling Xu, Mingshan Li, ChunFeng, Holger Schlichting, Tobias Erlbacher **Paper ID: P202-201911111204**	50-53
15	**Study on Preparation and Application of Nano-copper Powder for Power Semiconductor Device Packaging** Xu Pan, Jiacheng Zhou, Jingguo Zhang, Zhaohui Zhao, Minghui Liang, Huaiyu Ye, Qiang Hu, Huijun He, Limin Wang, Ligen Wang, Fengcai Qi, Youzhi Zhou **Paper ID: P202-201911221359**	54-58
16	**A Highly Integrated Multi-parameters RF Transceiver Module for Microwave Semiconductor Chip Testing** Guangshan Zhang, Miao Song, Rongbin Guo, Yahai Wang, Lei Liu,Jie Yang, Shichao Liu, Yisheng Yang **Paper ID: P203-201907301407**	59-62
17	**High Precision Model by Error Compensation Method based on the Angelov Model** Ziyue Zhao, Yang Lu, Hengshuang Zhang, Chupeng Yi, Yuchen Wang, Xiaohua Ma and Yue Hao **Paper ID: P203-201909122336**	63-66
18	**A Novel Scalable Series MIM Capacitor Model for MMIC Applications** Chupeng Yi, Yang Lu, Hengshuang Zhang, Ziyue Zhao, Xiaohua Ma, and Yue Hao **Paper ID: P203-201909151126**	67-70
19	**A Compact X-band Pallet Power Amplifier Using GaN MMIC and Discrete FETs with HMIC Technology** Wang Yi, Ni Tao, Yin Jun, Mo Jianghui, Yu Ruoqi, Li Jing, Dong Shiliang, Liu Ze, Chen Lei, He Jian, Huang Luoguang **Paper ID: P203-201910080737**	71-73
20	**Facet Formation of AlGaN/AlN-based Multiple Quantum Wells by Laser Scribing** Bin Xue, Jianchang Yan, Yanan Guo, Chunyan Liu, Yiping Zeng, Junxi Wang and Jinmin Li **Paper ID: P204-201909281853**	74-76
21	**Studies on primary lens for LED light source to enhance lateral emission intensity** Wenting Tang, Baojin Chen, Rui Zhang, Baoxing Wang, Yunfei Sun, Jiahua Min, Shuqi Li, Yong Cai **Paper ID: P205-201907020925**	77-80

2019 16th China International Forum on Solid State Lightning & 2019 International Forum on Wide Bandgap Semiconductors China (SSLChina: IFWS 2019)

Shenzhen, China
25 – 27 November 2019

IEEE Catalog Number: CFP19J64-POD
ISBN: 978-1-7281-5757-3

**Copyright © 2019 by the Institute of Electrical and Electronics Engineers, Inc.
All Rights Reserved**

Copyright and Reprint Permissions: Abstracting is permitted with credit to the source. Libraries are permitted to photocopy beyond the limit of U.S. copyright law for private use of patrons those articles in this volume that carry a code at the bottom of the first page, provided the per-copy fee indicated in the code is paid through Copyright Clearance Center, 222 Rosewood Drive, Danvers, MA 01923.

For other copying, reprint or republication permission, write to IEEE Copyrights Manager, IEEE Service Center, 445 Hoes Lane, Piscataway, NJ 08854. All rights reserved.

*** This is a print representation of what appears in the IEEE Digital Library. Some format issues inherent in the e-media version may also appear in this print version.

IEEE Catalog Number:	CFP19J64-POD
ISBN (Print-On-Demand):	978-1-7281-5757-3
ISBN (Online):	978-1-7281-5756-6

Additional Copies of This Publication Are Available From:

Curran Associates, Inc
57 Morehouse Lane
Red Hook, NY 12571 USA
Phone: (845) 758-0400
Fax: (845) 758-2633
E-mail: curran@proceedings.com
Web: www.proceedings.com

Sr. No.	Title	Page No.
22	**High color rendering index white LEDs fabricated using InP/ZnS green-emitting quantum dots and InP/ZnSe/ZnS red-emitting quantum dots** Doudou Zhang, Yuxian Yan, Fan Cao, Gongli Lin, Xuyong Yang, Wanwan Li, Luqiao Yin, Jianhua Zhang **Paper ID: P205-201909121659**	81-84
23	**Violet Chip Excited White LEDs for Sun-Like Lighting and Horticulture Lighting** Yue Zhuo, Hongyuan Zhu, Chongyu Shen, Guoxi Sun, Jay Guoxu Liu **Paper ID: P205-201909141235**	85-89
24	**High-efficiency GaN-based LED with patterned SiO₂ passivation layer and discontinuous current block layer** Jie Deng, Weiling Guo, Jianpeng Tai, Zehua Lu, Mengmei Li **Paper ID: P205-201909151348**	90-92
25	**Thermal Simulations of a UV LED module with nanosilver sintered die attach process on graphene-coated copper substrates** Pan Liu, Yong Li, Xiaobin Jian, Chen Jing, Min Li, Shurong Ding, Guoqi Zhang **Paper ID: P205-201909152037**	93-97
26	**The contrast ratio improvement of perovskite nanocrystals LEDs devices based on carbon nanotubes** Caiman Yan, Hanguang Lu, Zongtao Li, Jiexin Li, Jiasheng Li, Yong Tang, Binhai Yu **Paper ID: P205-201909171425**	98-102
27	**Preparation of Translucent Al₂O₃ Ceramic Substrates for LED Filament Bulb** Yizheng Zhang, Ling Gao, Fengpo Yuan, Yaguang Wu, Dongsheng Wang, Caihua Ren, Hongbo Bai **Paper ID: P205-201910121051**	103-106
28	**Study on high power density light-emting diodes light source** Huaiwen Zheng, Zhonghua Deng, Zhuguang Liu, Fei Yu, Yan Li, Yuming Yang, Qiao Liang, Hua Yang, Xiaoyan Yi, Junxi Wang, Jinmin Li **Paper ID: P205-201910281116**	107-110
29	**Review of High Power Phosphor-Converted Light-Emitting Diodes** Yan Li, Yuming Yang, Huaiwen Zheng, Fei Yu, Qiao Liang, Hua Yang, Xiaoyan Yi, Junxi Wang, Jinmin Li **Paper ID: P205-201910301118**	111-115
30	**Effects of Different Ratios of Red and Blue Light on the Morphology and Photosynthetic Characteristics of Anoectochilus roxburghii** Rui Li, Yinghui Mu, Hongyu Wei, Lixue Zhu, Wenqi Tang, Zhiyu MA **Paper ID: P206-201909152318**	116-120
31	**Effect of Different LED Light Sources on Growth and Development of Cherry Radish** Zhipeng Wen, Yumei Zhou, Shaoming Luo, Hongyu Wei, Xiaomin Li, Jiawei Liu, Zhiyu Ma* **Paper ID: P206-201909152325**	121-124
32	**Calculation Method for Chicken-Perceived Light Intensity** Zhichao Li, Xiaocui Wang, Baoming Li, Weichao Zheng, Zhengxiang Shi, Qin Tong **Paper ID: P206-201909252231**	125-129

Sr. No.	Title	Page No.
33	**Effects of LED Light Color and Intensity on Feather Pecking and Fear Responses of Layer Breeders in Natural Mating Colony Cages** Haipeng Shi, Baoming Li, Qin Tong , Weichao Zheng, Dan Zeng and Guobin Feng **Paper ID: P206-201909282235**	130-138
34	**Effects of Illumination and Color Temperature Distribution on Subjective Perception** Dandan Hou, Congshan Dai, Yan Lu, Yandan Lin **Paper ID: P207-201906302251**	139-143
35	**Research on train light environment evaluation method** Sijie He, Jinrong Liu, Shuo Jing, Yandan Lin **Paper ID: P207-201906302252**	144-148
36	**Effect of Illuminance and Light Strobe on Attention and Visual Fatigue in Indoor Lighting** Jin Yang, Tianchi Zhang, Yandan Lin, Wei Xu **Paper ID: P207-201906302253**	149-152
37	**Study of the Stroboscopic Effect Visibility Measure (SVM) based on Cognitive Performance** Xiaojie Zhao, Mengxin Li, Yandan Lin, Wei Xu **Paper ID: P207-201906302254**	153-156
38	**A summary to the personal consideration about the light's influence on myopia, procreation and even philosophy & society** Wenqing Fang, Chaopu Yang, Kaiqi Fang, Han Jin, Xu Zhang, Xiaojian Han, Chaolin Ma, Yuehui Zheng, Fan Yang, Jiang Fu, Tuanqing Fang, Guoqing Fang, Taiyang Chen, Shanxiao Huang, Youming Zhang, Zhenquan Lai **Paper ID: P207-201907271546**	157-162
39	**Influences of Blue Component in White Light on Visual Discomfort** Yin Zhang, Yan Tu, Lili Wang **Paper ID: P207-201907301427**	163-166
40	**Infer light diffuseness on light probes with different kinds of mesoreliefs** Yudi Wang, Ling Xia, Jinfeng Huang, Ruipeng Xu, Xiaofeng Liu **Paper ID: P207-201908031348**	167-170
41	**The Effect of Lighting on Stereo Vision Test with Random-Dot Stereogram** Jinfeng Huang, Tingting Zhang, Yudi Wang, Jinwei Xie, Xiaofeng Liu **Paper ID: P207-201908032049**	171-174
42	**Design of a multi - wavelength high irradiance LED phototherapy system for LLLT** Weimin Li, Zhiliang Jin,Jialin Liu, Liquan Guo,Haiyang Wang, Daxi Xiong **Paper ID: P207-201909051308 (Oral)**	175-179
43	**Multi-chip dynamic white light emitting diode with high level photobiological safety and good color fidelity** J.X. Nie, Z.Z. Chen, F. Jiao, C.C. Li, Y.F. Chen, J.L. Zhan, Y.Y. Chen, T.J. Yu, X.N. Kang, Y.Z. Wang, S.F. Li, G.Y. Zhang and B. Shen **Paper ID: P207-201909121908**	180-184

Sr. No.	Title	Page No.
44	**The Influence of Lighting on Human Circadian Rhythms** Hung-Wei Chen, Chien-Yu Chen, Pei-Jung Wu **Paper ID: P207-201909150203**	185-188
45	**The effect of Emel and CCT of Dynamic Lighting on Alertness, Fatigue and Circadian Responses** Yu Liu, Ming Ronnier Luo, Peijung Wu, Binyu Teng **Paper ID: P207-201909201744**	189-191
46	**Applying LEDs as Therapeutic Light Sources for Anti-microbial Treatment: An Experimental Study** Tianfeng Wang, Jianfei Dong and Guoqi Zhang **Paper ID: P207-201909271407**	192-195
47	**Influence of dynamic-brightness environment on ocular physiological parameters** Shanshan Zeng, Wentao Hao, Ya Guo, Xiangyu Qu, Ke Wei, Shanshan Tang, Rongrong Wen, Jianqi Cai **Paper ID: P207-201910081046**	196-200
48	**Study on 3D thermal transport in micro-LEDs on GaN substrate at the level of kW/cm^2** Zhizhong Chen, Chengcheng L, Fei Jiao, Jinglin Zhan, Yifan Chen, Yiyong Chen, Jingxin Nie, Tongyang Zhao, Xiangning Kang, Shiwei Feng, Guoyi Zhang, Bo Shen **Paper ID: P208-201908281659**	201-205
49	**Influence of the Charge Transfer on the Lifetime of Quantum-Dot Light-Emitting Diodes** Yue Liang, Chongyu Shen, Junfei Chen, Weiye Zheng, Zheng Xu, Jay Guoxu Liu **Paper ID: P208-201909152328**	206-209
50	**Effect of Mechanical Stress on the Electrical Characteristics of Different Type IGBT Chips** Yihui Zhang, Jinyuan Li, Yinghan Liu, Guanbin Wu **Paper ID: P209-201909171702**	210-213
51	**Optimal thermal design of LED automotive headlamp with the response surface method** Zhibin Tang, Jiajie Fan, Wei Chen, Yutong Li, Moumouni Guero Mohamed, Ru Li **Paper ID: P210-201907101533**	214-219
52	**Optical and Thermal Designs of LED Matrix Module used in Automotive Headlamps** Wei Chen, Jiajie Fan, Gaojin Qi, Chengzhong Sun, Weiqiao Yang **Paper ID: P210-201907290929**	220-224
53	**Comparison of Life Testing Standards for LED Lighting Products** Zhu Zhike, Cao Suming, Cai Shasha, Shi Tingting, Lian Yuanhui **Paper ID: P210-201907301518**	225-230
54	**A Comparative Study of the Lifetimes of High-End and Low-Cost Off-Line LED Drivers Under Accelerated Test Conditions** F. Keil, K. Hofmann **Paper ID: P210-201907310128 (Oral)**	231-234

Sr. No.	Title	Page No.
55	**Corrosion Failure Analysis and Coping Strategies of Light Reflecting Devices in Light-Emitting Diode Devices** Lyu Tiangang, Wang Yuefei, Lyu Henan, Wang Caixia, Chen Lihe, Xu Bingjian, Tang Leming, Ren Rongbin **Paper ID: P210-201909051759**	235-244
56	**Failure Analysis of Glass Transition Temperature of LED Insulation Layer** Yibin Wang, Fang Fang, Jing Wu, Kaixuan Lin, Tingting Xu, Weiqing Liang, Luqiao Yin **Paper ID: P210-201909121056**	245-248
57	**Intelligent Control Semiconductor Laser Reliability Test System** Xiaoling HU, Wensha LAN **Paper ID: P210-201909150809**	249-251
58	**Junction Temperature Prediction of the Multi-LED Module with the Modified Thermal Resistance Matrix** Fanny Zhao, Brian Shieh, Fangyun Zeng, Guoming Yang, S. W. Ricky Lee **Paper ID: P210-201909151656**	252-255
59	**Smart Lighting with Autonomous Color Tunability** Tianhang Zheng, Wujun, Zhixian Zhou, Wanghui Yan **Paper ID: P301-201909101118**	256-259
60	**Design of intelligent temperature control driving circuit for high power LED array** Fei Wang, Houda Zhou, Jingjing Liu, Luqiao Yin, Jianhua Zhang **Paper ID: P301-201909111644**	260-263
61	**Research on a Smart LED Lighting Based on Improved Flyback Driver** Wenran.Liu, Weiming.Lin **Paper ID: P301-201909151502**	264-268
62	**Analysis of Smart Wall Switch without Neutral Wire Compatibility Issue** Yang Hu, Wei Wen, Wanghui Yan and Ran Ding **Paper ID: P301-201909151630**	269-272
63	**A Wavelength Stabilized GaN based Laser Utilizing Distributed Bragg Reflector** Mingle Liao, Wuze Xie, Zejia Deng and Junze Li **Paper ID: P301-201909151659 (Oral)**	273-276
64	**Perovskite liquid quantum dots as a color converter for LD-based white lighting system for visible light communication** Shunming Liang, Zhou Lu, Xinrui Ding[*], Jiexin Li, Yong Tang, Zongtao Li Binhai Yu **Paper ID: P301-201909162207**	277-279

PROCEEDINGS
SSLCHINA & IFWS 2019

16th China International Forum on Solid State Lighting
2019 International Forum on Wide Bandgap Semiconductors China

November 25-27, 2019 Shenzhen, China

2019 16th China International Forum on Solid State Lighting &
2019 International Forum on Wide Bandgap Semiconductors China (SSLChina: IFWS)

25-27 November 2019
Shenzhen, Guangdong, China

Advanced Chemical Concentration Control for Fabrication of Devices Using SiC

J. Boecker, I. Kashkoush, and D. Waugh
NAURA Akrion Inc.
6330 Hedgewood Dr Suite #150, Allentown, PA 18106
ikashkoush@naura-akrion.com

Abstract

In conventional MEMS fabrication, relatively inert compounds such as Si_3N_4 are used as an etch stop or mask for creating patterns on wafers. However, materials such as this require insight as to their etch selectivity corresponding to that of the substrate material and are not suitable for high temperature devices. When developing these high temperature compatible devices, components composed of SiC are desired, and may be used as an etch stop due to it being chemically inert. For these applications, it is common for the substrate or sacrificial layer to be either Si or SiO_2. A technique for advanced chemical concentration control during processing is critical to be able to maintain a consistent etch rate, a controlled etch depth, and maintain the desired shape of the pattern. Using NIR technology it is possible to monitor both the concentration of chemicals in the bath as well as that of byproducts created from the etching of Si and SiO_2. The system can then increase bath life and the ability to etch consistently within and across batches. In the present paper, we present the mechanism of the advanced concentration control, the results of using either TMAH or KOH to etch Si, as well as its applications for the future of SiC integrated devices.

Introduction

For the last 20 years, SiC has been investigated as a replacement for traditional etch stops in bulk micromachining due to its properties of being chemically inert in conventional etching solutions.[1] It is also a desired material for high temperature devices due to its thermal properties. Applications for creating such devices as fuel atomizers, pressure sensors, and microfabricated molds can use typical wet etching techniques and take advantage of the properties of the SiC layer which make it chemically resistant.[1] Particularly for SiC-MEMS devices, large area substrates are essential. Due to the difficulty in manufacturing single growth crystal substrates, there has been much interest in epitaxial growth of single and poly-crystalline SiC layers on silicon. After the deposition, bulk etching is able to create microstructures and patterns suitable for the desired devices.

With bulk etching, one major concern is that the byproducts are released from the wafer into the bath. Depending on the open surface area of the patterned wafer, large amounts of silicates can be introduced. This can then create unwanted side effects such as decreased etch rate. By creating stable chemical conditions inside of the bath, it not only allows for consistent etch depth within and between lots, but it also allows for increased bath life which will lessen the cost of ownership for the process.

Naura-Akrion has developed a solution to the problem of consistent chemical control. The novel system tracks the concentration bands allowed within the process using NIR technology. When the concentration breaches those bands, a small volume will drain from the tank and the same volume of fresh chemical will be added to maintain the purity and concentration in the bath. These parameters are completely customizable to each process and situation, allowing for a wide variety of possible usage.

Experimental

Wet chemical processes were conducted on a fully-automated GAMA™ wet processing station using 200mm wafers with a typical cavity structure. TMAH was used as an alkaline etchant at a concentration suitable for achieving a maximum etch rate. KOH was used as an alkaline etchant during subsequent testing on Solar wafers following the same methods outlined. Silicon etching processes were conducted with the aid of Naura-Akrion's in-situ chemical concentration control system. Occasionally, samples of the baths were taken and titrated for comparison to the system's readings. The goal for the TMAH testing was to fully etch through the wafers by maintaining a consistent etch rate throughout the entire process. For the KOH testing, a consistent etch rate and reflectance were measured to ensure the versatility of the system.

Results and Discussion

The chemical reaction for the anisotropic alkaline etching of silicon is well known and a variety of etchants can be used for the process (KOH, NaOH, TMAH). For the purpose of this study, the starting setpoint concentration of TMAH was 5% and allowed to drop to a minimum level of 3% whereafter the etch rate was kept constant. Through the course of the experiment, it was discovered that the generally held overall reaction mechanism for the etching of silicon with TMAH, as is shown in equation 1, did not match the real data.

$$Si + 2TMAH + 2H_2O \rightarrow SiO_2(OH)_2^{2-} + 2TMA^+ + 2H_2 \quad (1)$$

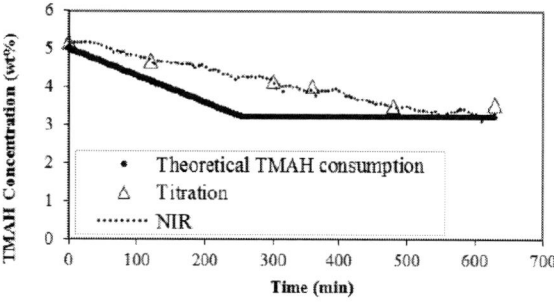

Figure 1. TMAH Concentration Model (2:1) vs Practice

As a result, a new theory was developed which fits the data much more accurately. The new theory supposes that only 1 mole of TMAH is consumed per 1 mole of Si, as opposed to

978-1-7281-5757-3/19 $31.00 © 2019 IEEE

the original theory of 2:1 (TMAH:Si, based on equation 1). The test was run again, and the results are shown in Figure 2. As can be seen, the new theory fits the data very closely. It can also be seen from Figure 1 that after the TMAH concentration hit the lower limit of 3%, the system initiated the steps of draining and refilling with fresh chemical, which in turn allows the concentration to maintain steady even during processing.

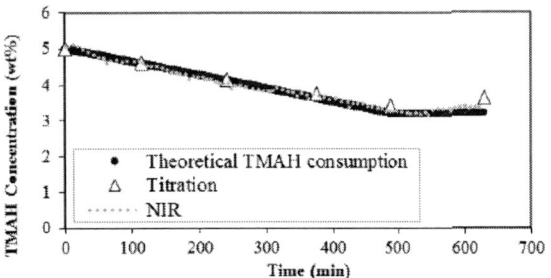

Figure 2. New TMAH Concentration Mode (1:1) vs Practice

As can be seen in Figure 3, when the system initiates the concentration control process, the bath is then left at a steady state position. Therefore, it is important to set all parameters for upper and lower bounds to allowable values for maintained chemical concentration. In Figure 3, when the Si concentration reaches 5 g/L, the balanced volume of chemical being drained from the tank and added to the bath allows the concentration to stay at 5 g/L. The system does not correct itself to lessen the amount of Si in the bath significantly, as this would provide too much disturbance to the process. The TMAH concentration during the process is also maintained at a steady concentration, thus allowing the bath to sustain itself in an equilibrium between TMAH concentration and Si concentration for an extended period of time. The constant concentration of TMAH also allows the system to keep a constant etch rate, which was monitored before and after the system was activated. The production etch rate was the same across the experiment, and the qual etch rate was comparable to that of normal conditions. This dynamic equilibrium can maintain itself for weeks or even months depending on the process and the needs of the fab, as shown in the chart where the data is collected for 11 days with consistent results. A benefit of the SiC etch processes is that the material itself is inert to the normal chemical concentrations used for bulk etching, therefore the models necessary to track the silicate byproducts in the bath for typical bulk Si etching have already been created.

For the KOH process, testing focused on the bath life capabilities and the ability to simultaneously maintain both the base chemical concentration and Si concentrations within their set allowable bands. From Figure 4, over the course of 77 hours the etch rate is held constant due to the steady state conditions of the process. The silicate maximum level was set at 4.2g/L. The Si reading reached the upper limit, and the system responded accordingly and decreased the amount of Si in the bath without compromising the KOH concentration or the etch rate. Due to the adjustable settings, the Si concentration in this case is adjusted to well below the upper limit each time the system corrects itself. Etch rate testing throughout the 77-hour

process indicated a consistent etch rate and reflectance. Both of these metrics represent the steady state of the bath once the chemical concentration monitor system is activated. Using not only KOH to etch but a mixture involving the surfactant IPA has important implications as well for SiC devices. Pressure sensors are being designed which use these two chemicals in various concentrations at high temperatures.[5] The process for designing these pressure sensors involves deposition of 3C-SiC on a bulk substrate of Si which undergoes backside etch. The NIR chemical concentration system is able to control these types of conditions just as well as if there was only one chemical involved.

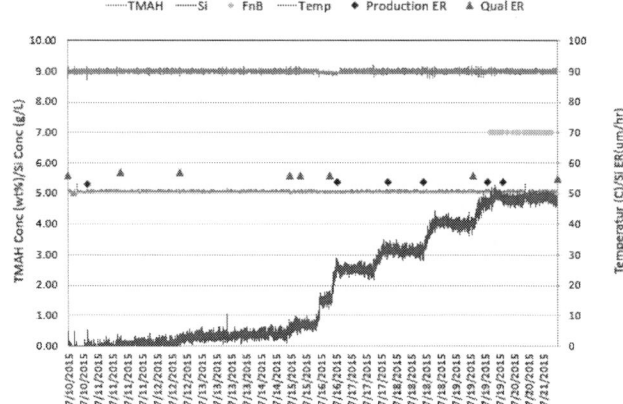

Figure 3. Concentration of TMAH and Si in bath with measured etch rates over 11 days

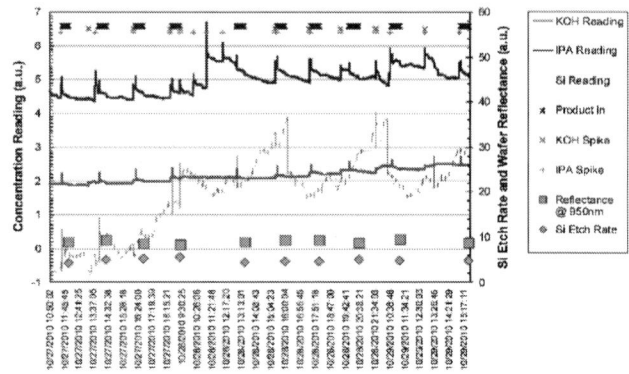

Figure 4. Concentration of KOH and Si with measured etch rates across 77 hours

This novel process can be especially useful when a pattern requires a long etch time. Molds to create some SiC devices require an etch depth of up to a few hundred microns, in which case the quantity of byproducts in the bath should be closely monitored so as to not negatively interfere with the etch.[2] This system has been shown to work with through etching 200mm sized wafers of 550um thickness. In this way, this system is capable of handling not only the front-side etching to create and potentially release devices from Si molds, but also the bulk back side etching to release the microstructures as pictured in Figure 5. This process will also aid in situations where oxide is

978-1-7281-5757-3/19 $31.00 © 2019 IEEE

required as a secondary etch stop for patterning as a constant concentration in the bath will directly relate to a constant selectivity.

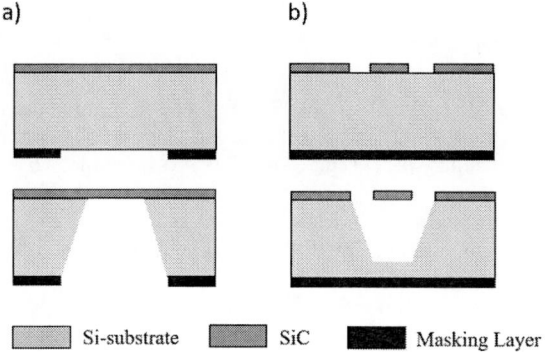

Figure 5. View of potential etching of micromachining for: (a) back side and (b) front side etching. [3]

For a similar process to this chemical concentration control system, conductivity may also be used to monitor the amount of chemical that is being used during etching. A prime example of a situation in which this design would be useful is when etching SiO_2 with HF. This is an important step for many SiC devices as the HF etch releases the device from the mold. During this process, etching by-products may affect the linear conductivity-concentration relationship seen for many acids. Therefore, a correction must be developed to correlate the amount of SiO_2 etched to the change in conductivity. Naura-Akrion calls this system of conductivity monitor and control ICE™. Through testing in a production tool, a correlation was found and implemented between the concentration of by-products, specifically H_2SiF_6, the concentration of HF, and the conductivity value over 23 runs with 50 oxide wafers. Based on this new formula for the expected conductivity, the setpoint would be renewed after each etching process is completed. Examples of oxide being etched without and then with this concentration control scheme are shown in Figures 6 and 7 respectively. From the figures it is apparent that the adjustment for conductivity from run to run results in a more uniform etch of SiO_2.

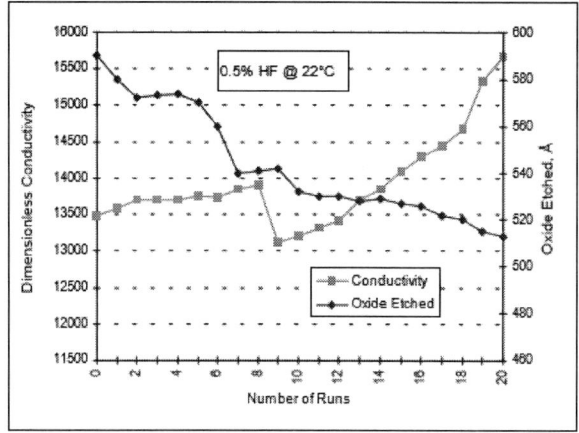

Figure 6. Effect of Dissolved SiO_2 on Etch Rate and Bath Conductivity

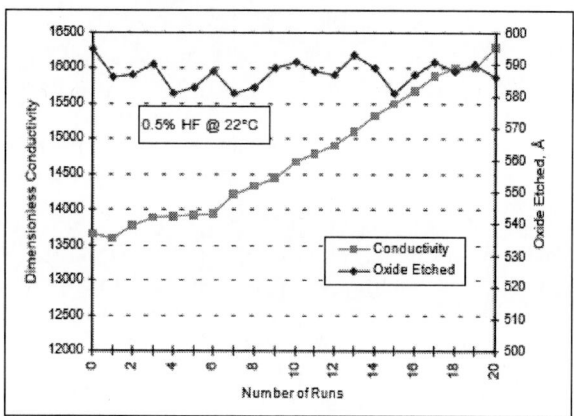

Figure 7. Stability of SiO_2 Etch rate with Modified Conductivity Algorithm

Conclusions

Results have shown that Naura-Akrion's novel closed loop concentration control system allows for a minimum usage of etchant, such as TMAH or KOH, while still maintaining the desired etch rate. This reduces the amount of chemical sent to the waste stream, decreasing the overall cost of ownership for the process. It also allows for consistent results within lots and between lots for wafer processing, effectively reducing the need to changeout baths after specific number of lots or time, depending on the demands of the process. This not only saves on more chemical cost, but also reduces the total time taken to qualify the baths in the long run after a chemical changeout occurs. The current state of SiC device manufacturing involves deposited films on bulk Si which must be etched. With closed loop concentration control, the process is more robust and results in less costly manufacturing.

References

1. Mehregany, M., Zorman, C., Rajan, N., & Wu, C. H. (1998). Silicon carbide MEMS for harsh environments. Proceedings of the IEEE,86(8), 1594-1609. doi:10.1109/5.704265

2. Mehregany, M., & Zorman, C. A. (1999). SiC MEMS: Opportunities and challenges for applications in harsh environments. Thin Solid Films,355-356, 518-524. doi:10.1016/s0257-8972(99)00374-6

3. Sarro, P. M. (2000). Silicon carbide as a new MEMS technology. Sensors and Actuators A: Physical,82(1-3), 210-218. doi:10.1016/s0924-4247(99)00335-0

4. Kashkoush, I., Rieker, J., Chen, G., & Nemeth, D. (2015). Process Control Challenges of Wet Etching Large MEMS Si Cavities. In Solid State Phenomena (Vol. 219, pp. 73-77). Trans Tech Publications.

5. Marsi, N., Majlis, B. Y., Mohd-Yasin, F., & Hamzah, A. A. (2014). The fabrication of back etching 3C-SiC-on-Si diaphragm employing KOH + IPA in MEMS capacitive pressure sensor. Microsystem Technologies, 21(8), 1651–1661. doi:10.1007/s00542-014-2267-8

The Synthesis of Large-size Silicon Carbide Powder for Crystal Growth

Fang Jiao[1], Dianpeng Cui[1], Mulong Yang[1], Zhouli Wu[1] and Boyu Dong[1,2,a]

[1] Beijing NAURA Microelectronics Equipment Co., Ltd.
[2] Beijing Engineering Research Center of MOCVD, China
No.8 Wenchang Avenue Beijing Economic-Technological Development Area, China
[a] E-mail: boyu.dong@naura.com

Abstract

A study on the synthesis of SiC powder with large size was presented. Special attentions were paid to synthetic temperature and ratio of large and small size silicon powder. Several experiments were designed to understand the effects of these conditions on phase composition, grain size and yield of SiC powder material by using a NAURA Advanced Physical Vapor Transport (PVT) System. It was found that high-quality SiC powder could be acquired at 1950°C for 10h. The rising proportion of large size silicon powder with 500~800μm was beneficial to synthesize large size SiC powder of >750μm. The results showed successful preparation of large size SiC powder and a significant reduction of carbon inclusions in SiC ingots by using SiC powder with large size.

1. Introduction

Silicon carbide (SiC) is playing an important role in current electronic device application due to its outstanding physical properties, such as high thermal conductivity, wide band gap, and high critical breakdown electric field [1-2]. In order to improve SiC wafer quality, it makes a demand for purity and size of SiC source powder. Conventionally, SiC powder is mainly prepared by Acheson method through a carbothermal reaction of silicon dioxide and carbon [3-4], while the purity of as-prepared SiC powder is not enough. Other SiC synthesis techniques include sol-gel [5], chemical vapor deposition [6], plasma[7], and so on. They are usually applied to prepare SiC nanopowder. For the size of SiC source powder, it's reported that large size SiC powder can reduce carbon inclusions, which were considered as the main cause of crystalline defects, such as polytypes, micropipes and dislocations[8-10]. Wang,et al. have investigated the influences of the C/Si ratio, reaction temperature and reaction time on the size, composition and structure of the SiC powder. The results indicate that longer reaction time causes larger size SiC grains[11], The effects of reaction temperature on the phase composition, particle size and the production yield are also discussed by Wang et al. It shows that the particle size and the production yield are first increased and then decreased with increasing the reaction temperature[12].

In this work, different from the above methods, combustion synthesis was applied for the preparation of SiC powder material using high-purity silicon and carbon powder. Emphasis was put on the synthesis of large size SiC powder from two aspects: synthetic temperature and particle size of silicon powder.

2. Experiments

SiC powder material was prepared by a NAURA Advanced PVT System APS180G. The system has a double-layer quartz chamber for better impurity background control and is equipped with a movable coil for medium-frequency induction heating, measured by a pyrometer. The synthesis process of SiC powder was described as follows: (i) the prebaking for graphite crucible and insulation felt at high temperature. (ii) High purity silicon (99.999%) and carbon (99.999%) powder were mixed uniformly in 1:1 molar ratio and transferred into the graphite crucible. (iii) The chamber was evacuated to a low pressure, injected argon gas when heated to 1000°C, and then reached a synthesis condition. (iv) After the reaction was completed, the chamber was cooled to room temperature. The prepared SiC powder material was used for crystal growth.

Table 1 Particle size of starting materials

Starting materials	Particle size (μm)
carbon powder (C)	50~100
silicon powder 1 (Si1)	50~100
silicon powder 2 (Si2)	500~800

In order to obtain large size SiC powder material, first of all, different synthetic temperatures were carried out from 1950°C to 2050°C. Then, changing the particle size of silicon powder to acquire large size SiC powder material at the suitable temperature. To be specific, the particle sizes of silicon and carbon powder were selected and listed in Table 1. The ratio of small size silicon powder 1 of 50~100μm and large size silicon powder 2 of 500~800μm was changed to get different mixtures of silicon powder materials. A series of silicon and carbon mixture samples were sintered to compare the particle size of SiC powder. As-grown SiC ingots using as-prepared SiC powder were sliced and polished with diamond slurries for wafer preparation.

SiC phase structure was analyzed by X-ray diffraction (XRD, SmartLab, Rigaku) with Cu Kα radiation. The morphology of as-prepared SiC powder was inspected by scanning electron microscopy (SEM, SU8030, Hitachi). SiC powder for different particle sizes was classified and weighed using different mesh size and electronic balance. The distribution of carbon inclusions was inspected by an optical microscope (OM, BX51, OLYMPUS).

3. Results and discussion

SiC has a great number of polytypes. Their structural stability is strongly dependent on temperature. At relatively low temperature, about 1700-1800°C, 3C-SiC can be prepared[11]. While 3C-SiC will be transformed into α-SiC polytypes, such as 6H-SiC, at 1900-2000°C[13]. When the temperature is above 2000°C, 6H-SiC, 4H-SiC and 15R-SiC polytypes are often observed[14].

In order to obtain α-SiC powder, SiC powder material was synthesized at 1950, 2000 and 2050°C, respectively. As shown in Fig. 1, the diffraction peaks of SiC powder at 1950°C matched well with 6H-SiC standard card, except minor Al_2O_3 peaks. Al_2O_3 phase originated from a grinding process that ground large SiC particles into small ones for XRD measurement, not from synthetic process. With the temperature increasing, the peaks of 6H-SiC coexisted with 15R-SiC, especially for SiC powder synthesized at 2050°C, in accordance with the above-mentioned description.

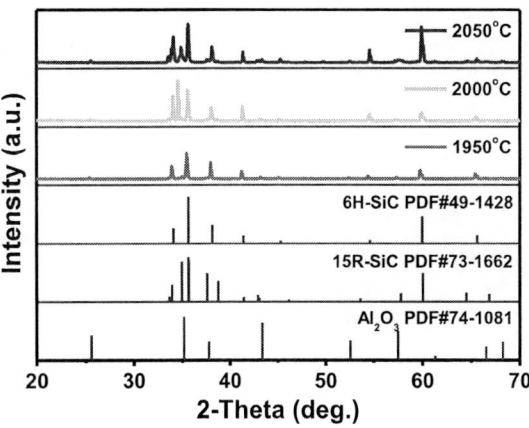

Fig. 1 XRD patterns of SiC powder material synthesized at 1950, 2000 and 2050°C.

Table 2 Different molar ratios of C, Si1 and Si2 powder for preparing SiC powder

Sample number	C:Si1:Si2 (molar ratio)
1#	1:1:0
2#	1:0.7:0.3
3#	1:0.5:0.5
4#	1:0.3:0.7
5#	1:0:1

In order to prepare large size SiC powder further, the molar ratio of Si1 and Si2 powder was changed and listed in Table 2. From sample number 1# to 5#, the ratio of large size Si2 powder increased gradually and reached the maximum for 5#. The XRD patterns were drawn in Fig. 2 for these five SiC powder samples synthesized at 1950°C for 10h. It indicated that SiC powder had been prepared successfully and very little Al_2O_3 could be seen in the patterns.

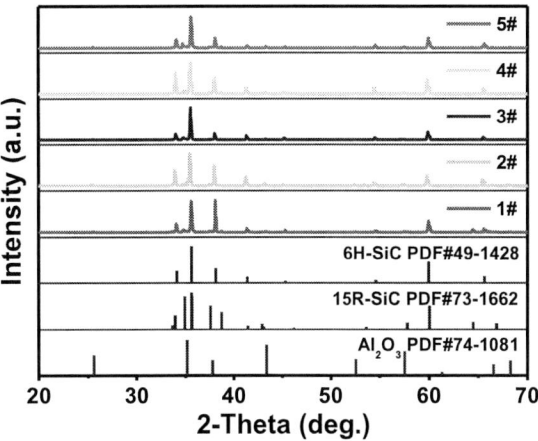

Fig. 2 XRD patterns of five SiC powder samples synthesized at 1950°C.

The particle size and morphology of five SiC powder material were characterized by SEM, as shown in Fig. 3. It was obvious that mean particle sizes of SiC powder gradually increased from 1# to 5# and large size SiC particles increased significantly, especially for 5# with large size Si2 powder only.

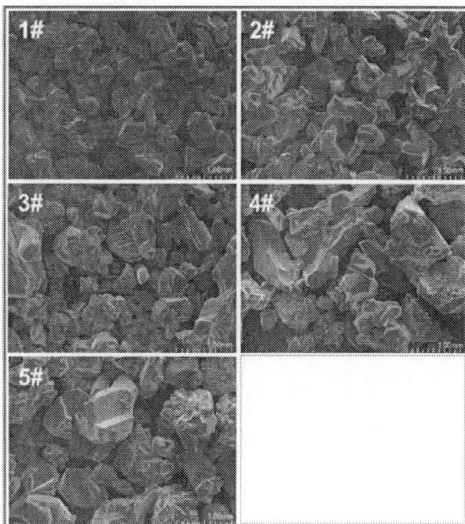

Fig. 3 SEM images of SiC powder for five samples.

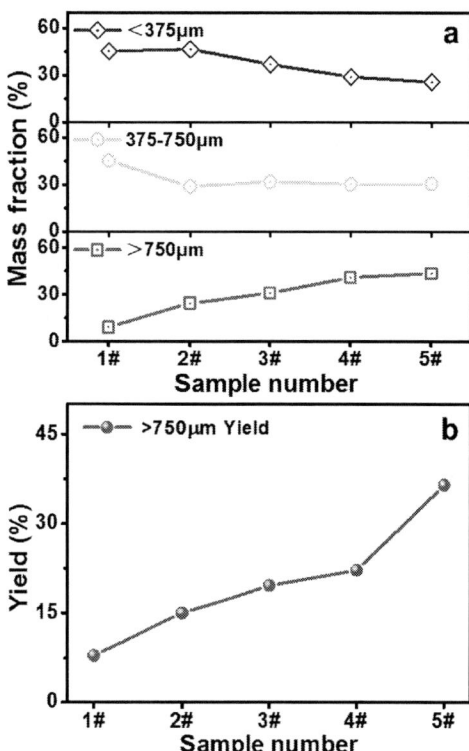

Fig. 4 (a) Mass fraction of SiC powder with different particle sizes; (b) Yield of SiC powder with large size of >750μm for five samples.

The mass fractions of five SiC powder material with different particle sizes were depicted in Fig. 4a. Here the mass fraction was calculated using the as-prepared SiC powder, and the yield of SiC was the ratio of as-prepared SiC powder to the raw materials. With the increase of large size Si2 powder from 1# to 5#, the mass fraction of large size SiC particles with >750μm gradually increased. When adding only C and Si2 powder, the mass percentage and yield of >750μm SiC particles reached the maximum, 43.6% and 36.5%, respectively. It showed that large size Si powder was beneficial to synthesize large size SiC powder material. The possible reason for this phenomenon was

978-1-7281-5757-3/19 $31.00 © 2019 IEEE

analyzed: During the synthetic process of SiC powder, when the temperature exceeded the melting point of Si, about 1410°C, the silicon was in the molten state. The carbon particles were fully diffused in the liquid silicon, and then reacted to form silicon carbide particles. The large size silicon was conducive to the diffusion of carbon particles in the liquid silicon, which could promote the agglomeration of silicon carbide particles, thus increasing the size of silicon carbide powder.

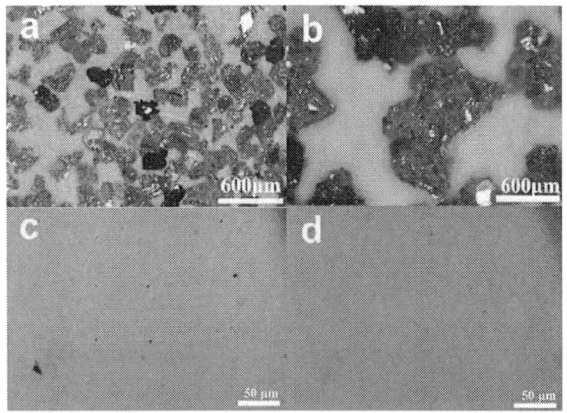

Fig. 5 OM images of (a) 1# SiC powder with small size, (b) 5# SiC powder with large size. The corresponding wafers growth using (c) 1# SiC powder with small size and (d) 5# SiC powder with large size, respectively.

SiC powder with different sizes was used as the starting material for the growth of 4H-SiC crystal. The SiC powder with large size was shown in Fig. 5b. It showed a significant reduction of carbon inclusions in grown crystals by using large size SiC powder in Fig. 5d, in comparison with that in Fig. 5c. During the process of SiC crystal growth, the SiC powder with small size was easier to graphitization than that of large size SiC powder. The carbon particles produced by graphitization could be transported to crystal growth interface in the interaction of drag force of fluid and thermophoretic force, and finally formed carbon inclusions in the crystal.

Conclusions

The synthesis of SiC powder material with large size was investigated. SiC powder was obtained at 1950°C for 10h. Its grains can grow up significantly by choosing large size silicon powder. As a result, the mass percentage and yield of large size SiC particles with >750μm reached 43.6% and 36.5%, respectively. High quality SiC wafers with less carbon inclusions were obtained by using SiC powder with large size.

References

1. Casady, J. B. *et al*, "Status of Silicon Carbide (SiC) as a Wide-Bandgap Semiconductor for High-Temperature Applications: a Review," *Solid State Electronics*, Vol. 39, No. 10 (1996), pp. 1409-1422.
2. Wellmann, P. J, "Review of SiC Crystal Growth Technology," *Semiconductor Science and Technology*, Vol. 33 (2018), pp. 103001 (21pp).
3. Martin, H. P. *et al*, "Synthesis of Nanocrystalline Silicon Carbide Powder by Carbothermal Reduction," *Journal of the European Ceramic Society*, Vol. 18, No. 12 (1998), pp. 1737-1742.
4. Lin, Y. J. *et al*, "The Effects of Transition Metals on Carbothermal Synthesis of β-SiC Powder," *Ceramics International*, Vol. 33, No. 5 (2007), pp. 779-784.
5. Najafi, A. *et al*, "Synthesis and Characterization of SiC Nano Powder with Low Residual Carbon Processed by Sol-Gel Method," *Powder Technology*, Vol. 219 (2012), pp. 202-210.
6. Gupta, A. *et al*, "The Influence of Diluent Gas Composition and Temperature on SiC Nanopowder Formation by CVD," *Journal of Materials Science*, Vol. 42, No. 13 (2007), pp. 5142-5146.
7. Cui, H. *et al*, "Nucleation and Growth of Silicon Nitride Nanoneedles Using Microwave Plasma Heating," *Journal of Materials Research*, Vol. 16, No. 11 (2001), pp. 3111-3115.
8. Wang, Z. Z. *et al*, "Study on Carbon Particle Inclusions during 4H-SiC Growth by Using Physical Vapor Transport System," *Materials Science Forum*, Vol. 954 (2019), pp. 46-50.
9. Dudley, M. *et al*, "The mechanism of micropipe nucleation at inclusions in silicon carbide," *Applied Physics Letters*, Vol. 75, No. 6 (1999), pp. 784-786.
10. Glass, R.C. *et al*, "SiC Seeded Crystal Growth," *physica status solidi (b)*, Vol. 202, No. 1 (1997), pp. 149-162.
11. Wang, L. H. *et al*, "Combustion Synthesis of High Purity SiC Powder by Radio-Frequency Heating," *Ceramics International*, Vol. 39, No. 6 (2013), pp. 6867-6875.
12. Y.M. Wang, X.R. Hou, W.Xu, M.Tian, "Effects of reaction temperature on the synthesis of high purity silicon carbide powder," Mater.Res.Innov., 19 (2015) S5-1338-S5-1343.
13. Yoo, W. S. *et al*, "Solid-State Phase Transformation in Cubic Silicon Carbide," *Japanese Journal of Applied Physics*, Vol. 30, No. 3 (1991), pp. 545-553.
14. Knippenberg, W. F, "Growth Phenomena in Silicon Carbide," *Philips Research Reports*, Vol. 18, No. 3 (1963), pp. 161-274.

Investigation of Defect Levels of Al/Ti 4H-SiC Schottky structures by Deep Level Transient Spectroscopy

Yawei He, Guoguo Yan*, Xingfang Liu, Zhanwei Shen, Wanshun Zhao, Lei Wang, Feng Zhang, Guosheng Sun, Yiping Zeng

Key Laboratory of Semiconductor Materials Science, Institute of Semiconductors, Chinese Academy of Sciences
No.A35, QingHua East Road, Haidian District, Beijing P R China
*ggyan@semi.ac.cn

Abstract

The performance of silicon carbide bipolar devices is limited by the material quality of substrates and epitaxial layers. There are many aspects, one of which is the deep level defects in the epitaxial layer, that affect the quality of materials. N-type 4°off-axis Si-face silicon carbide epitaxial layers were investigated by deep level transient spectroscopy (DLTS), which includes C-V/I-V measurement, DLTS signal spectrum measurement and Arrhenius fitting analysis processes. The energy level position, concentration and capture cross-section of the deep level defects were obtained accurately after the measurement and analysis. $Z_{1/2}$ defect was found to be a dominant deep level for all the as-grown samples because it existed in all the samples we characterized. So $Z_{1/2}$ defect is widely distributed on epitaxial layers. The largest defect concentration in all samples is $Z_{1/2}$ defect, which means that it was obviously the most influential defect type in this work. In addition to $Z_{1/2}$ defects, three other deep levels had been found which were $RD_{1/2}$, EH_5 and the one with defect energy level position near 1.209eV.

1 Introduction

4H-SiC (4H Silicon carbide) is appropriate for making power devices because of the larger band gap, higher electric breakdown field, higher thermal conductivity and larger electron saturation drift velocity [1-3]. 4H-SiC bipolar devices can sustain high forward blocking voltage and a low positive on-state resistance at the same time because of conductivity modulation [4]. Effective conductance modulation in the drift layer requires a high minority carrier lifetime with low deep level defect concentrations. However, the minority carrier lifetime of as-grown 4H-SiC epitaxial layer is limited, which cannot satisfy the requirements of high-voltage or ultra-voltage bipolar power devices. There are many reasons that lead to lower carrier lifetime in the as-grown epitaxial layer and the important one is the center of recombination caused by deep levels. The typical deep level in as-grown 4H-SiC epitaxial layer is $Z_{1/2}$, which has been clearly identified as originating from the carbon vacancy (V_C) defect [5-7]. Furthermore, $Z_{1/2}$ is considered as a dominant recombination center, which is well known lifetime killers in 4H-SiC [8]. Therefore, studying the types and concentrations of deep levels in the 4H-SiC epitaxial layer is conducive to improving bipolar device performance. The characterization of deep level defects in 4H-SiC epitaxial layer has been studied in recent years and many methods have been reported to reduce the concentration of $Z_{1/2}$ deep level defect. In 2004, Tawara et al. had shown that minority carrier lifetime limited by the $Z_{1/2}$ and $EH_{6/7}$ deep levels, located at ~0.7eV and ~1.5eV

below the conduction band edge (E_C), respectively [9,10]. In 2007, Danno et al. Unambiguously manifested that the $Z_{1/2}$ deep level defect works as an effective recombination center when the concentration is higher than an order of magnitude, but there are many other deep levels might affect minority carrier lifetime in epitaxial layer [11]. In 2016, Eiji et al decreased the density of $Z_{1/2}$ deep level defect from 2×10^{13}cm^{-3} to the detection limit (3×10^{10}cm^{-3}) in the region to a 130μm depth [12]. In 2017, Mitsuhiro et al. decreased the $Z_{1/2}$ deep level defect from 2×10^{13}cm^{-3} to 1.6×10^{11}cm^{-3} through the method of C^+ ion implantation/annealing [13]. Although the specific microstructure of the $Z_{1/2}$ defect is still unclear, there are many feasible methods to reduce the $Z_{1/2}$ defect concentration. In addition, the influence of other defect types on carrier lifetime is also not clear.

In this work, we investigated the information of deep level defects in the as-grown 4H-SiC epitaxial layers by the equipment of DLTS. Schottky structures were fabricated to be the measurement samples. What's more, the energy level position, concentration and capture cross section of the defects in the samples were obtained by the measurement of those samples.

2 Experiment

Nitrogen doped 4H-SiC (0001) epitaxial layers grown by chemical vapor deposition purchased from TYSiC, with a net donor doping concentration of 2×10^{14}cm^{-3}. In order to investigate the deep levels in the 4H-SiC epitaxial layer, a number of Al/Ti Schottky contact structures are fabricated.

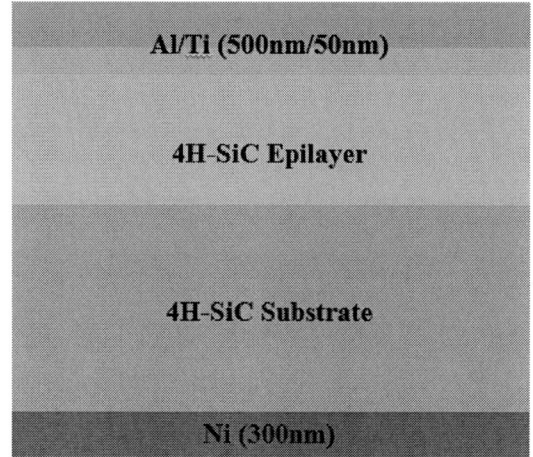

Fig. 1 A cross section of Schottky structures of the samples

978-1-7281-5757-3/19 $31.00 © 2019 IEEE

The 4-inch n-type 4H-SiC epitaxial wafer was cut into a number of 15mm×15mm pieces by laser. A Ni layer, whose thickness is 300nm, annealed at 999℃ for 1.5min was employed as a backside ohmic contact. After that, a Ti/Al (50nm/500nm) layer was evaporated on Si-face in turn to form Schottky contacts. The Schottky structure we fabricated is shown in Fig. 1.

The Schottky structures were electrically characterized by Deep Level Transient Spectroscopy (DLTS) [14], with the mode of Deep Level Transient Fourier Spectroscopy (DLTFS). The information about deep levels in the structures was obtained from DLTS, and the measurement was performed on the structures in two temperature ranges, which are 80-450K and 300-700K, respectively. The measuring parameters include Pulse width (Tp=1ms), Period width (Tw=5ms), Reverse bias voltage (Vr=-4v), Pulse voltage (Vp=-0.1v) were absolutely unchanged in two measurements.

3 Results and discussion

Capacitance-voltage (C-V) and current-voltage (I-V) are measured before the measurement of DLTS, and the curves,

which are indicated in Fig. 2 and Fig. 3, manifest that the structures have small leakage currents and effective Schottky contact. Through C-V test results, the effective doping concentration in epitaxial layer can be calculated by Eq. 1.

$$C = \sqrt{\frac{\varepsilon_s q N_D}{2(V_D - V)}} \qquad (1)$$

Here, ε_s is the dielectric constant of 4H-SiC, C and V are Schottky barrier capacitance and the applied voltage across the depletion width, respectively. V_D is the built-in potential barrier [15]. The functional relationship of C-V is shown in Fig. 1. So, the doping concentration N_D can be calculated for DLTS measurement and analysis.

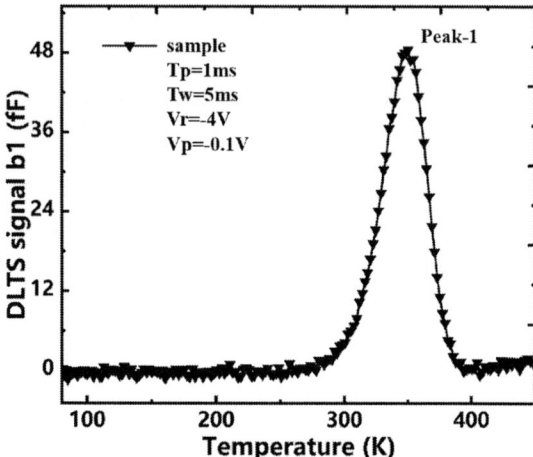

Fig. 3 DLTS signal spectrum of samples measured in 80-450K

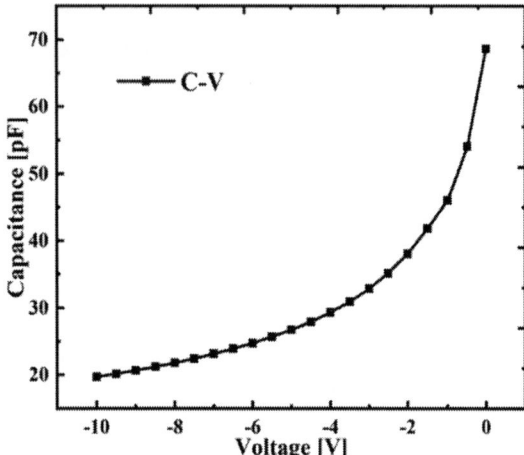

Fig. 2 C-V curve of n-type 4H-SiC Schottky structures

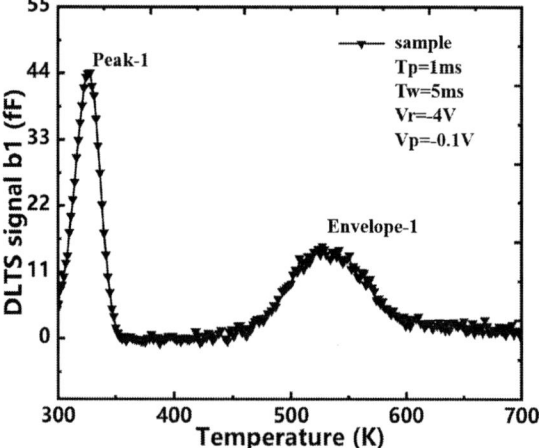

Fig. 4 DLTS signal spectrum of samples measured in 300-700K

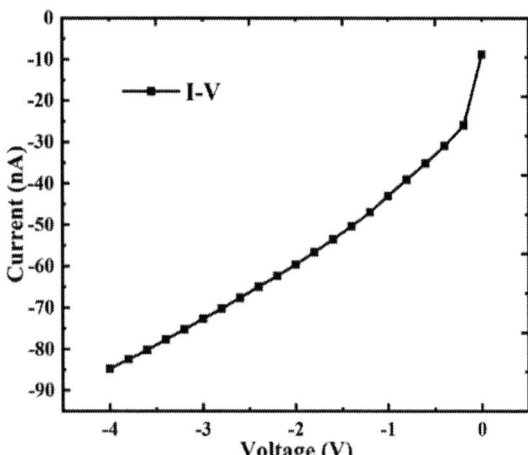

Fig. 3 I-V curve of n-type 4H-SiC Schottky structures

After C-V and I-V test, DLTS measurement was applied in all the samples we fabricated and the DLTS signal spectrums were acquired in the two temperature ranges. As shown in Fig. 3 and Fig. 4, peak-1 and envelope-1 were

978-1-7281-5757-3/19 $31.00 © 2019 IEEE

obtained in the whole temperature range from 80K to 700K through double tests.

In Fig. 3, only peak-1 can be fitted by a single signal peak around 350K, and there is no deep level at the lower temperature because of no peaks at the temperature range. Ti center is a typical deep level defect in n-type 4H-SiC epitaxial layer with the energy level around 0.14eV and the DLTS signal peak of Ti center around 100K [16,17]. The origin of Ti center may be Ti or Cr impurity [18], which means that our process for Al/Ti Schottky contact will not introduce deep level defects.

In Fig. 4, a number of deep level defects are detected including the one whose DLTS signal is Peak-1 appeared in Fig. 3. Peak-1 is the signal peak of a single defect, and the height of the signal peak reflects the concentration of the defect. Unlike Peak-1, the envelope-1 is the superposition peak of different defect signals, and there are three distinct signal peaks [peak-2~peak-4] are contained in the envelope-1. Through peak searching fitting of Peak-1 and envelope-1, Arrhenius plot which is shown as Fig. 5 can be obtained.

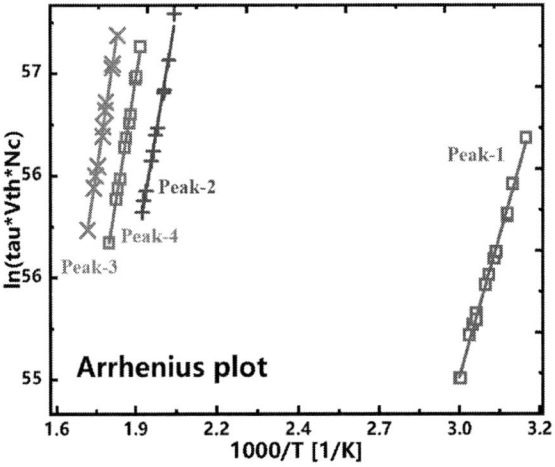

Fig. 5 Arrhenius plots for the deep level defects revealed in the Schottky structures. Peak-2~Peak-4 are fitting from envelope-1. The solid lines are the best least-squares fitting to the signal location.

The Arrhenius equation (Eq. 2) is the basis for calculating the information of deep level defects whose Arrhenius plots are shown as Fig. 5.

$$\ln(\tau_e v_{th} N_C) = \frac{E_C - E_T}{k}\frac{1}{T} - \ln(X_n \sigma_n) \qquad (2)$$

Here, τ_e and σ_n are the emission time constant and capture cross section for electrons, respectively. E_C-E_T is the energy level location of deep level defects in bandgap [19]. Combine Arrhenius equation (Eq. 2) and Arrhenius plots of the four peaks, the defect energy level and capture cross-section of deep level defects we detected are obtained and the defect information is shown as Table I.

In Table I, the energy level position, concentration and capture cross-section of the deep level defects in the epitaxial layer of the samples were shown and they were obtained by the signal fitting analysis. The energy level location and concentration of the deep level defect which is corresponding to Peak-1 are 0.657eV and 2.63×10^{12}cm^{-3}, respectively. It is concluded that Peak-1 is the signal spectrum of $Z_{1/2}$, which is usually appeared in n-type 4H-SiC. The cause of $Z_{1/2}$ defect has not been determined, but it is generally believed to be caused by carbon vacancy [20]. In order to obtain the best fitting results, envelope-1 is divided by Peak-2~Peak-4. Peak-2~Peak-4 are corresponding to different types of deep level defects in the epitaxial layer, and the concentration of all the three type of defects is a same level which is an order of magnitude below the defect of $Z_{1/2}$. So, it indicates that $Z_{1/2}$ is the main deep level defect of n-type as-grown 4H-SiC samples, which is one of the reasons why $Z_{1/2}$ is the typical killer of carrier lifetime. The energy level position of the defect corresponding to Peak-2 is 1.041eV, and we temporarily consider the defect to be $RD_{1/2}$ which is commonly associated with the vacancy pair V_{Si}-V_C [21]. The energy level position of the defect corresponding to Peak-4 is 1.13eV, which has been reported in the previous literature with exactly the same energy level [3]. We determined that Peak-4 is related to the deep level defect named EH_5, and the reason for the defect is a carbon cluster [22]. However, the deep level defect which is fitted by Peak-3 is not confirmed. The energy level of Peak-3 is 1.209eV which has not been reported before, and we will do further research on it in the future.

Table I. Deep level defects' parameters obtained from the fitting of DLTS curves in Fig. 4. These parameters include capture cross-section, defect energy level and defect concentration. The last column shows the possible types of defects.

Peak-#	σ cm^2	ΔE eV	Nt cm^{-3}	Possible trap
Peak-1	1.12E-14	0.657	2.63E+12	$Z_{1/2}$
Peak-2	5.09E-15	1.041	4.02E+11	$RD_{1/2}$
Peak-3	1.80E-14	1.209	3.92E+11	(?)
Peak-4	1.06E-14	1.130	4.21E+11	EH_5

978-1-7281-5757-3/19 $31.00 © 2019 IEEE

Conclusions

For all the as-grown samples in this work, the $Z_{1/2}$ defect is the dominant deep level, which was found exists in all samples we fabricated. It indicates that the $Z_{1/2}$ defect might be distributed throughout the wafer, and it is clearly that $Z_{1/2}$ defect concentration is larger than other deep level defects. $RD_{1/2}$ and EH_5 are also detected in the as-grown 4H-SiC epitaxial layers, and there is another deep level defect whose energy level location around 1.209eV cannot be determined yet. Although different deep levels will affect the properties of 4H-SiC epitaxial, $Z_{1/2}$ defect is obviously the most influential defect type in this work.

Acknowledgments

This work is supported by the National Key Research and Development Program of China (2016YFB0400400), the Science Challenge Project (TZ2018003) , the National Natural Science Foundation of China (Grant Nos. 61574140, 61804149 and 61604148).

References

[1] E. Janzen and O. Kordina, "SiC material for high-power applications," Materials Science and Engineering B-Solid State Materials for Advanced Technology, vol. 46, no. 1-3, pp. 203-209, Apr 1997.

[2] J. Camassel, S. Contreras, and J. L. Robert, "SiC materials: a semiconductor family for the next century," Comptes Rendus De L Academie Des Sciences Serie Iv Physique Astrophysique, vol. 1, no. 1, pp. 5-21, Jan-Feb 2000.

[3] C. Hemmingsson et al., "Deep level defects in electron-irradiated 4H SiC epitaxial layers," Journal of Applied Physics, vol. 81, no. 9, pp. 6155-6159, May 1 1997.

[4] "Fabrication of 4H-SiC n-channel IGBTs with ultra high blocking voltage," Journal of Semiconductors, vol. 39, no. 3, pp. 034005-1-034005-3, 2018 2018, Art. no. 1674-4926(2018)39:3<034005:Fo4snc>2.0.Tx;2-g.

[5] I. D. Booker, E. Janzen, N. T. Son, J. Hassan, P. Stenberg, and E. O. Sveinbjornsson, "Donor and double-donor transitions of the carbon vacancy related EH6/7 deep level in 4H-SiC," Journal of Applied Physics, vol. 119, no. 23, Jun 21 2016, Art. no. 235703.

[6] N. T. Son, W. M. Chen, J. L. Lindstrom, B. Monemar, and E. Janzen, "Carbon-vacancy related defects in 4H-and 6H-SIC," Materials Science and Engineering B-Solid State Materials for Advanced Technology, vol. 61-2, pp. 202-206, Jul 30 1999.

[7] J. Feng et al., "Characteristics and analysis of 4H-SiC PiN diodes with a carbon-implanted drift layer," Journal of Semiconductors.

[8] P. Hazdra, S. Popelka, and A. Schöner, "Local Lifetime Control in 4H-SiC by Proton Irradiation," Materials Science Forum, vol. 924, pp. 436-439, 2018.

[9] T. Tawara, H. Tsuchida, S. Izumi, I. Kamata, and K. Izumi, "Evaluation of free carrier lifetime and deep levels of the thick 4H-SiC epilayers," in Silicon Carbide and Related Materials 2003, Prts 1 and 2, vol. 457-460, R. Madar and J. Camassel, Eds. (Materials Science Forum, 2004, pp. 565-568.

[10] H. M. Ayedh, R. Nipoti, A. Hallen, and B. G. Svensson, "Thermodynamic equilibration of the carbon vacancy in 4H-SiC: A lifetime limiting defect," Journal of Applied Physics, vol. 122, no. 2, Jul 14 2017, Art. no. 025701.

[11] K. Danno, D. Nakamura, and T. Kimoto, "Investigation of carrier lifetime in 4H-SiC epilayers and lifetime control by electron irradiation," Applied Physics Letters, vol. 90, no. 20, May 14 2007, Art. no. 202109.

[12] E. Saito, J. Suda, and T. Kimoto, "Control of carrier lifetime of thick n-type 4H-SiC epilayers by high-temperature Ar annealing," Applied Physics Express, vol. 9, no. 6, Jun 2016, Art. no. 061303.

[13] M. Kushibe, J. Nishio, R. Iijima, A. Miyasaka, and K. Kojima, "Carrier Lifetimes in 4H-SiC Epitaxial Layers on the C-Face Enhanced by Carbon Implantation," Materials Science Forum, vol. 924, pp. 432-435, 2018.

[14] S. Weiss and R. Kassing, "DEEP LEVEL TRANSIENT FOURIER SPECTROSCOPY (DLTFS) - A TECHNIQUE FOR THE ANALYSIS OF DEEP LEVEL PROPERTIES," Solid-State Electronics, vol. 31, no. 12, pp. 1733-1742, Dec 1988.

[15] M. A. Mannan, K. V. Nguyen, R. O. Pak, C. Oner, and K. C. Mandal, "Deep Levels in n-Type 4H-Silicon Carbide Epitaxial Layers Investigated by Deep-Level Transient Spectroscopy and Isochronal Annealing Studies," IEEE Transactions on Nuclear Science, vol. 63, no. 2, pp. 1083-1090, 2016.

[16] N. Achtziger, J. Grillenberger, and W. Witthuhn, "Radiotracer Identification of Ti, V and Cr Band Gap States in 4H- and 6H-SiC," Materials Science Forum, vol. 264-268, pp. 541-544, 1998.

[17] T. Dalibor, G. Pensl, N. Nordell, A. Schöner, and W. J. Choyke, "Ground States of the Ionized Isoelectronic Ti Acceptor in SiC," Materials Science Forum, vol. 264-268, pp. 537-540, 1998.

[18] L. Gelczuk, M. Dabrowska-Szata, M. Sochacki, and J. Szmidt, "Characterization of deep electron traps in 4H-SiC Junction Barrier Schottky rectifiers," Solid-State Electronics, vol. 94, pp. 56-60, Apr 2014.

[19] W. Sieghard, "Semiconductor Investigations with the DLTFS Method," PhD Thesis, Department of Physics, University of Kassel, 1991.

[20] A. Galeckas, H. M. Ayedh, J. P. Bergman, and B. G. Svensson, "Depth-resolved carrier lifetime measurements in 4H-SiC epilayers monitoring carbon vacancy elimination," in European Conference on Silicon Carbide & Related Materials, 2017.

[21] L. Ottaviani, D. Barakel, V. Vervisch, and M. Pasquinelli, "Electrical Characterizations of Hydrogenated 4H-SiC Epitaxial Samples," Solid State Phenomena, vol. 108-109, pp. 677-682, 2005.

[22] C. G. Hemmingsson, N. T. Son, O. Kordina, and E. Janzén, "Metastable defects in 6H-SiC: Experiments and modeling," Journal of Applied Physics, vol. 91, no. 3, pp. 1324-1330, 2002.

978-1-7281-5757-3/19 $31.00 © 2019 IEEE

Polytype Transformation in 4H-SiC single crystals grown on on-axis Seeds

Xianglong Yang, Yan Peng, Xiufang Chen, Xuejian Xie, Jinying Yu, Xiaobo Hu, Xiangang Xu
State Key Laboratory of Crystal Materials, Shandong University, Jinan 250100, China
Email:cxf@sdu.edu.cn

Abstract

Polytype destabilization at the periphery of 4H-SiC single crystals during the initial stage of 4H-SiC bulk crystals grown on on-axis seeds by sublimation method were investigated. Optical microscopy and Raman spectroscopy, were used to study the distribution of polytypism. Three regions with different Raman peak intensity ratio (I_{150}/I_{204}) at the outer parts of the grown layer are distinctly observed, indicating the increase in parasitic polytype component towards the edge. The higher probability of 6H or 15R-SiC nucleation due to the higher supersaturation at the periphery could be responsible for the polytype transformation.

Introduction

Silicon carbide (SiC) is a promising semiconductor material for power device application due to its wide bandgap, high breakdown electric field and high thermal conductivity [1,2]. A key prerequisite to SiC device production is the availability of polytypic stable, low defect SiC substrate wafers of large diameter [3,4]. The physical vapor transport method (PVT) has been the most successful and common method for growing bulk SiC crystals [5]. One of the most important and interesting properties of SiC is that there are different polytypes in SiC single crystals which are easily formed during crystal growth, due to the very low formation energy differences among different polytypes[6]. Since the foreign polytype inclusions usually lead to serious quality deterioration in terms of nucleation of other defects such as micropipes, stacking faults and dislocations [7,8], it is important to completely eliminate the foreign polytypes throughout the entire growth process, including nucleation and subsequent growth, in further reduce defect density.

Many studies have been devoted to stabilize the desired polytype during PVT growth. The main growth parameters, such as the seed polarity [9,10], growth temperature [11,12], supersaturation [12,13], vapor phase stoichiometry [14] and impurity levels[14,15] that influence the polytype stability have been discussed. However, most of those parameters are not independent, giving rise to a very complex experimental problem. Although the occurrence of the polytypism in SiC crystals is still very common, the determining factor of polytype formation in SiC has not sufficiently been understood.

In this paper, we focus on the foreign polytype inclusions occurring at the seed periphery at the initial nucleation stage of on-axis SiC growth and investigate their formation and propagation through analysis of the growth initiation process.

Experiments

4-inch 4H-SiC crystal growth experiments were conducted in an inductively heated PVT reactor [16]. The growth system consists of SiC powder, a graphite crucible, an insulation shield, a graphite pedestal and induction coils. For all runs we used 4H (000-1) C-face on-axis orientated seeds and the crystals were nominally undoped. During the crystal growth, the lid temperature was set to be 2000-2300 °C and the temperature gradient along the axial direction was kept at 30-60°C/cm. To track the evolution of the polytypes, the grown crystals were horizontally sliced perpendicular to the growth direction into (0001) Si wafers, and also vertically cut parallel to the growth direction into (1-100) slices, which were then lapped and polished according to standard substrate processing technology. The polytype structures of the grown crystals were examined by micro-Raman spectroscopy. To explore the origin of 15R and 6H ploytype during the growth initiation, a short time growth experiment was run and the surface morphology was observed when the polytype inclusion probably appeared on the growth front.

Results and Discussion

In Fig. 1, typical polytype changes can be seen in an axial cut of a crystal. It is obvious that the 6H polytype transformation (yellowish color) starts near the seed-crystal interface and close to the seed edge. Besides the polytype inclusions, no other defects are visible at the beginning of the growth. Fig. 2 illustrates the Raman mapping of the polished wafers horizontally cut from another 4H-SiC crystals. The Raman mapping is fitted by the peak positions of the folded modes of transverse acoustic (FTA) bands which are related to the period of the SiC polytype and often used to easily identify the polytype structure [17]. From the FTA modes, 4H, 6H, and 15R polytypes were identified and their occupying areas are filled by red, green, and blue color, respectively. From Fig. 1 and Fig. 2, the polytypic lamellae were observed to be localized in the vicinity of the seed, which indicated that they were generated at the initial stage of the crystal growth and distributed at the periphery of the crystal. Following further growth, the inclusions were progressively overlapped by the main 4H polytype after approximately 3mm thick growth. It was noteworthy that a multitude of micropipes originated from the boundaries between foreign polytype and 4H-SiC. Although the polytype inclusions gradually disappeared during further growth, the micropipes caused by the poytype inclusions would penetrate through the whole crystal, leading to overall structural quality deterioration. Based on this key concept, it is essential for studying the generation of polytype inclusion at the seed periphery at the initial stage of crystal growth.

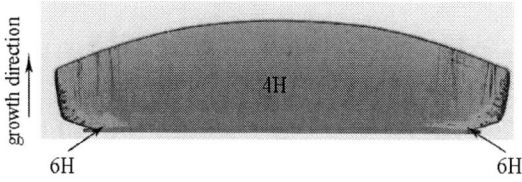

Fig. 1 Axial cut through a SiC bulk crystal in which polytype inclusions (6H parts on the edge of the crystal) appeared at the initial growth stage, and then were overgrown by the main 4H polytype.

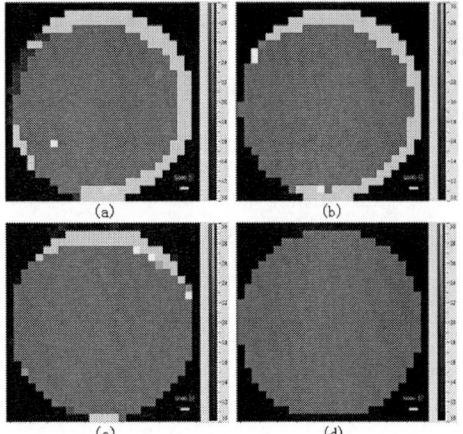

Fig. 2 Raman mapping of polished 4H-SiC wafers with 6H and 15R inclusions at the periphery at the beginning of growth. (a)、(b)、(c) and (d) are arranged according to growth sequence. The red, green, and blue areas belong to 4H, 6H and 15R polytypes, respectively.

Since parasitic nucleation of 15R and 6H generally occurred at the initial stage of crystal growth, we specially pay attention to the beginning stage of the growth process. The surface morphology of 4H-SiC after 2h growth at 2150℃ and 30 mbar was shown in Fig. 3(a). From this morphology, three significantly different regions at the periphery of the grown layer could be observed. Raman spectra for the corresponding three regions were shown in Fig. 3(b). It was noted that the peak intensity ratios (I_{150}/I_{204}) of the three regions were different.

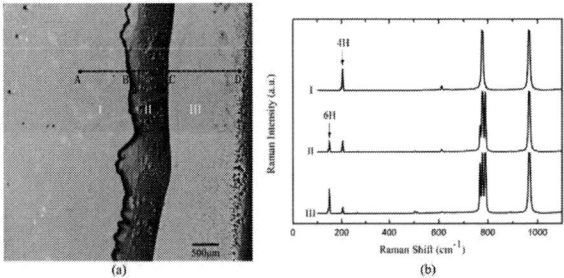

Fig. 3 The surface morphology of 4H-SiC a after 2h growth at 2150℃ and 30 mbar and Raman spectra for the three different regions in **a**, **b**.

To track the evolution of the 6H polytype along the basal plane, one dimensional Raman analysis along radial direction (A-B-C-D) was measured. Fig. 4 showed the peak intensity of FTA modes (150 and 204 cm^{-1}) as a function of the wafer position. The intensity ratio of Raman peaks at 150 and 204 cm-1 (I_{150}/I_{204}) was investigated to distinguish the polytypes at different regions. In region I, the ratios of 0 were obtained, which are typical of 4H-SiC {000-1} crystals. In region II, the I_{150}/I_{204} ratio increased monotonously with position, indicating the increase in 6H-SiC component towards the edge. The I_{150}/I_{204} ratio at region III took the highest value, and showed very little structure change with position. The component of 6H-SiC was completely dominated in this region. Thus, we proposed that 6H polytype inclusion nucleated at region III, where the intensity peak of the 4H (204cm^{-1}) still appeared

because of large laser beam focus in comparison with the thin lamellas. Once a foreign polytype originated at the periphery, it spread towards the center of the crystal (region II). At point B, the propagation of the 6H inclusion stopped by meeting the lateral growth on the facetted 4H-region.

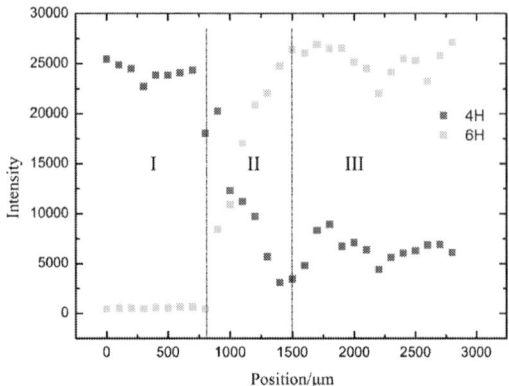

Fig. 4 The peak intensity of FTA modes as a function of position

Fig. 5 showed the surface morphology near the periphery of SiC crystal. The nuclei at the crystal edge were observed. The border between the periphery of nucleus at the periphery and the outer of facetted centre was clearly visible in Fig. 5(a).

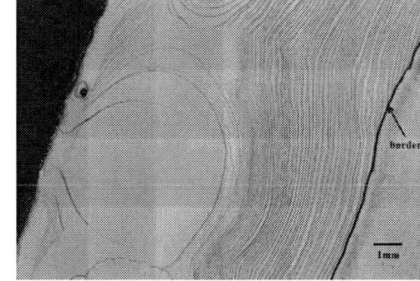

Fig. 5 The surface morphology near the periphery of SiC crystal. Arrow indicates the border between two crystallization centres.

Based on the above results, a sensible explanation of the polytype transformation close to the seed edge is the occurrence of random nucleation crystallization centres due to the high supersaturation at the periphery. Once polytype inclusion nucleated, their lateral expansion is driven by the higher growth rate at the initial stage of growth as seen in Fig. 1. Following further crystal growth, facetted centre in the central region became dominant and expanded into the periphery, resulting in foreign polytype overlapped by the main 4H polytype. The result is consistent with the thermodynamic analysis of SiC polytype stability by K. Kakimoto [18]. Fig. 6 shows the typical models of parasitic polytype generation and elimination.

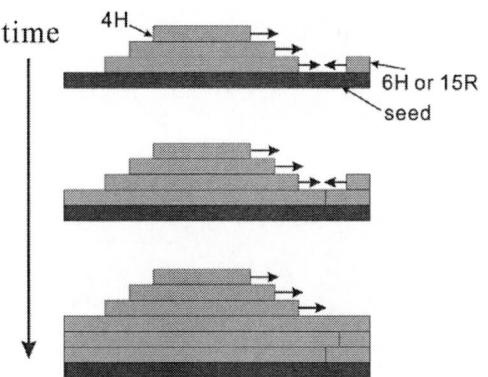

Fig. 6 Models of parasitic polytype generation at the outer parts of the crystal during the initial stage.

Conclusions

In this work, the polytype transformation during 4H-SiC PVT growth on on-axis seeds has been investigated. The initiation stage for a short-time growth is emphatically investigated to track the propagation and distribution of the polytype inclusion along radial direction. The results showed that the parasitic polytype originated at the periphery of the crystal. The lateral growth on the facetted 4H-region cannot expand to the periphery at initial stage of growth, thus making heterogeneous nucleation due to the high supersaturation more easily at the periphery of the crystal.

Acknowledgments

The work was supported by the National Key Research and Development Program of China (2016YFB0400501), the Shandong Outstanding Youth Fund (ZR2019JQ01), the Province Key R&D Program of Guangdong (2019B010126001), the Equipment pre-research Ministry of Education Joint Fund (6141A02022252), the Province Key R&D Program of Shandong (2017CXGC0412), the Fundamental Research Funds of Shandong University (2016JC037, 2018JCG01, 2019JCG010).

References

[1] A. Fissel, Journal of Crystal Growth, 212 (2000) 438.

[2] X.L. Wang, D. Cai, H. Zhang, Journal of Crystal Growth, 305 (2007) 122.

[3] X.R. Huang, D.R. Black, A.T. Macrander, J. Maj, Y. Chen, M. Dudley, Applied Physics Letters, 91 (2007) 231903.

[4] P. Wu, M. Yoganathan, I. Zwieback, Journal of Crystal Growth, 310 (2008) 1804.

[5] Y.M. Tairov, V.F. Tsvetkov, Journal of Crystal Growth, 43 (1978) 209.

[6] S. Limpijumnong, W.R.L. Lambrecht, Physical Review B, 57 (1998) 12017.

[7] D. Hofmann, E. Schmitt, M. Bickermann, M. Kolbl, P. J. Wellmann, A. Winnacker, Materials Science and Engineering B, 61 (1999) 48.

[8] E. Tymicki, K. Grasza, R. Bozek, M. Gata, Crystal Research and Technology 42 (2007) 1232.

[9] J. Takahashi, N. Ohtani, M. Kanaya, Japanese Journal of Applied Physics, 34 (1995) 4694.

[10] R.A. Stein, P. Lanig, S. Leibenzeder, Materials Science and Engineering B, 11 (1992) 69.

[11] R. Yakimova, M. Syvajarvi, T. Iakimov, H. Jacobsson, R. Raback, A. Vehanen, E. Janzen, Journal of Crystal Growth, 217 (2000) 255.

[12] M. Kanaya, J. Takahashi, Y. Fujiwara,A. Moritani, Applied Physics Letters, 58 (1991) 56.

[13] E.Y. Tupitsyn, A. Arulchakkaravarthi, R.V. Drachev, T.S. Sudarshan, Journal of Crystal Growth, 299 (2007) 70.

[14] E. Schmitt, T. Straubinger, M. Rasp, M.Vogel, A. Wohlfart, Journal of Crystal Growth, 310 (2008) 966.

[15] A. Itoh, H. Akita, T. Kimoto, H. Matsunami, Applied Physics Letters 65 (1994) 1400.

[16] Xiaobo Hu, Xiangang Xu, Xianxiang Li, Shouzhen Jiang, Juan Li, Li Wang, Jiyang Wang, Minhua Jiang, Journal of Crystal Growth, 292 (2006) 192.

[17] D.W. Feldman, J.H. Parker, W.J. Choyke, L. Patrick, Physical Review, 170 (1968) 698.

[18] K. Kakimoto, B. Gao, T. Shiramomo, S. Nakano, Shi-ichi Nishizawa, Journal of Crystal Growth, 324 (2011) 78.

Application of In-situ Pre-epi Clean Process for Next Generation Semiconductor Devices

Darian Waugh, Gim Chen, Jennifer Boecker and Ismail Kashkoush

NAURA Akrion Inc.

6330 Hedgewood Dr Suite #150, Allentown, PA 18106

ikashkoush@naura-akrion.com

Abstract

As electronic devices are evolving to more diversified and specifically function-oriented applications, silicon-based semiconductors have shown their limitation to unprecedented functionality requirements such as high power, high frequency, and high temperature operation. Growing utilization of IV-IV compounds (e.g. SiGe, SiC), III-V compounds (e.g. GaAs, GaN) as well as hetero-epitaxial structures with Si has become an inevitable trend. Due to cost and size of SiC and GaN wafers, epitaxial deposition on Si is utilized. However, this requires an efficient pre-epitaxial wet cleaning of the Si wafer yielding the lowest defects possible. In this study, different HF-last processes were tested yielding that an in-situ process with dilute chemicals gives the best results.

Introduction

Wide band-gap semiconductors show promise of improved performance at high frequencies, high power, and high temperature applications. Silicon Carbide (SiC) and Gallium Nitride (GaN) have shown an increase in uses due to this forward push to wide band-gap semiconductors versus the standard silicon.[1] Silicon Carbide has several properties such as: high electron saturated drift velocity, high thermal conductivity, high electric critical field, oxidative properties, and its chemical inertness, although, SiC has a lower electron mobility than that of Si. GaN on the other hand has a much higher electron mobility but lacks in thermal conductivity.[1]

Due to the size limitations of pure SiC wafers and GaN on sapphire as well as their high costs, epitaxial deposition of SiC and GaN on Si is desirable.[1,2] With further reduction of the dimensions of the microelectronic devices into the low nanometer scale, the cleaning procedures play an increasingly important role in the manufacturing of these new generation semiconductor devices. The standard approach for Si surface cleaning prior to epitaxial growth processes is a high temperature, usually greater than 1050°C, gas phase method to dissolve the native oxide along with any other contaminants on the surface of the wafer in order to prevent formation of any defects.[3] However, in the case that an advanced device with sensitive structures requires lower thermal budget treatments, a low-temperature pre-epitaxial cleaning process is needed. However, lowering the process temperature in turn causes an issue by lowering the desorption rate of SiO_2. This issue can be resolved by an HF-last process which converts the surface of the silicon wafer to a hydrogen-terminated surface, which when accomplished properly can yield a hydrophobic surface with the least defects.[3,6,7,8] The pre-epitaxial cleaning of Si wafers for SiC and GaN deposition can be approached in a wet bench at much lower temperatures than the gas phase method. The process chemicals, sequence and number of cleaning steps are becoming more critical in determining the desired end results.[4,5] The following study provides the data and process of proving that a one-step dilute in-situ-HF in the dryer is more effective than a traditional multi-tank HF-last process in a wet bench.[9]

Experimental

All experiments were conducted on Naura-Akrion's GAMA™ automated wet station which is capable of performing both a multi-tank sequence and single tank in-situ process. The silicon wafers are processed in the tool for the pre-epitaxial cleaning prior to the epitaxial growth step. Bare silicon wafers were processed with dummy oxide wafers, alternated or sandwiched, in order to simulate a situation with patterned wafers. The contamination levels from the oxide wafers on the bare wafers would be large due to the etch by-products from the oxide wafers depositing onto the bare wafers during processing. Multiple cleaning techniques were used in order to counteract the high level of contamination caused by the etching process. The conventional method consisting of SC1, Rinse, HF, Rinse, Dry was used first to remove any contaminants. Then two different techniques were tested in order to create the cleanest hydrophobic wafer. This consisted of either a surfactant being added to the HF tank or the in-situ HF process in the dryer. The experimental procedure and details of the materials used are shown in Figure 1. The materials used were: a GAMA™ wet bench equipped with a LuCID™ dryer (HF controlled injection), KLA-Tencor SurfScan (inspected at ≥ 0.12µm), bare Si wafers with low particle counts and thermal oxide wafers. Concentrations and parameters: 100:1 HF (23°C), 400:1 dHF (23°C), 1:2:50 dSC1 (50°C and 800W megasonic), DIO_3 rinse (~5-10ppm at 23°C).

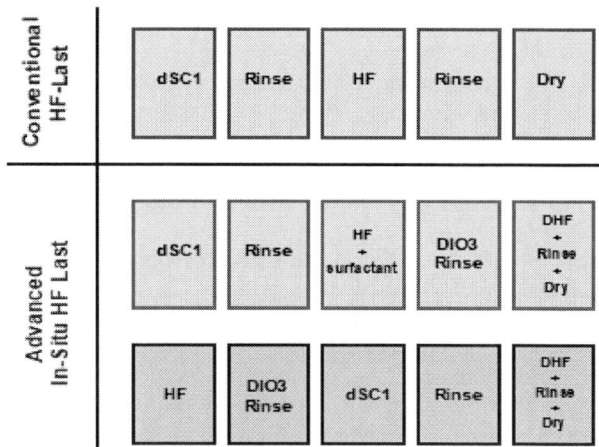

Figure 1: Experimental Procedures

Results and Discussion

Surface conditioning has proven to be the critical step in reducing the thermal budget for Silicon epitaxial growth. The typical standard process is to use a high temperature H_2 pre-bake to desorb the native oxide on the wafers to prepare the surface for an epitaxial layer deposition. However, lower temperatures are required to ensure isothermal processing for these advanced next-generation devices.[10,11] In IC manufacturing, wafers are typically mixed with oxide wafers or the wafers are patterned, and exposed silicon is typically adjacent to oxide or nitride areas. When the wafers are exposed to HF solutions, the by-products of the etched wafers will be removed from the hydrophilic surface and be deposited on the hydrophobic surface. This deposition results in high particle counts on the exposed silicon surface. The process contained herein was created to overcome this issue.

Before proceeding with the experimental procedures, tests to ensure particle neutrality within the GAMA™ wet bench were performed. The results of a conventional HF/Rinse/SC1/Rinse/Dry process yielded low particle addition even in the presence of oxide wafers; i.e. an average particle addition of − 6 (1 σ Stdev = 11, Figure 2). When only using bare silicon wafers, the conventional HF-last process yielded low particle addition as well; i.e. the average particle addition less than 40 particles at 0.12 µm (Figure 3). In addition, post epitaxial defects were also low (∼ 1.26 defects/cm²), as shown in Figure 4.

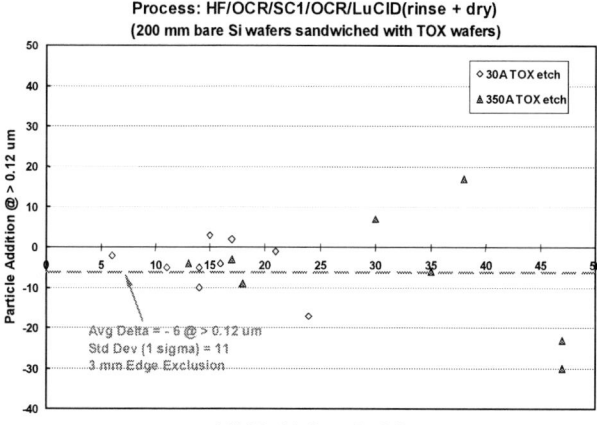

Figure 2: Particle Performance with simulated pattern oxide etch

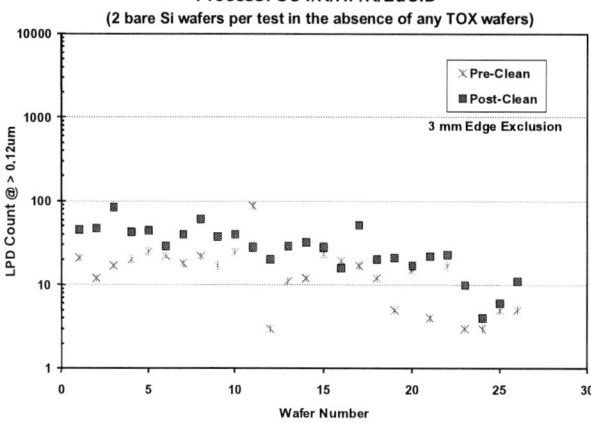

Figure 3: Conventional HF-last with bare Si wafers

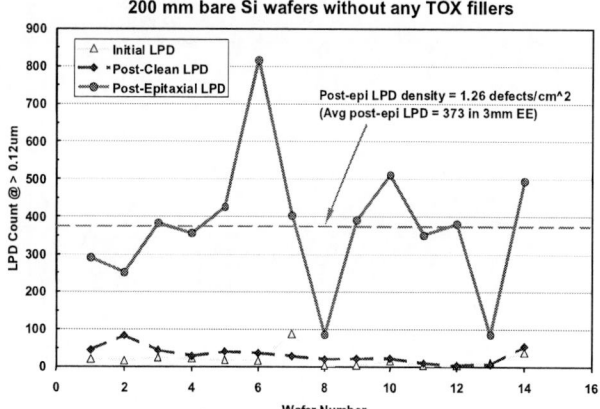

Figure 4: Post-Epitaxial LPD after conventional HF-last

Here silicon wafers were sandwiched between oxide filler wafers in order to simulate patterned wafers in a typical manufacturing environment. A conventional HF-last process (SC1/Rinse/HF/Rinse/Dry) resulted in high particle counts at 0.12µm (> 1,000). The high pre-epitaxial particle counts also caused high post-epitaxial defects (>30,000). The particulate defects are normally considered as nucleation sites of epitaxial defects during the epitaxial deposition process. Conventional methods of wafer transfer between tanks plays a significant role in increasing the deposition of silicate particles onto the silicon surface due to wafers crossing the liquid-to-air interface. To counteract the silicate deposition, two different approaches were tested.

The first tested method was to add a surfactant to the HF solution in the bath in order to improve the wettability of the wafers to reduce the particle adhesion. The process results proved to be slightly better than that of the conventional HF-last method. However, the presence of the surfactant on wafer surfaces in any trace amounts became problematic for the epitaxial growth process. In order to remove the surfactant, an additional step would be required in the process, e.g. ozone. Therefore, a new process to eliminate the use of the surfactant is desired. An in-situ process was thus developed in order to prevent the wafers from crossing the liquid-to-air interface in which the contaminants reside and deposit onto the wafer surface. HF chemical injection was used in the dryer to perform the in-situ process which yielded much lower particle deposition due to not crossing the liquid-to-air interface. Figures 5 and 6 show the results that the average particle adder of less than 50 particles.

An important note is that the use of ozonated rinse after HF and before going to the SC1 step is very critical in eliminating any potential for metal-induced pitting on the hydrophobic surface.[12] As reported by Knotter, Fe in the SC1 can induce pitting on hydrophobic wafer surface. The oxide chemically grown in the ozonated rinse is stable and thick enough (7-10 Å, as shown in Figure 7) to protect the silicon surface from any effects of metal roughening. The post epitaxial cleaning results for the in-situ method are shown in Figure 8, and the average LPD density per wafer is about 0.89 defects/cm². Figure 8 also indicates that the lower the HF-last defects is, the lower the post epitaxial deposition defects would be.

978-1-7281-5757-3/19 $31.00 © 2019 IEEE

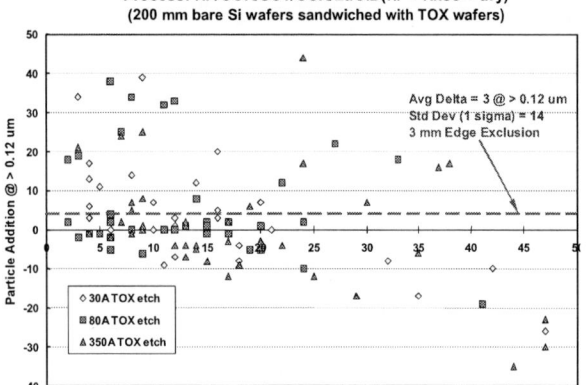

Figure 5: Particle performance with simulated pattern oxide etch (hydrophobic at the end)

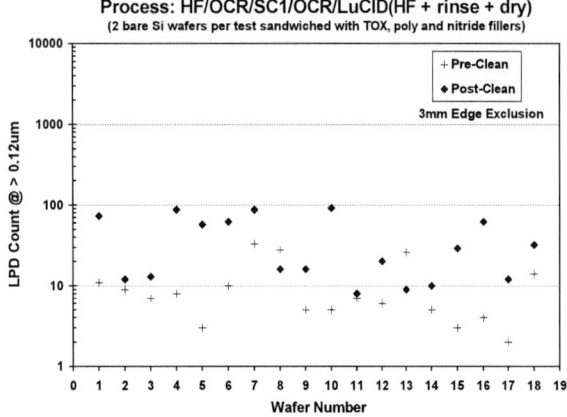

Figure 6: In-situ HF-Last with simulated pattern oxide etch

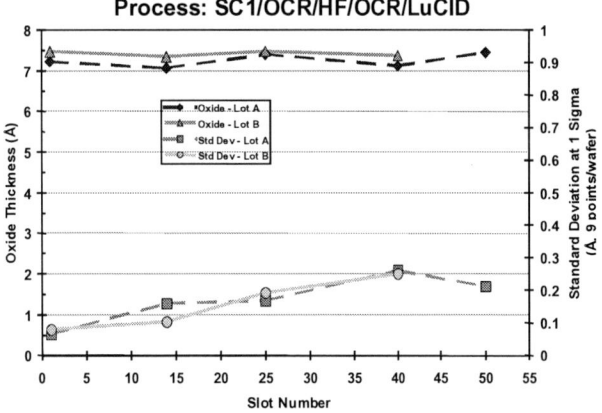

Figure 7: Oxide regrowth uniformity by OCR

The results of each of the different cleaning recipes are summarized in Table 1. Different types of filler wafers were also used, i.e. polymer and nitride, to investigate if the filler wafer type would have any negative impact on the results. The results from testing showed that the most critical step in order to achieve extremely low post epitaxial deposition defects is the in-situ process which requires no wafer transfer between steps.

In order to characterize the background oxide thickness, measurements were also taken as a measure of the oxygen content on the wafer surface. It is equally important to notice that the amount of oxygen present on the wafer surface could significantly increase the number of post-epitaxial defects on the wafer. The lower the oxygen content presents on an H-passivated surface, the lower amount of post-epitaxial defects is observed on the wafer surface, as shown in Figure 9.

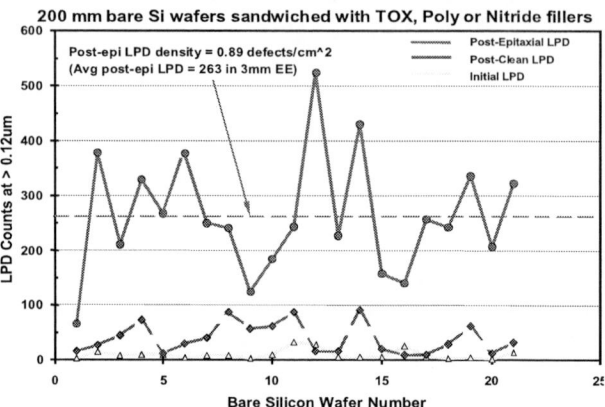

Figure 8: Post Epitaxial (900°C bake) LPD after In-situ HF-Last

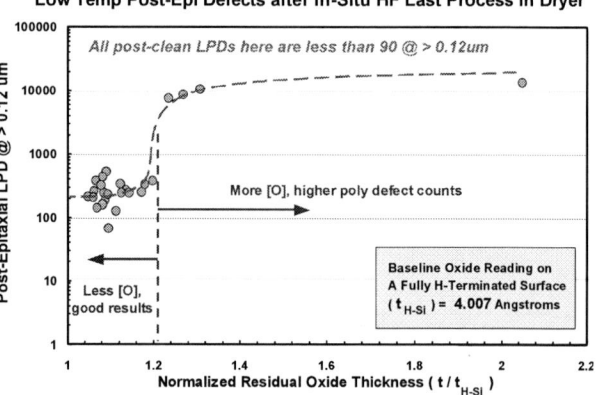

Figure 9: Low Silicate Particles and Low Residual Oxide Content on Surface are Crucial

Conclusions

With the increase in use of epitaxial materials such as GaN and SiC deposited on Si, an efficient pre-epitaxial cleaning resulting in low particle addition is necessary. An in-situ cleaning process in the dryer was developed and used for pre-epitaxial growth. Results for the in-situ cleaning showed a significant improvement over the standard HF-last process. The reason for the in-situ process having a great impact on the elimination of particle deposition is due to the Si wafers not crossing the liquid-to-air surface between the etch, rinse, dry process. The experiments conducted in the study proved that the use of dilute chemicals and the in-situ HF, rinse, dry process yield lower defects than that of the standard multi-tank HF-last process. The lower defects after the pre-epitaxial clean (post-clean) are reflected on the epitaxial growth yielding lower defects after the product deposition.

978-1-7281-5757-3/19 $31.00 © 2019 IEEE

Cleaning Recipe	48 Filler Wafs	LPD Sum @ > 0.12um			Rudolph post-cln oxide (A)	Rudolph post-cln 1 Sigma	Post-Epi LPD >0.12um
		Pre-Clean	Post-Clean	Delta			
SC1/OCR/HF(+surft)/OCR/LuCID(inj HF+R+Dry)	TOX	3	16	13	4.390	0.256	66
	TOX	15	27	12	4.275	0.188	379
	TOX	8	45	37	4.201	0.234	211
SC1/OCR/LuCID(HF+R+Dry)	TOX	9	73	64	4.728	0.414	329
	TOX	10	12	2	4.553	0.294	267
OCR/LuCID(HF+R+Dry)	TOX	4	30	26	4.805	0.133	377
	TOX	8	40	32	4.7	0.117	250
HF/OCR/SC1/OCR/LuCID(inj HF+R+Dry)	TOX	8	87	79	4.51	0.191	241
	TOX	3	57	54	4.459	0.107	125
	TOX	10	62	52	4.362	0.142	185
HF/OCR/SC1/OCR/LuCID(stg HF+R+Dry)	TOX	33	87	54	4.351	0.095	243
	TOX	28	16	-12	4.37	0.124	524
	poly & ntrd	5	16	11	4.389	0.095	227
	poly & ntrd	5	91	86	4.341	0.129	430
	poly & ntrd	6	20	14	4.334	0.112	158
	poly & ntrd	26	9	-17	4.288	0.087	141
	poly & ntrd	5	10	5	4.258	0.134	257
	TOX	3	29	26	4.584	0.18	242
	TOX	4	62	58	4.501	0.231	335
	TOX	2	12	10	4.252	0.113	207
	TOX	14	32	18	4.325	0.126	322

Table 1: Cleaning recipe result data

References

1. Microsemi, Gallium Nitride (GaN) versus Silicon Carbide (SiC)In The High Frequency (RF) and Power Switching Applications
2. Golecki, I., et al., Appl. Phys. Let., Vol. 60 (1992) pp. 1703-1705.
3. Caymax, M. et. al, Solid State Phenomena Vols. 65-66 (1999) pp. 237-240, 1999 Scitec Publications, Switzerland.
4. Besson, P. et al, UCPSS '2000, Vols. 76-77 (2001) pp. 199-202.
5. Kashkoush, I., et al., Mat. Res. Soc. Symp. Proc., Vol. 477, 1997, pp. 311-316.
6. Golecki, I., et. al., Appl. Phys. Lett., Vol. 69 (1992) p. 1730.
7. Fissel, A., et. al., Appl. Phys. Lett., Vol. 66 (1995), p. 3182.
8. Patruno, P., Fleury, A., Andre, E., and Tardif, F., UCPSS '94 Proc., pp. 247-250.
9. Kashkoush, I., et al., Elec. Soc. Clean. Symp. Proc., Vol. 26, 2001, pp. 345-351.
10. Mouche, M., et al, UCPSS '96 Proc. Pp 269-272.
11. Verhaverbeke, S and Pagliaro, B., Electrochem Soc. Proc. Vol. 99-36, pp. 445-451.
12. Knotter, M. and Dumensil, Y., UCPSS '2000, Vols. 76-77 (2001) pp. 255-258.

An Improved Composite JTE Termination Technique for Ultrahigh Voltage 4H-SiC Power Devices

Rui Hu[1,3*], Xiaochuan Deng[1,3], XiaoJie Xu[1,3], Xuan Li[1,3], Juntao Li[2], Zhiqiang Li[2], Yourun Zhang[1], and Bo Zhang[1]

[1.] State key Laboratory of Electronic Thin Films and Integrated Devices, University of Electronic Science and Technology of China, Chengdu 610054, China.

[2.] Microsystem and Terahertz Research Centre, China Academy of Engineering Physics, Mianyang 621900, China

[3.] Institute of Electronic and Information Engineering in Guangdong, University of Electronic Science and Technology of China, Dongguan 523808, China

ruihu2018@yeah.net

Abstract

This paper presents a novel and efficient multiple-step-modulated JTE (MSM-JTE) termination technique for ultrahigh voltage (>10 kV) silicon carbide (SiC) devices, to extend the ultrahigh voltage JTE dose window and increase the breakdown voltage. MSM-JTE takes advantage of ring assisted JTE, etched JTE and space modulated JTE, to relief local electric field concentration and form a gradual decrease of effective charges overall. This is similar to lateral variation doping (VLD) technique which is widely used in silicon. A practical fabrication processes is also described. Compared with conventional TZ-JTE, MSM-JTE requires only one extra etching process and is insensitive to doping dose and energy of ion implantation. The MSM-JTE is applied to 15 kV PiN rectifier and simulated by Silvaco TCAD. The simulation result shows MSM-JTE could reach a nearly ideal maximum efficiency of 99 % and keep an efficiency of 95 % in a doping interval of 7×10^{12} cm^{-2}. Tolerance to etching depth uncertainties is also high enough for process reliability and repeatability.

1. Introduction

Modern high-performance power electronic systems such as sophisticated military systems and the next generation DC distribution networks are surpassing the power density, efficiency and reliability limitations set by the inherent properties of widely employed silicon-based devices [1], [2], [3]. To make a breakthrough, new device technologies are being explored intensely during the past decades such as silicon carbide (SiC) device. SiC devices have exceeded previously constraining restrictions imposed by silicon-based devices especially on ultrahigh voltage and extreme high temperature as shown in Fig. 1.1, thus becoming a promising candidate to reduce size, weight and thermal management requirements of future equipment [4], [5], [6], [7]. However, the edge termination problem limits the breakdown voltage of practical SiC devices and poorly terminated junction decreases 10 % to 20 % of ideal breakdown voltage, which is intolerable for ultrahigh voltage devices. To solve the problem, termination techniques are developed to reduce electric filed crowding at junction edge. Typical termination techniques include field limiting ring (FLR), junction termination extension (JTE), bevel-edge termination and field plate (FP), but they all show obvious drawbacks in ultrahigh voltage application. FLR is completed simultaneously with the main junction and such a simple fabricating process makes it widely used. But the disadvantage is that for ultrahigh voltage (≥10 kV) device more than 100 FLRs are needed which is quite a consumption for chip area and design cost [8], [9], [10]. JTE termination covers much less area than FLR, but its sensitivity to dose and energy of ion implantation leads to a narrow optimum JTE-dose window. The interfacial charge between Si-to-SiO2 contact intensifies the problem by leading the shift of doping dose curve. Moreover, JTE might encounter crystal damage and surface roughness caused by the ion implantation and high temperature annealing. All these obstacles cause unacceptable decline in yield in practice for JTE termination technique. Although deep bevel-edge termination is area-efficient, it causes interfacial defects and high leakage current. FP is rarely used in ultrahigh voltage applications due to oxide reliability problem. According to Gauss' Law ($\varepsilon_{oxide}E_{oxide} = \varepsilon_{sic}E_{sic}$), the electric field in oxide dielectric is about 3 times higher than that in SiC substrate under voltage stress, leading to early breakdown of oxide layer. In order to achieve optimal electric field modulation effect, new termination techniques are trying to combine conventional termination techniques to complement each other and reach the optimal electric field modulation effect. Among them, JTE based termination structures show the best comprehensive performance. Composite JTE structures like space-modulated JTE (SM-JTE) [11], ring assisted JTE (RA-JTE) [12], [13]and hybrid JTE types [14], [15], [16] all manage to expand the optimum JTE-dose window with small increase in process cost.

In this paper, a further improved composite JTE structure having larger optimum JTE-dose window is proposed and referred to as multiple-step-modulated JTE (MSM-JTE). By device simulation and comparison with two zone JTE (TZ-JTE), we try to introduce the new termination technique in detail.

2. Device Structure

The cross-sectional structures of 15 kV PiN rectifier with conventional TZ-JTE and proposed MSM-JTE are shown in Fig. 2.1. The structures are studied by Silvaco TCAD, and simulation parameters of active area are listed in Table 1. For termination area, doping is defined by process simulation framework, Athena from Silvaco. A box profile with a depth of about 500 nm is formed by three step Al-ion implantation with 500 keV, 250 keV and 120 keV implant energy respectively. Furthermore, in practical production, the steps will be patterned using a thick photoresist mask which causes incline of step sidewall. This angle between the horizontal surface and the sidewall is also taken into consideration in our simulation. The JTE length are 500 μm for both with 3:2 JTE1 and JTE2 ratio. the overlap of JTE1 and JTE2 is 1 μm. The bevel-edge has a height of 2 μm and an angle of 45°. Except the bevel edge, the etching depth are the same 0.2 μm for the whole MSM-JTE termination. The steps in each group are evenly spaced at 7 μm interval, and the width of the steps within one group are 21 μm, 17 μm and 13 μm respectively.

978-1-7281-5757-3/19 $31.00 © 2019 IEEE

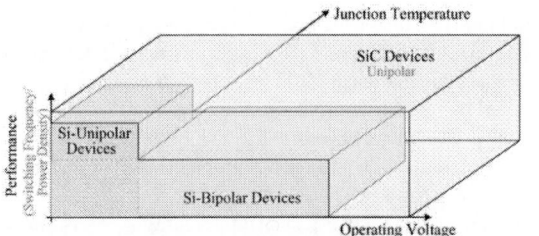

Fig. 1.1 The performance under high temperature and voltage of SiC and Si devices for general systems

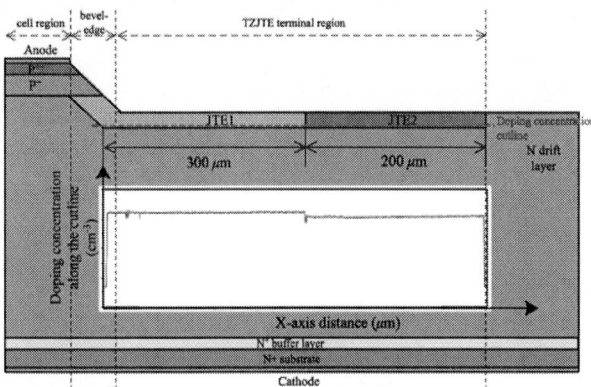

Fig. 2.1 (a) conventional TZ-JTE structure

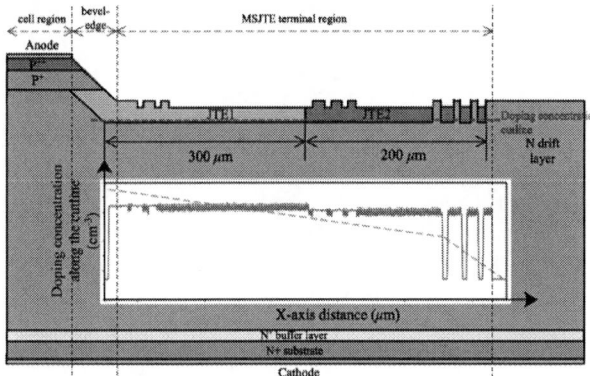

Fig. 2.1 (b) MSM-JTE structure

Fig 2.1 Schematic cross sections of 15 kV 4H-SiC rectifier with conventional TZ-JTE and proposed MSM-JTE and their termination doping concentration variation

Table 2.1 15kV PiN rectifier active area simulation parameters

Layer	Thickness (μm)	Doping dose (cm^{-2})
P^{++}	0.5	2×10^{19}
P^{+}	1	2×10^{18}
N$^-$ drift area	130	5×10^{14}
N$^+$ buffer	2	1×10^{19}
N$^+$ substrate	3	1×10^{18}

Compared with TZ-JTE, MSM-JTE attained a laterally tapered distribution of effective JTE dose without multiple JTE-dose injection as shown in Fig. 2.1(b). The red line is effective doping concentration and the dashed redline is its fitting line. The doping concentration has both local consistency and overall gradual trend. This lateral variation

doping (VLD) effect can enhance the protection of the main junction and reduce the peak electric field at termination edge. Also, the modulation steps are added at easy breakdown areas trying to strengthen these local areas.

The step width and step interval affect the lateral doping concentration change gradient. To keep a relative smooth slope, the steps in each group are evenly spaced and the ratio of three step widths are kept 8:7:5. The step interval is designed refer to maximum FLR interval which can be calculated by [17]

$$L_{spacing} = (r_i^2 + 2r_i \frac{\varepsilon_{SiC}E_C}{qN_D})^{\frac{1}{2}} \qquad (1)$$

where i is the number of rings, r_i is the radius of curvature of the i-ring.

A practical fabrication process for MSM-JTE is illustrated in Fig. 2.2. Compared with conventional TZ-JTE process, there is only one extra etching process. Both the bevel-edge and step shapes can be done by inductively coupled plasma etching.

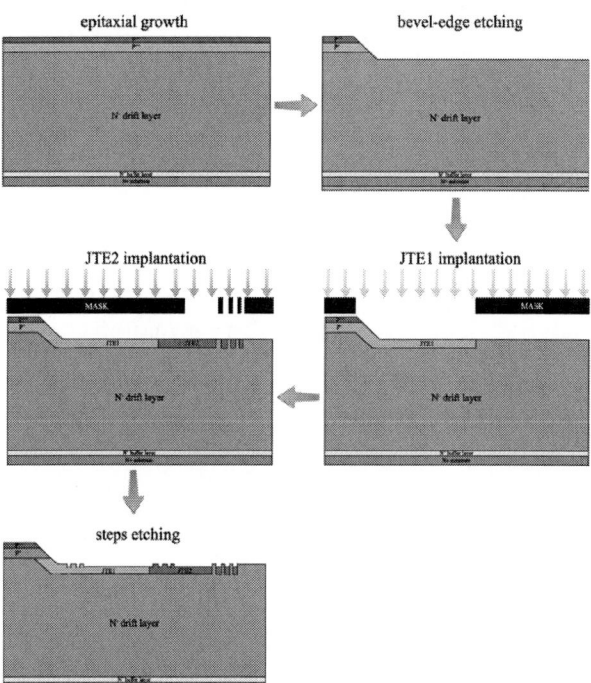

Fig. 2.2 Major process flow of 15kV 4H-SiC PiN rectifier with MSM-JTE termination

3. Simulation Result and discussion

Fig. 3.1 and Fig. 3.2 shows MSM-JTE dose dependence of simulated breakdown voltage. The breakdown voltages in both figures are obtained when leaking current reaches 1 nA for a single simulation unit. Fig. 3.1 is the comparison among optimal JTE dose window of MSM-JTE, MRM-JTE and TZ-JTE. Under 15 kV, our previously reported MRM-JTE [15] has successfully extend the optimal JTE dose window to about 2 times larger than that in TZ-JTE. MSM-JTE further enlarges the optimal JTE dose window to more than 2 times larger than MRM-JTE dose window and about 5 times larger than TZ-JTE dose window. The optimal JTE dose plateau of MSM-JTE as the red dashed ling in Fig. 3.1 shows is long enough for high process reliability in mass production.

978-1-7281-5757-3/19 $31.00 © 2019 IEEE

Fig. 3.2 shows optimal MSM-JTE-dose window, along with effect of SiO_2-SiC surface charge to MSM-JTE-dose curve. At 1.90×10^{13} cm^{-2} JTE1 doping and 1.27×10^{13} cm^{-2} JTE2 doping, MSM-JTE termination has a nearly ideal efficiency up to 99%. An efficiency of 95% could be maintained in a doping interval of 7×10^{12} cm^{-2}. The shift amount of doping curve is approximately equal to the SiO_2-SiC surface charge amount, and the ultra-wide optimal MSM-JTE-dose window reduces the breakdown voltage degradation caused by interface charge even in high interface charge up to 2×10^{12} cm^{-2}. Besides benefiting fabricating process, this advantage is of great help for devices to withstand extreme environments such as aerospace, where long-term cosmic radiation introduces SiO_2-SiC surface charges and lead to the degradation of device reverse bias performance.

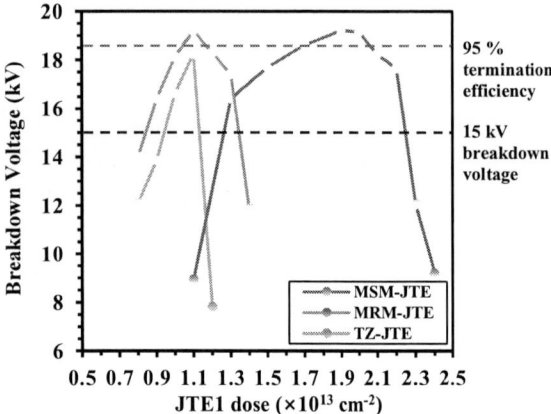

Fig. 3.1 Breakdown voltage versus the implant doses for 15kV PiN rectifier with MSM-JTE, MRM-JTE and TZ-JTE termination

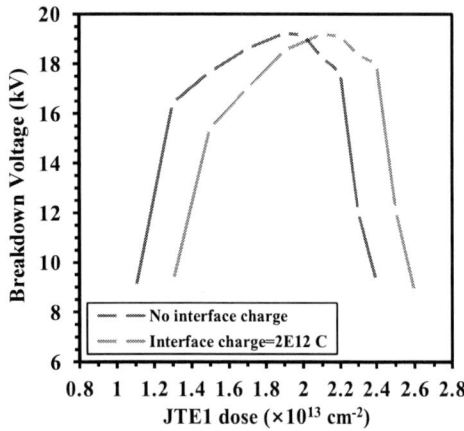

Fig. 3.2 Impact of SiO_2-SiC surface charge on breakdown voltage

The uniformity of electric field in termination area directly affects reverse breakdown voltage. From low doping to high, the breakdown region extends from bevel-edge bottom to JTE1 and JTE2 overlap area and finally shrinks to the JTE2 edge as shown in Fig. 3.3. The depletion layer edge curvature keeps a radius much larger than drift area under 1.5×10^{13} cm^{-2} to 2.2×10^{13} cm^{-2} doping dose at off-state breakdown. Beyond the

doping interval, the curvature of depletion layer sharply shrinks, causing electric field concentrating to its edge and breakdown voltage abrupt decreasing. This is because each JTE area is protected by its following JTE part except the JTE end who has no extra protection. Therefore, the JTE2 dose is designed to be much lower than JTE1 to lock high electric field at JTE1 and increase robustness.

Local modulation effect of the etching steps is shown in Fig. 3.4. Both MSM-JTE and TZ-JTE are at their optimal doping dose. The surface electric field is obtained from lateral cutline 3.3 μm below the surface, trying to pass as much breakdown area as possible. The MSM groups are inserted at bevel-edge bottom, JTE1 edge and JTE2 edge where electric field crowds and they effectively cut down local electric field peaks. In consider of fabrication process uncertainties, special attention is paid to breakdown voltage sensitivity to etching depth. Simulation result of etching depth effect on the 15 kV PiN rectifier breakdown voltage is shown in Fig. 3.5. The etching depth is able to tolerant ±0.2 μm deviation from the designed value, which is quite large compared to 0.1 nm etching process accuracy.

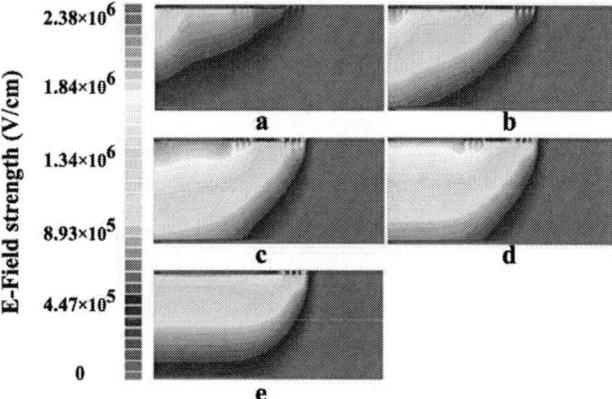

Fig. 3.3 Electric field distribution of MSM-JTE under different termination injection dose. The JTE1 doping dose are respectively (a) 1.1×10^{13} cm^{-2}, (b) 1.5×10^{13} cm^{-2}, (c) 1.9×10^{13} cm^{-2}, (d) 2.2×10^{13} cm^{-2}, (e) 2.3×10^{13} cm^{-2}

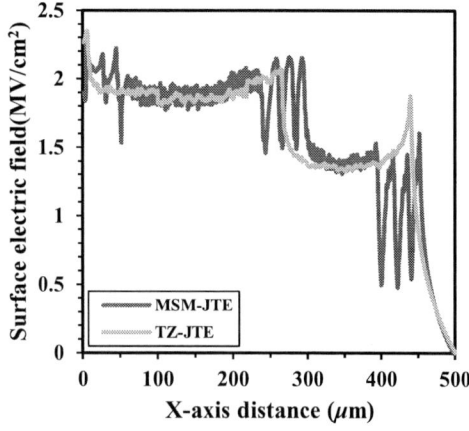

Fig. 3.4 Surface electric field distribution of TZ-JTE and MSM-JTE with different parts.

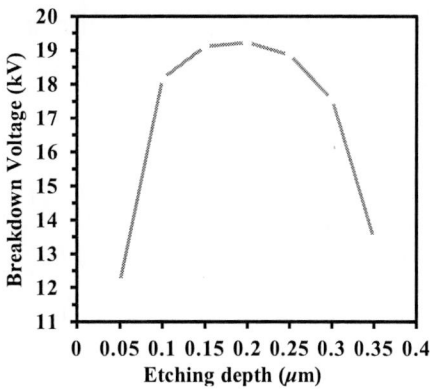

Fig. 3.5 Etching depth effect on breakdown voltage. The JTE1 doping dose is kept at 1.9×10^{13} cm^{-2}

4. Conclusions

MSM-JTE termination technique with ultrahigh reverse blocking voltage, extra-large optimum JTE-dose window and high tolerance to etching depth uncertainties has sufficiently improved process reliability and mitigated SiO_2-SiC surface charge effect. The simulation results demonstrate that the MSM-JTE termination provides a nearly ideal breakdown voltage at optimal doping concentration and keeps 95 % efficiency in a doping interval of 7×10^{12} cm^{-2}. These performances show MSM-JTE has great potential to enable high processability, device yield and robustness which are urgently required for ultrahigh voltage SiC device technology.

Acknowledgments

This work was supported in part by Science Challenge Project under Grant TZ2018003-1-201, and Natural Science Foundation of Guangdong under Grant 2019A1515012085.

References

1. Matsumoto M, Umeda Y, Masui K, *et al*, Design for Innovative Value Towards a Sustainable Society, Springer (Dordrecht, 2012).
2. Costabeber A, Tenti P, Caldognetto T, *et al*, "Selective compensation of reactive, unbalance, and distortion power in smart grids by synergistic control of distributed switching power interfaces," *EPE*, Lille, September. 2013, pp. 1-9.
3. Rothmund D, Ortiz G, Guillod T, *et al*, "10kV SiC-based isolated DC-DC converter for medium voltage-connected Solid-State Transformers," *IEEE APEC*, Charlotte, NC, March. 2015, pp. 1096-1101.
4. Hornberger J, Lostetter A, McNutt T, Magan Lal S, and Mantooth A, "The Application of Silicon Carbide (SiC) Semiconductor Power Electronics to Extreme High-Temperature Extraterrestrial Environments," *Proceedings of the 2004 IEEE Aerospace Conference*, Big Sky, MT, March. 2004, pp. 2538-2555.
5. Mustain H A, Lostetter A B and Brown W D, "Evaluation of gold and aluminum wire bond performance for high temperature (500 /spl deg/C) silicon carbide (SiC) power modules," *ECTC'05.*, Lake Buena Vista, FL. 2005, pp. 1623-1628.

6. Ilyin V A, Afanasyev A V, Ivanov B V, *et al*, "High-Voltage Ultra-Fast Pulse Diode Stack Based on 4H-SiC," *Mater. Sci. Forum*, Vol. 4, No.858 (2016), pp. 786-789.
7. Biela J, Schweizer M, Waffler S, *et al*, "SiC versus Si—Evaluation of Potentials for Performance Improvement of Inverter and DC–DC Converter Systems by SiC Power Semiconductors," *IEEE Trans. Industrial Electronics*, Vol. 58, No. 7 (2011), pp. 2872-2882.
8. Onose H, Oikawa S, Yatsuo T and Y. Kobayashi, "Over 2000 V FLR termination technologies for SiC high voltage devices," *12th International Symposium on Power Semiconductor Devices & ICs. Proceedings*, Toulouse, France, June. 2000, pp. 245-248.
9. Wada K , Uchida K , Kimura R, *et al*, "Blocking characteristics of 2.2 kV and 3.3 kV-class 4H-SiC MOSFETs with improved doping control for edge termination," *Mater. Sci. Forum*, vols. 778–780 (2014), pp. 915–918.
10. Ben Tan, Xiao-Li Tian, Jiang Lu *et al*, "Design and Optimization of Four-Region Multistep Field Limiting Rings for 10kV 4H-SiC IGBTs," *IEEE ICSICT*, Qingdao, October. 2018, pp. 1-3.
11. Feng G, Suda J and Kimoto T, "Space-Modulated Junction Termination Extension for Ultrahigh-Voltage p-i-n Diodes in 4H-SiC," *IEEE Trans. Electron Devices*, Vol. 59, No. 2 (2012), pp. 414-418.
12. Dheilly N, Planson D, Paques G, *et al*, "Light triggered 4H–SiC thyristors with an etched guard ring assisted JTE," *Solid-State Electronics*, Vol. 73 (2012), pp. 32-36.
13. Kaji N, Niwa H, Suda J and Kimoto T, "Ultrahigh-Voltage SiC p-i-n Diodes With Improved Forward Characteristics," *IEEE Trans. Electron Devices*, Vol. 62, No. 2 (2015), pp. 374-381.
14. Zhou C N, Wang Y, Yue R F, *et al*, "Step JTE, an Edge Termination for UHV SiC Power Devices With Increased Tolerances to JTE Dose and Surface Charges," *IEEE Trans. Electron Devices*, Vol. 64, No. 3 (2017), pp. 1193-1196.
15. Xiaochuan Deng, Lijun Li, Jia Wu *et al*, "A Multiple-Ring-Modulated JTE Technique for 4H-SiC Power Device With Improved JTE-Dose Window," *IEEE Trans. Electron Devices*, Vol. 64, No. 12 (2017), pp. 5042-5047.
16. W. Sung and B. J. Baliga, "A Near Ideal Edge Termination Technique for 4500V 4H-SiC Devices: The Hybrid Junction Termination Extension," *IEEE Electron Dev. Lett.*, Vol. 37, No. 12 (2016), pp. 1609-1612.
17. X. B. Chen, "A simple theory of floating field limiting rings," *Acta Electron Sinica*, Vol. 16, No. 3 (1988), pp. 6-10.

Electro-thermal Analysis of 1.2kV-100A SiC JBS Diodes Under Surge Current Stress

Wei Zhong[1,2], Yidan Tang*[1,3], Chengzhan Li[4], Hong Chen[1], Yourun Zhang[2], Yun Bai[1], Xinyu Liu[1]

[1] High-Frequency High-Voltage Devices and Integrated Circuits R&D Center, Institute of Microelectronics of Chinese Academy of Sciences, Beijing, China

[2] University of Electronics Science and Technology of China, Chengdu, China

[3] University of Chinese Academy of Sciences, Beijing, China

[4] Zhuzhou CRRC Times Electric Co. Ltd., Zhuzhou, Hunan, China

E-mail: tangyidan@ime.ac.cn (corresponding author)

Abstract

The reliability of power diode under large surge current stress is crucial to the applications. In this paper, the electro-thermal simulation and experimental evaluation of surge failure mechanisms of high current 1200V SiC JBS diodes were conducted. Different from thermal analysis from the perspective of thermal network, this paper combines TCAD simulation with 3D-thermal simulation for electrothermal simulation, which makes the electrical and thermal characteristics more intuitively combined. Through simulation, it is found that the contact thermal resistance between bonding wire and pad and the heating of bonding wire itself will lead to the temperature concentration near the bonding wire. The mechanism is verified by a forward nonrepetitive surge test, and the results are consistent with the simulation results.

Introduction

High-power junction barrier Schottky (JBS) diodes, manufactured by third-generation semiconductor material SiC, have the advantages of high frequency, low power consumption and high switching frequency, and have been widely used in various high current and frequency switching power supply fields. The ability of diode to resist the surge overload current is the guarantee of its reliability in practical application. To analyze the thermal stress on a power device, the device is simulated using DC steady state and transient conditions. One of the standards that relate to the surge current testing of power semiconductor rectifiers is the forward nonrepetitive surge test, which consists of longer, lower current pulses in the range of 10 ms [1]. The purpose of this test is to test the surge current capability of the device chip itself, internal and external wires and contacts under high current stress conditions and to verify the nonrepetitive surge current grade of a device. However, the surge test is destructive to the device. Therefore, it is necessary to calibrate the simulation model to guide the test and reduce the number of tests.

Establishing a thermal-network model for SiC JBS devices is a common thermal analysis method, and using 3D thermal simulation software will make the thermal network model more specific. The combination of TCAD simulation and 3D thermal simulation can successfully demonstrate the complete heating in the working process of the power device. We use Mixed Mode to build a surge test circuit in TCAD simulation, and build a package model of the device in 3D thermal simulation. Regarding the failure mechanism, most of the existing researches focus on the surge failure analysis of SiC diodes with a rated current of about 10A, the failure mechanisms they proposed are mostly metal melting near the termination, which leads to device failure[2]-[4]. In these small current devices, the

size of the bonding wire is so small that its heat dissipation of the chip is basically ignored[5],[6].

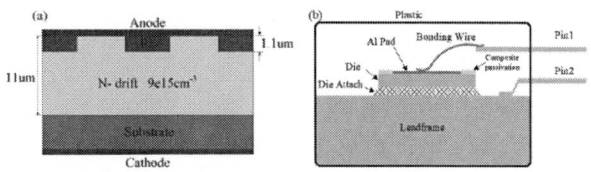

Fig. 1 Schematic cross-section structure of (a) the JBS diodes used in simulation and (b) the TO-247 package.

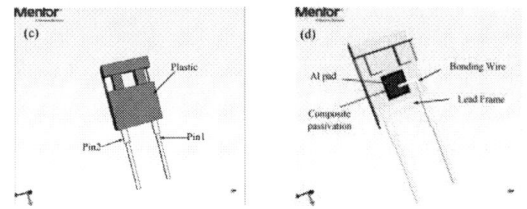

Fig. 2 (c) and (d) are 3D numerical models of the package.

Nevertheless, in the high current device, as the effect on the size of the bonding wire increases, the heat becomes apparent, which dissipates heat to the device.

In this work, we analyze the surge failure of a high-current 1.2kV SiC JBS diode with a rated current of 100A. By combining TCAD simulation with Flotherm thermal simulation to establish an integrated solution of electro-thermal simulation from the chip to the package. Also, structure function was used to calibrate the electro-thermal simulation model, which made the simulation results more accurate. In the simulation, we added the effect of bonding wire on the heat dissipation of the chip and simulated the position of device failure point. Finally, we conducted relevant experimental verification.

Device and package characteristics

The 1200V/100A SiC JBS diode fabricated in this experiment has a single-chip area of 6.1 mm × 6.1 mm. The material portion mainly includes a n+ substrate, a n-type buffer layer, and a n-type epitaxial layer. The epitaxial layer has a concentration of 9×10^{15}cm^{-3} and a thickness of 11 μm. The p+ region has a width of 2 μm and a pitch of 3 μm. Active region of the device is introduced into the p+ doping structure, and the junction termination adopts a multi-field limiting ring structure. The active region p+ doping is formed simultaneously with the p+ doping of the field limiting ring.

978-1-7281-5757-3/19 $31.00 © 2019 IEEE

The multi-step implantation of different ions and different doses of Al ions is performed in SiC. A box-shaped distribution is having a junction depth of 1.1μm and an average doping concentration of 1×10^{19} cm^{-3} was formed on the epitaxial layer; the terminal passivation was a composite passivation (Fig. 1(b)) layer structure of silicon dioxide (SiO$_2$) and polyimide.

(a)

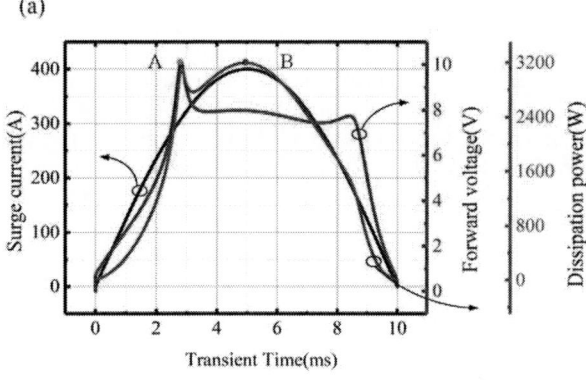

(b)

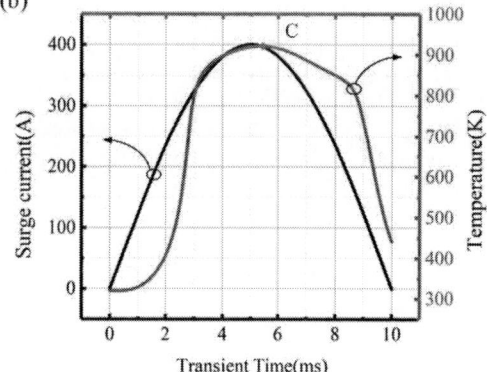

Fig. 3 Forward voltage, surge current and dissipation power (a) and surge current and lattice temperature (b) waveforms for the pulsed simulation of the JBS diode.

TABLE I. SPECIFICATIONS OF THE TEST MODULE COMPONENTS

Component	Specifications
Die	Material: SiC Size: 6.1 × 6.1 mm², Thickness: 0.350 mm
Composite passivation	Silicon dioxide (SiO$_2$) and polyimide (PI) Thickness: 8μm
Al pad	Thickness: 4 μm
Lead frame	Material: Cu Size: 15 × 20 mm², Thickness: 1.5 mm
R$_{\theta JC}$	thermal resistance from junction to case: 0.41K/W
Bonding wire	380 μm diameter aluminum wires

The cathode electrode is formed of a Ni metal system and forms a superior ohmic contact with the SiC material.

For the high temperature application, the fabricated 4H-SiC JBS dies were packaged in TO-247-2L package with high temperature solder material on direct-bond-copper board and bonded using the 380 μm diameter aluminum wires. The TO-247-2L package (Fig. 1(b)) from top to bottom is: plastic material, aluminum wire, chip, solder paste, tin-plated copper board. The average thickness of 50 μm solder paste containing lead was served as die-attach materials, which are capable of operating at temperature about 280 °C. Fig.2 (c) and (d) are 3D numerical models of the package established by the thermal CFD simulator-Flotherm, the package parameters are determined by the standard TO-247-2L package, the chip area is the same as the chip design size, and the Al bonding wire cannot be ignored because this diode has a high rated current. Detailed specifications of the components are listed in Table 1.

Device simulation

The static parameters of the devices studied in this paper were mentioned in previous studies[8]. In this study, we performed a two-dimensional electrothermal mixed-mode device simulation and the Shockley-Read-Hall (SRH) recombination was included[9]. Considering high current density in a power SiC JBS, the Auger recombination

(AUGER) was also applied[9]. The incomplete ionization model (INCOMPLETE) was selected since doping impurities in 4H-SiC material were not activated fully at room temperature. The band gap narrowing (BGN) model was also used. The Schottky contacts contained the Universal Schottky Tunneling (UST) Model and Parabolic Field Emission Model. The breakdown calculation was implemented with anisotropic impact model and the self-heating effect was also considered which was calculated by using the Wachutka's thermodynamics model[9]. The cross-sectional surface area $x \times z$ act as the heat sink and is used as the thermal boundary condition in the simulation, the thermal boundary conditions are determined by the measured R$_{\theta JC}$ (thermal resistance from junction to case).

The heat distribution of the device simulator ATLAS generated by solving a two-dimensional heat equation, where C is the heat capacity per unit volume, κ is the thermal conductivity, lattice temperature is T_L, H is the heat generated.

$$C\frac{\partial T_L}{\partial t} = \nabla(\kappa \nabla T_L) + H \qquad (1)$$

Thermal conductivity (κ) and heat capacity (C) of 4H-SiC were expressed as a function of lattice temperature (T_L) using polynomial expressions shown in (1) and (2), respectively. Experimentally obtained coefficients were used for the parameters (TC. A, TC. B, and TC. C for thermal conductivity and HC. A, HC. B, HC. C, and HC. D for heat capacity) in (1) and (2) [10]. During the simulation, the device structure file is saved periodically, and the values of various electrothermal parameters are recorded at discrete time intervals.

$$\kappa(T_L) = \frac{1}{TC \cdot A + (TC \cdot B) * T_L + (TC \cdot C) * T_L^2} \qquad (2)$$

$$C(T_L) = HC \cdot A + (HC \cdot B) * T_L + (HC \cdot C) * T_L^2 + \frac{HC \cdot D}{T_L^2} \qquad (3)$$

Because the surge current is about four times the rated current, so a 400A half-sine wave with a pulse width of 10ms is applied to the SiC JBS diodes by ATLAS. The results obtained for JBS diode pulsed simulation are shown in Fig. 3. It can be seen from the figure that the forward voltage of the diode drops suddenly at 2.3ms (point A) because the diode enters the bipolar conduction. Diode power dissipation up to 3250W at 5ms (point B). The maximum temperature of the diode reaches 920K at around 5.5ms. Fig. 4 show the lattice temperature profile of the JBS diode structure at 5.5ms (point C). A maximum temperature of 921 K was observed in the diode surface region which was also the area of localized heating.

The result obtained by TCAD simulation is that the junction temperature of the device has reached 921K, this temperature is far from the intrinsic excitation temperature of the material. However, the melting point of aluminum metal is generally 873K~933K. In order to simulate the heat distribution of the aluminum metal pad and verify that the failure point after the surge is near the bonding wire, we performed a visual analysis by establishing a 3D finite element thermal model.

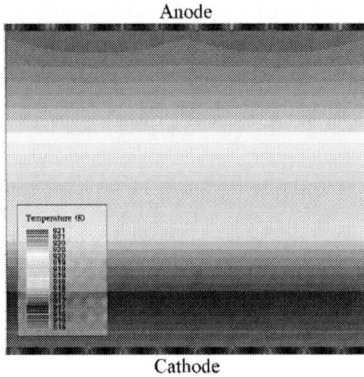

Fig. 4 2-D simulation of the temperature distribution for cross section inside the structure at time 5.5ms (C position in Fig.2).

Thermal simulation

The accuracy of the simulation model needs to be guaranteed before the simulation of the surge test. The parameters that have a greater impact on the surge test are the heat producing area (active region and location (Fig.5(a))) and the thermal resistance at the die attach. Calibrate the structure function(Fig.5(b)) can improve the accuracy of these parameters effectively.

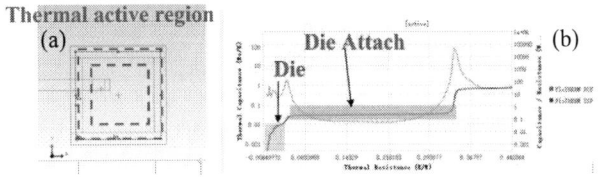

Fig. 5 (a)Thermal active region of device,(b)the structure function.

According to the actual working conditions of the power JBS device, the boundary conditions of the device are as follows: 1. The heat source is the power consumption of the chip, and the transient heat generating power of 3250 W(Half

sine wave with a pulse width of 10ms approximate by Figure 2) is applied to the chip entity (the thickness of the chip heat producing region is 8μm); 2. Ignore the radiation dissipation of the substrate is only considered in the form of their heat conduction heat dissipation. 3. The ambient temperature is 25 °C, and the device is subjected to steady-state temperature field analysis.

In the simulation model we set up a monitoring point on the surface of the chip. After a 10ms half-sine wave pulse, the temperature curve at the monitor point is shown in Fig. 6. The surface temperature of the chip reaches a maximum value of about 650 ° C (923 K) at 7ms, and the surface temperature decreases to about 460 ° C (733 K) at 10ms.

Fig. 7 shows the heat distribution of the device at 7ms. It can be seen that the temperature of the Al pad near the bonding wire has reached nearly 650 ° C (923 K). At this point, the temperature has not spread to the package, and the heat is mainly concentrated on the surface of the chip. Fig. 9 shows the heat distribution of the device at 10ms. It can be seen that the temperature of the substrate under the chip rise significantly, which means that the heat of the chip has begun to be transmitted to the package direction at 10ms. At this point, the heat at the top of the chip is concentrated near the bonding line, and the temperature is still maintained at 460 ° C (733K), which

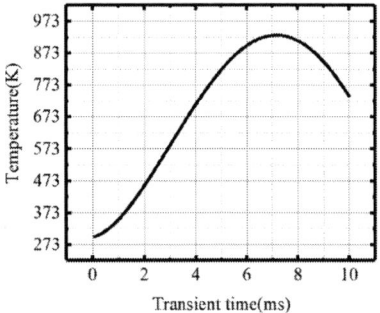

Fig. 6 The temperature curve at the surface monitor point

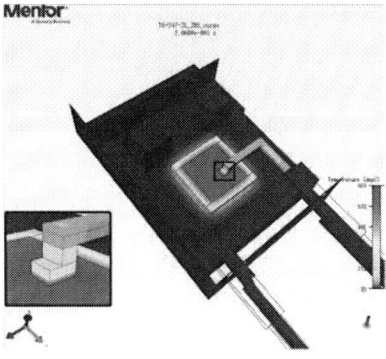

Fig. 7 Cross package (TO-247-2L) plane temperature distribution at 7ms

978-1-7281-5757-3/19 $31.00 © 2019 IEEE

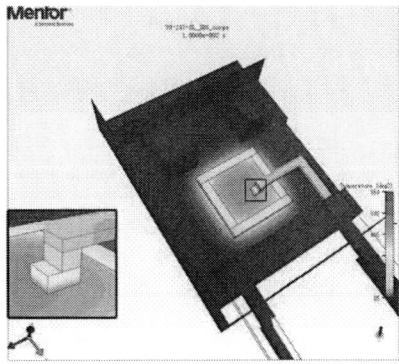

Fig. 8 Cross package (TO-247-2L) plane temperature distribution at 10ms

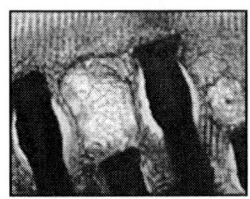 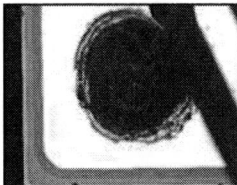

Fig. 9 The failure point of the chip after a 10ms half sinusoidal surge test

is due to the contact thermal resistance between the bonding wire and the Al pad, and the bonding wire itself has relatively large heat.

In order to verify the accuracy of the simulation, we performed a forward surge test using a half-sine wave with a period of 10ms. When the peak current reaches 400A, the device fails. Fig. 8 shows the failure point of the decaped chip after a 10ms half sinusoidal surge test. It can be seen that the Al pad near the bonding wire has experienced thermal fatigue or even burned. This is consistent with the simulation we have done before.

Conclusions

The failure mechanism of a 1.2kV-100A high current SiC JBS diode under non-repetitive surge tests was investigated in this paper. In the process of mechanism analysis, TCAD simulation and 3D thermal simulation are combined. The TCAD simulation can approximate the heating power of the device to a half-sine wave of 10ms, and then visually display the position of the hot spot in 3D thermal simulation. During the entire surge pulse, the surface temperature of the chip reached the maximum temperature at 7ms. The thermal concentration is obviously near bonding wire at 10ms. From the simulation results, it can be seen that the contact thermal resistance of bonding wire and pad and a heating of bonding wire itself will cause the temperature to be concentrated near the bonding wire. After a half-sine wave surge test, it was found that the device failed after a non-repetitive surge test of a 400A/10ms half sine wave pulse, and the point of failure is concentrated near the bond wire which is consistent with our simulation results and provides an experimental basis for the failure of high current devices near the bonding wire.

Acknowledgments

This work was supported by the National Key Research and Development Program of China under Grant No. 2016YFB0100601 and Institute of Microelectronics of Chinese Academy of Sciences.

References

[1]. JEDEC JSD282-B Rev. .01, "Silicon Rectifier Diodes," Section 4.2,2002

[2]. Radhakrishnan R , Cueva N , Witt T , et al. Analysis of Forward Surge Performance of SiC Schottky Diodes[C]// 2018:621-624.

[3]. Huang X , Wang G , Lee M C , et al. Reliability of 4H-SiC SBD/JBS diodes under repetitive surge current stress[C]// Energy Conversion Congress & Exposition. IEEE, 2012

[4]. Ruggedness of 1200V SiC MPS diodes[J]. Microelectronics Reliability ,2015 , 55 (s9-10): S0026271415001808 .

[5]. Jianfeng W, Siyang L, Weifeng S . Thermal Characteristic Calibration and Optimization of TO-220 Package Power Device Based on ANSYS Software[J]. Chinese Journal of Electron Devices, 2015.

[6]. Zhao Heran, Kang Xi'e, Ma Yanyan . Methods to Fix Simulation Value and Measurement Results of Thermal Resistance of Electronic Device Package[J]. MicroProcessors,2017

[7]. Treu M , Rupp R , Tai C S , et al. A Surge Current Stable and Avalanche Rugged SiC Merged pn Schottky Diode Blocking 600V Especially Suited for PFC Applications[C]// Materials Science Forum. Trans Tech Publications, 2006:1155-1158.

[8]. Yidan Tang, Chengzhan Li, Jingjing Shi, et al. Development of 1200V/100A High Temperature High Current 4H-SiC JBS Diodes [J]. Semiconductor Devices, 2018.

[9]. Atlas User 's Manual. Santa Clara, CA, USA: Silvaco, Inc., 2018

[10]. Snead L L , Nozawa T , Katoh Y , et al. Handbook of SiC properties for fuel performance modeling[J]. Journal of Nuclear Materials, 2007, 371(1-3):329-377.

978-1-7281-5757-3/19 $31.00 © 2019 IEEE

Simulation Study on Current Collapse Effect of E GaN HEMT

Guo Weiling[1*], Ma Qijing, Du Shuai, Lin Tianyu, Zhu Yanxu

(Key Laboratory of Optoelectronic Technology under the Ministry of Education ,Beijing University of Technology, Beijing 100124, China)

Beijing University of Technology, No.100, Pingleyuan, Chaoyang District, Beijing, China

Corresponding Author, E-mail:guoweiling@bjut.edu.cn , phone:13701298554

Abstract

GaN high electron mobility transistor(GaN HEMT) has high channel electron concentration, high electron mobility and high breakdown voltage, which makes it have great development prospects in high frequency, microwave and other fields. In the development of GaN HEMT, suppressing current collapse has always been the focus and difficulty of research. Based on SILVACO TCAD simulation software, the current collapse suppression by field plate structure is studied and discussed in this paper. Firstly, the model of enhanced GaN HEMT device is established and its correctness is verified. Then, the phenomenon of current collapse and virtual gate model are introduced. From the theory of improving breakdown voltage of device by field plate structure as the starting point, field plate structure is added to the device model and simulation experiments are carried out. According to the results, the principle of suppression current collapse by field plate structure is analyzed, and the correctness of virtual grid model is verified.

Introduction

As the third generation semiconductor material, GaN is considered to be the most promising semiconductor material after the first generation semiconductor materials Si, Ge and the second generation semiconductor materials GaAs, InSb, etc [1]. In recent decades, GaN HEMT devices have been widely used in the field of high frequency and high power electronic devices due to their high electron mobility, high current density and high temperature tolerance[2]. Current collapse is mainly divided into two types: strong field current collapse and radio frequency current collapse. The current collapse described in this paper is strong field current collapse. It was first found by Khan et al. [3] when testing GaN HEMT, It means the phenomenon that saturated output current decreases and transconductance, knee-point voltage and on-resistance increase after high voltage impulse. Aiming at this phenomenon, based on SILVACO TCAD, the device model is optimized by using field plate structure. According to the simulation results, the principle of current collapse suppression by field plate structure is analyzed, and the virtual gate theory is verified.

Device Model

Compared with conventional GAN HEMT devices, enhanced GaN HEMT devices have the advantages of high threshold voltage and low static power consumption. Considering the difficulty of process and the effect of implementation, P-type GaN gate structure is adopted to realize enhanced devices. The principle of P-type GaN gate structure is to insert P-type GaN structure between gate and AlGaN barrier layer to make ohmic contact between gate and barrier layer. When the gate voltage is zero, the electrons in the channel below the gate are consumed under the effect of P-type doping increasing energy band, thus forming an enhanced device.

The model design refers to the GaN HEMT device structure with P-GaN gate proposed by O. Hilt et al. and is modified and optimized. [4] The detailed model is shown in Figure 1. From bottom to top, it is 2μ m AlGaN buffer layer, 35nm GaN channel layer and 15nm AlGaN barrier layer. The surface of the device is 1μ m source, 1.4μ m gate and 1μ m drain from left to right. The distance between gate sources is 1 μ m and between gate drains is 6μ m. A P-type GaN material with doping concentration of $3*10^{17}/cm^3$ was added below the gate to form ohmic contact with the gate electrode.

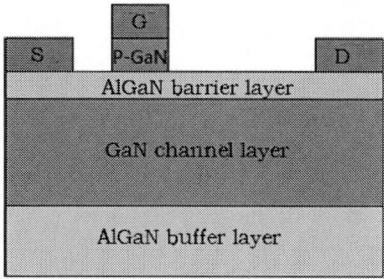

Figure1 Device model

Silvaco TCAD is used to simulate the model shown in Figure 1. The results show that the electron concentration in the channel below the gate of the device under zero gate voltage is significantly lower than that in the rest of the channel, and the highest electron concentration is about 107/cm2, which is not enough to make the channel open.Compared with zero gate voltage, the channel carrier concentration of the device has changed greatly at positive gate voltage. The lowest concentration reached 10^{12} cm^{-2}., the peak concentration increases from 10^7 cm^{-2} to 10^{18} cm^{-2}, and the channel is conducting.

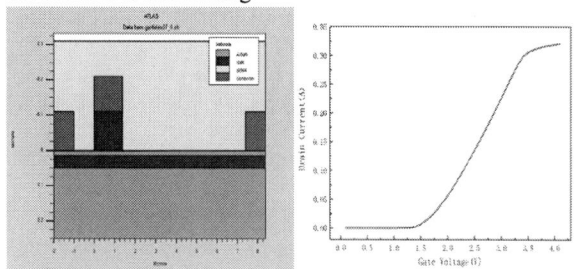

Figure2 (a)Structure Diagram of P-GaN Gate Enhanced GaN HEMT (b)Transfer Curve of P-GaN Gate Enhanced GaN HEMT

Figure 2 is a P-GaN gate-enhanced GaN HEMT structure and its transfer curve. The structure is consistent with the enhanced device shown in Figure 1. When the gate voltage is zero, the device is not turned on, and when the gate voltage reaches the threshold voltage of 1.3V, the device is turned on,

978-1-7281-5757-3/19 $31.00 © 2019 IEEE

which proves the feasibility of making enhanced GaN HEMT with P-type GaN gate structure.

Theoretical Principle and Simulation Analysis of current collapse

For the phenomenon of current collapse, the virtual gate model is more accepted[5-6]: as shown in Figure 3, Under the action of the built-in electric field and the electric field formed by the Drain voltage, Channel electrons move toward the buffer layer and device surface, and are trapped by defects in the buffer layer and interface states on device surface, forming a virtual gate similar to the gate. The virtual gate reduces the electron concentration in the channel, which results in the decrease of the output current density. When the device is turned on, it will take time for the current of the device to exchange with the electrons in the trap and surface states, which further aggravates the current collapse effect.

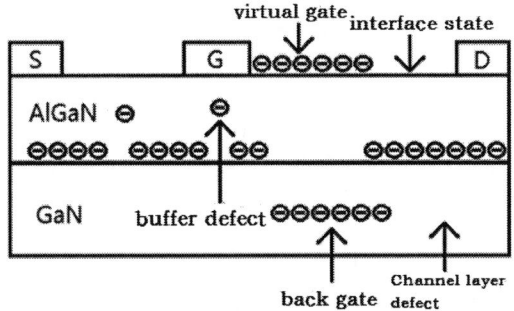

Figure3 Schematic diagram of virtual grid and back grid model

Field plate structure is an effective method to improve the breakdown voltage of devices. Field plate structure can reduce the peak value of electric field by changing the electric field distribution between gate and drain to improve the breakdown voltage of devices. As shown in Figure 4, the grid field plate and the grid have the same potential, which makes some electric field lines originally pointing to the grid field plate point to the grid field plate and make the electric field lines more dispersed. At the same time, some electric field lines pointing to the right of the grid plate and the electric field lines originally pointing to the grid cancel each other. This makes the electric field distribution more uniform and the peak value of the electric field lower, thus increasing the breakdown voltage of the device. [7]

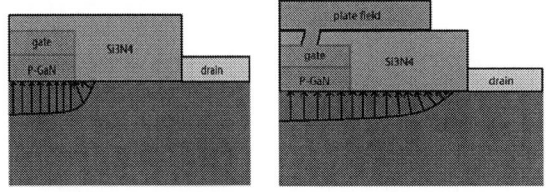

Figure4 Schematic diagram of electric field with or without field plate

Usually the breakdown of devices is caused by the high electric field, and the breakdown point is the highest point of the electric field. Therefore, it is of great significance to discuss the peak value of the electric field for studying the breakdown voltage of devices. Figure 5 shows the channel electric field distribution of the same device with or without the field plate. It can be seen that the peak value of the channel electric field of the device with the field plate is two orders of magnitude lower than that of the device without the

field plate structure,which proves that the field plate structure can effectively balance the channel electric field distribution of the device and improve the breakdown voltage of the device.

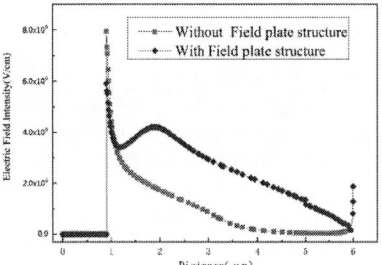

Figure5 an electric field distribution diagram with a field free plate device

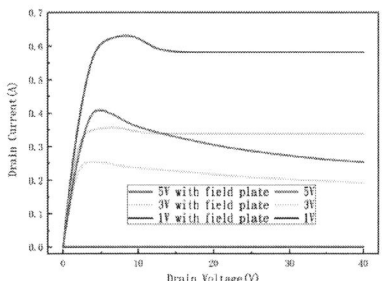

Figure6 Contrast of Current Collapse Curves of Field Plate Structure

The field plate structure is an effective structure to reduce the peak electric field of the device and to suppress the current collapse. The effect of the field plate structure to suppress the current collapse effect is simulated. Figure 6 is a comparison of output curves of devices with and without field plates. It can be seen that the effect of field plate structure on current collapse suppression is obvious, and the peak value of output current is also greatly improved compared with the original structure.

The structure of this comparison is based on the structure shown in Figure 2. A 3μ m grid field plate structure is added. For ordinary enhanced devices, the peak output current is 0.25A at 3V gate voltage, 0.19A at 40V and 0.41A and 0.25A at 5V gate voltage respectively. It can be seen from the Figure 6 that the two curves increase with drain voltage after the peak value. For convenience of statistics, the ratio of current value to peak value of current value at 40V leakage voltage is defined as K coefficient, and the K coefficient at 3V and 5V gate voltage is 0.76 and 0.61 respectively. Current Peak of Field Plate Devices and Current Value at 40V Drain Voltage are higher than those of conventional enhanced devices. When the gate voltage is equal to 3v, the peak current is 0.36A and the leakage current is 0.34A at 40V drain voltage.When the gate voltage is equal to 5v, the peak current is 0.63A and the leakage current is 0.58A at 40V drain voltage. The K coefficients of the two gate voltages are 0.94 and 0.92, respectively.

It is believed that the field plate structure reduces the peak electric field near the gate of the device, thus reducing the number of electrons crossing the barrier layer to the surface of the device, weakening the effect of the virtual gate and suppressing the current collapse of the device. It can be seen from the foregoing that the field plate structure can effectively reduce the peak electric field of the device, so its transport capacity for electrons will also be weakened by the field plate structure. Figure 7 is a comparison of electron concentration with or without field-plate structure. It can be seen that the ultra-low concentration electron region under the field-plate structure is wider than that without field-plate structure, and the electron concentration is significantly lower in the region closer to the drain, which proves the suppression effect of field-plate structure on virtual grid. Figure 8 is a comparison of the electron concentration in the barrier layer with or without the field plate structure at 0.4 and 0.6 microns on the right side of the gate. It can be seen that the electron concentration in the barrier layer varies greatly with the field plate structure, especially when the gap between the side near the gate and the side near the channel layer reaches 10^9 times and 10 times respectively. And for the area near the grid, the effect of the field plate structure is more obvious, and the difference is about ten times by calculation. In view of the more obvious effect near the grid, the simulation results of the field plates with the length of 1.5, 3 and 4.5 μ m are compared. The results show that there is no difference. As shown in Figure 9, the correctness and importance of the virtual grid model are proved, and the conclusion that the virtual grid is located near the right side of the grid is verified.

The field plate structure can balance the channel electric field distribution of the device, which is an effective way to improve the breakdown voltage of the device. At the same time, the field plate structure is also an effective structure to suppress the current collapse effect of enhanced GaN HEMT. From the simulation results, it can be seen that the field-plate structure has obvious suppressing effect on current collapse under different gate voltages. It can be inferred that the principle is that the field-plate structure reduces the peak electric field near the gate, thus reducing the number of electrons crossing the barrier layer to the surface of the device, weakening the role of the virtual gate and suppressing the current collapse of the device. The comparison of electron concentration at different distances on the right side below the gate can confirm the inference of this paper from the side, and also verify the correctness of the virtual gate model.

Acknowledgments:

We would like to thank National Key R&D Program of China (Grant No. 2017YFB0402800, 2017YFB0402803).

References

[1].Karmalkar S , Mishra U K . Enhancement of breakdown voltage in AlGaN/GaN high electron mobility transistors using a field plate[J]. IEEE Transactions on Electron Devices, 2001, 48(8):1515-1521.

[2].Li H , Yao C , Fu L , et al. Evaluations and applications of GaN HEMTs for power electronics[C]// Power Electronics & Motion Control Conference. IEEE, 2016.

[3].Khan M A , Shur M S , Chen Q C , et al. Current/voltage characteristic collapse in AlGaN/GaN heterostructure insulated gate field effect transistors at high drain bias[J]. Electronics Letters, 1994, 30(25):2175-0.

[4].Hilt O . Normally-off AlGaN/GaN HFET with p-type GaN Gate and AlGaN Buffer[C]// International Conference on Integrated Power Electronics Systems. IEEE, 2010.

[5].Sabuktagin S , Dog?An S , Baski A A , et al. Surface charging and current collapse in an AlGaN / GaN heterostructure field effect transistor[J]. Applied Physics Letters, 2005, 86(8):083506.

[6].Simin G , Koudymov A , Tarakji A , et al. Induced strain mechanism of current collapse in AlGaN/GaN heterostructure field-effect transistors[J]. Applied Physics Letters, 2001, 79(16):2651.

[7].Ambacher O , Foutz B , Smart J , et al. Two dimensional electron gases induced by spontaneous and piezoelectric polarization in undoped and doped AlGaN/GaN heterostructures[J]. Journal of Applied Physics, 2000, 87(1):334-0.

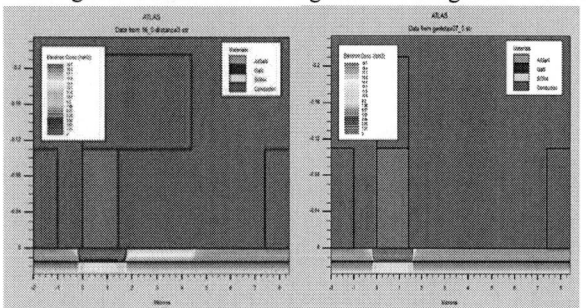

Figure7 Comparison of Electronic Distribution Diagrams (a) Field Plate Structure (b) Primary Structure

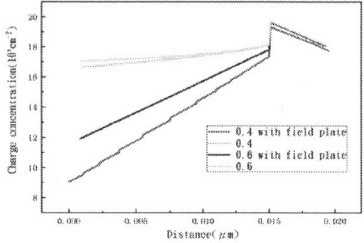

Figure8 Contrast Chart of Electron Concentration at Virtual Gate Position

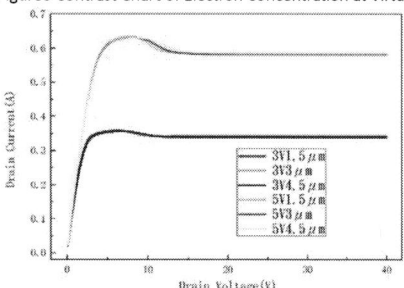

Figure9 Suppression effect of different length field plate on current collapse

Conclusions

A New SiC Split-gate MOSFET Structure With Protruded P-base and the Mesa above JFET for Improving HF-FOM

Kunlin Li, Yourun Zhang, Wei Zhong, Xiaochuan Deng, Xiao Yang, Hang Chen, Bo Zhang
University of Electronics Science and Technology of China, Chengdu, China
E-mail: yrzhang@uestc.edu.cn (corresponding author)

Abstract

A novel 4H-SiC MOSFET （PM-MOSFET） for rated 3.3 kV applications is proposed, which features the protruded P-base and the mesa above JFET. Numerical simulation based on Silvaco is carried out to investigate the benefits of the proposed structure. The on-state resistance of PM-MOSFET is 11.9 m$\Omega \cdot$cm^2, which is dramatically lower compared to on-resistance of 18.2 m$\Omega \cdot$cm^2 of the traditional split-gate MOSFET (SG-MOSFET). The C_{rss} of SG-MOSFET extracted at $V_d = 1800$ V is 17.5 pF/cm^2, while the C_{rss} of PM-MOS extracted is 6.5 pF/cm^2, which is three times lower than that of the SG-MOSFET. It is demonstrated that the PM-MOSFET structure is superior to the SG-MOSFET. More importantly, the benefits above are achieved without degradation of other performances of MOSFET. As a result, the PM-MOSFET presents superior figure of merit (HF-FOM) ($R_{on} \times C_{rss}$) than that of the SG-MOSFET. The PM-MOSFET achieves much faster switching speed than the SG-MOSFET.

Keywords: *Silicon carbide, MOSFET, Split gate, C_{rss}, HF-FOM*

Introduction

Wide bandgap semiconductor SiC MOSFETs are considered as promising choices in high voltage field owing to their superior material characteristics, such as high breakdown electric field and high thermal conductivity. Nowadays SiC MOSFETs have been commercialized because of small on-resistance, high switching speed and low on-resistance[1]. With the pursuit of efficiency, more efficient and better performance MOSFETs are required. The switching loss is the prime factor limiting utilization of much higher switching frequency. The value of HF-FOM ($R_{on} \times C_{rss}$) is widely used to measure the high frequency performance. Switching loss is directly affected by Miller capacitance which depends on the overlapping area of the gate to drain electrode. Therefore, the value of HF-FOM should be reduced as much as possible to increase the high frequency characteristics of the device[2]. Lots of novel methods have been proposed in the past decades to reduce the on-resistance and C_{rss}, such as reduction of the channel length and novel structure designs[3,4]. Among these works, the Split-Gate MOSFET structure is considered as a promising solution, which allows C_{rss} to be reduced by decreasing its gate to drain area and improving HF-FOM. The 4H-SiC SG-MOSFET structure[5] has been recently proposed owing to superior HF-FOM compared to conventional MOSFETs.

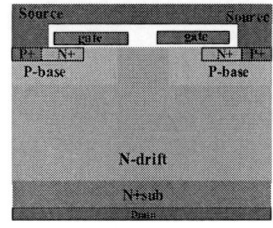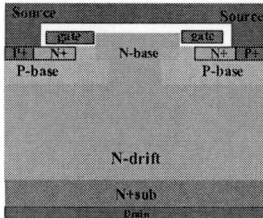

(a)SG-MOS　　　　　　　　(b) PM-MOS

Figure. 1 Schematic structures of (a) SG-MOSFET and (b) PM-MOSFET.

In this work, a novel MOSFET structure(PM-MOSFET), where the P-base region is extended beyond split-gate electrode and mesa is beyond on the JFET, is designed to improve figures of merit without sacrificing device performances. An additional N-base region is required in the PM-MOSFET structure to provide the current passage, because the inversion layer can not be formed in the prominent part of the p-type base. A lower R_{on} is maintained in the PM-MOSFET by decreasing the resistance of the JFET region R_{JFET} and the resistance of the channel region R_{CH}. A lower C_{rss} is maintained by reducing of overlapping area of the gate-to-drain. The switching performance of the PM-MOSFET is also significantly improved owing to the reduction in C_{rss}.

TCAD is applied here to investigate the characteristics of the proposed structure which are based on semiconductor theory. In this paper, the simulated physical models include Schockley–Read–Hall(SRH), the auger recombination model (AUGER) and the velocity-dependent mobility model(CVT). Because of high doping concentration in the substrate and source region, the bandgap narrowing model(BGN) is applied. The incomplete ionization model is performed because the doped impurities in 4H-SiC material have larger activation energy. The selberherr's impact model is also implied to calculate the breakdown voltage.

Device Structure And Simulation Setup

Fig. 1 shows the fabricated SG-MOSFET and PM-MOSFET structures. The PM-MOSFET features a mesa and protruded P-base, which benefits to the reduction of C_{rss} and R_{on}. The devices are designed for 3300 V and the structure parameters of the device are as follows . The concentration of the drift region is 2.82×10^{15}cm^{-3}. The thickness of the drift region is 30 μm. The oxide thickness is 50 nm. The channel mobility is set to be 20 cm^3/v · s.

978-1-7281-5757-3/19 $31.00 © 2019 IEEE

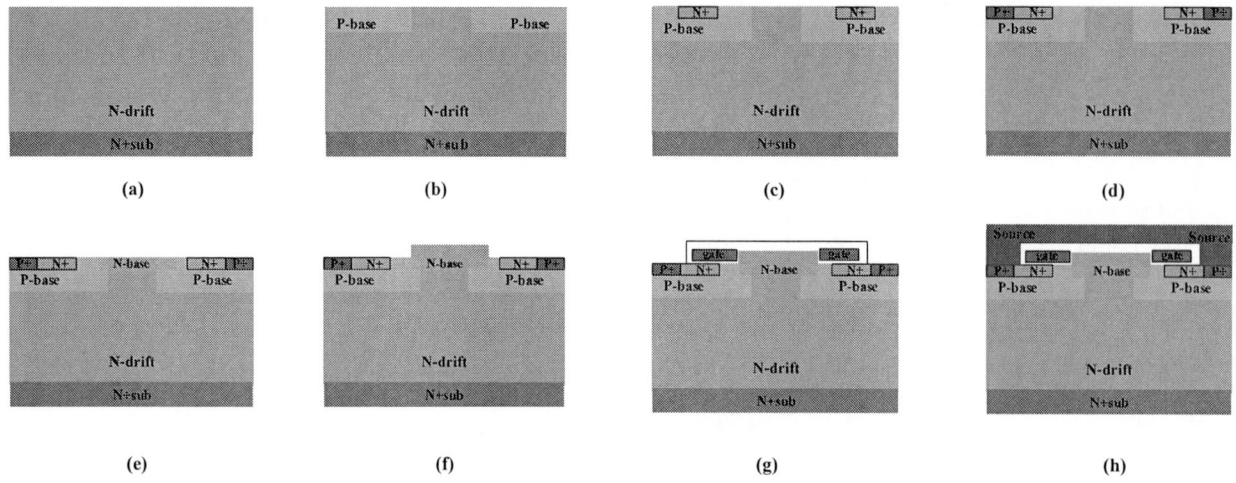

Figure. 2 The proposed process flow for the PM-MOSFET

A layer of N-base area is applied in the PM-MOSFET, owing to that the channel is not completely covered by the gate. The N-base doping concentration must be used to provide current passage with caution. According to the numerical simulation with TCAD, the concentration of $2 \times 10^{17} cm^{-3}$ of the N-base region is rightly appropriate. In addition, the height of the mesa is 0.5 μm which is critical for providing a current pathway and reducing JFET effect.

Figure 2 presents the proposed processing method for the PM-MOSFET. The part of the P-base protrude beyond the gate plays an important role in the performance of the PM-MOSFET, which can shield the gate and reduce the overlapping area of the gate and source. So the C_{rss} has an obvious drop.

The whole process mainly includes ion implantation of N+ region, P+ region, P-well region and N-base region, the secondary epitaxy, the deposition of polysilicon, SiO_2 and metal.

The forward IV characteristic curves of the devices are shown in Fig 3. The area of the device is 1 mm². The R_{on} of the SG-MOSFET is 18.2 mΩ·cm². The R_{on} of the PM-MOSFET is 11.9 mΩ·cm². Compared with the SG-MOSFET, the on-resistance of PM-MOSFET decreases by 18%.

In the conventional SG-MOSFET, on-resistance mainly consists of the resistance of the channel region of inverse layer R_{CH}, the resistance of the accumulation region R_A, the resistance of the JFET region R_{JFET} and the resistance of the drift region R_{drift} and other resistance R_{others}.

$$R_{on} = R_{CH} + R_{drift} + R_A + R_{others} \tag{1}$$

$$R_{CH} = \frac{L_{CH}}{Z\mu_{ni}C_{OX}(V_G - V_{th})} \tag{2}$$

$$R_A = \frac{L_A}{Z\mu_{nA}(V_G - V_{th})} \tag{3}$$

$$R_{JFET} = \frac{\rho_{JFET}x_{JP}Z}{Z(W_G - 2x_{JP} - 2W_0)} \tag{4}$$

where L_{CH} is the length of channel region, L_A is the length of accumulation area, Z is the width of the device cell and C_{ox} is the oxide capacitance, ρ_{JFET} is resistivity in JFET region, W_0 is the width of JFET depletion region at zero bias.

According to the above formulas, It is demonstrated that reduction of R_{on} of PM-MOS is attributed to the decrease of the resistance of the channel region R_{CH} and the resistance of the JFET region R_{JFET}. The channel length of PM-MOSFET is lower than that of SG-MOSFET. For PM-MOSFET, the resistance of the channel region is greatly reduced. Thus as shown Figure 4(a). In the results of traditional SG-MOSFET, electrons in the N source region are injected into the N drift region by injection through the JFET region. In order to make the current distribution more uniform and minimize the current congestion in JFET region, as shown Figure 4(b), a mesa beyond the JFET region in PM-MOS is designed to reduce R_{JFET}. It provides a current passage for carriers. Carriers can enter the mesa from the channel then spread in the JFET region uniformly. The resistance in the JFET area is reduced accordingly

The PM-MOSFET presents a similar breakdown voltage (4600 V) to that SG-MOSFET(4680 V). The PM-MOSFET presents superior static parameters than those of the SG-MOSFET.

Figure 5 shows C_{rss} of the PM-MOSFET and that of SG-MOSFET influenced by V_D. As shown Figure 6, C_{rss} is composed of the semiconductor capacitance C_s beneath oxide and the oxide capacitance C_{ox}. The formulas below show that the reducing of overlapping area of gate-to-drain is conducive to the decreasing of C_{rss}. As shown in Figure 1, it is obvious that the overlapping area of the gate and drain in PM-MOSFET is much smaller than that in SG-MOSFET. Owing to this, the C_{rss} of SG-MOSFET extracted at $V_d = 1800\,V$ is 17.5 pF/cm²., While the C_{rss} of PM-MOSFET extracted at $V_d = 1800$ V is 6.5 pF/cm² ,which is three times lower than that of the SG-MOSFET

978-1-7281-5757-3/19 $31.00 © 2019 IEEE

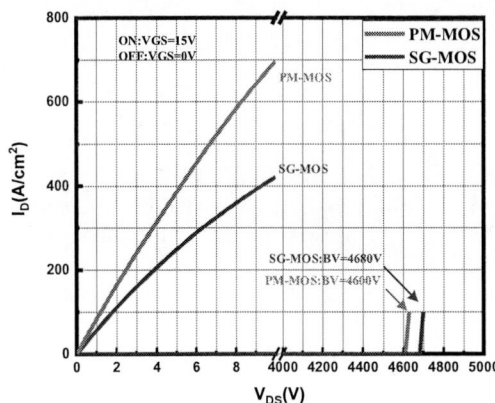

Figure. 3 The I-V characteristics of the PM-MOSFET and SG-MOSFET

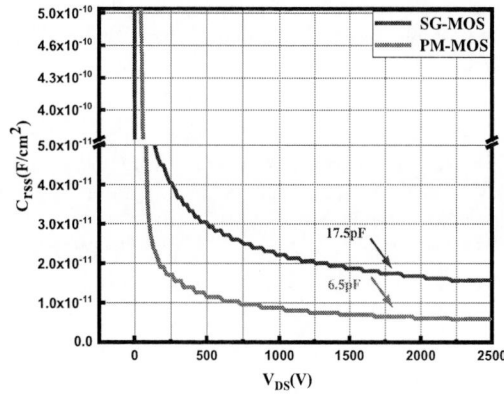

Figure. 5 The C-V characteristics of the PM-MOSFET and SG-MOSFET

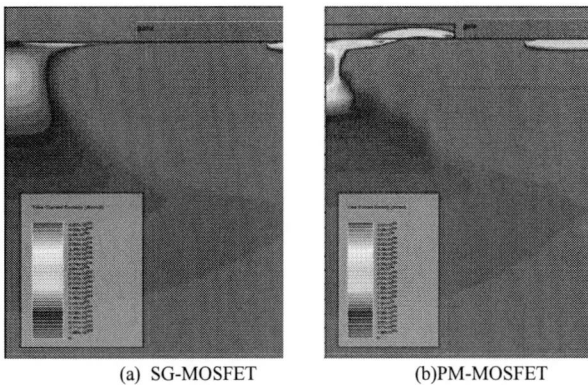

(a) SG-MOSFET　　　　(b)PM-MOSFET

Figure. 4 The total current density contours of (a) SG-MOSFET and (b) PM-MOSFET

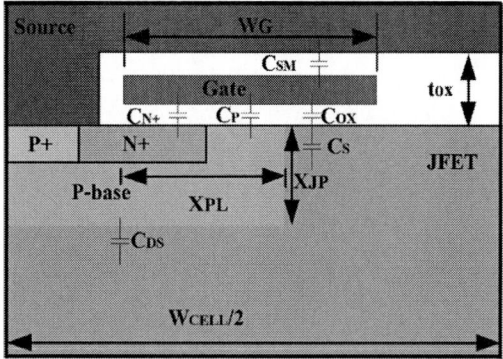

Figure. 6 Main capacitances of SG-MOSFET.

It is demonstrated that the PM-MOSFET structure is superior to the SBG-MOSFET.

$$C_{rss} = \frac{2W_G - 2X_{PL}}{W_{CELL}} \left(\frac{C_{ox}C_s}{C_{ox} + C_s} \right) \quad (5)$$

(5)

$$C_s = C_{ox} \left(\sqrt{1 + \frac{2V_D}{q\varepsilon_s N_D}} - 1 \right) \quad (6)$$

In Figure 6, W_G is the length of the polysilicon gate, X_{PL} is the distance between the boundary of P-body region and the edge of the gate, and $Wcell$ is the width of device cell. C_{ox} is decided by the oxide thickness usually has a constant value.

The PM-MOSFET has superior HF-FOM $C_{rss} \times R_{on} = 77.35$ pF×nC . Compared to $C_{rss} \times R_{on} = 318.5$ pF×nC for SG-MOSFET. For better comparison, the key electrical characteristics of the PM-MOSFET and SG-MOSFET are summarized in Table I.

Table I SUMMARY OF SIMULATED RESULTS FOR PM-MOSFET AND SG-MOSFET.

Parameters	PM-MOSFET	SG-MOSFET	Unit
Breakdown voltage, BV	4600	4680	V
The threshold voltage, V_{th}	4.3	4.4	V
On-state resistance, R_{on}	11.9	18.2	mΩ·cm
Miller capacitance, C_{rss}	6.5	17.5	pF/cm²
$C_{rss} \times R_{on}$	77.35	318.5	pF×nC

CONCLUSION

A SiC Split-gate MOSFET with the protruded P-base and the mesa above JFET (PM-MOSFET) has been proposed. Compared to the conventional Split-gate MOSFET (SG-MOSFET), the PM-MOSFET achieves a dramatically lower R_{on}, owing to the decrease of resistance of the channel region R_{CH} and the resistance of the JFET region R_{JFET}. And it also exhibits a smaller C_{rss}, which is beneficial to improve the frequency characteristic. It has been demonstrated that the PM-MOSFET has improved HF-FOM. The PM-MOSFET has superior static and dynamic characteristics. Therefore, the PM-MOSFET can be applied to high frequency applications.

Acknowledge

This work was supported by the National Key Research and Development Program of China under Grant under Grant 2017YFB0102302.

References

1. Palmour, John W. , et al. "Silicon carbide power MOSFETs: Breakthrough performance from 900 V up to 15 kV." IEEE International Symposium on Power Semiconductor Devices & Ics IEEE, 2014.
2. Sakai, T. , and N. Murakami . "A new VDMOSFET structure with reduced reverse transfer capacitance." IEEE Transactions on Electron Devices 36.7(1989):1381-1386.
3. Jiang, Huaping , et al. " [IEEE 2017 29th International Symposium on Power Semiconductor Devices and IC\"s (ISPSD) - Sapporo, Japan (2017.5.28-2017.6.1)] 2017 29th International Symposium on Power Semiconductor Devices and IC\"s (ISPSD) - SiC MOSFET with built-in SBD for reduction of reverse recovery charge and switching loss in 10-kV applications." (2017):49-52.
4. Han, Kijeong , B. J. Baliga , and W. Sung . "A Novel 1.2 kV 4H-SiC Buffered-Gate (BG) MOSFET: Analysis and Experimental Results."IEEE Electron Device Letters39.2(2018):248-251
5. Han, Kijeong , B. J. Baliga , and W. Sung . "Split-Gate 1.2 kV 4H-SiC MOSFET: Analysis and Experimental Validation." IEEE Electron Device Letters (2017):1-1.

6.5kV Silicon Carbide Discontinuous Trenched Junction Barrier Schottky Diode

Zhiyu Chen[1], Xuan Li[1], Xiaochuan Deng[1], Juntao Li[2], Zhiqiang Li[2], Yourun Zhang[1], and Bo Zhang[1]

[1.] State Key Laboratory of Electronic Thin Film and Integrated Devices, University of Electronic Science and Technology of China, Chengdu, China

[2.] Microsystem and Terahertz Research Center, China Academy of Engineering Physics, Mianyang 621900, China

13540789909@163.com

Abstract

In this paper, a new affordable trenched junction barrier Schottky (TJBS) called discontinuous TJBS (DTJBS) diode is proposed and investigated for its better current conduction capability than TJBS. TJBS diode has a poor area of Schottky contact, which is the main cause of the increment of forward voltage (V_f) compared with DTJBS diode. In order to optimize the V_f, physical insights into 6.5kV silicon carbide (SiC) DTJBS diode is carried out by TCAD Silvaco, then shielding effect and current conduction capability are compared with TJBS and analyzed to verify the advantages of DTJBS's lower V_f. Eventually, The V_f of the device achieves 1.89V while the current is 15A.

0. Introduction

Nowadays, SiC power device has attracted many attentions for its superlative electrical prospects, Compared with silicon (Si). For instance, SiC has ten times higher critical field, three times higher electron drift velocity and three times higher thermal conductivity than Si, [1] with almost equivalent dielectric constant concurrently. [2] Meanwhile, with a wider band gap, the intrinsic carrier concentration of SiC is only ~10^{10}cm^{-3} at 800K, which is nearly the same with that of Si at room temperature. [3] Due to distinctive material characteristics, 6.5kV SiC diode is extensively utilized in power and energy system,[4, 5] such as wind power, rail, regarding traction, grid systems, and pumped hydro. [6, 7]

It must be noticed that large V_f is an Unavoidable problem for diodes Under high voltage primarily due to balance the V_f and leakage current. Planar JBS diode meets challenges when applying more than 1.7kV. Depth of P$^+$ region in planar JBS diode is extremely limited for 6.5kV applications, which limits the shielding effect of the P$^+$ region on the Schottky junction under the reverse voltage. Reducing the width of the Schottky junction suppresses leakage current, but weakens diode's unipolar current capability. On the contrary, conventional TJBS diode can increase the depth of P$^+$ region, [8] so TJBS diode improves deficiencies of planar JBS diode. However, P$^+$ region nearing the sidewall of the trench weakens V_f. Therefore, it is of significance to improve this contradiction for 6.5kV SiC diode. There have been some researches to alleviate the pivotal contradiction, such as trench MOS barrier Schottky (TMBS) and dual TMBS, [9] which also complicates the manufacturing process.

To address this problem, an idea of DTJBS is proposed, and it is found that this structure can greatly lower V_F with slight decline in shielding effect.

1. Structure and Simulation Preparations of the Device

Schematic cross-section of TJBS and DTJBS is shown in Fig. 1-1(a) and Fig. 1-1(b) respectively. Traditional TJBS has P$^+$ regions nearing sidewall of the trench, which serves a key role in enhancing the shielding effect. [10] Thus, the current flow can only pass through a small portion of anode area under forward bias, stringently restricting the current capability. Therefore, it is necessary to improve this defect by discontinuous trenching, which allows the current passing through the sidewall of the trench. Meanwhile, common P$^+$ region and trenched P$^+$ region can also suppress leakage current well.

To sustain the high voltage, thickness and concentration of drift layer are 62μm and 1×10^{15}cm^{-3} respectively. Models

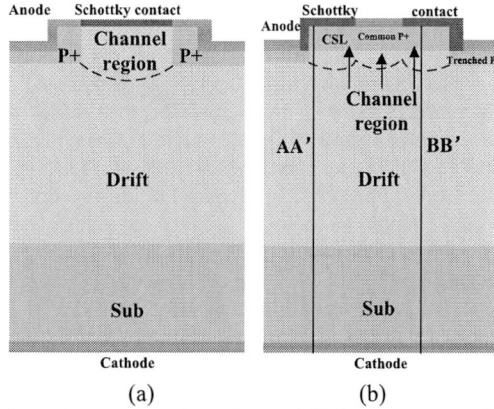

Fig. 1-1. Cross-section views of diode structures. (a) TJBS diode (one cell). (b) DTJBS diode. The maximum electric field (E_m) is located at line AA', while the maximum current density (J_m) is located at line BB'.

used in simulation contain SRH, Auger, CVT, fldmob, BGN, etc. Through Silvaco, the V_f, E_m and J_m is demonstrated, compared with a conventional TJBS. Electric field distribution and current density distribution are revealed to explain differences between TJBS and DTJBS.

2. Simulation Results

Depth of trench, the improvement about concentration of CSL and comparison between TJBS and DTJBS are discussed in this part.

To optimize the DTJBS, depth of the trench should be concerned firstly. The static characteristics are shown in Fig. 2-1.On the one hand, deeper trench mitigates the overlap of depletion region due to the increased distance between common P$^+$ region and trenched P$^+$ region shown in Fig. 1-1 (a), which is adverse for shielding effect. As a consequence, more leakage current is produced and lowers the BV. On the other hand, due to the Schoktty contact at the sidewall, deeper trench brings larger areas of Schoktty contact, especially when the depth varies from 1μm to 2μm, which may turn on the sidewall's Schottky junction (not shielded by trenched P$^+$ region) and decrease the V_f.

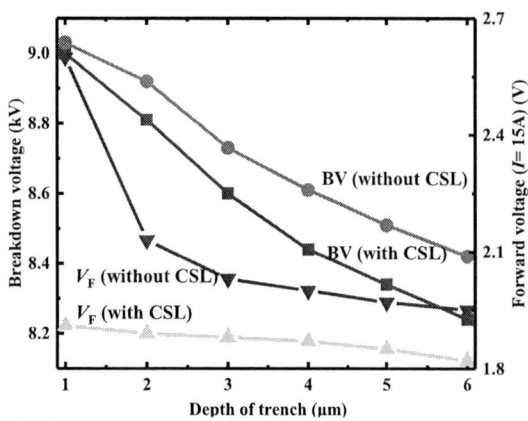

Fig.2-1. BV and V_f varied with different depths of trench in DTJBS. (V_f is achieved when I=15A)

In addition, to verify the necessity of CSL, different concentrations of CSL (1×10^{15}cm^{-3} and 5×10^{16}cm^{-3}) are set for comparison, shown as without CSL and with CSL respectively in Fig. 2-1. Higher concentration brings trouble to the overlap of depletion layer in channel region and hence BV is lower in DTJBS with CSL. However, the decline in BV is still acceptable. Meanwhile, with a narrow width, resistance in channel region (R_{ch}) makes an essential contribution to the total resistance. Properly high concentration in CSL can narrow the depletion layers and enhance the current flow .Though resistance in drift layer (R_{drift}) is a main part of total resistance, higher drift layer concentration severely decreases BV, which cannot be ignored. Therefore, it is an effective way to lower V_f by rising the concentration in CSL instead of drift layer.

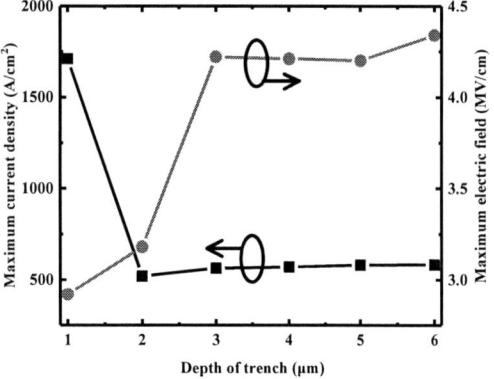

Fig. 2-2. E_m and J_m varied with different depths of trench.

E_m and J_m are also factors affected when optimizing the trench depth, shown in Fig. 2-2. Large E_m and J_m are both notorious features for DTJBS, which influences the stability of the device, but there is always an unavoidable contradiction. As a consequence, it is necessary to compromise E_m and J_m. When depth of trench is insufficient such as 1μm, high level of current density may cause huge energy loss. Deeper trench also brings about larger electric field. Combining the above two points, 2μm depth of trench is considered the best parameter for its low enough E_m, J_m, and acceptable shielding effect.

With the CSL concentration rising, there is also a compromise between E_m and J_m, shown in Table 1.

Table 1. Characteristics varied with different concentration of CSL for 6.5kV DTJBS

Concentration of CSL (cm^{-3})	V_F (V) (when I=15A)	E_m (MV/cm)	J_m (A/cm^2)
5.00×10^{15}	1.93	2.9	551
1.00×10^{16}	1.91	2.9	537
2.00×10^{16}	1.89	2.97	527
5.00×10^{16}	1.89	3.18	520
8.00×10^{16}	1.88	3.48	521

Concentration of CSL is about two orders of magnitude higher than that of drift layer. With concentration growing, there is a more uniform current distribution, though effect of pinch-off process weakens. In addition, narrower depletion layer causes a larger electric field, which is harmful for the device due to the limit of SiC critical electric field. Therefore, it is appreciative to control CSL concentration at 2×10^{16}cm^{-3}, and V_f achieves 1.89V when the current is 15A.

In addition, with optimized parameters, J_m is 527A/cm^2, while the electric field of spherical junction nearing P$^+$ region (E_m) reaches 2.97MV/cm (Fig. 2-3). On the one hand, the E_m (when breakdown occurs) occurs nearing the trenched P$^+$ region due to curve effect. However, electric field around common P$^+$ region is much lower, leading to a dispersion of electric field. On the other hand, J_f is caused by the narrowed path for current flow and preference of carriers to pass through the way that needs the least energy.

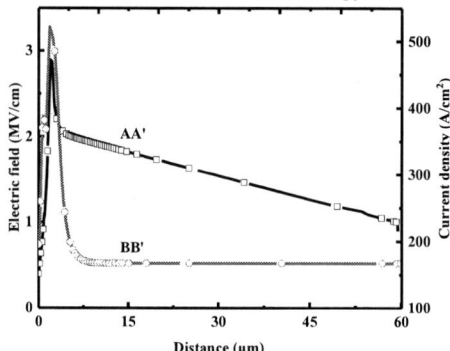

Fig. 2-3. Electric field along AA' and current density along BB' (shown in Fig. 1-1 (b)) with optimized parameters.

Furthermore, there is a trouble that higher concentration (5×10^{16}cm^{-3}) of near interface between CSL and drift layer (about 1μm thick) weakens effect of shielding, reaching around 88% value of BV compared with original concentration (1×10^{15}cm^{-3}). With thickness increasing, shielding effect even possibly fails. Thus, area of high concentration is supposed to be limited in CSL.

After optimizing parameters (depth of trench and concentration of CSL) of DTJBS, comparison between TJBS and DTJBS (without CSL) is made to verify the advantages of DTJBS. Concentration of CSL is set the same with drift layer. When P$^+$ region causes a wide depletion layer that can prevent carriers passing, TJBS needs wider channel region to maintain the same current scale as DTJBS, or close P$^+$ regions in TJBS may even cut off the current flow.

Therefore, 7μm cell width is set in the simulation, *I-V* characteristics of DTJBS and TJBS are shown in Fig. 2-4. It is obvious that DTJBS has lower V_f compared with TJBS due to a larger Schottky contact, and the common P$^+$ region and trenched P$^+$ region can result in the overlap of depletion layer and sustain 6.5kV high voltage, which can protect the Schottky contact well. That is to say Common P$^+$ region in DTJBS has a similar function with P$^+$ region in TJBS. However, the shielding effect of TJBS in simulation (shown in Fig. 2-5) cannot even work. Since the depletion layer cannot overlap, the leakage current increases dramatically and passes through the channel region into drift layer. As the result, only half BV of 6.5kV can this TJBS achieve. Narrowing width of CSL can lead to the shielding effect in TJBS, but it can also worsen V_f of TJBS, which is already worse in Fig. 2-4.

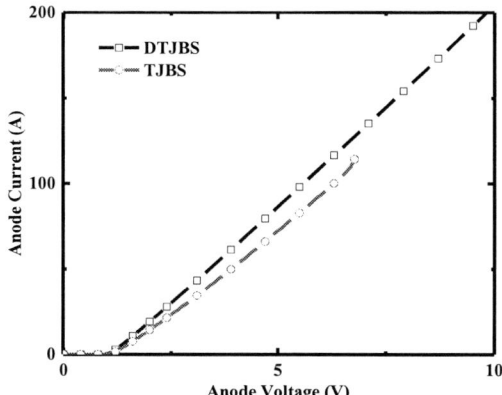

Fig. 2-4. *I-V* characteristics of DTJBS and TJBS.

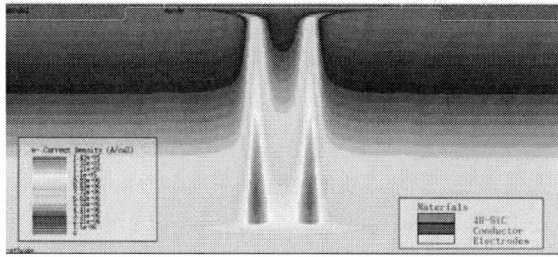

Fig. 2-5. 2D-leakage current distribution of TJBS when the breakdown occurs. (TJBS has the same width with DTJBS)

Otherwise, since DTJBS has two current paths for carriers and larger areas of Shocttky contact, it can effectively avoid centralized current and make current distribution more uniform to decrease the maximum of current in channel region (about 200A/cm^2 lower compared with TJBS), as are shown in Fig. 2-6 (a) and (b). Current is mainly resulted from the sidewall of trench, bypasses depletion layer nearing P$^+$ regions and goes into the drift layer, which is different from TJBS. It is also obvious that current density in drift layer is higher in DTJBS, which means injection of more carriers in drift layer for DTJBS. Above the two reasons, lower V_f can be found in Fig. 2-4.

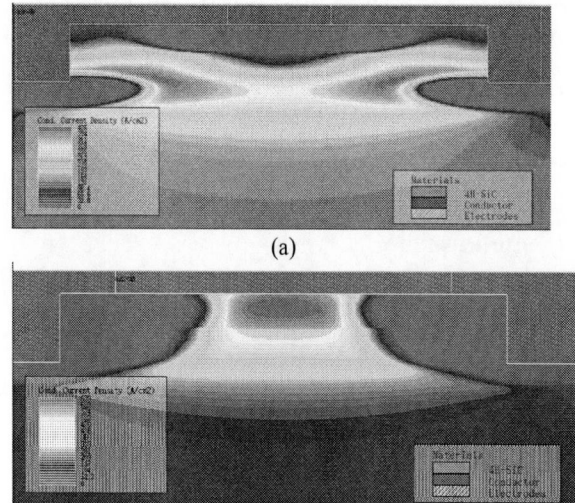

(a)

(b)

Fig. 2-6. 2D-current density distribution of (a) DTJBS; (b) TJBS when V_{anode}=7.0V.

To further illustrate this phenomenon, potential distribution of DTJBS and TJBS are shown in Fig. 2-7 (a) and Fig. 2-7 (b) respectively. Drops of potential in channel region are about 1V and 2.5V respectively in DTJBS and TJBS, and equipotential lines in DTJBS are sparser in channel region, from which it can be known that resistance of DTJBS's channel region is lower. Meanwhile, areas with high potential mean the high concentration areas where carriers are injected into SiC and have neglected resistance, which may be called effective Schottky area (S_e) to measure the current conduction capability of the devices. It can be seen that S_e in DTJBS is

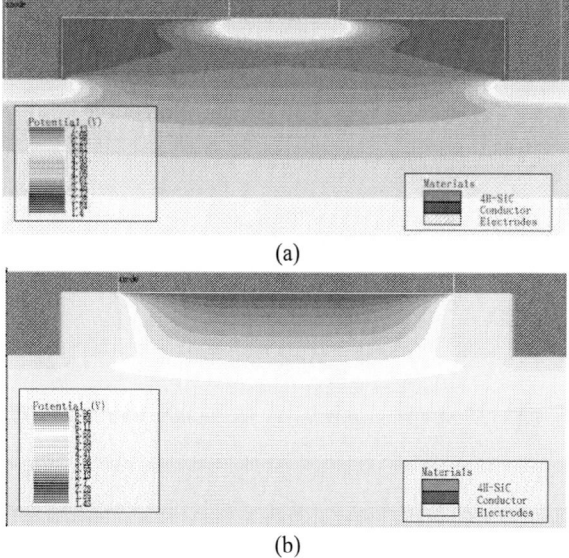

(a)

(b)

Fig. 2-7. 2D-potential distribution of (a) DTJBS; (b) TJBS when V_{anode}=7.0V.

larger than TJBS, so DTJBS has advantages of better current conduction capability.

Overall, DTJBS can operate with narrower channel region compared with TJBS, which can also minimize the weakness

978-1-7281-5757-3/19 $31.00 © 2019 IEEE   35

of shielding effect in DTJBS. With a same width of CSL in simulation, TJBS cannot even sustain 6.5kV voltage due to the high leakage current under high voltage. Meanwhile, DTJBS has an outstanding performance in current conducting, leading to a much higher current density in drift layer where is fundamental for the whole current path, which means the lower V_f in DTJBS under the same condition. Therefore, DTJBS surpasses TJBS in terms of V_f.

Conclusions

Depth of trench and concentration of CSL can both affect the characteristics in DTJBS, such as shielding effect, V_f, E_m, J_m. It is then demonstrated that DTJBS has advantages of better current conduction capability and lower V_f when compared with TJBS, while it also has a similar shielding effect. In the future work, optimizing width of channel region and P^+ regions are planned to finish.

Acknowledgements

This work was supported by Science Challenge Project under Grant TZ2018003-1-201.

References

1. R. D. e. al, "Replacing silicon IGBTs with SiC IGBTs in medium voltage wind energy conversion systems," in *2016 7th India International Conference on Power Electronics (IICPE)*, Patiala, India, 2016, pp. 17-19 Nov. 2016.
2. M. U. e. al, "Specific features of SiC-IGBT with 13kV switching," in *2012 24th International Symposium on Power Semiconductor Devices and ICs*, Bruges, Belgium, 2012, pp. 261-264.
3. A. K. e. al, "Operation of ultra-high voltage (>10kV) SiC IGBTs at elevated temperatures benefits & constraints," in *2019 31st International Symposium on Power Semiconductor Devices and ICs (ISPSD)*, Shanghai, China, China, 2019, pp. 175-178.
4. Y. C. e. al, "Development of 6.5kV 50A 4H-SiC JBS Diodes," presented at the 2018 15th China International Forum on Solid State Lighting: International Forum on Wide Bandgap Semiconductors China (SSLChina: IFWS), Shenzhen, China, 23-25 Oct. 2018, 2018.
5. A. M. e. al, "Experimental investigation of SiC 6.5kV JBS diodes safe operating area," in *The 29th International Symposium on Power Semiconductor Device*, Sapporo, Japan, 2017, pp. 53-56.
6. T. S. e. al, "Large area (150mm) high voltage (6.5kV) reverse conducting IGCT," in *2017 19th European Conference on Power Electronics and Applications (EPE'17 ECCE Europe)*, Warsaw, Poland, 2017, pp. 1-8.
7. L. K. e. al, "Dynamic switching and short circuit capability of 6.5kV silicon carbide MOSFETs," in *2018 IEEE 30th International Symposium on Power Semiconductor Devices and ICs (ISPSD)*, Chicago, IL, USA, 2018, pp. 451-454.
8. N. e. a. Ren, "An Analytical Model With 2-D Effects for 4H-SiC Trenched Junction Barrier Schottky Diodes," *IEEE Transactions on Electron Devices*, vol. 61, No. 12 (2014), pp. 4158-4165.
9. N. R. e. al, "Design and Experimental Study of 4H-SiC Trenched Junction Barrier Schottky Diodes," *IEEE Transactions on Electron Devices* vol. 61, No. 7 (2014), pp. 2459-2465.
10. L. Di Benedetto *et al.*, "Analytical Model and Design of 4H-SiC Planar and Trenched JBS Diodes," *IEEE Transactions on Electron Devices*, vol. 63, No. 6 (2016), pp. 2474-2481.

Design and Characteristics of an Etching Field Limiting Ring for 10kV SiC Power Device

Yi Wen [1,2*], Xiaojie Xu [1], Hao Zhu [1], Xuan Li [1,2], Xiaochuan Deng [1,2], Fei Yang [4], Juntao Li [3] and Bo Zhang [1]

(1. University of Electronic Science and Technology of China, Chengdu 610054, China;

2. Institute of Electronic and Information Engineering in Guangdong, University of Electronic Science and Technology of China, Dongguan 523808, China;

3. Microsystem and Terahertz Research Center, China Academy Of Engineering Physics, Chengdu 610000, China;

4. Global Energy Interconnection Research Institute, Beijing 102209, China)

wenyi169@163.com, 18708185608

Abstract

For ultra-high voltage SiC devices of 10kV and above, the length of conventional FLR even reaches up to millimeters, which impedes the miniaturization and development in ultra-high voltage applications. In this paper, a novel termination structure of the Etching Uniform Field Limiting Ring (EU-FLR) for 10kV SiC power device is proposed and analyzed based on the theory of charge field modulation. The blocking capability achieves at 14.2kV and the EU-FLR exhibits a reduction of more than 30% in size compared with the conventional FLRs. The voltage efficiency factor η_1 of EU-FLR is 90% and the area efficiency factor η_2 is 17.8V/μm, respectively. The influence of the pivotal structural parameters of EU-FLRs on termination protection efficiency has been analyzed and researched by TCAD Silvaco.

1. Introduction

Ultra-high voltage devices that refer to power devices with blocking capability over 6.5kV have exhibited promising application prospects in smart grid application, rail transit, electric vehicle, and industrial frequency conversion. Traditional Si-based IGBTs, GTOs, ETOs, and other ultra-high voltage devices are encountering challenges such as relatively low theoretical upper limit voltage, large losses due to stacking in series and parallel, and serious thermal effects. Silicon Carbide (SiC) has become the new option for the improvement of power applications based on its characteristics of the wide bandgap, excellent thermal conductivity, high electron saturation velocity, and high critical breakdown field. Nowadays, SiC has been particularly suitable for ultra-high power applications and promoted revolutionary changes in power system equipment and grid structure [1-2].

The commercial SiC power device rated voltage levels are concentrated in the 650V ~ 1700V medium and high voltage field, including diodes and MOSFETs throughout the industry. Ultra-high voltage power devices of 6.5kV or even more than 10kV are still in the process of research and development. In 2001, Sugawara reported 12~19kV SiC PiN diodes and developed the first SiC power rectifier with a blocking voltage over 10kV [3]. In 2015, Kimoto developed a SiC PiN diode with a breakdown voltage of 26.9kV and an on-resistance of 9.7mΩ·cm² by means of SM-JTE and DZ-JTE termination structures and thermal oxidation processes with enhanced carrier lifetime [4], which has been the highest breakdown voltage in SiC power devices reported so far. The withstand voltage capability of SiC power devices is much lower than its theoretical value for the existence of the curvature effect, which limits the development in the field of ultra-high voltage and miniaturization of devices. Thus, the

junction termination technology to optimize the curvature effect has become a considerable part of the design of ultra-high voltage SiC power devices. The field limiting ring (FLR), junction termination extension (JTE) and the hybrid combination of the two are mainly adopted for the terminal structure of the ultra-high voltage SiC power device. The mechanism of the JTE is to introduce the additional charges directly by depletion in OFF state. The electric charge field generated by the introduced charge is opposite to the electric potential field at the main junction, which weakens the electric field at the main junction to contribute to increasing the breakdown capability. The FLRs achieve the purpose of increasing the breakdown voltage by extending the depletion region of the main junction through highly doped P-type rings. In the view of actual fabrication, the JTE structure requires a relatively narrow range of P-type ion implantation doping concentration. The impurity activation level is sensitive to the activation conditions and is not easy to reach the optimal doping concentration response to the maximum breakdown voltage. However, FLR structure is formed at the same time with the p-type region of the main junction. The process is simple without additional injection and the tolerance of doping concentration is wide, which make the FLRs structure widely used. However, a larger number of rings are required to obtain a higher blocking voltage, resulting in a large area of junction termination and a waste of the dedicated chip eventually. Therefore, terminal efficiency has become a new important issue in the field of ultra-high voltage devices.

A linearly graded 4H-SiC JBS rectifier field limiting ring was researched in [5]. The area of the linearly graded field limiting ring is reduced by 37% compared with the traditional field limiting ring under the premise of the same breakdown voltage. In 2016, a trench multiple field limiting rings termination was used in 4H-SiC JBS rectifier which obtained a breakdown voltage of 6.7kV at a terminal length of 350μm [6]. In 2017, a single-zone JTE designed for 3.3kV SiC JBS rectifier with the terminal voltage efficiency at 81% of the parallel plane junction [7]. In the same year, a three-zone junction termination extension was demonstrated for a 13kV SiC JBS rectifier with a thickness of 150μm and a junction terminal length of 710μm [8].

2. Design of the EU-FLR structure

a) Termination efficiency factor

The conventional expression for judging the efficiency of the termination relies on the ratio of the breakdown voltage with the terminal structure to that of the parallel plane junction. However, for ultra-high voltage devices of 10kV and above, the termination length of the device is usually more than 5~8 times of the epitaxial thickness. It is comprehensive to judge the device termination efficiency simply by the ratio of the actual breakdown voltage to the theoretical voltage. Under the circumstance of the specified blocking capability, it has become one of the most considerable research fields to reduce the terminal area, increase the chip utilization rate and improve the terminal voltage withstanding capability for ultra-high voltage power devices. In this paper, a new terminal efficiency expression η is expressed as:

$$\eta = \eta_1 \times \eta_2 \qquad (1)$$

$$\eta_1 = \frac{BV}{BV_0} \qquad (2)$$

$$\eta_2 = \frac{BV}{L_{JTT}} \qquad (3)$$

The terminal efficiency η of the ultra-high voltage power device is determined by two factors, η_1 and η_2. η_1 is the voltage efficiency factor. BV is the blocking voltage. BV_0 is the blocking voltage of the parallel plane junction. η_2 is the area efficiency factor, which represents the voltage assumed by every unit length of the terminal structure. L_{JTT} is the terminal length.

b) Conventional FLR

For the conventional FLR structure, as the reverse bias voltage increases, the field limit ring introduces additional charge ΔQ and additional electric field $\Delta E(r)$ to improve the electric field distribution of the main junction and other rings [9]. The depletion region in the N⁻ epitaxial layer expands and gradually punches through to the rings. By solving the Poisson equation, the FLR parameter satisfying the blocking voltage can be obtained [10]. In the case where $i+1$ rings exist, the voltage V_{Fi} and maximum electric field $E_{max,Fi}$ on the i th ring are described in the expression (4) and (5), respectively.

$$V_{F_i} = \frac{qN_d}{2\varepsilon\varepsilon_0}\left(\frac{r_j^2}{2} + r_j^2 \ln\left(\frac{r_j + l_{j+1}}{r_j}\right) - \frac{(r_j + l_{j+1})^2}{2}\right) \quad (4)$$

$$E_{max,F_i} = \frac{qN_d}{2\varepsilon\varepsilon_0}\left(\frac{r_i^2 + r_j^2}{r_i}\right) - \frac{qN_d}{2\varepsilon\varepsilon_0} \qquad (5)$$

r_j is the depth of the FLR. l_{i+1} is the distance between the rings. r_i is the depletion radius of the i-th ring. In terms of actual fabrication conditions and high precision requirements, the parameters obtained by (4) and (5) are not usable for design and fabrication process for the blocking voltage of 10kV and above.

Using the linearly graded FLR termination mentioned in [5], the rings required for devices over 10kV tends to be more than 200 in number and 1000µm in length, respectively. As the ring space increases to a certain extent, the depletion region will be interrupted and the latter ring will no longer function to expand the depletion region into the drift region.

c) EU-FLR structure

Based on the manufacturing advantage of the uniform FLR and the charge field modulation principle, the etching uniform FLR structure proposed in this paper is shown in Fig. 2.1. The basic simulation parameters are as follows: $d_{Epi} = 100$ µm, $N_{Epi} = 5 \times 10^{14} \text{cm}^{-2}$, $N_{Sub} = 1 \times 10^{19} \text{cm}^{-2}$, and ring width $W = 2$µm. The distance between the ring and the main junction, the ring space are both S. The length of the termination Zone I and Zone II are L_I and L_{II}, respectively. D is the etching depth. A 1µm deep P⁺ doped region of Gaussian distribution was defined by multiple Al ion implantation processes. The

mechanism of EU-FLR structure is to increase or reduce the additional charge ΔQ by adjusting the region of Zone I and Zone II in order to extend the depletion region outward and modulate the electric field in the blocking state so that the electric field peak of the main junction or the certain ring can be impaired. Thereby the breakdown voltage could be increased and the termination area could be reduced. Compared with the etching JTE structure, the advantage of the EU-FLR is that when there is a large amount of positive charge at the SiC/oxide interface, the surface of lightly doped N regions between the rings function as thin heavily doped N layers and the generated charge filed would avail to reduce the original electric field applied to the main junction and each rings.

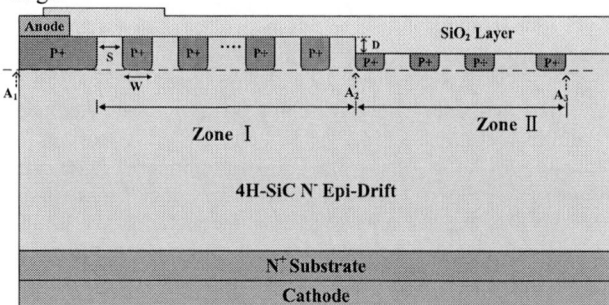

Fig. 2.1 Schematic cross-section of the EUFLR

Through TCAD Silvaco, it is found that the EU-FLR structure achieves a breakdown voltage of 14.2kV, which reaches 90% of the parallel plane junction breakdown voltage and shows a reduction of 30.4% in termination length compared with the conventional FLR. As shown in Table 1, The EU-FLR has a voltage efficiency factor η_1 of 90% and an area efficiency factor η_2 of 17.8 V/µm. The general efficiency is superior to conventional FLR.

Tab. 1 Performance comparison between EU-FLR and Con-FLR

Structure	Length/µm	BV/kV	voltage efficiency /η_1	area efficiency /η_2 (V/µm)	Breaking point
Cell	/	15.8	/	/	/
Con-FLR	1150	14.2	90%	12.4	Main junction
Con-FLR	800	9	56%	11.3	Last ring
EU-FLR	800	14.2	90%	17.8	Main junction

It is necessary to constantly increase the number of field limiting rings to keep the depletion region continuous in the conventional FLR design. For ultra-high voltage devices of 10kV and above, the length of conventional FLR even reaches over 1 mm. The EU-FLR structure could manage the additional charges amount of Zone I and Zone II through modulating the etching depth D and the position $L_{zone\ I}$. The reduction of the additional charge in Zone II leads the electric field to be pulled up at A_2, effectively flattening the electric field distribution and increasing the voltage assumed by the terminal depletion region. The electric field distribution in the proposed structure and the conventional FLR along A_1A_3 is shown in the figure. 2.2.

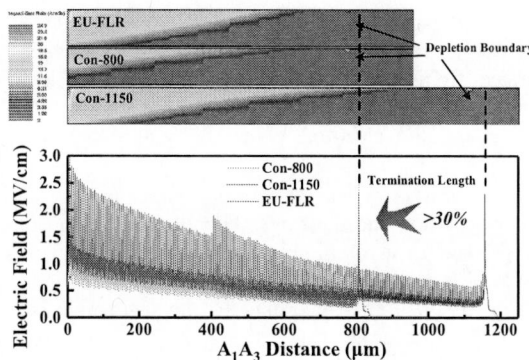

Fig. 2.2 Comparisons of the electric field distribution in EU-FLR and conventional FLR at breakdown

3. Characteristics and Analysis of EU-FLR

a) The impact of etching depth

The mechanism of the EU-FLR is to increase or reduce the additional charge ΔQ by adjusting the region of Zone I and the Zone II, thereby modulating the electric field in blocking state to mitigate the electric field concentration at the main junction. As shown in Fig. 3.1, when the etching depth D varies among 0.35-0.45μm, the breakdown voltage can be maintained at 14.2kV. When D changes from 0.35μm to 0.55μm, the breakdown voltage loss is kept within 5%, exhibiting an etching depth tolerance of 0.2μm. As the domestic etching precision is several tens of nanometers at present, the EU-FLR proposed is very suitable for the actual manufacture process level.

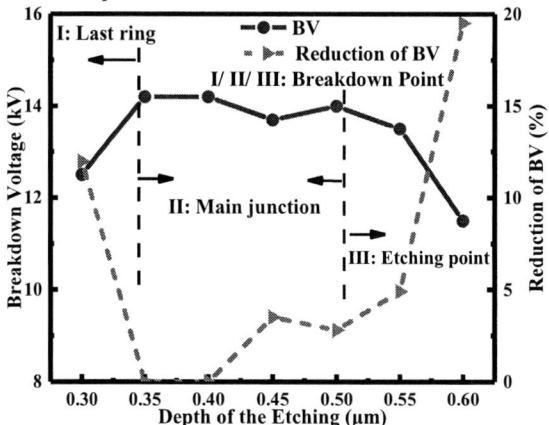

Fig. 3.1 Impact of etching depth D on the breakdown

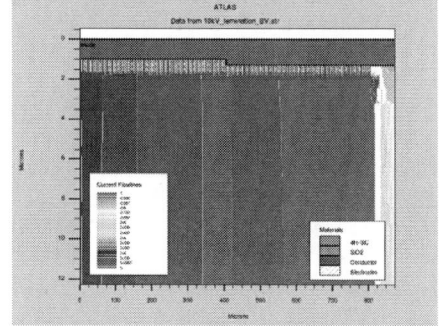

(A) D_{etch}=0.3μm: breakdown occurs in the last ring

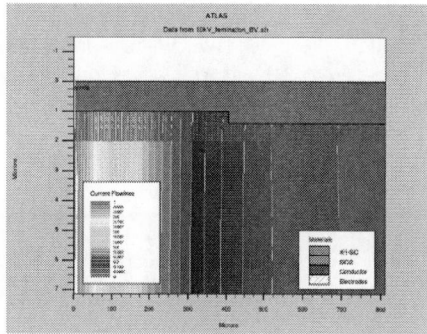

(B) D_{etch}=0.4μm: breakdown occurs in the main junction

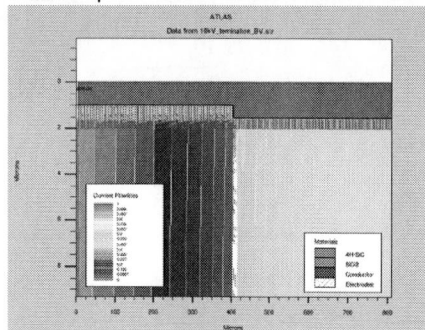

(C) D_{etch}=0.5μm: breakdown occurs in the etching point

Fig 3.2 Current path in the situation of breakdown

For various etching depths, the breakdown points of the EU-FLR in the blocking state are mainly concentrated in three places: the main junction, the etching portion, and the last ring. Figure 3.2 shows the path of breakdown current under three different breakdown conditions respectively. When the etching depth D_{etch} is between 0.35-0.45μm, the EU-FLR has the optimal electric field modulation effect on the main junction and breakdown occurs in the main junction, as shown in Fig. 3.2(B). When D_{etch} is less than 0.3μm, according to the charge field modulation theory, the additional negative charge ΔQ introduced by Zone II of P$^+$ under reverse blocking state increases and the generated charge field ΔE outside the last ring is in accordance with the original electric field E_0, which leads to the electric field concentration in the last ring. The breakdown current distribution is shown in Fig. 3.2(A). Fig. 3.2(C) denotes that when D_{etch} exceeds 0.5μm, the breakdown occurs at the etching portion because the additional negative charge ΔQ introduced in Zone II decreases with the area of Zone II enhances, which results in a decrease in the charge field ΔE generated by ΔQ. Comparing to Fig. 3.2(A), the total electric field of the last ring is alleviated. However, on the side of the etching ring close to the main junction, the charge field ΔE generated by ΔQ is opposite to the original electric field introduced by reverse bias. Therefore, when ΔE decreases, the total electric field increases relatively, causing breakdown occurs at A$_2$.

b) The impact of ring space

The etching depth D is fixed at 0.4μm and the EU-FLR width is kept at 800μm in the optimization. The distance between the main junction and the first ring is equal to the ring space. It is found that when the ring space S is less than 1.9μm, the avalanche breakdown occurs at the last ring and the loss of blocking capacity exceeds 10%. When the ring space S is greater than 2.1μm, although breakdown occurs at the main junction, the breakdown voltage drops rapidly and the voltage loss is over 10% as well. Therefore, the ring space S has a high requirement and the terminal protection effect is optimal when S=2μm, as shown in Fig. 3.3. The actual manufacture is not affected by the process mask making owing to the equally ring space design.

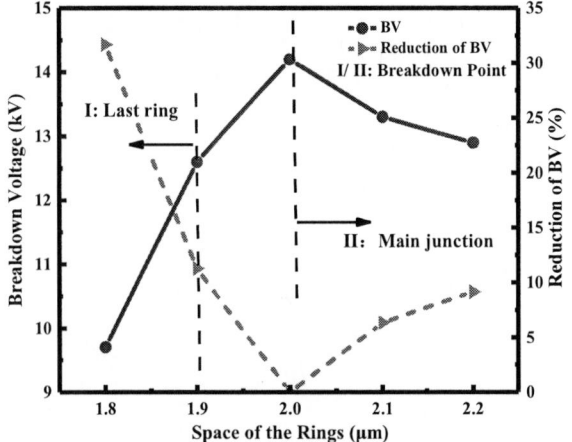

Fig. 3.3 Effect of ring spacing on breakdown

c) The impact of etching position

Another way to modulate the amount of ΔQ is to change the etching position, specifically by adjusting the area of Zone I and Zone II regions to optimize the electric field concentration of the main junction and the rings. The design parameters are as follows: the length of termination is kept at 800μm, etching depth D=0.4μm and the ring space S=2.0μm, respectively.

The tolerance of the etching position range is relatively excellent. As shown in Fig. 3.4, when $L_{zone\ I}$ varying from 200μm to 400μm, the breakdown voltage is kept above 14kV and the voltage loss is within 2% while breakdown occurs at the main junction and the electric field distribution achieves the optimization. When $L_{zone\ I}$ exceeds the optimal range, the effect of ΔE on the last ring and the etching portion becomes significant and the breakdown voltage drops rapidly where the voltage loss exceeds 14%.

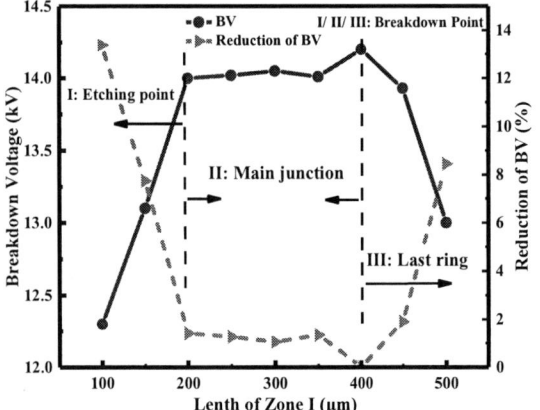

Fig. 3.4 Effect of different etching positions on breakdown

d) The impact of interface charge

The blocking characteristics of SiC power devices are affected by traps at the SiC/SiO$_2$ interface, which are generated from the holes captured by the deep interface energy level, the fixed oxide charge at the P-SiC/SiO$_2$ interface and the electrons captured by the shallow body energy level [11][12].

The SiC/SiO$_2$ interface charges Q_f induce charges at both the main junction and the rings. The electric field generated from Q_f remains zero at the main junction with no reverse voltage. If Q_f is positive, positive charges will be induced at the edge of the ring. According to the theorem of charge conservation, negative charges are induced at the main junction edge. The direction of the electric field generated by the negative charges is pointing to the main junction, which is the same as the original electric field by reverse. The peak field at the main junction is concentrated, leading to lower breakdown voltage. However, if Q_f is negative, the total peak electric field will decrease at the main junction while increase at the outer ring.

The results are shown in Fig. 3.5, the blocking capability of the power devices decreases for the exist of Q_f. When Q_f is positive, the breakdown occurs at the main junction with a 2%~4% blocking voltage loss. On the other hand, the breakdown occurs at the last ring with a blocking voltage loss up to 12% when Q_f is negative.

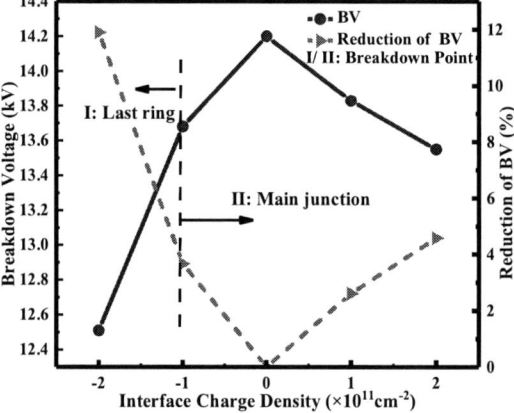

Fig. 3.5 Effect of interface charge on breakdown

Conclusions

The EU-FLR proposed in this paper satisfies the design requirements of 10kV SiC power devices. The blocking capacity of the EU-FLR termination is up to 14.2kV with the voltage efficiency factor η_1 = 90% and the area efficiency factor η_2 = 17.8V/μm, respectively. The structure exhibits an area decrease of more than 30% compared with the conventional FLRs. According to the charge electric field modulation theorem, the effects of the etching depth, etching position, ring space and the interface charge on the blocking characteristics of EU-FLR are analyzed in this paper, providing some reference for design and development of ultra-high voltage SiC power devices.

Acknowledgments

This work was supported in part by Science Challenge Project under Grant TZ2018003-1-201, Natural Science

Foundation of Guangdong under Grant 2017A030313344 and State Grid Corporation Foundation under 52020118000A.

References

1. Casady J B, Pala V, Lichtenwalner D J, et al. New generation 10 kV SiC Power MOSFET and diodes for industrial applications[C]// Pcim Europe , International Exhibition & Conference for Power Electronics. VDE, 2015.

2. Li X, Jiang J, Huang. A. Q, et al. A SiC Power MOSFET Loss Model Suitable for High Frequency Applications[J]. IEEE Transactions on Industrial Electronics, vol. 64, no. 10, pp. 8268-8276, Oct. 2017.

3. Sugawara Y, Takayama D, AsanO K, et al. 12-19 kV 4H-SiC pin diodes with low power loss[C]//Proceedings of the 13th International Symposium on Power Semiconductor Devices & ICs. ISPSD'01 (IEEE Cat. No. 01CH37216). IEEE, 2001: 27-30.

4. Kaji N, Niwa H, Suda J, et al. Ultrahigh-voltage SiC pin diodes with improved forward characteristics[J]. IEEE Transactions on Electron Devices, 2015, 62(2): 374-381.

5. Deng X, Wen Y, Wang X, et al.[C]// IEEE International Conference on Solid-State and Integrated Circuit Technology. IEEE, 2014:1-3.

6. Hao Yuan, Qingwen Song, Trench Multiple Floating Limiting Rings Termination for 4H-SiC High-Voltage Devices, IEEE Electron Device Letters,vol. 37, no. 8, AUGUST 2016.

7. Pan Y, Tian L, Wu H, et al. 3.3 kV 4H-SiC JBS diodes with single-zone JTE termination[J]. Microelectronic Engineering, 2017, 181: 10-15

8. Kitai H, Hozμmi Y, Shiomi H, et al. Demonstration of 13-kV class junction barrier Schottky diodes in 4H-SiC with three-zone junction termination extension[C]// European Conference on Silicon Carbide & Related Materials. IEEE, 2017:451-454.

9. Zhang B, Luo X R, Li Z J. Electric field optimization technology for power semiconductor devices [M]. Chengdu: University of Electronic Science and Technology of China, 2015: 12-15.

10. He J, Chan M, Zhang X, et al. A new analytic method to design multiple floating field limiting rings of power devices[J]. Solid-state electronics, 2006, 50(7-8): 1375-1381.

11. M. Noborio, J. Suda, and T. Kimoto, P-channel MOSFETs on 4H-SiC {0001} and nonbasal faces fabricated by oxide deposition and N_2O annealing, IEEE Trans. Electron Devices, vol. 56, no. 9, pp. 1953–1958, Sep. 2009.

12. D. C. Sheridan, Guofu Niu, J. Neil Merrett, J. D. Cressler, J. B. Dufrene, J. B. Cassady, and I.Sankin, Comparison and optimization of edge termination techniques for SiC power devices, In Proc. International Symposium on Power Semiconductor Devices and ICs, Osaka, Japan, Jun. 2001, pp. 191–194.

Improvement of Clamped Inductive Turn-Off Ruggedness of Trench IGBT at Overcurrent Condition with Optimized Split Gate Structure

Jiang Lu[1], Jiawei Liu[1], Xiaoli Tian[1*], Hong Chen[1], Fei Liang[2], Yun Bai[1]

1. Institute of Microelectronics of Chinese Academy of Sciences, Beijing, China
2. Sichuan Huacan Electronics Co., Ltd., Sichuan, China
*tianxiaoli@ime.ac.cn

Abstract

In this paper, the clamped inductive turn-off ruggedness of a novel Trench Insulated Gate Bipolar Transistor (TIGBT) at overcurrent condition is studied by numerical simulation. This proposed structure is optimized by using split gate structure to improve the turn-off reliability. Simulation result shows that the maximum turn-off critical current of the proposed TIGBT increases for about 42.8% compared with the conventional TIGBT structure. The main reason is that the electric field of the proposed structure is redistributed by RESURF effect at the trench bottomed area, resulting to attenuate the current filament at local area. In addition, the hole which concentrated at the trench bottom area is less because the bottom part of split gate polysilicon is shorted with the emitter. Therefore, the local dynamic avalanche effect which triggered by the excessive carriers concentration and high peak electric filed accumulation is weakened. The turn-off ruggedness of the proposed structure is enhanced effectively and the electrical parameters are not compromised but even better compared with conventional structure.

Keywords: Trench Insulated Gate Bipolar Transistor (TIGBT), dynamic avalanche, split gate, turn-off ruggedness.

1. INTRODUCTION

The Trench Insulated Gate Bipolar Transistor (TIGBT) is widely used in power electronic area due to a great trade-off performance between the on-state saturation voltage and switching losses. Typically, the TIGBT is used as the power switch application under the clamped inductive condition. It is inevitable to work at the harsh circumstances with high current and voltage simultaneously and the safe turn-off ability is important for the long-time working requirement. During the overcurrent turn-off process, the peak electric field is accumulated at the local small area and the current filament is formed easily after the electron current form the MOS channel disappeared and then the dynamic avalanche phenomenon is triggered at the trench area [1]. This dynamic avalanche behavior is related to the excessive holes which cumulated at the trench bottom area, leading to the large local peak electric filed. In order to suppress this behavior, some optimizing strategies are investigated, such as using the large gate resistance, changing the cell pitch size or using negative gate drive voltage at the turn-off process [2-4]. However, these methods will sacrifice other electric parameters to a certain extent or increase the fabrication cost due to an additional gate drive design.

In this paper, an optimized TIGBT with split gate structure to enhance the turn-off reliability is proposed by numerical simulation. As we known, the split gate structure is originally from the low voltage power MOSFET design skill to reduce the gate-drain charge [5]. Moreover, it also brings an improved RESURF effect on the breakdown ability by local

electric filed optimization [6]. We use these structure advantages to enhance the turn-off reliability under clamped inductive condition. In order to analysis the performance of the proposed structure, the turn-off behavior is compared with the conventional TIGBT structure by investigating the inner electric field distribution and carrier movement.

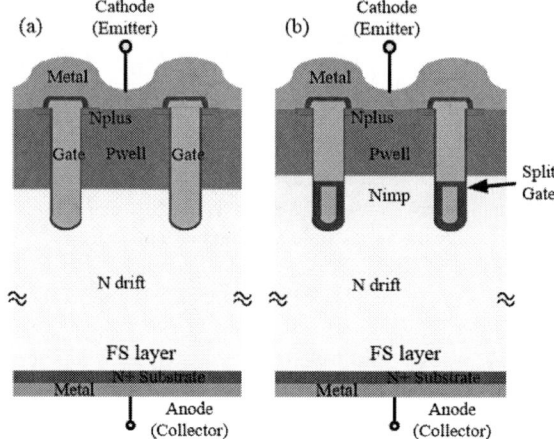

Fig. 1. Cross-sectional view (not to scale) of (a) the conventional structure and (b) the proposed structure.

2. DEVICE STRUCTURE AND SIMULATION SETUP

Fig. 1 shows the structure of the conventional TIGBT and the proposed TIGBT. Both of structures are designed with the 120μm wafer thickness, low concentration Ndrift region and high implant dose FS layer to achieve 1.2kV targeted breakdown voltage. The cell pitch and trench depth are 4μm and 5 μm, respectively. Furthermore, two structures are designed with the same doping profile and structure size parameter except the trench gate area. As can be seen from the Fig. 1(b), the gate polysilicon is split into top and bottom parts. The bottom gate polysilicon is enclosed by thick oxide layer and connect with the emitter through the cell periphery contact. In order to balance the electric field of the trench bottom area, the N enhanced layer with higher doping concentration, namely Nimp, is used at the local split gate area. And it also increases the conduction modulation effect as the hole blocking layer. The simulations are performed by Synopsis Sentaurus Technology Computer Aided Design (TCAD) software. Basic semiconductor models are included to analyze the device's physical behavior in simulation. The generation and recombination models include the Shockley-Read-Hall (SRH) recombination, auger recombination and avalanche generation by Van Overstraeten model. The mobility models include the high-field saturation, carrier-carrier scattering and mobility degradation by normal electric field (Enormal) model. In order to save the simulation time, we select the single cell pitch 4μm and the 1 mm^2 die area factor in the simulation.

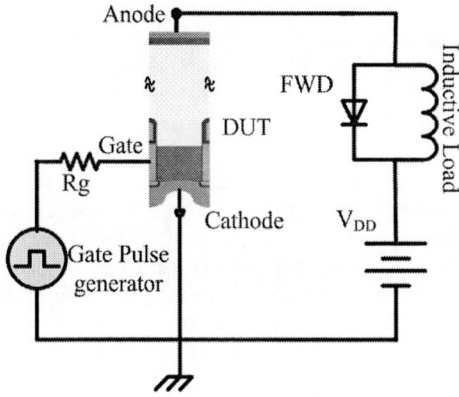

Fig. 2. Circuit used in mixed-mode transient TCAD simulations for investigating the turn-off behavior of TIGBT at clamped inductive condition.

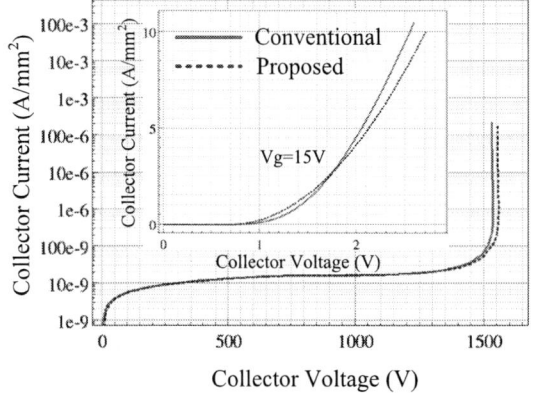

Fig. 3. Simulation results of the breakdown voltage and I-V characteristic. The curve of I-V characteristic is set at V_g=15V.

Fig. 2 shows the simulated circuit setup of the device with clamped inductance by freewheeling diode (FWD). During the turn-off process, the bias voltage was fixed at the rated voltage 1.2kV and several times current over the rated current was set to flow in the device. Here the gate pulse generator is used to control the on-off state and the current value which flowed through the device and the inductance. Once the current reached to the critical current, the device can't turn off normally. We capture the critical conditions to compare the different performance of two structure devices. Through this procedure, the device turn-off ruggedness at the clamped inductance and overcurrent condition is evaluated.

3. RESULTS AND DISCUSSIONS

Before analyzing the clamped inductive turn-off ability of the proposed structure, the basic electric parameter is compared with the conventional TIGBT. Fig. 3 shows the curve of the breakdown voltage and I-V characteristic for each structure. It can be seen that the breakdown voltage are all over 1.5kV and the proposed structure is much higher due to the local RESURF effect. From The I-V characteristic comparison of the conventional structure and the proposed structure, it can be captured that the saturation voltage at Vge=15V with current density 200A/cm² are 1.63V and 1.60V, respectively. Therefore, the basic electric parameters

of the proposed structure are better through this gate structure adjustment.

Then, the turn-off ability of two structures is compared by using the circuit setup based on the Fig. 2. As we mentioned before, the bias voltage is fixed at 1.2kV and the gate pulse width is set to increase gradually until the device out of control. Once the current reaches to the critical value, it can be seen in Fig. 4 that the collector current cannot down to zero although the gate voltage already disappeared. Here we capture the maximum critical current of two structures. From the results, the maximum peak current density for the conventional TIGBT and the proposed TIGBT are 72.5 A/mm² and 103.6A/mm², respectively. It indicates that the maximum critical current which the device can turn off safely for the proposed structure increases for about 42.8% compared with the conventional structure. That is to say, the proposed structure achieves great turn-off reliability and better basic electric parameters compared with the conventional structure.

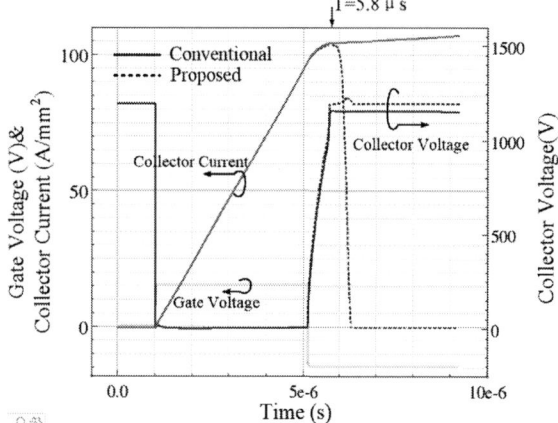

Fig. 4. Turn-off waveform at the same bias condition: (a) the conventional TIGBT (solid line) and (b) the proposed TIGBT (dotted line).

In order to analyze the inner difference of electric field distribution and carrier profile at the turn-off process, we capture the turn-off waveform of the device with the same bias condition. Fig. 4 shows the comparison waveform of two structures with the same gate voltage pulse width. It can be seen that there is a turning point at the time 5.8μs, where the current and voltage reached to the maximum value for each structure. After this time point, the collector current of the proposed structure (dark red dotted line) drops to zero instantly and that means the device turns off safely. However, the collector current of conventional TIGBT (red solid line) still exists and flows continually although the gate voltage already disappeared. It indicates that the conventional TIGBT cannot turn off by gate control at this current density. At this time point, the electric current disappears because the gate voltage is turned off and the huge holes accumulate at the bottom of trench area. If the peak electric field at the local area is too high, the electric-hole pairs will generate by local micro dynamic avalanche effect continually [2].

Then, the inner physical behavior of the device is analyzed at this time point. Fig. 5 shows the 3D electric field profile of two structures at time 5.8μs. It can be seen in Fig. 5(a) that the peak electric field of the conventional structure accumulates at

978-1-7281-5757-3/19 $31.00 © 2019 IEEE   43

the trench bottom corner area. According to the enlarged electric filed profile (right side in Fig. 5), the peak electric field locates near the corner of trench bottom and the maximum peak value is about $5×10^5$V/cm. However, for the proposed structure in Fig. 5(b), the electric field distributes along the split gate area with two peak locations, which can be seen in the enlarged electric field profile. This phenomenon is related to the local RESURF effect and the maximum peak electric field is only $4×10^5$V/cm. Obviously, the peak electric field of the proposed structure is lower compared with the conventional structure at the same overcurrent and voltage bias condition. Thus, it helps to attenuate the local dynamic avalanche effect which related to the local electric field accumulation and carrier generation.

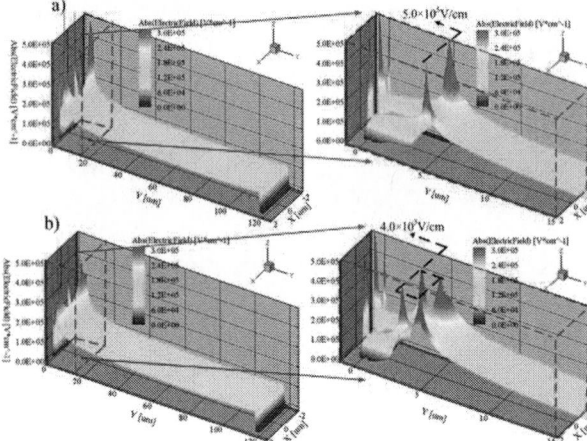

Fig. 5. TCAD simulated 3D electric field profiles during the turn off situation: (a) the conventional TIGBT and (b) the proposed TIGBT.

During the turn-off process, the hole in plasma area is pushed to the cathode by the electric field stress. Due to the negative gate voltage, some holes accumulate easily at the trench bottom area and the local electric field is intensified continuously by this influence [7-8]. Fig. 6 shows the 2D hole density profile at the time 5.8μs and the compared distribution of the hole density by lateral cutline beneath the trench bottom 0.05μm. The cutline on the conventional structure and the proposed structure are the A-A' (red solid line) and the B-B' (blue dotted line), respectively. It can be seen in Fig. 6. that the hole density of conventional structure is much higher at the trench bottom area. But for the proposed structure, lots of holes accumulate at the center of the two adjacent split gate areas due to the gate at bottomed area is shorted with emitter contact. That means that holes which concentrated at the bottom are less for the proposed structure.

By analyzing the turn-off behavior of the proposed TIGBT with clamped inductance at overcurrent bias conditions, the physical failure mechanism can be concluded as following reasons. Firstly, it can be known that the peak electric field profile of proposed structure is smaller and smother by RESURF effect at trench bottom area compared with the conventional structure. For the proposed structure, the electric field doesn't concentrate at the trench bottom area but distribute more smoothly. On the contrary, the electric field in the conventional structure is easy to accumulate at the corner of bottom because of the cylindrical structure. Thus, the

conventional structure induces higher peak electric field at local area and the dynamic avalanche behavior is much severe inevitability. The electron-hole pairs which generated by the high electric field continue increasing and the latch-up is triggered easily.

Secondly, the hole is not easy to accumulate at the trench bottom area for the proposed structure. During the turn-off process, the excessive holes are pushed by the electric field to the cathode side. The bottom part polysilicon of the split gate connects with the emitter contact (ground) instead of the negative gate voltage which appealed the hole at turn-off condition. As we already known, latch up is the destruction effect in the IGBT [9]. If the fewer holes are concentrated at the trench bottom, the less chance is to trigger the latch up behavior. Otherwise, the electric filed increase with the holes concentrating continuously. This feedback behavior causes the intensifying of local peak electric field and increasing of the local current density, which can cause the device burnout at local area finally.

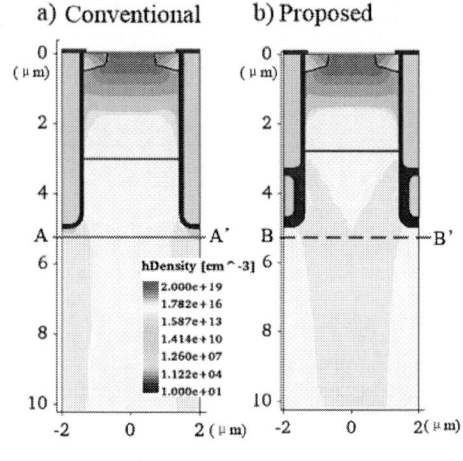

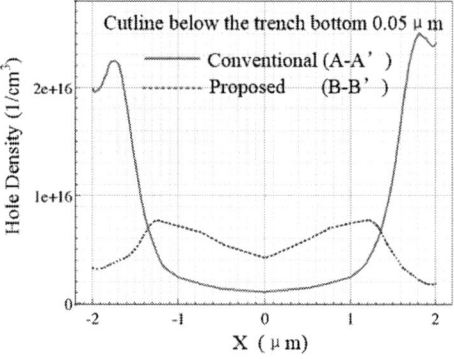

Fig. 6. Simulated 2D hole density distribution at the time 5.8μs and the comparison of hole density by lateral cutline. (a) the conventional TIGBT and (b) the proposed TIGBT.

Conclusions

The clamped inductive turn-off reliability of the optimized TIGBT with split gate structure is compared with the conventional TIGBT by TCAD simulation. From the simulation results, the maximum turn-off critical current of the proposed structure increases for about 42.8% and the basic parameters becomes better simultaneously. The inner electric field distribution and current density profile are analyzed at

978-1-7281-5757-3/19 $31.00 © 2019 IEEE

the same overcurrent condition. The electric field distribution of the proposed structure is smaller and smoother at the trench bottom area by RESURF effect. It helps to attenuate the peak electric field stress and the dynamic avalanche behavior. Furthermore, the hole which concentrated at trench bottom is less because the split gate polysilicon is shorted with emitter contact (ground). Therefore, it indicates that this proposed structure shows a great performance in turn-off ruggedness without compromising other parameters. Through these structure advantages, the proposed structure displays a great potential in the clamped inductive application.

Acknowledgments

This work was supported in part by the National Key Research and Development Program of China 2017YFB1200902, in part by the National Natural Science Foundation of China under Grant 51490681, in part by the Sichuan Science and Technology Program, and in part by the National Key Research and Development Program of China under Grant No. 2017YFB0102302.

References

1. C. Toechterle, F. Pfirsch, C. Sandow, and G. Wachutka, "Evolution of current filaments limiting the safe-operating area of high-voltage trench-IGBTs," in *Proc. 26th ISPSD*, 2014, pp. 135-139

2. S. Machida, K. Ito, and Y. Yamashita, "Approaching the limit of switching loss reduction in Si-IGBTs," in *Proc. 26th ISPSD*, 2014, pp.107-111

3. M. Riccio, L. Maresca, A. Irace, G. Breglio, and Y. Iwahashi, "Impact of gate drive voltage on avalanche robustness of trench IGBTs," *Microelectronics Reliability,* vol. 54, 2014, pp. 1828-1832

4. R. Baburske, V. van Treek, F. Pfirsch, F. J. Niedernostheide, C. Jaeger, H. J. Schulze*, et al.,* "Comparison of critical current filaments in IGBT short circuit and during diode turn-off," in *Proc. 26th ISPSD*, 2014, pp. 47-51

5. P. Goarin, G. E. J. Koops, R. van Dalen, C. Le Cam, and J. Saby, "Split-gate Resurf Stepped Oxide (RSO) MOSFETs for 25V applications with record low gate-to-drain charge," in *Proc. 19th ISPSD*, 2007, pp. 61-65

6. G. E. J. Koops, E. A. Hijzen, R. J. E. Hueting, and M. A. A. in 't Zandt, "RESURF stepped oxide (RSO) MOSFET for 85V having a record-low specific on-resistance," in *Proc. 16th ISPSD*, 2004, pp. 185-189

7. H. Tao, F. Pfirsch, B. Reinhold, J. Lutz, and D. Silber, "Transient avalanche oscillation of IGBTs under high current," in *Proc. 26th ISPSD*, 2014, pp. 43-47

8. C. Toechterle, F. Pfirsch, C. Sandow, and G. Wachutka, "Analysis of the latch-up process and current filamentation in high-voltage trench-IGBT cell arrays," in *Simulation of Semiconductor Processes and Devices (SISPAD), 2013 International Conference on*, 2013, pp. 296-300

9. Baliga, B. Jayant. *Fundamentals of power semiconductor devices*. New York, NY: Springer Science & Business Media, 2008, pp. 974-975

Optimized Design of 4H-SiC VDMOSFET for Low ON-resistance

Defu Yin, Zhiming Wu*, Xian Zou, Yongqiang Sun, Yaping Wu, Weiping Wang, Xu Li, and Junyong Kang*
Department of Physics, OSED, Fujian Provincial Key Laboratory of Semiconductor Materials and Applications, Jiujiang
Research Institue, Xiamen University
Xiamen, 361005, China
E-mail: zmwu@xmu.edu.cn, jykang@xmu.edu.cn

Abstract

In this work, we develop an optimized VDMOSFET cell structure based on 4H-SiC material. In the optimized structure, two high n-doped regions are added at both sides of the JFET region. Simulation results reveal that the additional n-doped regions not only effectively limit the depletion width in JFET region at ON-state, but also could protect the oxide layer at OFF-state due to depletion expansion. As a result, the optimized structure reduces the specific ON-resistance by 18% while keeping breakdown voltage as roughly high as the conventional structure; meanwhile, the value of figure of merit increases by 22%, which exhibits a significant improvement in device performance.

Keywords-Silicon Carbide, VDMOSFETs, Breakdown Voltage, ON-resistance, Figure of merit

Introduction

Electronic devices based on wide bandgap semiconductors have extensive applications in the fields such as consumer electronics, automotive, renewable energies, transportations, etc [1-3]. As a kind of wide bandgap semiconductor, 4H silicon carbide (4H-SiC) has superior physical and electrical properties, such as high thermal conductivity, high critical electric field, which make it the most promising material for the applications of high temperature and high power semiconductor devices [4-8].

Currently, vertical double-implanted MOSFET (VDMOSFET) is one of major structures in power MOSFETs, and 4H-SiC VDMOSFETs are considered as an ideal choice to replace Si bipolar power devices owing to their low ON-resistance and absent minority carrier storage [9]. Compared with U-shaped trench-gate MOSFETs (UMOSFETs), 4H-SiC VDMOSFETs exhibits a high reverse breakdown voltage due to the existence of the JFET region, but it possesses a relatively large ON-resistance. To reduce the ON-resistance, many optimized structures based on conventional VDMOSFETs were proposed [10-12]. For example, A. Saha et al. developed a VDMOSFET with a specific ON-resistance of 6.95 m$\Omega \cdot$cm^2 via adding a current spreading layer (CSL) under the P-Well and increasing the concentration of the JFET region [10]. S. Howell et al. proposed a 10 kV/50 A 4H-SiC power VDMOSFET by growing a 100 nm thick epitaxial layer with a donor concentration of 5.0×10^{14} cm^{-3} on the surface of P-Well [11]. In addition, Q. C. Zhang et al. developed an optimized VDMOSFET structure named CIMOS which exhibited only 4 m$\Omega \cdot$cm^2 specific ON-resistance at room temperature via an additional p-type ion implant at the center of the JFET region [12].

In this work, we propose an optimized VDMOSFET structure that has high n-doped regions at both sides of the JFET region. The electric properties of the optimized device are investigated in detail. The simulation results show that the optimized structure reduces the specific ON-resistance ($R_{on,sp}$) by 18% while keeping breakdown voltage roughly equaling to conventional structure, resulting in a 22% improvement on figure of merit (FoM).

Device structure and mechanism

Fig. 1 (a-c) show the schematic cross-sectional views of the conventional, doped-JFET, and optimized proposed VDMOSFET devices, respectively. The JFET region in conventional device has the same concentration as the drift layer does, while doped-JFET device has a high doped JFET region with the concentration of 4×10^{16} cm^{-3}. The proposed optimized device has two additional high n-doped regions with the thickness same as P-Well regions at both sides of the JFET region. The length of the additional n-doped region (L_A) is 0.2 μm and the donor concentration is 3×10^{17} cm^{-3}.

In the conventional VDMOSFET structure, high doped P-Well region is formed by Al ion implantation on the surface of lightly n-doped epitaxial layer for avoiding device punch through in the forward blocking mode of operation, and then a p$^+$-n junction is formed between P-Well and epitaxial layer. Drift region locates under the P-Well and JFET region exists between two continuous P-Well regions. When the device is in the forward blocking mode of operation, the depletion region in the n-doped drift layer expands as the drain voltage increases and undertakes a high reverse voltage. The depletion region in JFET region could effectively protect the gate oxide layer from breakdown at high reverse voltage. However, for conventional structure, the expansion of the depletion mainly occurs in the lightly n-doped JFET region when the device is at ON-state. As a result, the $R_{on,sp}$ of the device increases due to the reduced actual width of the JFET region.

A heavy constant n-doping in the JFET region is one way to reduce the $R_{on,sp}$ caused by the JFET region, as shown in Fig. 1 (b). The heavy n-doping increases electron concentration in the JFET region, but it decreases the electron mobility. As a result, the resistance of JFET region significantly decreases compared with that of the conventional device. In the case of the optimized structure (seeing Fig. 1 (c)), the additional high n-doped region could limit the expansion of the depletion in the JFET region at ON-state and inject electrons into the lightly doping JFET region, which is beneficial for the low $R_{on,sp}$. When the VDMOSFET is in the forward blocking mode of operation, as the drain voltage is much higher than that in ON-state, the expansion of the depletion region could still extend to the whole JFET region and well protects the gate oxide layer.

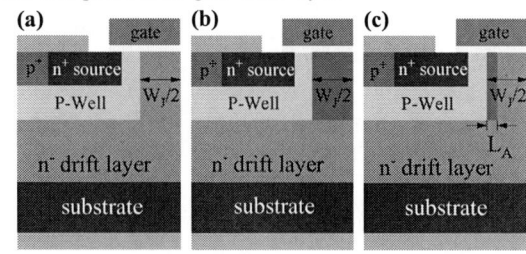

Fig. 1 Cross-sectional views of (a) conventional VDMOSFET structure, (b) doped-JFET VDMOSFET structure, and (c) optimized VDMOSFET structure.

978-1-7281-5757-3/19 $31.00 © 2019 IEEE

Table 1 Detailed structure parameters.

Parameter	Value
Gate oxide thickness	50 nm
n⁺ source junction depth	0.25 μm
P-Well junction depth	0.60 μm
Added high n-doped region thickness	0.60 μm
n⁻ drift region thickness	11.00 μm
Channel length	0.50 μm
Added high n-doped region length (L_A)	0.20 μm
JFET region width (W_J) in conventional device	2.20 μm
JFET region width (W_J) in doped-JFET device	1.70 μm
JFET region width (W_J) in optimized device	2.00 μm
n⁻ drift region concentration	9×10^{15} cm⁻³
JFET concentration in doped-JFET device	4×10^{16} cm⁻³
Added high n-doped region concentration	7×10^{17} cm⁻³

Results and discussion

We comprehensively compare the device performance of the three structures above. The numerical simulations of all three structures were performed using Silvaco ATLAS [13]. Several physical models were considered including BGN, FERMI-DIRC, INCOMPLETE, SRH, AUGER, FLDMOB, ALTCVD, and OKUTO [13]. SRH and AUGER are carrier generation-recombination models. FLDMOB is a parallel electric field-dependent mobility model which describes the phenomenon that the velocity of the electron is asymptotic to the saturation velocity at high electric field. ALTCVD is a comprehensive mobility model at low electron field and has been used successfully in the simulation of 4H-SiC MOSFETs [13-15]. OKUTO is an empirical model with temperature dependent coefficients, where the parameters for 4H-SiC material are fitted according to Ref. [16]. The 4H-SiC/SiO₂ interface state distribution is extracted from experimental data in Ref. [17]. Detailed structure parameters are listed in Table 1. In the simulation, device breakdown occurred at which the following conditions are true: (1) semiconductor avalanche breakdown occurs, or (2) the electric field in gate oxide layer reaches 4 MV/cm, whichever is less [10].

Fig. 2 shows the transfer characteristic curves of the studied devices. The drain voltage is set to 0.1 V, which

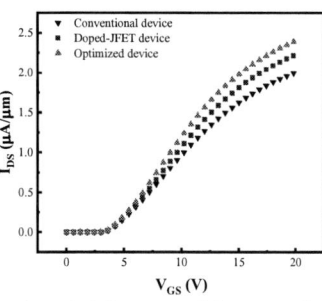

Fig. 2 Transfer characteristic curves of the conventional, doped-JFET, and optimized devices at a constant drain voltage of 0.1 V.

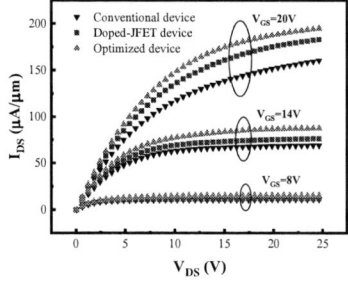

Fig. 3 Output curves of the conventional, doped-JFET, and optimized

devices.

makes sure that the device works at linear region. The cutline with the maximal slope of the transfer characteristic curve is extracted and the threshold voltage V_t is defined as its horizontal axis intercept. The obtained V_t for the conventional, doped-JFET, and optimized devices are 4.23 V, 4.33 V, and 4.26 V, respectively. As the additional high n-doped region has little impact on the surface concentration of P-Well region, the V_t of the optimized device is almost same as that of conventional one. Notably, the current density I_{DS} of the optimized device is obviously higher than those of the other two devices, and an improvement of 20% is achieved at the gate voltage of 20 V compared with the conventional one.

Fig. 3 shows the output curves of the conventional, doped-JFET, and optimized devices under different gate voltages. As the additional high n-doped region effectively depress the expansion of depletion in JFET region at the ON-state and the effect injection of electrons, the I_{DS} of optimized device is larger than those of the other two devices at same drain and gate voltages. The optimized device has the lowest specific ON-resistance of 2.93 mΩ·cm² (the values of conventional and doped-JFET devices are 3.56 mΩ·cm² and 3.09 mΩ·cm², respectively.).

To analyze the essence of the variation of $R_{on,sp}$ in the JFET region, we extract the band diagram and electron concentration distribution at thermal equilibrium state and the condition of VDS = 10 V (seeing Fig. 4 and Fig. 5). Fig. 4 shows band structures of the P-Well/JFET junction for all three devices at thermal equilibrium state. As for the conventional device, the depletion expansion mainly occurs in the lightly n-doped JFET region, and the depletion width reaches up to 500 nm in current case. By contrast, the doped-JFET device decreases the depletion width to 300 nm due to the higher n-doped concentration. However, for

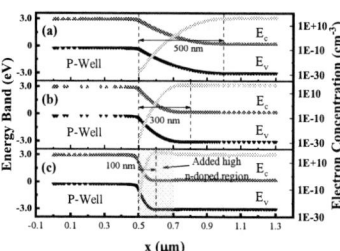

Fig. 4 Energy band diagram at thermal equilibrium for (a) conventional device, (b) doped-JFET device, and (c) optimized device.

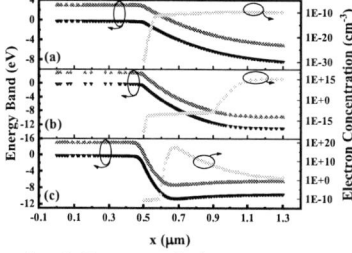

Fig. 5 Energy band diagram and electron concentrations at drain voltage of 10 V for (a) conventional device, (b) doped-JFET device, and (c) optimized device.

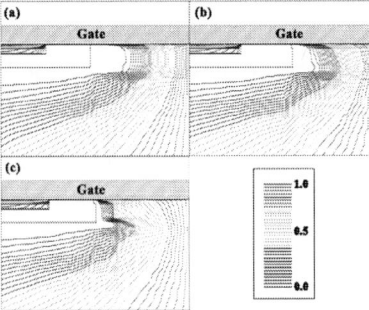

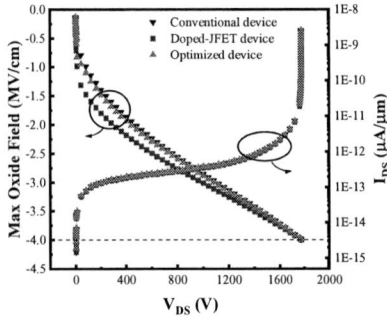

Fig. 6 Current flowline distribution of (a) conventional device, (b) doped-JFET device, and (c) optimized device at On-state.

Fig. 7 Maximal electric fields on y-axis in gate oxide layer and breakdown characteristics curves for the conventional, doped-JFET, and optimized devices.

the optimized device, the depletion mainly expands in the added high n-doped region, and the depletion width in JFET region is even less than 100 nm in each side.

Fig. 5 shows the band structures of P-Well/JFET junction and electron concentrations for all three devices at the condition of VDS = 10 V. Compared with the results in Fig.4, it can be seen that the depletion width in the JFET region increases and electron concentration decreases when the P-Well/JFET junction works at the reverse bias. Notably, the depletion width is still less than the added n-doped width of 200 nm owing to the high doped concentration in the optimized device. In this case, the injected electron concentration reaches up to a maximal value of 1.79×10^{17} cm^{-3}, almost two orders of the magnitude higher than that in doped-JFET device, which is beneficial to current expansion in the JFET region. Fig. 6 shows the current flowline distributions in the three kinds of devices. As shown in Fig. 6, the current only passes part of the JFET region in conventional and doped-JFET devices, whereas the current flows through the full JFET region in the optimized device. Consequently, the optimized device has the lowest $R_{on,sp}$.

Table 2 Breakdown Voltage, specific ON-resistance, and FoM at LA=0.2 μm.

Concentration (cm^{-3})	V_{BR} (V)	$R_{on,sp}$ (mΩ·cm^2)	FoM (MW/cm^2)
1×10^{17}	1772.0	3.13	1003
2×10^{17}	1771.5	2.99	1050
3×10^{17}	1770.9	2.93	1071
4×10^{17}	1712.8	2.89	1013
5×10^{17}	1646.3	2.87	943
6×10^{17}	1579.6	2.86	873
7×10^{17}	1503.2	2.85	793
8×10^{17}	1078.5	2.84	410
9×10^{17}	508.4	2.83	91

Table 3 Breakdown Voltage, specific ON-resistance, and FoM at n-doped concentration of 3×10^{17} cm^{-3}.

L_A (μm)	V_{BR} (V)	$R_{on,sp}$ (mΩ·cm^2)	FoM (MW/cm^2)
0.05	1772.1	3.44	912
0.10	1771.7	3.07	1021
0.15	1771.3	2.98	1051
0.20	1770.9	2.93	1071
0.25	1688.5	2.88	991
0.30	1610.9	2.85	912

The avalanche breakdown curves of all three simulated devices are shown on the right y-axes of Fig. 7. The avalanche breakdown voltage ($V_{br, ava}$) corresponds to the drain voltage V_{DS} with a leakage current density of 1×10^{-10} μA/μm. All three curves are almost overlapped, and the estimated $V_{br, ava}$ are 1771 V, 1773 V, and 1771 V for the conventional, doped-JFET, and optimized devices, respectively. This means that the additional n-doped regions with a suitable JFET width do not affect the avalanche breakdown voltage.

Generally, the performance of MOSFET is estimated by the parameter FoM which is defined as following formula:

$$FoM = \frac{V_{BR}^2}{R_{on,sp}},$$

where V_{BR} denotes the breakdown voltage. In our cases, according to above results, the FoMs of the conventional, doped-JFET, and optimized devices are 881 MW/cm^2, 1018 MW/cm^2, and 1071 MW/cm^2, respectively, indicating a 22% improvement for the optimized device. The left y-axes in Fig. 7 shows the dependence of the maximal electric field on y-axis in the oxide layer on drain voltage. The value of the electric field in all three devices is still less than the threshold of 4 MV/cm. This means that the JFET region in the optimized device still can well protect the oxide layer.

In addition, the optimization of the length L_A and concentration of the additional high n-doped region were performed for the optimized device. Table 2 and Table 3 show the values of $R_{on,sp}$, V_{BR}, and FoM of devices with the different L_A and concentration of the additional high n-doped region, respectively. The maximal FoM value of 1071 MW/cm^2 is achieved when the length L_A and the concentration are 0.2 μm and 3×10^{17} cm^{-3}, respectively.

Table 2 lists the stimulated results of V_{BR}, $R_{on,sp}$, and FoM with different concentrations and a constant L_A of 0.2 μm. As the concentration increases, the V_{BR} shows a slow decrease. It is worth noting that the V_{BR} is determined by the avalanche breakdown voltage in the three cases, demonstrating the weak influence of concentration on electric field in the oxide layer. In addition, the $R_{on,sp}$ also slowly decreases with the increase of the concentration. This is because higher concentration in the additional n-doped region causes the smaller depletion expansion and higher electron concentration of JFET region. As a result, the actual width that electrons flow through in JFET region and electron concentration increase, and then the $R_{on,sp}$ decreases. FoM increases initially then decreases with a peak value of 1071 MW/cm^2 at the concentration of 3×10^{17} cm^{-3}. Table 3 lists the stimulated results of V_{BR}, $R_{on,sp}$, and FoM with different L_A and a constant concentration of 3×10^{17} cm^{-3}. When the L_A is beyond 0.2 μm, the wide additional n-doped region causes a weak depletion expansion in JFET region, resulting in the breakdown of the oxide layer in a low voltage. Similar to the results in Table 2, the $R_{on,sp}$ decreases with the increase of L_A. The FoM slightly increases at initial step then decreases due to the reduced V_{BR}. The maximal FoM is achieved for L_A of 0.2 μm.

Conclusion

In this work, we propose an optimized 4H-SiC VDMOSFET structure, in which two high n-doped regions are added at both sides of JFET region. The simulation results indicate a significant improvement in device performance. The specific ON-resistance reduces by 18% for the optimized device while keeping the breakdown voltage essentially unchanged. As a result, the value of FoM increases by 22%. This work provides a potential to develop high-performance SiC MOSFETs.

Acknowledgements

This work was supported in part by National Key Research and Development Program of China (Grant No. 2016YFB0400801), Science and Technology Key Project of Xiamen (Grant No. 3502ZCQ20191001), National Natural Science Foundations of China (Grant Nos. 61774128, 61974123, 61674124, 61874092, U1405253, and 11604275), and Outstanding Youth Foundation Project of Jiujiang (Grant No. 2018042).

References

1. Roccaforte, F., et al., *Emerging trends in wide band gap semiconductors (SiC and GaN) technology for power devices.* Microelectronic Engineering, 2018. **187**: p. 66-77.
2. Srdic, S., et al. *A SiC-based power converter module for medium-voltage fast charger for plug-in electric vehicles.* in *2016 IEEE Applied Power Electronics Conference and Exposition (APEC).* 2016.
3. Furuhashi, M., et al., *Practical applications of SiC-MOSFETs and further developments.* Semiconductor Science And Technology, 2016. **31**(3): p. 9.
4. Willander, M., et al., *Silicon carbide and diamond for high temperature device applications.* Journal of Materials Science Materials in Electronics, 2006. **17**(1): p. 1.
5. Lee, H., et al., *1200-V 5.2-m$\Omega\cdot$cm^2 4H-SiC BJTs With a High Common-Emitter Current Gain.* IEEE Electron Device Letters, 2007. **28**(11): p. 1007-1009.
6. Salemi, A., et al. *Conductivity modulated on-axis 4H-SiC 10+ kV PiN diodes.* in *2015 IEEE 27th International Symposium on Power Semiconductor Devices & IC's (ISPSD).* 2015.
7. Zhang, Q.J., et al., *Next Generation Planar 1700 V, 20 mΊ© 4H-SiC DMOSFETs with Low Specific On-Resistance and High Switching Speed.* Materials Science Forum, 2017. **897**: p. 521-524.
8. Deng, X.C., et al., *Numerical analysis on the 4H-SiC MESFETs with a source field plate.* Semiconductor Science And Technology, 2007. **22**(7): p. 701-704.
9. Kimoto, T., *Material science and device physics in SiC technology for high-voltage power devices.* Japanese Journal of Applied Physics, 2015. **54**(4): p. 040103.
10. Saha, A. and J.A. Cooper, *A 1-kV 4H-SiC Power DMOSFET Optimized for Low on -Resistance.* IEEE Transactions on Electron Devices, 2007. **54**(10): p. 2786-2791.
11. Howell, R.S., et al. *Demonstration of 10 kV, 50A 4H-SiC DMOSFET with stable subthreshold characteristics across 25-200 ºC operating temperatures.* in *2007 International Semiconductor Device Research Symposium.* 2007.
12. Zhang, Q.C.J., et al. *CIMOSFET: A New MOSFET on SiC with a Superior Ron·Qgd Figure of Merit.* in *Materials Science Forum.* 2015.
13. Silvaco, *Atlas User's Manual.* 2017.
14. Uhnevionak, V., et al., *Comprehensive Study of the Electron Scattering Mechanisms in 4H-SiC MOSFETs.* IEEE Transactions on Electron Devices, 2015. **62**(8): p. 2562-2570.
15. Uhnevionak, V., *Simulation and Modeling of Silicon Carbide Devices.* 2015.
16. Loh, W.S., et al., *Impact Ionization Coefficients in 4H-SiC.* IEEE Transactions on Electron Devices, 2008. **55**(8): p. 1984-1990.
17. Shen, H.-J., et al., *Fabrication and Characterization of 1700 V 4H-SiC Vertical Double-Implanted Metal-Oxide-Semiconductor Field-Effect Transistors.* Chinese Physics Letters, 2015. **32**(12): p. 127101.

Design and Fabrication of 3300V 100mΩ 4H-SiC MOSFET with Stepped p-body Structure

Weijiang Ni[1,2,3,4*], Xiaoliang Wang[1,2,3*], Miaoling Xu[4], Mingshan Li[4], Chun Feng[1,2], Hongling Xiao[1,2], Wei Li[1],
Quan Wang[1], Holger Schlichting[5], Tobias Erlbacher[5]

1. Key Lab of Semiconductor Materials Science, Institute of Semiconductors, Chinese Academy of Sciences, Beijing, China
2. Center of Materials Science and Optoelectronics Engineering, University of Chinese Academy of Sciences, Beijing, China
3. School of Microelectronics, University of Chinese Academy of Sciences, Beijing, China
4. Power device department, Beijing Century Goldray Semiconductor Co., Ltd, Beijing, China
5. Fraunhofer Institute for Integrated Systems and Device Technology IISB, Erlangen, Germany
*niweijiang@semi.ac.cn; xlwang@semi.ac.cn

Abstract

In this paper, a 3300 V 100 mΩ 4H-SiC planar MOSFET with stepped p-body structure was designed and fabricated. 2D TCAD tool was used to design the MOSFET cell and field limiting ring junction termination structure. The junction field-effect transistor (JFET) region was optimized to get better trade-off between on-resistance and maximum gate oxide electric field in off-state. The stepped p-body was formed by two step ion implantations to transfer the avalanche point from the edge of the p-body to the deep p+ area. Finally, the threshold voltage of 1.7 V, subthreshold swing of 188 mV/decade and the average interface state density of 5.35E11 cm^{-2}eV^{-1} were obtained from the measured transfer curve.

Introduction

Wide-band gap semiconductor material SiC is very suitable for fabricating high power and high efficiency electronic devices[1]-[4]. The on-resistance of SiC unipolar devices is about 400 times lower than that of Si devices at the same blocking voltage [5]. Therefore, high voltage SiC unipolar devices such as MOSFET are very suitable for high power application up to 10 kV voltage [6]. In SiC MOSFET, the on-resistance is greatly affected by the high MOS interface state density and low channel mobility [7][8]. Compared to low and medium voltage SiC MOSFET, the low channel mobility is not very importance in high voltage (3.3 kV or above) SiC MOSFET because of high drift resistance. The propotion of channel resistance to drift resistance is less than 25%. The 3.3 kV rating SiC MOSFET studied in this paper is suitable for main and auxiliary converters of rail traction, wind turbines, smart power grid and other fields, and it has important application prospects[9].

In this paper, 2D TCAD tool was used to simulate electrical properties of the 4H-SiC MOSFET. A 3.3 kV rating SiC MOSFET with stepped p-body and optimized JFET region was designed and fabricated on 4-inch n type 4H-SiC epi-wafer. By optimizing fabricaiton process, the blocking voltage of larger than 3.3 kV and the on-resistance of 100 mΩ were obtained.

Device Design

The Sentaurus tool from Synopsys Company is used to simulate 2D devices. Drift-diffusion equation and continuity equation of electrons and holes are solved self-consistently. Impact ionization rate, Shockley-Read-Hall recombination, incomplete ionization and material anisotropy models are used.

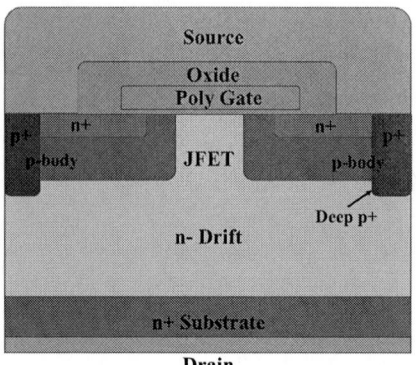

Fig.1 Cross-sectional schematic of 4H-SiC MOSFET

Fig. 1 is a schematic of the cell structure of the planar 4H-SiC MOSFET with stepped p-body structure. The thickness of epitaxy layer is 30 μm with actual doping concentration of 1.8E15 cm^{-3}. The designed channel length is 0.9 μm and the whole cell pitch is 12 μm. The surface doping concentration in p-body region is 1E17 cm^{-3}. The thickness of gate oxide is 50 nm. Because the relationship between short-circuit current withstand time and saturation current of MOSFET devices can be approximated by [10]

$$t_{SC} = 4\rho C_P \, \Delta T_{MAX}/(E_C * J_{D,sat}) \qquad (1)$$

where $J_{D,sat}$ is the saturation current, ρ is the material density, C_P is the specific heat, ΔT_{MAX} is the maximum allowable temperature rise, E_C is the critical field. The long channel length which will cause small saturation current is good for short-circuit current capability.

Fig. 2 shows the relationship between on-resistance and breakdown voltage with various of JFET width by TCAD simulation. Fig. 3 shows the trade-off between on-resistance and maximum gate oxide electric field at drain voltage of 3.3 kV and gate voltage of 0 V in off-state with various of JFET width by TCAD simulation. The JFET region was additionally implanted by N+ ions to form retrograde doping profile to improve the trade-off between on- resistance and maximum gate oxide electric field [11]. The avalanche breakdown voltage was almost same when JFET width W_{JF} less than 4.0 μm. The design window of JFET width is larger compare to lower voltage device with high doping drift layer [12]. When W_{JF} was smaller than 2.0 μm, R_{ON} increased drastically because the whole JFET region was depleted. The maximum gate oxide electric field (E_{MOX}) monotonically increased with increasing of W_{JF} and was designed to less than

978-1-7281-5757-3/19 $31.00 © 2019 IEEE

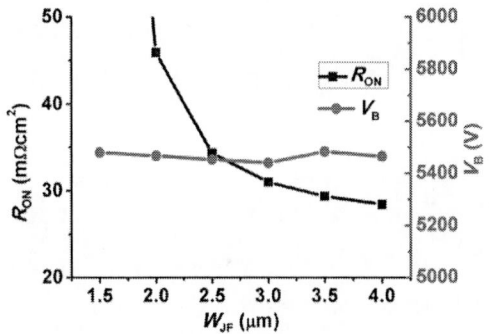

Fig.2. Relationship between on-resistance (R_{ON}) and avalanche breakdown voltage (V_B) with various of JFET width (W_{JF}) based on modelling of the unit cell.

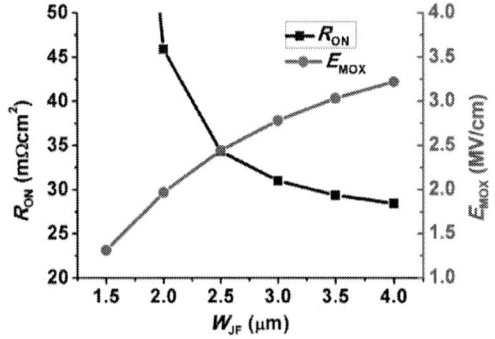

Fig.3. Trade-off between on-resistance (R_{ON}) and maximum gate oxide electric field (E_{MOX}) with various of JFET width (W_{JF}) based on modelling of the unit cell.

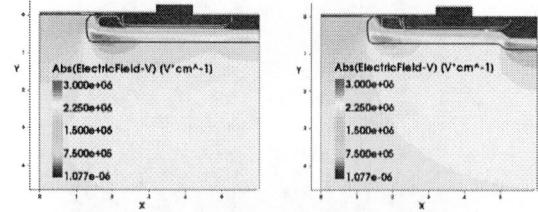

Fig.4. Electric field distribution in SiC at the avalanche voltage (Left: conventional MOSFET. Right: MOSFET with stepped p-body.

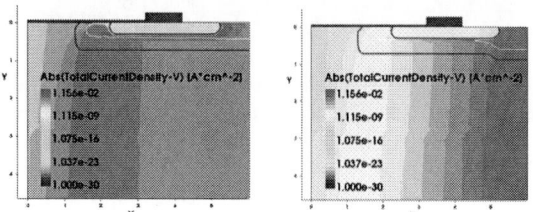

Fig.5. Current distribution in SiC at the avalanche voltage (Left: conventional MOSFET. Right: MOSFET with stepped p-body.

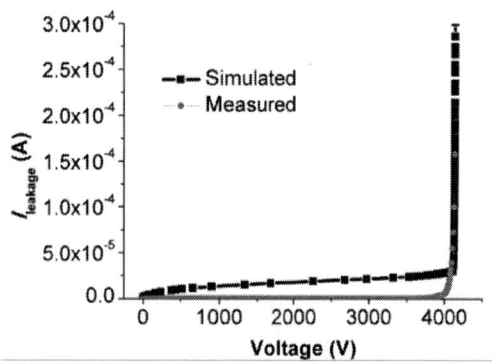

Fig.6. Simulated and measured breakdown voltage of pn diode with designed FLR.

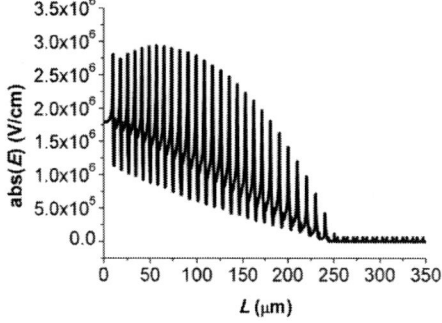

Fig.7. Absolute electric field at PN junction from main junction to edge.

3 MV/cm for long time reliability [13]. The specific on-resistance of 31 mΩcm^2, breakdown voltage of 5440 V and maximum gate oxide electric field of 2.78 MV/cm were obtained when W_{JF} = 3.0 µm. Compared to the conventional planar MOSFET, the stepped p-body structure with deep p+ region transfers the avalanche point from the edge of the p-body to the deep p+ area which is show in fig.4-5. The new structure can shorten the avalanche current path to the source which will suppress turning on of the parasitic BJT and help to improve the UIS capability.

The designed junction termination used a uniformly increased field limiting ring structure (FLR). Fig. 6 is the breakdown *I-V* curve of FLR termination. The breakdown voltage of 4153 V was obtained which was much lower than the cell's. Fig. 7 shows the electric field distribution at the depth of the PN junction at the breakdown voltage. The design of field limiting ring structure was limited by the critical dimensions available in the manufacturing process. The breakdown voltage can be improved by reducing the space between the field limiting rings. The breakdown voltage of junction termination will be further improved by optimizing the FLR's structure or use other termination structures [14].

Device Fabrication and Characteristics

The MOSFET was fabricated on N-type 4 inch 4◦-off 4H-SiC (0001) Si-face epilayers. Substrate thickness was 350 µm and resistivity was 0.02 Ωcm. The total chip size is 0.55 cm*0.6 cm and the active area is about 20.9 mm2. The p-body and JFET region was multiply implanted by Al+ and N+ ions respectively to form retrograde doping profile. A more detailed fabrication process is described in the paper [11]. Fig.8 shows the cross section of the fabricated MOSFET.

978-1-7281-5757-3/19 $31.00 © 2019 IEEE

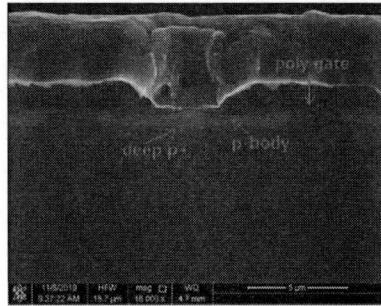

Fig.8. SEM cross-sectional view of the fabricated MOSFET.

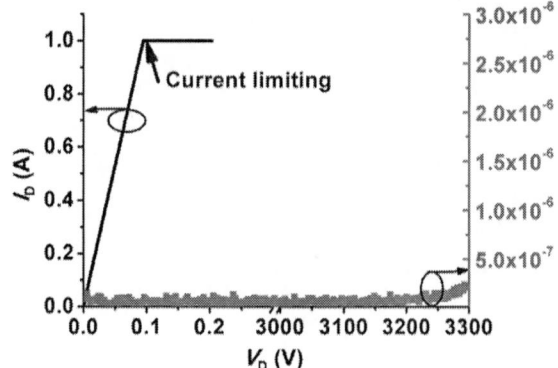

Fig.9. Output and blocking characteristics of the fabricated 4H-SiC MOSFET.

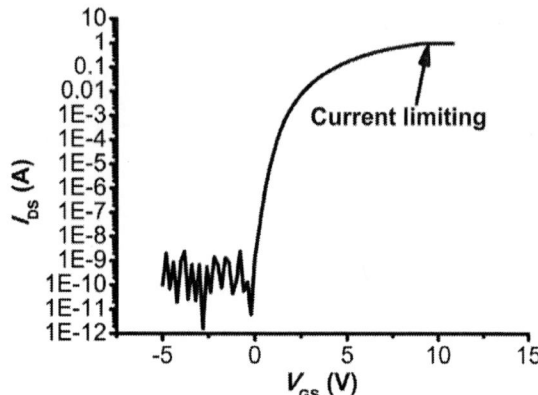

Fig.10. Transfer characteristics of the fabricated 4H-SiC MOSFET. The current was limited to 1 A by the measurement equipment.

Fig.6 shows the breakdown voltage of fabricated pn diode with designed FLR. The breakdown voltage of the pn diode is almost same as the simulated results. The high leakage current of the simulation results compares to the measurement results is due to the simulation method and doesn't represent actual value.

Fig. 9 shows the output and blocking I-V characteristics of the fabricated 4H-SiC MOSFET. Blocking voltage was measured up to 3300 V at V_{GS} = 0 V and the output characteristic was obtained at V_{GS} = 20 V. The current was limited to 1 A by the measurement equipment. The on-resistance of 100 mΩ at I_D = 1 A was obtained which contains the parasitic resistance of the probe. Considering the active area, specific on-resistance ($R_{ON,SP}$) of the fabricated MOSFET is about 20.9 m$\Omega\cdot$cm^2 which is much smaller than the simulation results. This is mainly due to the hexagonal cell structure that was used in actual MOSFET while the stripe cell was used in simulation [15].

Figure 10 shows the transfer characteristics of the MOSFET at V_{DS} = 0.2 V. The threshold voltage V_{TH} of 1.7 V was obtained. V_{TH} is defined as the gate voltage when the drain current is 1 mA. According to the sub-threshold swing curve, the sub-threshold swing S = 188 mV/decade was obtained. The interface state density D_{it} can be obtained by the relationship between subthreshold swing and D_{it} [16][17]

$$S \text{ (with } D_{it}) = (\ln 10)(kT/q)(C_{OX} + C_D + C_{it})/ C_{OX} \quad (2)$$

The interface state density of 5.35E11 cm^{-2}eV^{-1} is obtained. The channel mobility of 15 cm^2/Vs which is obtained from the LDMOS is consist with the interface state density[7][18][19]. Although the gate oxide process with NO POA improves the channel mobility, the low V_{TH} will be a problem to use the SiC MOSFET to replace Si IGBT. To develop a new gate oxide process to improve the trade-off between the channel mobility and the threshold voltage is very necessary [20].

Conclusions

We have developed a 3300 V 100 mΩ 4H-SiC planar MOSFET. The stepped p-body with deep p+ region was designed to alleviate the electric field at the edge of the p-body and transfer the avalanche point to the deep p+ area. The MOS interface state density of 5.35E11 cm^{-2}eV^{-1} and the threshold voltage of 1.7 V was obtained. The on-resistance can be further reduced by using higher doping drift layer. The breakdown voltage of the MOSFET is limited by the FLR termination which is much lower than the designed breakdown voltage of the cell. Next work will focus on optimizing junction termination structure to obtain much more efficiency breakdown voltage.

References

[1] M. Imaizumi and N. Miura, "Characteristics of 600, 1200, and 3300 v planar SiC-MOSFETs for energy conversion applications," *IEEE Trans. Electron Devices*, vol. 62, no. 2, pp. 390–395, 2015.

[2] P. Godignon *et al.*, "New trends in high voltage MOSFET based on wide band gap materials," *Proc. Int. Semicond. Conf. CAS*, vol. 2017-Octob, pp. 3–10, 2017.

[3] J. Millan, P. Godignon, X. Perpina, A. Perez-Tomas, and J. Rebollo, "A survey of wide bandgap power semiconductor devices," *IEEE Trans. Power Electron.*, vol. 29, no. 5, pp. 2155–2163, 2014.

[4] Q. J. Zhang and A. K. Agarwal, "Design and technology considerations for SiC bipolar devices: BJTs, IGBTs, and GTOs," *Phys. Status Solidi Appl. Mater. Sci.*, vol. 206, no. 10, pp. 2431–2456, 2009.

[5] J. J. A. Cooper, M. R. Melloch, R. Singh, A. Agarwal, and J. W. Palmour, "Status and prospects for SiC power MOSFETs," *IEEE Trans. Electron Devices*, vol. 49, no. 4, pp. 658–664, 2002.

[6] J. W. Palmour *et al.*, "Silicon carbide power MOSFETs: Breakthrough performance from 900 v up to 15 kV," *Proc. Int. Symp. Power Semicond. Devices ICs*, pp. 79–82, 2014.

[7] T. Kimoto, Y. Kanzaki, M. Noborio, H. Kawano, and H. Matsunami, "Interface Properties of Metal-Oxide-Semiconductor Structures on 4H-

978-1-7281-5757-3/19 $31.00 © 2019 IEEE

SiC {0001} and (11-20) Formed by N2O," *Japan Soc. Appl. Phys.*, vol. 44, no. 3, pp. 1213–1218, 2005.

[8] G. Liu, B. R. Tuttle, and S. Dhar, "Silicon carbide: A unique platform for metal-oxide-semiconductor physics," *Appl. Phys. Rev.*, vol. 2, no. 2, 2015.

[9] Hong Lin, "Power SiC 2017: Materials, Devices and Application," 2017.

[10] J. A. Cooper, D. T. Morisette, and M. Sampath, "Increased Short-Circuit Withstand Time and Reduced DIBL by Constant-Gate-Charge Scaling in SiC Power MOSFETs," in *International Conference on Silicon Carbide and Related Materials*, 2019, p. Mo-P-39.

[11] W. Ni *et al.*, "1700V 34mΩ 4H-SiC MOSFET With Retrograde Doping in Junction Field-Effect Transistor Region," *2019 IEEE Int. Conf. Electron Devices Solid-State Circuits*, pp. 1–3, 2019.

[12] A. Saha and J. A. Cooper, "A 1-kV 4H-SiC power DMOSFET optimized for low ON-resistance," *IEEE Trans. Electron Devices*, vol. 54, no. 10, pp. 2786–2791, 2007.

[13] R. Singh and A. R. Hefner, "Reliability of SiC MOS devices," *Solid. State. Electron.*, vol. 48, no. 10-11 SPEC. ISS., pp. 1717–1720, 2004.

[14] W. Sung, B. J. Baliga, and L. Fellow, "A Comparative Study 4500-V Edge Termination Techniques for SiC Devices," *IEEE Trans. Electron Devices*, vol. 64, no. 4, pp. 1647–1652, 2017.

[15] P. A. Losee *et al.*, "SiC MOSFET design considerations for reliable high voltage operation," *IEEE Int. Reliab. Phys. Symp. Proc.*, pp. 2A2.1-2A2.8, 2017.

[16] D. Okamoto, H. Yano, K. Hirata, T. Hatayama, and T. Fuyuki, "Improved inversion channel mobility in 4H-SiC MOSFETs on Si face utilizing phosphorus-doped gate oxide," *IEEE Electron Device Lett.*, vol. 31, no. 7, pp. 710–712, 2010.

[17] S. M. Sze and K. K. Ng, *Physics of Semiconductor Devices*, Third Edit. Wiley, 2007.

[18] S. Suzuki, S. Harada, R. Kosugi, J. Senzaki, W. J. Cho, and K. Fukuda, "Correlation between channel mobility and shallow interface traps in SiC metal-oxide-semiconductor field-effect transistors," *J. Appl. Phys.*, vol. 92, no. 10, pp. 6230–6234, 2002.

[19] A. Agarwal *et al.*, "Critical Technical Issues in High Voltage SiC Power Devices," *Mater. Sci. Forum*, vol. 600–603, no. 3, pp. 895–900, 2009.

[20] M. Furuhashi, S. Tomohisa, T. Kuroiwa, and S. Yamakawa, "Practical applications of SiC-MOSFETs and further developments," *Semicond. Sci. Technol.*, vol. 31, no. 3, p. 34003, 2016.

Study on Preparation and Application of Nano-copper Powder for Power Semiconductor Device Packaging

Xu Pan[1,2], Jiacheng Zhou[1,2,3], Jingguo Zhang[1,2]*, Zhaohui Zhao[4]*, Minghui Liang[5], Huaiyu Ye[6],
Qiang Hu[1,2], Huijun He[2], Limin Wang[2], Ligen Wang[7], Fengcai Qi[1,2], Youzhi Zhou[1,2,3]

(1. GRIPM Research Institute Co., Ltd, Beijing 101407, China; 2. GRIPM Advanced Materials Co. Ltd Beijing 101407, China; 3. General Research Institute for Nonferrous Metals, Beijing 100088, China; 4. Beijing COMPO Advanced Technology Co.,Ltd, Beijing 101407, China; 5. National Center for Nanoscience and Technology, Beijing 100190, China; 6. School of Microelectronics, Southern University of Science and Technology, Shenzhen Institute of Wide-bandgap Semiconductors, Shenzhen 518055, China; 7. Materials Computation Center, GRIMAT Engineering Institute Co., Ltd, Beijing 101407, China)

zjg@gripm.com, zhaozhaohui@composolder.com

Abstract

Nano-copper powder is prepared by liquid phase reduction method in ethylene glycol solvent system. Nano-copper paste is prepared by mixing with organic carriers and subjected to pressureless sintering. The influence of different reducing agent systems on the particle size and morphology of nano-copper powder are studied. The sintering properties of the copper paste are tested by TEM, XRD and SEM. The results indicate that the synthesized product of using ethylene glycol as the reaction solvent and reducing agent is pure copper power and particle size from 100 nm to 200 nm at 160℃;The nano-copper paste can achieve metallurgical bonding at 300℃ and the density of the sintered layer gradually increases with the sintering time prolonged, the porosity is gradually reduced, and the sintering of the nano-copper powder for 30 minutes is basically stable.

Keywords-- Nano-copper powder; Nano-copper paste; The sintering properties;

1. INTRODUCTION

Power semiconductor devices have been widely used in various fields of modern society, including power electronics, electric vehicles, smart grid, automation systems and home appliances, many domestic and foreign research institutes and semiconductor companies have conducted extensive research and in-depth industrialization. Due to the development process in the fields of 5G and new energy ,there is an urgent need for high power, low cost, high heat dissipation and high reliability for power semiconductor devices. In addition to chip technology, the package, as a bridge between connecting materials, chips, devices and applications, directly provides a stable electromagnetic, mechanical and thermal environment for the chip, which makes the chip work stably and stably, especially for the reliability of the device, and can be realized through package integration. The integration and expansion of the chip's functions and applications through package integration provides a good foundation for further improving device performance and giving full play to the superior characteristics of third-generation semiconductors.

Faced with the stringent requirements of the third-generation semiconductor working environment, the traditional brazing package is very difficult to improve the temperature resistance of the solder itself. The temperature of the high-temperature solder currently developed is lower than 350 ℃, which is far from the operating temperature (500 ℃) that can be achieved by the new generation of power chips. In

recent years, new interconnect materials are evolving from welding to sintering technology. Sintering is a process in which atoms on the surface of a material are mutually diffused by high temperature to form a dense crystal. By reducing the size of the sintered particles, the sintering temperature can be lowered, and thus the nano silver particles are applied to the interconnection of power devices [1-3]. The advantages of nano silver mainly include low processing temperature (about 250 ℃), high operating temperature (melting temperature of 960 ℃) and low modulus of elasticity. However, since silver has a serious electromigration phenomenon and a high production cost, silver is limited as a connecting material in practical applications, and it is necessary to add other intermediate metal layers in consideration of differences in thermal expansion coefficients of silver and semiconductor chip back materials. Compared with silver, copper has good thermal conductivity and electrical conductivity, is relatively inexpensive, and has good resistance to electromigration. Therefore, the new nano-copper has received extensive attention [4].

Internationally, Lockheed Martin (USA) [5] and Intrinsiq (UK) have developed the conductive paste prepared by nano-copper with organic coating. In 2013, the C. S. Tan team of Nanyang Technological University in Singapore used the nano-copper paste of Lockheed Martin Company of the United States for low-temperature sintering [6-9]. The sintering results are improved by surface deoxidation treatment, doping with micron copper particles, and the like. Zürcher et al. used copper nanopaste developed by Intrinsiq of the United Kingdom to weld copper stud bumps to achieve full copper interconnection [10]. In China, in recent years, for the nano-copper synthesis and semiconductor device interconnection technology, Fudan University has carried out research work on the synthesis of nano-copper paste since 2012 [11,12]. In 2018, Qipeng Liu et al. studied the interconnectivity of nano-copper in high-temperature devices [13]. At present, the sintering mechanism of nano-copper particles coated with organic materials, the optimization of sintering parameters, the influence of pressure sintering on the device, and the influence with the interface have not been thoroughly explored. Therefore, there is still a lot of research space in the nano copper paste synthesis process, sintering and semiconductor power device interconnection applications.

In this paper, the preparation of nano-copper powder is carried out by chemical reduction method. The morphology of nano-copper powder under different reaction systems is studied. One of the preparation schemes is optimized. The

978-1-7281-5757-3/19 $31.00 © 2019 IEEE

nano-copper powder is mixed with organic carrier to prepare nano-copper paste. The sintering properties of the nano copper paste are tested.

2. EXPERIMENT

2.1. Preparation of nano-copper powder

The main raw materials used in the experiment are: cupric chloride dihydrate, polyvinylpyrrolidone (PVP), ethylene glycol, sodium hydroxide, hydrazine hydrate, ascorbic acid and K6 antioxidant.

Specific experimental steps: prepare 500ml of 0.2mol/l copper chloride glycol solution, take 50ml of the solution into three beakers A, B and C respectively, add about 1.8g of PVP K-30 into each beaker, add 100ml of glycol to dilute, drop 0.5mol/l NaOH glycol solution, adjust the pH to 12 left and right, the solution changes from green to dark blue, stir for 30min, add 6ml of 85% hydrazine hydrate into beaker A, When the temperature is raised to 80 ℃ for 3 hours, the dark red dispersion A1 is obtained; when the beaker B is added with about 3 grams of ascorbic acid and heated to 80 ℃ for 3 hours, the dark red dispersion B1 is obtained; when the beaker C is heated to 160 ℃ for 3 hours under the protection of argon, the dark red dispersion C1 is obtained. Add equal volume of water and twice volume of acetone to the above-mentioned copper nano dispersion to obtain PVP coated copper nano particles precipitation; A small amount of water is used again to dissolve and precipitate, and acetone was added to precipitate. The process was repeated for 3 times to get the washed copper nano particles.; disperse the copper nano particles in water and add antioxidant for antioxidant treatment; finally, the nano-copper powder is rapidly frozen by liquid nitrogen, and dried by a lyophilizer to obtain dried copper nano powders A2, B2, C2.

2.2. Preparation of nano-copper paste

(1) Selection of organic carriers

The organic carrier consists of five components: resin, protective agent, thixotropic agent, active agent and solvent. The resin plays the role of adhesion, and the protective agent prevents the nano-copper powder from oxidation and corrosion during storage. The thixotropic agent can make the nano-copper paste have good operability. The active agent can remove the oxide layer on the surface of the nano-copper powder during sintering. The solvent is the carrier of all organic components.

(2) Preparation of nano-copper paste

Weigh the organic carrier in proportion and put it into the mortar; weigh the nano-copper powder, add 1/3 to the mortar, grind to fineness, add 1/3 nano-copper powder, grind to fine, and then add the remaining nano copper powder, ground for 20 min; the ground nano copper paste was placed under vacuum of -0.1 Mpa for 10 min to remove bubbles in the paste.

2.3. Characterization

Scanning electron microscope and transmission electron microscope are used to observe the micro morphology of nano copper powder and the cross section of copper paste after sintering; X-ray diffraction (XRD) is used to determine the composition of nano-copper powder; The carbon content of sintered copper paste is measured by CS-901 high frequency infrared carbon-sulfur analyzer; The density balance is used to test the density of the sintered layer of copper paste, and the porosity of the sintered layer is calculated.

3. RESULTS AND DISCUSSION

3.1. Influence of preparation conditions on the morphology of nano-copper powder

In this paper, the effects of three different reducing agent systems of hydrazine hydrate, ascorbic acid and ethylene glycol on the morphology of nano-copper are investigated. Figure 1 is a photomicrograph of a transmission electron micrograph of copper nanoparticles prepared by different reducing agents in an ethylene glycol reaction system. It can be seen from Fig. 1(a)(b) that most of the copper powder prepared by using hydrazine hydrate as a reducing agent has a dendritic structure distribution. It can be seen from Fig. 1(c)(d) that the copper nanoparticles prepared using ascorbic acid have a cubic structure with a size ranging from 100 to 300 nm. Fig. 1(e)(f) are copper nanoparticles obtained by using ethylene glycol as a reducing agent, exhibiting an irregular structure, and has a size ranging from 100 to 200 nm, and the powder dispersion effect is good. Figure 2 is a scanning electron micrograph and XRD spectrum of nano-copper powder. The XRD spectrum shows strong diffraction peaks at 43°, 50° and 74°, belonging to Cu face-centered cubic structures (111), (200) and (220). Crystal plane, no other impurities were detected.

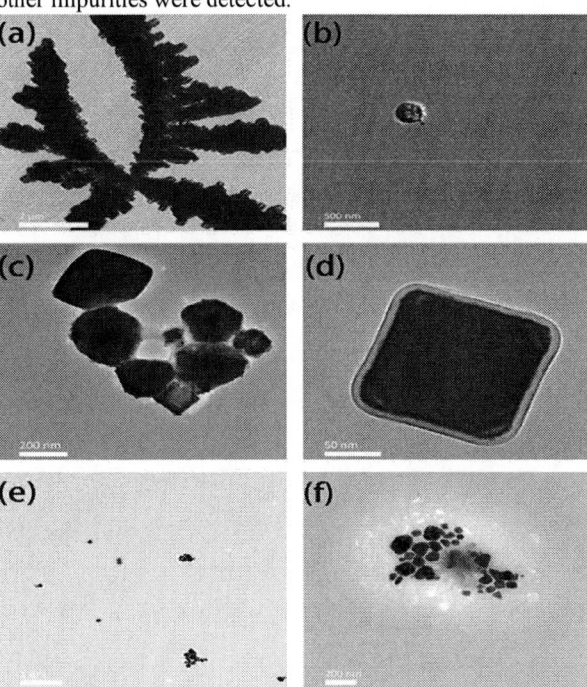

Fig.1 TEM of nano-copper powder prepared with different reducing agents (a) (b) hydrazine hydrate, (c) (d) ascorbic acid, (e) (f) ethylene glycol

From the above experimental results, it is established that nano-copper powder having a particle diameter of about 200 nm was prepared as a raw material of the nano copper paste by using ethylene glycol as a reducing agent. The size and morphology of copper nanoparticles prepared by using ethylene glycol as a reducing agent are easy to control. It is ideal for preparing nano copper paste, and ethylene glycol as a reducing agent is relatively inexpensive, and no other solvent is introduced, the recovery of glycol is relatively easy and environmentally friendly.

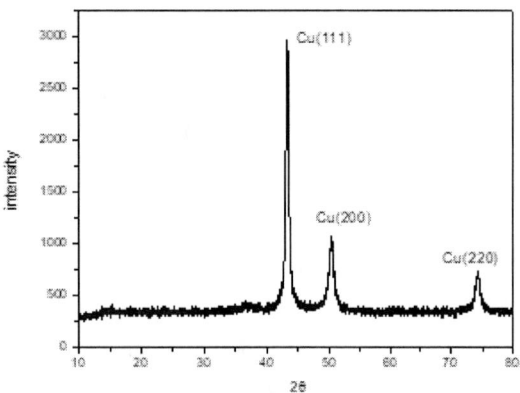

Fig.2 XRD diffraction pattern of nano-copper

3.2 Comparison of stability of nano-copper paste

The organic carrier is prepared, and the nano-copper paste is prepared by using 200 nm copper powder. The state of the copper paste is observed after being placed at 40 ℃ for 60 hours, and the distribution ratio of the organic carrier is initially determined. Table 1 shows the stability comparison of nano-copper paste prepared by different organic carriers. It can be seen from Table 1 that the organic carrier A has no obvious change after being placed at 40 ℃ for 60 h, and the performance is better than that of the organic carrier B and the organic carrier C, Therefore, organic carrier A is selected as the organic carrier for the next experiment.

Table 1 Comparison of stability of nano-copper paste

Organic carrier	Copper powder content / %	Initial state of copper paste	40℃，60h
A	85%	Exquisite	No significant change
B	85%	Exquisite	Viscosity increase
C	85%	Exquisite	Surface crust

3.3 Effect of sintering process on nano-copper paste

The nano-copper paste is sintered in a hydrogen atmosphere at 300 ° C for 15 min, 30 min, and 60 min. The results are shown in Table 2. As can be seen from Table 2, the original nano-copper paste shows turbid brown and semi-solid properties; after sintering, the copper paste turns into a solid state. With the increase of time, the nano-paste gradually appeares metallic luster, showing the form of a metal block and high strength. After sintering for 15 minutes, the metal block is formed. The sintered block has poor metallic luster, weak strength and powder drop on the surface. From 30 min to 60 min, the metallic luster of the sintered body becomes more and more obvious, and the strength of the sintered body is improved.

Table 2 appearance of nano-copper paste after different sintering time

Organic carrier	Copper powder ratio	Copper paste state	Sintering time / min		
			15min	30min	60min
A	85%				
	90%				
State		Exquisite	Poor metal luster, poor strength of sintered layer	Good metal luster, high strength of sintered layer	Good metal luster, high strength of sintered layer

Observing the cross section of the sintering layer, as shown in Fig. 3 (a), (b) and (c), the cross section is relatively flat when sintering for 15min, and the cross section gradient increases gradually when sintering for 30min and 60min, which shows that metallurgical bonding can be formed between nano-copper powder at 300 ℃, and the sintering process degree increases gradually with the increase of sintering time. When sintering for 15 minutes, the degree of sintering is weak, and a small number of particles appear sintering neck; when sintering for 30 minutes and 60 minutes, the degree of sintering is increased, and sintering neck appears. The sintering layer of nano-copper paste looks dense, and fracture traces can be seen clearly from the fault, indicating that there is a good sintering between copper powders. In addition, because the copper paste is not pressurized in the sintering process, there is randomness in the combination of nano-copper powder in the sintering layer, so the cross-section is uneven, and the cross-section is not smooth. The roughness of the cross section also shows that the sintered layer has certain toughness.

978-1-7281-5757-3/19 $31.00 © 2019 IEEE

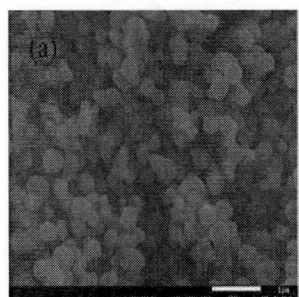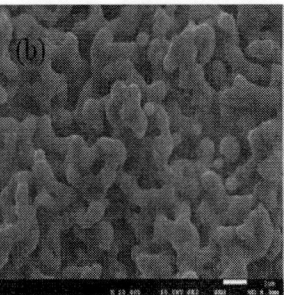
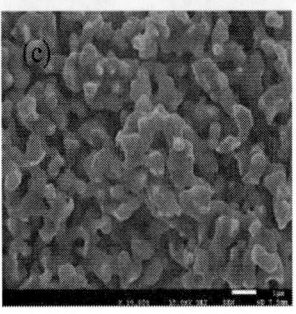

Fig.3the perspective of section
(a)15min(b)30min(c)60min

Table 3 porosity and density of nano-copper paste at
different sintering time

Porosity %			Density g/cm^3		
15min	30min	60min	15min	30min	60min
39.58	33.29	29.51	5.377	5.937	6.274

The results of sintering layer density and porosity are shown in Table 3. It can be seen that with the extension of sintering time, the density of sintered layer gradually increases and the porosity gradually decreases. After 30min, the density of sintered layer increases slowly and the porosity decreases slowly, indicating that the sintering of nano-copper powder is basically stable at 30min. This is consistent with SEM morphology of sintered copper paste.

The carbon content in the sintered layer is measured after different sintering time. The results are shown in Table 4. It can be seen that with the increase of sintering time, the carbon content in the sintering layer decreases gradually, and the decrease speed of carbon content slows down after 30 minutes, which is consistent with the density test results. It shows that the sintering structure is basically stable after 30 minutes.

Table 4 carbon content of nano-copper paste at different
sintering time

Copper powder content %	carbon content %		
85%	15min 0.54	30min 0.30	60min 0.23

Conclusions

(1) Compared with hydrazine hydrate and ascorbic acid, the size distribution of copper nanoparticles prepared by ethylene glycol is easier to adjust and the particle size distribution is more uniform. In this study, ethylene glycol is used as reducing agent to prepare nano-copper particles at 160 ℃, with uniform size distribution and good dispersion, with an average size of 100-200nm.

(2) With the increase of sintering time, the density of the sintering layer increases gradually, the porosity and carbon content decrease gradually, and the increase speed of the density of the sintering layer and the decrease speed of the porosity and carbon content of the copper paste slow down after 30min, which shows that the sintering between the nano-copper powders in the copper paste is basically stable and the removal of organic matters is best under the hydrogen environment at 300 ℃.

Acknowledgments

This work is supported by the Key-Area Research and Development Program of GuangDong Province No.2019B010131001.

References

1. Jung I , Jo Y H , Kim I , et al. A Simple Process for Synthesis of Ag Nanoparticles and Sintering of Conductive Ink for Use in Printed Electronics[J]. Journal of Electronic Materials, 2012, 41(1):115-121.
2. Makrygianni M , Kalpyris I , Boutopoulos C , et al. Laser induced forward transfer of Ag nanoparticles ink deposition and characterization[J]. Applied Surface Science, 2014, 297:40-44.
3. Sohn J H , Pham L Q , Kang H S , et al. Preparation of conducting silver paste with Ag nanoparticles prepared by e-beam irradiation[J]. Radiation Physics and Chemistry, 2010, 79(11):1149-1153.
4. E. Beyne, "3D system integration technologies," in VLSI Technology, Systems, and Applications, 2006 International Symposium on, 2006, pp. 1-9.
5. A. Zinn, R. Stoltenberg, A. Fried, J. Chang, A. Elhawary, J. Beddow, et al., "nanoCopperbased solder-free electronic assembly material," Nanotech, vol. 2, pp. 71-74, 2012.
6. D. F. Lim, J. Wei, K. C. Leong, and C. S. Tan, "Cu passivation for enhanced low temperature (⩽300 °C) bonding in 3D integration," Microelectronic Engineering, vol. 106, pp. 144-148, 6// 2013.
7. B. H. Lee, M. Z. Ng, A. A. Zinn, and C. L. Gan, "Evaluation of copper nanoparticles for low temperature bonded interconnections," in 2015 IEEE 22nd International Symposium on the Physical and Failure Analysis of Integrated Circuits, 2015, pp. 102-106.
8. Y. Y. Dai, M. Z. Ng, P. Anantha, Y. D. Lin, Z. G. Li, C. L. Gan, et al., "Enhanced copper micro/nano-particle mixed paste sintered at low temperature for 3D interconnects," Applied Physics Letters, vol. 108, p. 263103, 2016.
9. Y. Y. Dai, M. Z. Ng, P. Anantha, C. L. Gan, and C. S. Tan, "Copper micro and nano particles mixture for 3D interconnections application," in 3D Systems Integration

Conference (3DIC), 2015 International, 2015, pp. TS8. 9.1-TS8. 9.5.

10 Jonas Zürcher, Luca Del Carro, Gerd Schlottig, Daniel Nilsen Wright, Astrid-Sofie B. Vardøy, Maaike M. Visser Taklo, et al., "All-Copper Flip Chip Interconnects by Pressureless and Low Temperature Nanoparticle Sintering," presented at the 2016 IEEE 66th Electronic Components and Technology Conference, Las Vegas, USA, 2016.

11 D. Deng, T. Qi, Y. Chen, Y. Jin, and F. Xiao, "Preparation of antioxidativenano copper pastes for printed electronics application," in Electronic Packaging Technology and High Density Packaging (ICEPT-HDP), 2012 13th International Conference on, 2012, pp. 250-253.

12 D. Deng, Y. Jin, Y. Cheng, T. Qi, and F. Xiao, "Copper nanoparticles: aqueous phase synthesis and conductive films fabrication at low sintering temperature," ACS applied materials & interfaces, vol. 5, pp. 3839-3846, 2013.

13 Liu, Qipeng& Chen, Xianping& Zhu, Jie& Zhang, HuanKun& Zhang, Jiang & Zhang, Jing & Wang, Li & Ye, Huaiyu&Koh, S.W. & Zhang, G.O. & Zhang, G.Q.. (2018). The performance of sintered nanocopper interconnections for high temperature device. 1476-1478. 10.1109/ICEPT.2018.8480591.

A Highly Integrated Multi-parameters RF Transceiver Module
for Microwave Semiconductor Chip Testing

Guangshan Zhang[1], Miao Song[1], Rongbin Guo[1], Yahai Wang[1], Lei Liu[1], Jie Yang[2], Shichao Liu[1], Yisheng Yang[1]

1. The 41st Research Institute of CETC. 98 Xiangjiang Road, Qingdao, CHN 266555

2. College of Arts, China University of Petroleum (East China). 66 Changjiang West Road, Qingdao, CHN 266555

Abstract

The test of microwave semiconductor chip is currently a hot topic. A miniaturized and highly integrated RF (Radio Frequency) module is designed to test microwave semiconductor chip. Similar to the design concept of synthetic instrument which has an open structure with standard bus and software defined radio, the newly designed integrated RF channel technology realize a variety of RF testing functions. It provides a small and highly integrated common hardware platform with a flexible software platform for re-development of control software. This platform is highly integrated with vector signal generation, vector signal analysis, vector network analysis and noise figure analysis, which can achieve a variety of functional test through dynamic reconfiguration and real-time loading. Multi-stage switch and coupler network are used to route reference signals and reflected signals and transmit signals for vector network analysis. The module receives and generates signals with frequency covering 100k~18GHz. It shares baseband FPGA which makes received signals simultaneously replay in generated channels. The bandwidth reaches 500MHz and supports a variety of vector modulated standards. This module provides a perfect solution to the testing system for microwave semiconductor chip.

1. Introduction

With the rapid development of new technologies such as 5G and vehicle anti-collision radar, the market demand for microwave semiconductor chip is exploding [1]. Due to the highly- integrated multiple channels and complicated functions of the microwave semiconductor chips, the testing of microwave semiconductor chips faces great challenges. At the same time, the extremely short cycle for research, development, and launching of the chip will inevitably require extremely fast design and product testing [2-3].

The test of microwave semiconductor chip is different from that of digital circuit chip and low frequency circuit chip. It needs to test different parameters such as frequency, power, gain, 1dB compression point, third-order intercept point, and phase noise. It must provide high-speed and weak-radio frequency signal transmission. If noise interference exists in testing circuit and testing load board, it will weakly affect the quality of high-speed RF signal transmission. It also needs to support the testing of more complex protocols. In addition to the current mainstream Wlan, GPS and 5G, testing suitable for more non-standard protocols is required.

The representative state-of-the-art manufacturers of radio and microwave semiconductor chip testing are Advantest Company, Teradyne Company and National Instruments Company [4-5]. The representative product of Advantest is the T2000 semiconductor test system. The system has a frequency up to 12GHz, multi-channel parallel test capability, high test-efficiency, full features, and module upgrading support [6]. Due to the use of the closed bus, it is hard to introduce new technologies on such a system. Teradyne's UltraFLEX series of integrated circuit test systems own test capabilities of power, gain, spectrum, noise figure, network parameter. Singe-channel patrol method for testing multiple microwave semiconductor devices, microwave testing capability and efficiency is relatively low. NI's NI_STS microwave semiconductor testing system adopts open standard bus and modular architecture; thus, it is of good versatility and scalability, and strong compatibility.

At present, the development of microwave semiconductor multi-parameter test has the following trends and characteristics. First, multi-channel parallel test capability is built to improve efficiency. Second, it uses common standard bus architecture to improve system platform compatibility. Third, improved integration and reliability is adapted to satisfy production line test requirements.

In this paper, we present a multi-parameters RF transceiver module for microwave semiconductor chip testing which is highly integrated with vector signal generation, vector signal analysis, vector network analysis and noise figure analysis. It provides power, gain, spectrum, noise figure, network parameter and vector testing function. This RF transceiver-testing module adopts advanced open-standard bus architecture and software radio platform which share the FPGA to generate digital baseband vector signals and process digital received signals. It applies dynamical reconfiguration to load a series of required test functions, such as transceiver real-time interaction including data collection and replay. The generating channel and receiving channel uses multi-stage routing switch to realize network parameter analysis. This module is very small and the rich RF test functions make it ideal for integrated automated-test system especially for microwave semiconductor chip testing. It can be easily extended to test multi-channel parallel RF parameters to improve test efficiency for microwave semiconductor chip testing.

2. Design of the Multi-parameter RF Transceiver Module

The RF transceiver module adopts standard bus and modular instrument. Its communication interface uses PXIe bus mode for high-speed data transmission and command control. Its structure is 3U 5Slot, easy to install at PXIe standard chassis. A standard PXIe chassis can insert 3 RF transceiver modular to realize multi-channel parallel RF parameters testing.

The RF transceiver module mainly includes a digital signal-processing unit, a direct down-conversion receiving unit, a direct up-conversion generating unit and a multi-stage routing switch unit, and it is presented in Fig. 2.1.

978-1-7281-5757-3/19 $31.00 © 2019 IEEE

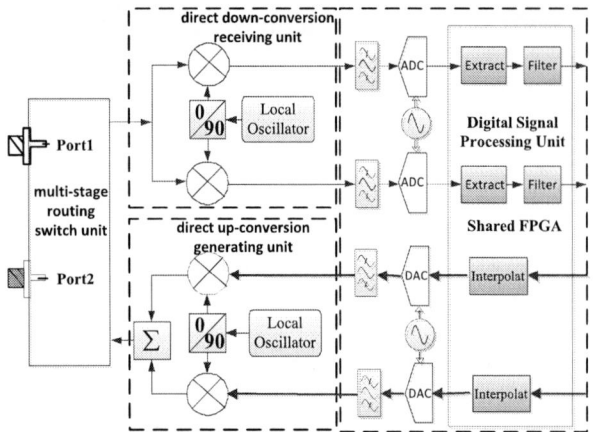

Fig. 2.1 Schematic diagram of the RF transceiver module

The digital unit of signal processing is mainly comprised of ADC acquisition, DAC generation, shared FPGA and filter circuit. The shared FPGA receives baseband signals and generates baseband signals, which directly interact with each other for signal playback. When implementing vector signal generation, vector signal analysis, vector network analysis or noise figure analysis, it only needs to load the corresponding FPGA program through dynamic reconfigurable mode, and execute the corresponding software codes to achieve the corresponding functions. Baseband reception and transmission can be performed in a common time base mode including co-sampling rate and co-trigging to achieve coherent signal transceiver.

When the digital signal processing unit receives the baseband signal from the direct down-conversion receiving unit, it performs low-pass filtering on the input baseband signal, and the digital baseband signal is sent to FPGA after the ADC acquisition. The digital baseband signal is divided into two in-phases and quadrature signals for digital filtering and extraction processing. Then the in-phase and quadrature signals such as FFT are processed, computing gain and phase according to different required functions.

The direct down-conversion receiving unit, which includes local oscillator, mixer circuit, band-pass filtering and amplification, amplifies and attenuates the input RF signal. Then it directly mixes the zero-IF frequency signal with local oscillator. The zero-IF frequency signal is divided into in-phase and quadrature signals which can adjust the zero-IF gain flexibility. Local oscillator adopts multiple VCO stitching mode to generate pure wide-band RF signals which provide the in-phase and quadrature signal for mixer.

The direct up-conversion generating unit directly up-converts the baseband quadrature and in-phase signals to the RF signals, and then performs band-bass filtering and amplification. It includes local oscillator, modulation circuit, band-pass filtering and amplification.

The multi-stage routing switch unit routes receiving signals and generating signals to port 1 or port 2. It includes microwave couplers, microwave switches, amplification circuit and attenuator. Port 1 and the port 2 both can receive or generate signals through the coupler and switch in Fig. 2.2. It connects the direct down-conversion receiving unit or the direct up-conversion generating unit. The transmission signals

pass through the coupler 3, switch 4, switch 2, switch 1 and coupler 1 to port 1 when port 1 needs to generate RF signals. The input RF signals pass through the coupler 2, switch 1, switch 2, switch 3 to receive RF signals when the port 2 needs to be the input port. This module has the flexibility to change the port attributes to generate or receive the RF signals.

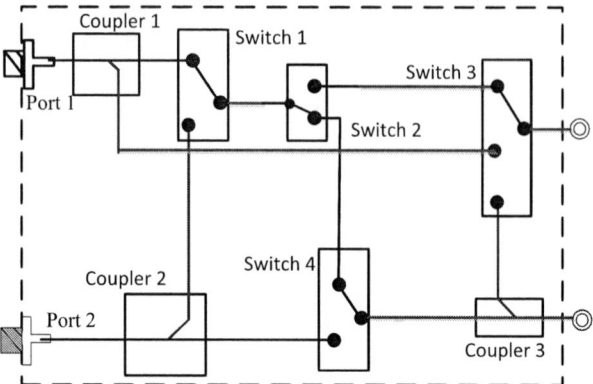

Fig. 2.2 Schematic diagram of the multi-stage routing switch unit

When the RF transceiver module is used to generate vector signals, the digital signal processing unit generates pseudo-random sequence, interpolates the digital signals and digital filters etc. by the shared FPGA. The digital baseband signal outputs to the DAC and then it is filtered by low-pass filter. The two in-phase and quadrature signals mixing with the local oscillator are directly modulated to the microwave signal. The modulated RF signal is conditioned with band-pass filtering, amplification and attenuation. Then it passes though coupler 3, switch 4 and coupler 2 to the port 2 for generating vector signals in the multi-stage routing switch unit. It also transmits by coupler 3, switch 4, switch 2, switch 1 and coupler 1 to the port 1 for generation.

When the RF transceiver module performs vector signal analysis or noise figure analysis, the input RF signal passes through coupler 1, switch 1, switch 2 and switch 3 to the direct down-conversion receiving unit. The RF signal is mixed with the local oscillator and is turned to zero IF signal after the band-pass filtering, amplification, and attenuation. It is divided into the in-phase and quadrature signals. And then they are input to the digital signal processing unit. They are digitalized by ADC after low-pass filtering. The digital in-phase and quadrature signals are extracted and filtered in the shared FPGA. It also does some other digital processing such as FFT, computing gain and phase, window function calculation and multi-stage filtering. The input RF signal can also pass though coupler 2, switch 1, switch 2, switch 3 to the direct down-conversion receiving unit from port 2. The noise figure analysis shares the same way with the vector signal analysis except for the software calculation.

When the RF transceiver module is used for vector network analysis, port 1 or port 2 can be set as an input or output. It can achieve dual port network test by the multi-stage routing switch. The signal generation and reception have the similar way to the vector signal generation and analysis. Port 1 is set to generate microwave signals, and it uses coupler 3, switch 4, switch 2, switch 1 and coupler 1 to generate incident signal for network parameter analysis, and then it is input to port 2 passing though coupler 2, switch 1, switch 2 and switch 3. Coupler 3 and switch 3 can be used to generate reference signals. The reflected signal is generated by coupler 1 and switch 3. When the port 2 is set to generate microwave signal, it uses coupler 3, switch 4 and coupler 2 to generate incident signal for network parameter analysis, and then it inputs to the port 1 passed though coupler 1, switch 1, switch 2 and switch 3. It also uses coupler 3 and switch 3 to generate reference signals. The reflected signal is generated by coupler 2, switch 1, switch 2 and switch 3. The digital signal processing of vector network analysis is achieved by the shared FPGA.

To achieve high performance of vector network analysis, the directional bridge device uses Ceyear 80622B to isolate or split signals with high quality in Fig. 2.3. It has wide-band frequency, high directionality and low loss. It can be used for coupling or isolating signals at the transmitting or receiving terminal.

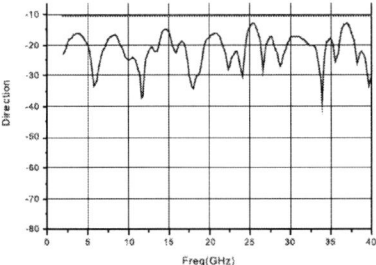

Fig. 2.3 Performance of the directional bridge device

3. Implementation of the Multi-parameter RF Transceiver Module

The multi-parameter RF transceiver module uses standard bus modular instrument architecture which has a size of 100mm*140mm*215mm. Its size is very small and suitable for automatic microwave test system construction. It can also build the multi-channel generating or receiving microwave signals for paralleling test by multiplexing the RF transceiver depicted in Fig. 3.1. Therefore, it can be applied to test the microwave semiconductor chips because of its paralleling testing characteristics.

Fig. 3.1 The Ceyear 6955DA multi-parameter RF transceiver module

In this RF transceiver module, it provides digital I/O function for expanding its application which can trigger the different channels. It outputs the local oscillator for signal generation and signal receiving which can provide coherent signals for multi-parameter transceiver system equipment. It also input local oscillator for generating channel or receiving

channel while the multi-parameter RF transceiver module's local oscillators are shut down. At the same time, the 10MHz reference can be used to input or output to different modules or different testing devices to synchronize time base. It is very useful for coherent receiving or generating test.

The multi-parameter RF transceiver module has highly pure signal generation which is depicted in Fig. 3.2. The generating signal has low spurious and high output power whose harmonic is under the average noise level. It can also provide the analog modulation such as AM, PM, FM etc.

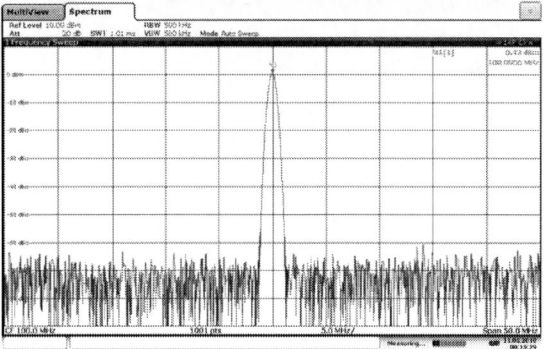

Fig. 3.2 Spectrum of the multi-parameter RF transceiver module's generating signal

The multi-parameter RF transceiver module can perform high-quality vector signal generating and receiving, which is depicted in Fig. 3.3. It supports a variety of vector modulation such as ASK, PSK, FSK, QAM etc. It can achieve high performance of error vector magnitude which is under 1% at the QPSK modulation and 10MHz symbol rate.

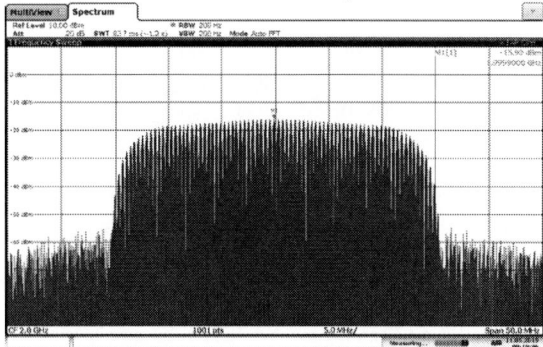

Fig. 3.3 Spectrum of the multi-parameter RF transceiver module's vector signal

In this paper, we also introduce the Ceyear 9801C microwave semiconductor device multi-parameter tester, which is depicted in Fig. 3.4. It can realize 32-channel RF parallel signal test by using 8 sets of the proposed multi-parameter RF transceiver modules. One chassis can include 3 sets of the multi-parameter RF transceiver modules. Therefore, it just needs two chassises to perform 32-channel RF parallel signal test. It embeds 2 sets of microwave switch modules to route 32-channel microwave signals. At last it can perform 16-channel transmission and 16-channel reception individually. According to these transceiver channels, it can be flexibly matched to vector signal generating, vector signal analysis, vector network analysis and noise figure analysis. For the different RF transceiver module they shares the same reference signal of 10MHz and have the same trigger for each

978-1-7281-5757-3/19 $31.00 © 2019 IEEE

module. They also have the same time base for ADC clocking or DAC clocking.

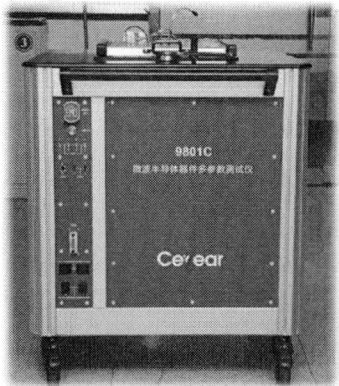

Fig. 3.4 The Ceyear 9801C microwave semiconductor device multi-parameter tester

The multi-parameter RF transceiver module adopts the software defined radio platform and open FPGA digital platform which can load a variety of communication protocol such as 4G, 5G, Wlan etc. It can be widely used in fields including microwave semiconductor chip test, vehicle collision avoidance radar test, internet of things protocol test.

Conclusions

We developed a multi-parameters RF transceiver testing module for microwave semiconductor chip testing. It is a high integrate with vector signal generation, vector signal analysis, vector network analysis and noise figure analysis. Our experiment results show that it can perform high performance on radio signal parameter testing including low spurious and high error vector magnitude. The power, gain, spectrum, noise figure, network parameter and vector testing function can be easily worked out. It adopted advance open standard bus architecture and software radio platform which share the FPGA to generate digital baseband vector signal and process received digital signal. It used dynamical reconfiguration to load a variety of required communication protocol testing. This module was very small and rich of RF test functions making it ideal for integrated automated test system especially for microwave semiconductor chip testing. It can be easily extended to multiple-channel parallel RF parameters testing to improve test efficiency for microwave semiconductor chip testing.

Acknowledgments

The authors acknowledge researchers from the 7th Research Department of China Electronics Technology Instrumentation Co., Ltd. The authors also acknowledge the financial support from "National Key Research and Development Plan, Major Scientific Equipment Development Project" (Project No. 2017YFF0106702) of National Science and Technology.

References

1. Stefan R. Vock & Omar J. Escalona *et al*, "Challenges for Semiconductor Test Engineering: A Review Paper", *J Electron Test*, Vol. 28, No. 10 (2012), pp. 365-374.
2. Alessandra Di Paola, Mario Sannino, "A Novel Noise Figure and Gain Test Set for Microwave Devices", *IEEE Trans. Instrum. Meas.*, Vol. 48, No. 5 (1999), pp. 921-926.
3. Tyler Junyao T. *et al*, "An loT Inspired Semiconductor Reliability Test System Integrated with Data-Mining Applications", *2th International Conference on Cloud Computing and Internet of Things*, Dalian, China, May. 2016, PP. 111-114.
4. Z. Zhang, *et al*, "Capacity planning with reconfigurable kits in semiconductor test manufacturing", *International jouruanl of Production Research*, Vol. 44, No. 13 (2006), pp. 2625-2644.
5. Young Min J. *et al*, "Multi-Pass Lot Scheduling Algorithm for Maximizing Throughput at Semiconductor Final Test Facilities", *27th International Conference on Flexible Automation and Intelligent Manufacturing*, Modena, Italy, June, 2017, PP. 1992-1996.
6. A. Di Paola and M. Sannino, "Determination of gain of microwave amplifying devices through noise figure measurements," *in Proc. IMTC'95*, Waltham, MA, Apr. 1995, pp. 526–529.
7. T. He, *et al*, "Dispatching optimization with sequence dependent setup times in semiconductor final testing scheduling", *Proceedings of the 26th International Conference on Flexibile Automation and Intelligent Manufacturing*, Seoul, Korea, June, 2016, PP. 83-87

High Precision Model by Error Compensation Method based on the Angelov Model

Ziyue Zhao, Yang Lu, Hengshuang Zhang, Chupeng Yi, Yuchen Wang, Xiaohua Ma and Yue Hao
Key Laboratory for Wide Band-Gap Semiconductor Materials and Devices,
Xidian University, Xi'an 710071, China
18691874570@163.com

Abstract

With the development of the microelectronics, circuit design also becomes more and more important. The circuit needs to be designed by the model, and the accuracy of the model will also determine the quality of the circuit design. In order to improve the accuracy of circuit design, it is necessary to improve the accuracy of the circuit model. This paper mainly discusses a more accurate method basing on the Angelov Model of the Gallium Nitride (GaN) based high electron mobility transistors (HEMTs). For the Angelov Model, the fitting of DC curve is the most critical. There are many parameters to fit and modify the fitting curve. Some of these parameters are of certain physical significance, while others are mainly used to improve the accuracy of curve fitting. Each parameter does not change with the bias, which will lower the accuracy of the model. In this paper, the sensitive parameters in the model are changed into a function of gate voltage and drain voltage by formula. In this way, the fitting accuracy of the model is improved. However, the improvement of the model accuracy is very limited by changing the sensitive parameters. Next, the curve is compensated by error function to improve the accuracy of the whole model. It also makes up for the defect that the Angelov Model cannot fit when the drain current is negative. In order to further improve the accuracy of the model, the weight function is used to correct the error function, which improves the fitting accuracy of the error function and makes the fitting of the output curve more accurate. Finally, the accuracy of the model was improved by 57%, which satisfied the needs of the circuit design.

Introduction

At present, with the development of technology and the continuous improvement of industrial demand for power and frequency of devices, GaN devices show good characteristics and meet people's requirements for power devices in all aspects [1]. With the increase of frequency, the monolithic microwave integrated circuit (MMIC) will become the main technology of circuit design. For MMIC, the design is largely determined by the accuracy of the model. The ideal model is ensure the accuracy of the model, but also need to be able to combine with the software [2]. In order to achieve these goals, many kinds of models have been proposed one after another in recent decades. Among these models, the Angelov Model shows good performance. The parameters of the model are relatively few, and most of them can be obtained through basic tests of GaN HEMT. However, the Angelov Model also has the common problems of the empirical base model. The model lacks feedback and guidance on device manufacturing process, and its accuracy needs to be further improved.

This paper mainly aims at improving the accuracy of the traditional Angelov Model. The parameters that are more sensitive to curve fitting are replaced by formulas to improve the accuracy of dc characteristics under each bias, and the errors between fitting values and measured values are compensated by formulas to once again improve the accuracy of the model. For these sensitive parameters, it is difficult to accurately describe under any bias, so in this paper, the commonly used bias is described. This paper focuses on the accurate fitting of the output curve of the device.

The Small Signal Model

The Angelov Model can well incorporate in commercial Harmonic Balance (HB) simulators, meanwhile, it can accurately predict gain, intermodulation distortion, generation of harmonics [3]. Because of these advantages of the model, it has been well developed and revised in recent years.

First, the performance of the device needs to be tested. The device was tested which was on wafer. S parameters of device were measured by the Vector Network Analyzer (E8363B) and the Semiconductor device analyzer (B1500A) which were made in Agilent Technologies. The frequency of the test is from 0.1GHz to 40GHz. The equivalent circuit of the transistor shown in Fig.1 was used to model the GaN HEMT on small signals. There are 17 parameters in this circuit model [4]. The parameter values of each parasitic parameter are obtained after matrix transformation and calculation through S parameter test under specific bias. The intrinsic parameters are obtained by embedding parasitic parameters and then by mathematical calculation. In this paper, for example, the bias of the device is Vds=5V and Vgs=1V. And the small signal simulation results are shown in the fig.2. In the figure, the red line indicates the test value and the blue line indicates the simulation value.

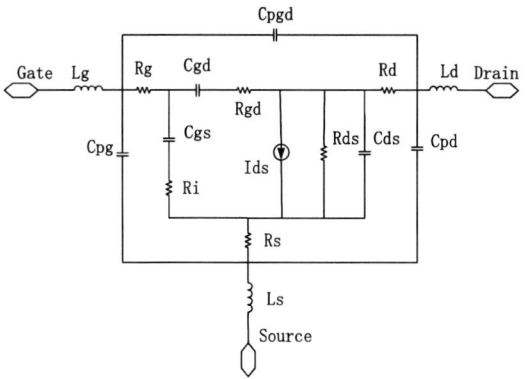

Fig. 1. Equivalent circuit of the transistor

978-1-7281-5757-3/19 $31.00 © 2019 IEEE

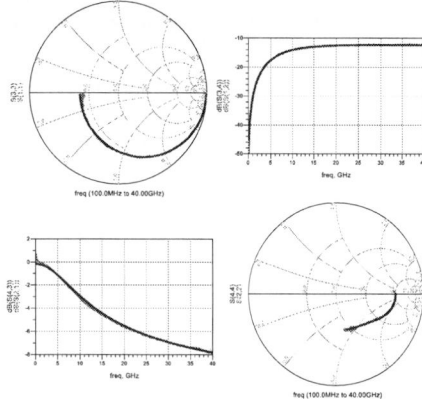

Fig. 2. Simulation results for small signals

As can be seen from the simulation results of small signal, the circuit topology and parameter extraction method can be used to fit the small signal performance of GaN HEMT well. The jitter at the lower frequencies is caused by test errors, which can be ignored here.

The Tradition Angelov Model

DC performance is the core nonlinear part of every model. Drain current Ids is also the DC feature that circuit designers are most concerned about. The Angelov Model is no exception, and the fitting error of Ids in the Angelov Model is generally about 10%. The tanh function was chosen since this function can describe the curve exactly and the tanh fuction can accurately characterize the derivatives of the Ids curve.

Many parameters in the Angelov Model have their specific physical significance, while some parameters are only for the fitting of curves. Some parameters are relatively sensitive to bias. Under different bias, slight change of parameter value will cause large change of Ids curve. For traditional Angelov Models, each parameter is constant under any bias. In this way, due to the existence of sensitive parameters, the fitting difficulty is greatly increased, and the fitting error is also large.

By adjusting the value of each parameter through the optimization software, the accuracy of the model can be improved, but there is still an error of nearly 5% or more. For the traditional Angelov Model, since the Ids fitting formula is certain, there is no room for improvement. Then, without changing the original Ids formula, the parameters need to be changed to further improve the accuracy of the model.

The Modified Angelov Model

In this paper, there are three parameters that need to be changed. They are IPK0, α s and λ. They all have their own physical meanings, and with the change of bias, these three parameters have obvious changes in the shape of the curve. The physical significance of IPK0 mainly represents the Ids current value corresponding to the maximum transconductance [3]. In the fitting process, this parameter mainly affects the saturation current value of Ids under different gate voltage. For α s and λ respectively represent the slope of Ids curve at the knee point voltage under large gate voltage and Ids curve in the saturation area under small gate voltage. Because these parameters are sensitive to bias,

the accuracy of traditional model is greatly limited. Here, converting these parameters into variables changing with the bias will lose their original unique physical significance, but it can improve the accuracy of the model, and each parameter value under different bias can be easily obtained.

When the parameter value of each parameter is obtained, the formula is needed to fit the parameter value. The formula you choose should be as simple as possible, and the derivative should be as simple as possible. If the fitting of sensitive parameters is carried out under all possible bias, then the formula form is difficult to be simple and the fitting is difficult to be more accurate. Therefore, for circuit design, the most commonly used work range is selected for fitting. In this way, not only the accuracy of Ids curve fitting can be guaranteed, but also the complexity of parameter formula can be significantly simplified. For gate voltage, the range selected in this paper is -2V to 2V, and the drain voltage range is 0 to 10V.

The gate voltage range of the test is -2V to 2V, and the step size of the gate voltage is 1V. So for the output curve, there are 5 curves that need to be fitted. For each curve, we use the formula of the traditional Angelov Model to fit, IPK0, α s and λ can be equal to any value, and make the error between fitting curve and test curve as small as possible. The parameter values are shown in the table. 1.

Table.1. Values of parameters

Vgs	-2V	-1V	0V	1V	2V
I_{PK0}	3.0e-6	3.8e-3	1.0e-1	1.6e-1	1.8e-1
α s	5.82	5.47	5.40	5.34	5.34
λ	1.44	1.04	0.01	-0.003	-0.008

Next, we need to select a function of gate voltage to fit these three sensitive parameters. In order to facilitate calculation and derivative, the following formula is selected to fit the parameters.

$$I_{PK0} = \sum_{n=0}^{\infty} A_n V_{gs}^n \qquad (1)$$

$$\alpha_s = \sum_{n=0}^{\infty} B_n V_{gs}^n \qquad (2)$$

$$\lambda = \sum_{n=0}^{\infty} C_n V_{gs}^n \qquad (3)$$

In the above formula, An, Bn and Cn are arbitrary constants. Moreover, the larger the value of n is, the more accurate the parameter fitting will be. But at the same time, this will bring about more complicated formulas. In order to have both accuracy and simplicity, the value of n is finally determined to be 4.

978-1-7281-5757-3/19 $31.00 © 2019 IEEE   64

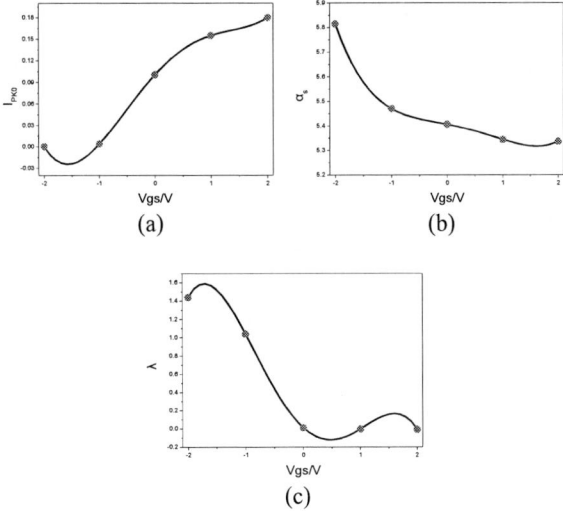

(a)　　　　　　　　　(b)

(c)

Fig. 3. Fitting curve for the parameters

The parameter fitting curve is shown in fig.3. The fitting curve is represented by a black line and the parameter values of the five bias points are represented by a red point. It can be seen that the three sensitive parameters can be well fitted by the formula.

The measurement curves and the simulation curves are compared by the fig.4. In the picture, the measurement curves are represented by lines, and the fitted curves are represented by points. It can be seen from the curve in the figure that when the drain voltage is large, the fitting is better. But for the small drain voltage, there is a very obvious error. When the gate voltage is increased, the drain current Ids is negative when the drain voltage is 0. For the Angelov Model, the current at this point cannot be fitted. Therefore, the accuracy of the Angelov Model cannot be further improved.

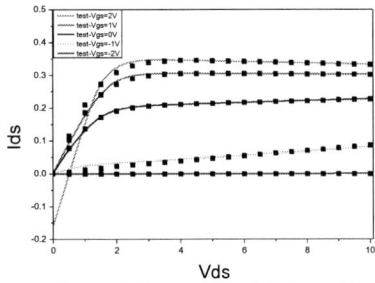

Fig. 4. Fitting results of different bias

In order to improve the accuracy of the model, error function is needed to modify the model. The error function is obtained by subtracting the current value obtained by the test and the current value fitted by the formula under the same bias. And then you have a bunch of curves. In this way, the accuracy of the model can be greatly improved and the shortcomings of the Angelov Model can be made up.

The error function is an equation of gate voltage and drain voltage. Therefore, in the process of error fitting, it is necessary to consider not only the influence of gate voltage on the error function, but also the influence of drain voltage on the error function. Here, the higher order function is still

selected to fit the error. After many experiments and comparisons, the highest order item finally selected was 3. On the premise of ensuring accuracy, the complexity of the formula is reduced. However, it is difficult to accurately fit the error under any bias. According to the fitting results, the fitting is more accurate when the leakage pressure is smaller, and the fitting error is larger when the leakage pressure is larger. In order to reduce the impact of error function on Ids fitting accuracy when drain voltage is large, the error function is modified by introducing weight function.

There are two parameters in the weight function, V_p and k. V_p mainly determines the position of the inflection point of the curve, and k mainly determines the slope of the curve. The specific curve form of the weight function is shown in Fig. 5.

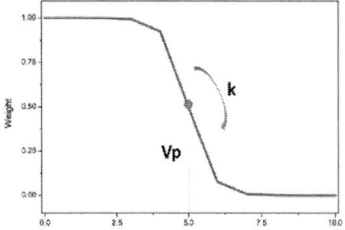

Fig. 5. The curve of the weight function.

The error function is modified by the weight function. Finally, the error function is compensated into the model formula, and the resulting curve is shown in the figure below.

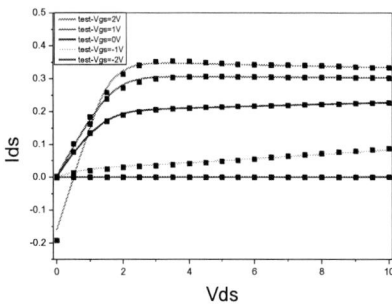

Fig. 6. Fitting results of different bias with error function

Through the analysis of the data in the figure, it can be obtained that the curve with the gate voltage of 2V has the most obvious improvement, and the error has been reduced by 72%. For the whole curves, the accuracy has also been significantly improved, with the error reduced by 57%.

Conclusions

In the Angelov Model, some parameters are relatively sensitive. If it stays constant all the time, it is difficult to improve the accuracy of the model. Then, replacing these parameters by formulas will obviously improve the accuracy of the model. The Angelov model also has its own defects for modeling the GaN HEMT. It is impossible to fit the situation that the drain current is negative when the drain voltage is low. Therefore, error function should be introduced to modify the model. However, the inaccuracy of the error fitting will also increase the overall error of the model, so the weight function

978-1-7281-5757-3/19 $31.00 © 2019 IEEE

is introduced to correct the error function and selectively compensate the error. In the end, the model can fit well with the test value and reduce the overall error by 56.05%.

References

1. Khan M A, Bhattarai A, Kuznia J N, et al. "High electron mobility transistor based on a GaN-AlxGa1-xN heterojunction," Applied Physics Letters, pp.1214-1215. 1993.

2. Angelov I, Thorsell M, Kuylenstierna D, et al. "Hybrid measurement-based extraction of consistent large-signal models for microwave FETs," 2013 European Microwave Conference, IEEE, Nuremberg, Germany, Oct. 2013, pp. 267-270.

3. Angelov I, Zirath H, Rorsman N. "A new empirical nonlinear model for HEMT-devices," 1992 IEEE MTT-S Microwave Symposium Digest, Albuquerque, USA, August. 1992, pp. 1583-1586.

4. Wang C, Xu Y, Yu X, et al. "An Electrothermal Model for Empirical Large- Signal Modeling of AlGaN/GaN HEMTs Including Self-Heating and Ambient Temperature Effects," IEEE Transactions on Microwave Theory and Techniques, Vol. 62, (2014), pp. 2878-2887.

A Novel Scalable Series MIM Capacitor Model for MMIC Applications

Chupeng Yi, Yang Lu*, Hengshuang Zhang*, Ziyue Zhao, Xiaohua Ma+,*, and Yue Hao+,*

School of Adanced Materials and Nanotechnology, Xidian Univ., Xi'an 710071, China,
*School of Microelectronics, Xidian Univ., Xi'an 710071, China
+ Key Laboratory of Wide Band-Gap Semiconductor Technology, Xidian Univ., Xi'an 710071, China
yicp905@163.com

Abstract

With the development of the wireless communication and integrated circuits, Integrated passive devices (IPDs) have gained more and more application in radio frequency (RF) integrated circuit and monolithic microwave integrated circuits (MMICs). Among IPDs, MIM capacitor is a key passive component, widely used in DC-bias circuit, decoupling, etc. Therefore, the accuracy and practicability of the capacitance model is one of the key factors for the success of circuits design.

However, since the MIM capacitor has many parasitic effects at high frequencies, it no longer exhibits pure capacitor characteristics anymore, thus, in high frequency circuit design, a simplified equivalent circuit model that can accurately characterizes the components is essential for the successful design of the circuit.

As SiC-based GaN devices exhibit the advantages of high temperature and high power in millimeter wave applications, more and more integrated circuit use SiC-based GaN devices and components. With regards to this we produced a series of MIM capacitors of different sizes with air-bridges have been fabricated on 4H–SiC substrate. Also, we proposed a novel of simple broadband, based on the physical parameter equivalent circuit model of MIM capacitor structure.

Based on the traditional equivalent circuit model, we introduce the corresponding components that characterize the skin effect and loss of the metal. The lumped element model is a simplified physical based model with broadband and high precision, and parameter extraction is given. To verify the accuracy and versatility of the model, we compared the measured and model simulated results of a series of MIM capacitors of different sizes. The results show that the model can accurately fit the S-parameters between measured and simulated in operating frequency range. In addition, the model shows good scalability as well.

1. Introduction

In MMICs and RF power amplifier, a large number of passive components such as spiral inductors, thin film resistors, and MIM capacitors are important components of circuit design. Among them, MIM capacitor is the main component in MMIC design, widely used in DC-bias circuit, decoupling, impedance matching and DC-bias network. Therefore, the accuracy and practicability of the capacitance model is one of the key factors for the success of MMIC design. [1]

As the frequency of use of MMICs continues to increase, the accuracy and frequency applicability of passive components are increasingly being studied and focused. However, since the MIM capacitor has many parasitic effects at high frequencies, it no longer exhibits pure capacitor characteristics anymore, thus, in high frequency circuit design,

a simplified equivalent circuit model that can accurately characterizes the components is essential for the successful design of the circuit. [2]

As SiC-based GaN devices exhibit the advantages of high temperature and high power in millimeter wave applications, more and more MMICs use SiC-based GaN devices and components. Recently, there have been a large number of researches on the modeling, optimization design and performance evaluation of MIM capacitors, but mainly focused on CMOS [3], GaAs and Si substrates [4]. Few modeling studies on SiC-based MIM capacitors have been reported.

At the same time, in the previous research, the traditional equivalent circuit model [5] failed to accurately characterize the skin effect and frequency characteristics such as metals, so the traditional model has obvious disadvantages in the face of increasingly high precision modeling requirements.

For this reason, in this paper, we proposed a new novel of simple broadband, based on the physical parameter equivalent circuit model of MIM capacitor structure, and give the corresponding parameter extraction process. At the same time, in order to verify the accuracy and versatility of the model, we also produced a series of MIM capacitors of different sizes with air-bridges have been fabricated on 4H–SiC substrate. The experimental results show that the model can accurately describe the frequency characteristics of the series capacitor in the frequency range of 0.1-40 GHz. In addition, the model demonstrates good scalability in addition to different sizes of MIM capacitors.

2. MIM capacitance structure and manufacturing

Fig. 2.1 shows the cross-sectional diagram and 3D view of the MIM capacitor. Where, the upper plate of the capacitor connects the feed line through the air bridge and the metalized through-hole. The filling medium in the middle is SiN, and the bottom plate is directly connected to the feed line on the other side.

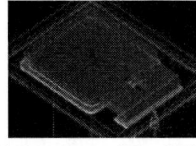

Fig. 2.1 the cross-sectional diagram and 3D view of the MIM capacitor

The MIM capacitor made by GaN process line in Xidian University, which is the standard MMIC process. Fig. 2.2 shows the micrograph of the MIM capacitor, which is made on a 4H-SiC substrate. The thickness of the substrate is 100nm, and the dielectric constant is 9.6. Firstly, a layer of Ti/Au with a thickness of 20nm/400nm was evaporated on a 400nm SiC

978-1-7281-5757-3/19 $31.00 © 2019 IEEE

substrate with AlGaN/GaN HEMT epitaxial layer. The Ti/Au layer is formed by the metal evaporation device. And a 200nm thick SiN dielectric layer was deposited via plasma-enhanced chemical vapor deposition (PECVD). Then, through ICP, the metal interconnected through-holes for connecting the upper plate to the Feed line are etched, and air Bridges are formed and thickened by electroplating, as well as Feed line to about 3um.

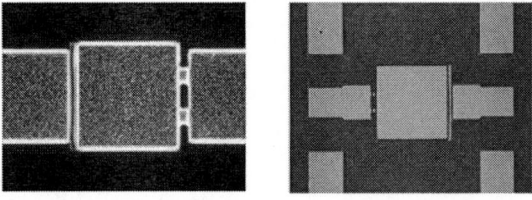

Fig. 2.2 The microphotograph of the fabricated on-chip MIM capacitor on SiC substrate

3. MIM capacitance model and parameter extraction method

The EM electromagnetic simulator in Keysight ADS is used to simulate the microwave characteristics of the MIM capacitors. In order to better understand the performance of MIM circuits and the loss mechanism in millimeter wave band, a simplified equivalent circuit model with certain physical significance is needed. This paper presents a simplified capacitance model which takes into account the loss of metal conductivity and skin effect. The proposed new model is shown in fig. 3.1.

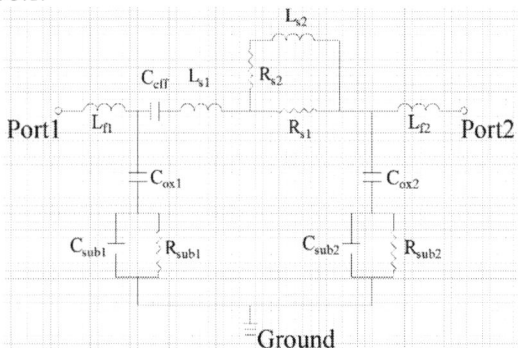

Fig. 3.1 The lumped element enhanced model for the MIM capacitor

The topology structure of the traditional 9 elements make improvements on the basis of the lumped model. By contrast, two lumped inductance Lf1/Lf2 used to describe the microstrip line which is connecting the input port and the output port. Ceff is the size of the two plate between the up and down the main capacitor. Ls1 and Rs1 represent the parasitic resistance and metal loss and parasitic inductance of the top plate metal respectively, while Ls2 and Rs2 are used to represent the parasitic inductance and resistance of metal due to skin effect. Cox1 and Cox2 are equivalent dielectric capacitance (caused by epitaxial layer) between MIM capacitor metal and SiC substrate. Csub and Rsub are the corresponding loss and parasitic parameters of SiC substrate.

Lf1/Lf2 can be obtained by fitting the S-parameters between the aggregate inductance and the Feed line with CAD tools. After Lf1/Lf2 is embedded, the equivalent circuit

consists of a π type circuit, as shown in fig. 3.2. This circuit can be divided into three parts, through transfering of the S parameter for Y parameter, and the y-parameter components of part A, B and C are calculated by using formula 1 [6]. The parameter values of each lumped element in the circuit can be calculated by the following formula, but different parameters need to be obtained at different frequencies.

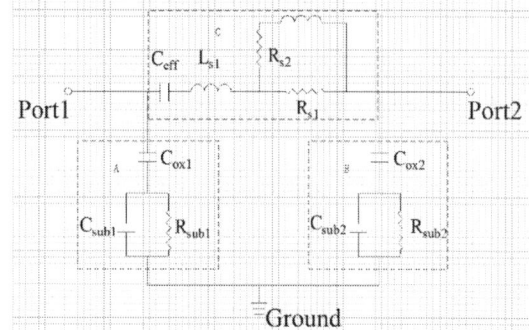

Fig. 3.2 The equivalent circuit after Ls1 and Ls2 are de-embedded

$$Y = \begin{bmatrix} A + C & -C \\ -C & C + B \end{bmatrix} \quad (1)$$

$$A = \frac{1}{1/_{jwC_{ox1}} + \left(1/_{R_{sub1}} + jwRC_{sub1}\right)^{-1}}$$

$$B = \frac{1}{1/_{jwC_{ox2}} + \left(1/_{R_{sub2}} + jwRC_{sub2}\right)^{-1}} \quad (2)$$

$$C = \frac{1}{1/_{jwC_{eff}} + jwL_{s1} + \left(1/_{R_{s1}} + 1/_{(R_{s1} + jwL_{s2})}\right)^{-1}}$$

$$Ceff = \frac{\varepsilon_{eff}*A}{d} \quad (3)$$

With the calculated initial value, the model is then optimized by gradient algorithm until the local minimum error is reached.

4. Measurement results and model verification

In order to study the Scalability of the new model, we fabricated a series of square capacitors with different side lengths (side-lengths from 40um-140um) with coplanar waveguides to microstrip line conversion structures, and used the vector network analyzer to measure the on chip MIM capacitor's S-parameter from 0.1 GHz to 40 GHz. And then, the S parameter of the conversion structure is embedded by the corresponding TRL calibration standard, thereby obtaining the actual S parameter of the capacitor. Fig. 4.1 shows the S-parameter with a side-length of 80um. After converting the S-parameter to the Y-parameter, in order to characterize the performance of the fabricated capacitor, we introduce the Q-quality factor and the effective capacitance Ceff from the Y-parameter, as shown in the following formula.

978-1-7281-5757-3/19 $31.00 © 2019 IEEE   68

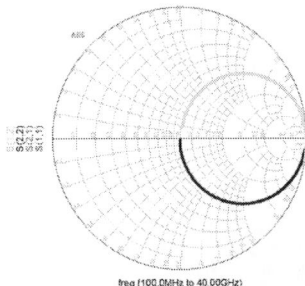

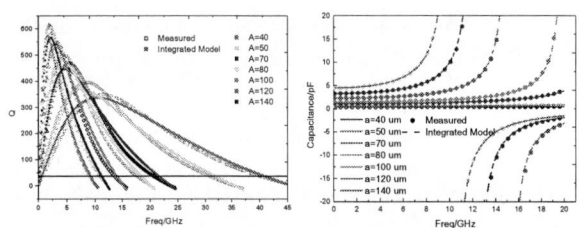

Fig. 4.3 The measured Q values and Ceff of different sizes of capacitance

Fig. 4.1 Smith Chart results of S11, S21 measurement of 80x80 um^2 MIM capacitor in the frequency range of 100MHz-40GHz.

Then we use the parameter extraction method in the third section to extract the model parameters of the side-length 80um capacitor. Based on the mentioned lumped circuit component values, Fig. 4.2 shows the capacitance Q value and the effective capacitance value ratio obtained from the model simulation and the measurement. It can be seen from the figure that the simulation value fits well with the measured value in the whole frequency band. The S parameter error and the Q root mean square error are 5% and 3%, respectively.

Similarly, the root mean square error percentage of the S and Q values is used to test the quality of the model. The S parameter percentage error shown in Fig. 4.4 is the average over the entire measurement frequency range. It can be seen from the figure that the error of the S parameter is less than 5% over the entire frequency range, and the same Q value error is also used to verify the accuracy of the model, and the Q value error is 10%.

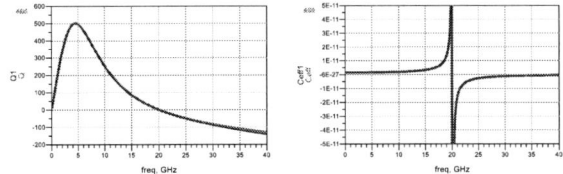

Fig. 4.2 Q values and Capacitance values comparison between measurement and model simulation

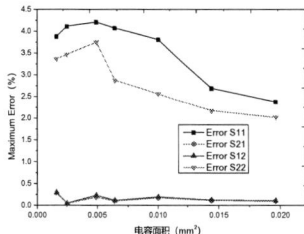

Fig. 4.4 Average Percentage Error (%) of S-parameters between model simulation and measurement varying to size of MIM area

The model proposed in this paper is further verified by extracting model parameters of capacitors of different side length dimensions. Table 4.1 lists various typical model parameters for capacitors of different sizes.

5. The scalability of the Lumped Model

Comparing the parameters of the different size MIM capacitor models above, we find that the main parameters and the effective area A of the capacitor show a good linear or nonlinear relationship, which can be expressed by different formulas. Fig. 5.1 shows a plot of the relationship between the partial model parameters and the effective area A of the capacitor.

A (um²)	Cs (pF)	Ls (pH)	Rs (Ω)	Csi1 (fF)	Csi2 (fF)	Rsi1 (KΩ)	Rsi2 (MΩ)	Fres (GHz)
40×40	0.379	31.22542	0.067553	2.813175	3.267758	24.7536	141.292	46.2
50×50	0.587	34.296596	0.04267	3.7469404	4.1880796	24.801836	129.628004	35.4
70×70	1.135	39.6356	0.031005171	5.81762	6.11202	25.4424	130.495	23.6
80×80	1.474	42.72491	0.028303067	6.72544	7.7966906	25.1485	128.228	20
100×100	2.28962	47.6641	0.02143135	9.21516	10.69391	24.9971	132.452	15.3
120×120	3.28764	50.9897	0.0192864	11.962	13.5315	24.9984	133.3933	12.3
140×140	4.4582	54.4421	0.01675517	14.9674	17.1064185	24.933145	134	10.2

Table 4.1 Extracted MIM capacitor model parameters

By extracting the parameters of each model, Fig. 4.3 shows the measured Q values and Ceff of different sizes of capacitance compared with the model simulation values.

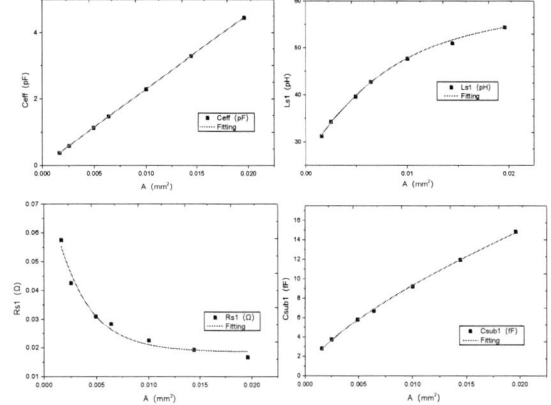

Fig. 5.1 Fitting curve of different parameters versus MIM capacitor plate area

978-1-7281-5757-3/19 $31.00 © 2019 IEEE

6. Conclusions

The new MIM series capacitor model for MMIC design application is presented. Introduced the parasitic inductance and resistance of the metal due to the skin effect, which greatly improved the accuracy of the model. Meanwhile, The method of parameters extraction and formulas are introduced. The MIM capacitor model is suitable for MMIC design application in wide frequency range with acceptable quality of error evaluation.

References

1. Mondal, J. P, "An Experimental Verification of a Simple Distributed Model of MIM Capacitors for MMIC Applications," *IEEE Transactions on Microwave Theory and Techniques*, Vol. 35, No. 4 (1987), pp. 403-408.

2. Sul W S , Pyo S G., "RF Characteristic Analysis Model Extraction on the Stacked Metal-Insulator-Metal Capacitors for Radio Frequency Applications," *IEEE Transactions on Electron Devices*, Vol. 61, No. 8 (2014), pp. 3011-3013.

3. Ando T , Cartier E , Jamison P , et al, "CMOS compatible MIM decoupling capacitor with reliable sub-nm EOT high-k stacks for the 7 nm node and beyond," *IEEE Electron Devices Meeting*, 2017.

4. Lintao L , Jinxiang W , Lai F C., "A New Equivalent Circuit Model of MIM Capacitor for RFIC," *International Conference on Microwave & Millimeter Wave Technology*, 2007.

5. Esa M , Subramaniam K , Victor Kordesch A,. "Nine-element lumped Metal-Insulator-Metal(MIM) capacitor model for RF applications," *International Conference on Electronics Manufacturing & Technology*, 2006.

6. Zheng T , Han M , Xu G , et al., "Design and fabrication of wafer level suspended high Q MIM capacitors for RF integrated passive devices," *Microsystem Technologies*, Vol. 23, No. 1 (2017), pp. 1-7.

A Compact X-band Pallet Power Amplifier
Using Gallium Nitride MMIC and Discrete FETs with HMIC Technology

Wang Yi[a], Ni Tao[a], Yin Jun[a], Mo Jianghui[a], Yu Ruoqi[a], Li Jing[a], Dong Shiliang[a], Liu Ze[a],

Chen Lei[b], He Jian[b], Huang Luoguang[a]

[a] The 13th research Institute, CETC, *Shijiazhuang* 050051,*China*

[b]*TianJian TECH*，*Chengdu* 610041，*China*

Abstract

This paper describes the design and characterization of a highly integrated pallet power amplifier using Gallium Nitride MMIC and discrete FETs. The amplifier is realized by the combination of a driving stage which consists of two Gallium Nitride MMICs and a pair of Lange couplers and a power stage which consists of two bare Gallium Nitride FETs and ceramic matching/biasing circuits. The complete pallet power amplifier is assembled with hybrid microwave integrated circuits (HMICs) technology with has an overall size of 20mm×12mm. When biased under pulsed condition, the pallet power amplifier has a saturated output power up to 130W, an associated power gain larger than 33dB, and a drain efficiency greater than 35%.

1.Introduction

As the representative of the third generation semiconductor materials , Gallium Nitride (GaN) has various advantages, such as high energy gap, 2DEG structure of AlGaN/GaN, which makes it a suitable candidate for active devices of high frequency, high power and high voltage applications. The GaN MMIC and IMFET have attracted much focus of different research institutes and companies in recent years. X-band amplifiers are widely used for radar system, air traffic control networks, and so on. Developments show that X-band GaN MMIC is of 48 dBm[1] saturated output power, and IMFET is of 500W output power and 7.7dB associated gain[2]. This article shows the design and characterization of a pallet power amplifier which compromises of GaN MMICs and discrete FETs. The amplifier is realized using HMIC technology and has a hundred-watt level output performance. The second part of the article gives a detailed circuit design procedure, and the third part of it covers measurement results.

2.Circuit Design

The target performances of the amplifier are given below．
Frequency: 9 GHz~10 GHz.
Voltage: -5V/28V.
Pulse Width: 500us.
Duty Cycle: 10%.
Saturated Output Power: >51dBm.
Power Gain : >33 dB.
Efficiency: >30%.
Dimensions：20mm×12mm.

The design of the pallet power amplifier is based on the comprehensive consideration of different parameters, such as size, gain and power level. And the power stage is the core of the entire chain. First, the topology and realizing format of the power stage are determined between monolithic implementation and discrete GaN transistor. The latter one is chosen because it has the following advantages, such as reduced cost by small GaN area, good tuning flexibility and low network losses. And then, much research work has been done to integrate the biasing, matching and blocking structure. A simplified schematic of the final stage is shown in Figure 1.

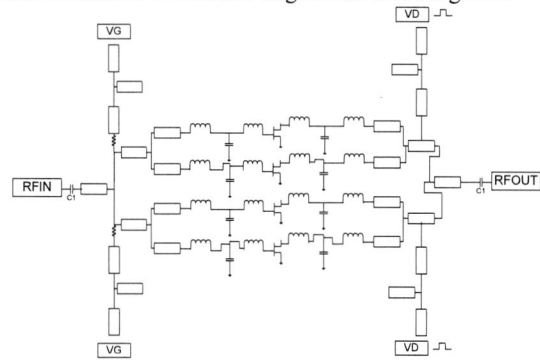

Figure 1 Simplified Schematic of the Power Stage

The overall size of the power stage is 13mm×10mm. The power divider and combiner are realized on 15mil thick Al_2O_3 ceramic substrate, and the shunt capacitors are realized on 5mil thick ε=40 board material. Besides ，the series inductances are realized by bonding wires which can have different shapes, heights and lengths.

Given the power density of 28V, the working voltage is 4W/mm. A total gate width of 33mm is enough to produce an output power greater than 130W, but we chose two 20mm gate width power bars for the final stage considering the loss of the combination network. Each power bar has 16 unit FET cells and a unit gate width of 125um. The power bar is manufactured under a standard and stable process.

As for the bias circuit, an innovative design method is used. The working principle is illustrated in Figure 2. Three quarter-wavelength transmission lines TL 1, TL 2 and TL3 are connected at Point B. At point A, the gate & drain voltages are feed，and Point C is connected to the power divider & combiner networks. TL 2 transmits RF open to RF short at point B, which 'shorts' TL 1, and then TL3 transmits RF short at Point B to RF open at Point C, which makes the bias network a RF high impedance loop which can be seen from Point C. The widths of TL 1, TL 2 and TL3 are optimized to give a best frequency response.

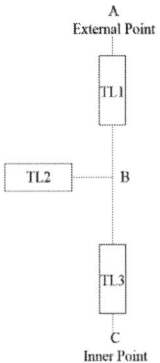

Figure 2 Simplified Schematic of the Bias Network

As for the DC blocking capacitor, 1pF single layer ceramic capacitor is used. This capacitor is home-made and tested one by one to ensure it has a breakdown voltage of greater than 300V and RF loss less than 0.2dB.

The simulation of the large signal model is done when all the circuits used in the power stage are modeled by an FEM simulator. And simulation results show that the power stage can have a transducer power gain of 10dB, an output saturate power of 52dBm and power added efficiency greater than 48% in Figure 3.

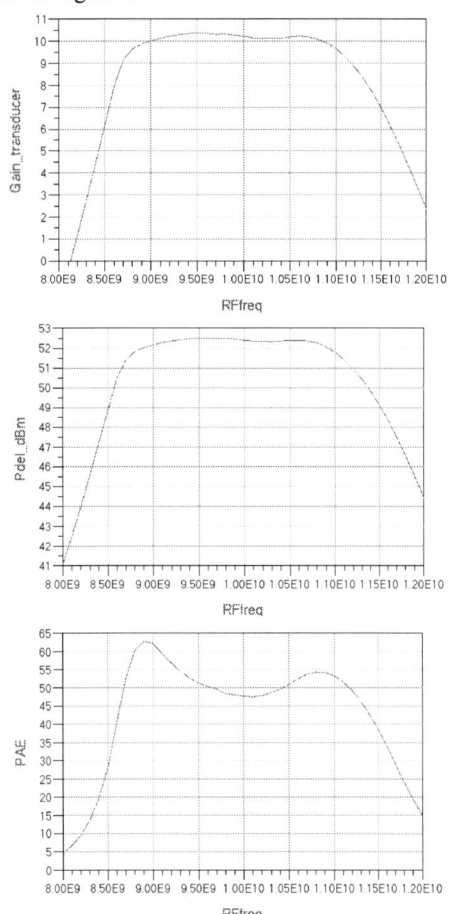

Figure 3 Simulation results of the power stage.

Based on the simulation performance of the power stage , we can draw the chain gain arrangement in Figure 4. An input power of 18 dBm is fed into the power amplifier chain , and then it is divided into two part by a Lange coupler. An insertion loss of 0.5dB is estimated for it. It is not a difficult task to find suitable driver MMIC amplifiers that can have an output power of 5 watts ,which can be combined in pairs to have an enough power to drive the power stage into saturation .

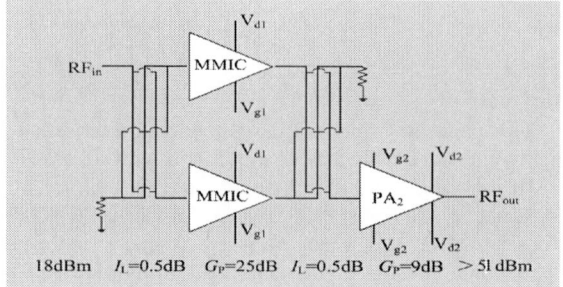

Figure 4 Chain Gain Arrangement.

As we all know the performance of an amplifier lies greatly on the output VSWR, an output mismatch performance simulation has been done to investigate the amplifier chain's anti-mismatch ability . With the criteria of output power greater than 130W, the power stage amplifier can withstand VSWR up to 1.5 in figure 5：1. At the same time, there is no significant degradation of efficiency （~5%）.The simulation results show the power stage amplifier have a good load mismatch compatibility.

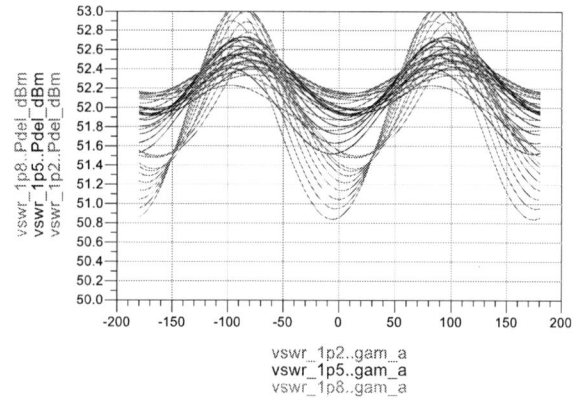

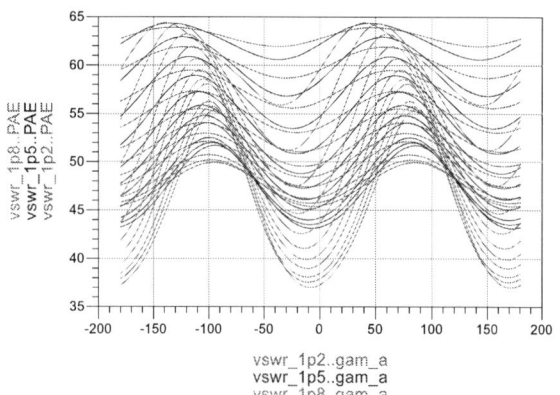

Figure 5 Simulation results of the amplifier's anti-output mismatch ability.

3. Measurement results

The assembled amplifier is shown in Figure 6. The various components are soldered on 40mil thick 20mm×12mm CuMo carrier plate and connected by gold wires with a diameter of 38um and gold ribbons with a width of 125um, respectively. The gate voltage of the driving stage and power stage is -1.8V and -2.2V, which are divided from the outer -5V by different resistance chains.

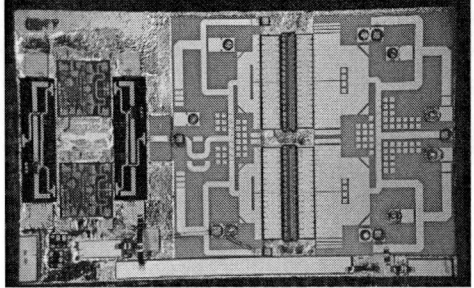

Figure 6 Photograph of the Pallet Power Amplifier.

Measured power data under required pulsed conditions is shown in Figure 7. As can be seen the output power exceeds 130W with a peak point of 162.1W in the entire bandwidth from 9.0GHz to 10GHz, and the efficiency is larger than 33%. Besides, the associated power gain is greater than 33.5dB. From the data listed above, we can draw the conclusion that the performance of the developed pallet power amplifier meets the design target.

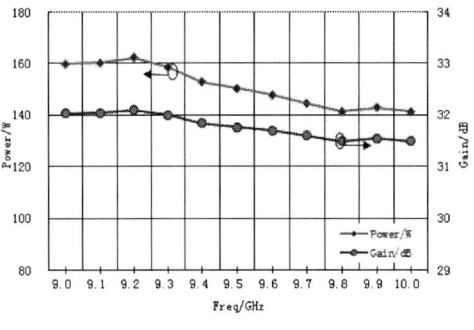

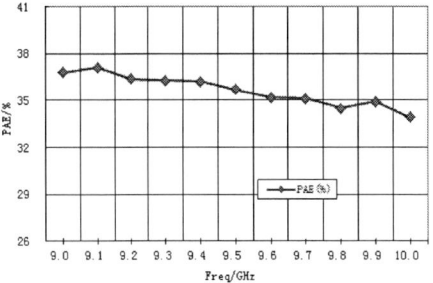

Figure 7 Measurement results of the Pallet Power Amplifier

Conclusions

Detailed design process and measurement of an X-band pallet power amplifier are given in this article. The amplifier consists of the Ga MMIC driving stage and discrete FETs power stage. An associated power gain of 33dB and hundred-watt level output power are obtained within a small area of just 20mm×12mm. Measurements confirm the topology a promising solution to integrate amplifiers for electronic systems.

References

1. Junhyung. J, "Wafer-Level-Packaged X -Band Internally Matched Power Amplifier Using Silicon Interposer Technology," *IEEE Microwave and Wireless Components Letters*, Vol. 29, No. 10 (2019), pp. 665-668.
2. Neil .C. *et al*, "Method of extraction of virtual X-parameters for a 500W internally matched device," *Wireless and Microwave Technology Conference (WAMICON),* Cocoa Beach, FL, USA,April. 2017.

Facet Formation of AlGaN/AlN-based Multiple Quantum Wells by Laser Scribing

Bin Xue[1], Jianchang Yan[1*], Yanan Guo[1], Chunyan Liu[1], Yiping Zeng[1,2], Junxi Wang[1,2] and Jinmin Li[1,2]

1. Institute of Semiconductors, Chinese Academy of Sciences
2. Youwill Hitech. Co. Ltd.
No.35A Qinghua East Road, Haidian District, Beijing, China
E-mail: yanjc@semi.ac.cn

Abstract

We proposed a laser-assisted method to obtain smooth Fabry-Perot cavity facet of AlGaN-based UV laser structure grown on *c*-plane sapphire. We demonstrated that laser scribing parameters has significant influence on the morphology of the cleavage plane. Smoothness and uniformity of the facet was improved by choosing appropriate cleaving direction, distance and depth. Therefore, laser scribing is an effective and controllable process to fabricate facets of AlGaN-based laser.

Introduction

AlGaN/AlN-based ultraviolet (UV) lasers are ideal candidates for next generation UV coherent light source owning to a list of advantages, such as low power consumption, compact size and eco-friendly nature, etc. [1]

One of the key steps of laser diode fabrication is to obtain high-quality facets[2]. Key parameters of a laser diode, such as threshold current density and external quantum efficiency are dependent on the facet quality. A tilt angel or roughness of the facet have substantial influence on the reflectivity. Therefore, it is essential to optimize the morphology and increase the smoothness.

For GaAs and InP-based laser diodes operating at red and infrared region, the heterostructure and substrate material have cubic zincblende structure with identical orientation. In this case, formation of high quality resonator facets through cleaving is simple. Unfortunately, for UV laser diodes based on group-III nitride, the material has wurtzite structure and it is impossible to cleave rectangularly shaped laser bars[3]. Moreover, group-III nitride epitaxial structure are commonly grown upon c-plane sapphire substrate due to cost and availability, and the existence of the 30° rotation between hexagonal nitride unit cell and unit cell of *c*-plane sapphire prevents a common cleavage plane[3, 4].

In order to circumvent the issue, different approaches have been presented to circumvent the issue[2, 5-12]. Dry etching is an alternative option to fabricate laser facets[2, 10] though ion bombardment damage during dry etching will increase the facet surface roughness[4]. Fortunately, it has been noticed that tetramethylammonium hydroxide (TMAH) and potassium hydroxide (KOH) based aqueous solution are useful in making highly anisotropic feature for c-direction wurtzite III-nitride semiconductors[8, 12-14]. Hence, surface morphology of the etched facet can be further improved through wet chemical etching. However, the combination of plasma etching and chemical wet etch increases the complexity of the fabrication process. Since laser scribing and dicing has been adopted to make nitride-based light emitting diodes grown on sapphire substrate. In this letter, we present a laser assisted method for obtaining smooth resonator facets of AlGaN/AlN-based laser structure grown on *c*-plane sapphire.

Experimental procedures

As illustrated in Fig.1, a typical 2-inch AlGaN multiple quantum wells (MQWs) laser structure grown on (0001) sapphire substrate by using metal-organic chemical vapor deposition (MOCVD) system was used in the experiment.

The AlN template and *n*-AlGaN has a thickness of 1500nm and 3000nm, respectively. AlN/AlGaN super-lattices (SLs) were placed between the template and *n*-AlGaN to block dislocation penetration. The MQWs consists of 5 pairs of 3nm AlGaN wells and 12nm barriers. After the growth process, the wafer was soaked in acetone and isopropanol solvent respectively for chemical surface cleaning.

The laser scribing and cleaving process for making Fabry-Perot cavity of the AlGaN-based UV laser used in this work is initially developed on nitride-based LED fabrication. Before scribing, wafer thinning process was performed and the wafer thickness was kept at 160nm to avoid cracks. A commercial scriber with nanosecond-pulsed 355nm laser was used in this work. During the fabrication process, backside of the wafer was scribed before the facets were manually cleaved by pressing a knife-edge against the sapphire substrate. In addition, the sample was cleaved along the *a*-direction in order to obtain *m*-plane facet before applying mechanical force on the sample to obtain laser bars. Scribing parameters were adjusted to optimize the facet smoothness and uniformity.

Fig.1 Schematic of AlGaN-based MQWs UV laser structure.

Results and discussions

Since laser ablation would cause material vaporization, redeposition and damage to the epitaxial surface, especially when the scanning speed of the laser is reduced at the end of the scribe lines, the backside of the wafer (sapphire substrate) was chosen to be laser-scribed to avoid the potential risks mentioned above. It needs to be pointed out that material redeposition caused by laser ablation may not be an issue for sapphire-based lasers, considering both metal pads are on the epitaxial side. While for laser diodes with vertical-injection structure, the redeposited materials would complicate the fabrication of metal contacts and heat sink soldering. This could be improved by chemical etching to remove the redeposited material.

Typical resonator facets, as obtained with non-optimized cleaving process along *a*-plane, cracks and terraces were observed and has profound impact on the surface smoothness. In some cases, the terraces could even reach up to the active layers, as illustrated in Fig.2. In such case, reflectivity of the facet will be decreased and therefore severely restricted the performance of the emitter. The large stress difference in the epitaxial layers and substrate are believed to cause the undesired facet feature. The suppression of the cracks and terraces would require strain engineering during epitaxial growth, while optimization of scribing and cleaving parameters could also prevent the formation of undesired morphology.

As illustrated in Fig.3, surface roughness of the laser facet is greatly improved by increasing the laser beam scanning speed from 15mm/s to 35mm/s along *m*-plane. Laser ablation length and distance between scribing line is kept at 3000um and 1000um, respectively. In this case, the laser ablation depth is around 40um and it can be seen that that the formation of cracks or grooves is initially absent at the beginning of the cleaving process. However, the cracks and grooves start to generate along the laser beam scanning line. In addition, the length of the cracks is further deepened with increasing cleavage distance. Fortunately, undesired extension of the cracks did not propagate further into the active region of the laser structure. The formation of the cracks and grooves could be attributed to the existence of the 30° rotation between hexagonal nitride unit cell and unit cell of *c*-plane sapphire prevents a common cleavage plane. It could be helpful to further increase the laser ablation depth by decreasing the laser beam scanning speed.

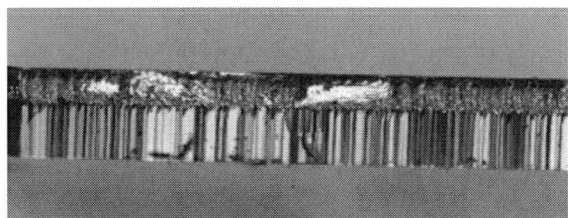

Fig.2 Initial facet morphology after laser scribing without optimization.

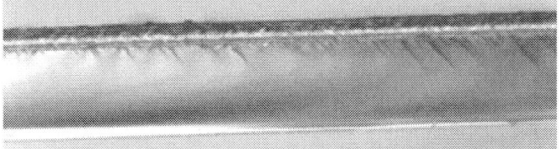

Fig.3 Cleavage plane after laser scribing with preliminary optimization.

Fig.4 Cleavage plane morphology after further optimization.

In order to prevent the formation of the cracks by increasing the laser ablation depth, scribing speed was decreased to 25mm/s. In addition, the distance between each scribe line is kept at 1mm. In this case, laser ablation depth is nearly 50um, as illustrated in Fig.4. It can be seen that with further adjustment of the scribing parameters, smoothness of the cleaved facet is greatly improved. A scanning electron microscope (SEM) micrograph of such facet is shown in Fig. 5 and it can be seen that the cleaving process successfully fabricated laser facets with mirror-like surface.

Fig.5 SEM image of the obtained facet after scribing optimization.

Therefore, it can be found that the laser ablation depth has to be chosen appropriately to minimize the formation of imperfections on the cleavage plane. It the depth was too small, the sapphire substrate would not break exactly along the scribe line and could lead to a high density of cracks and terraces. If the laser ablation depth is too deep, it may damage the AlGaN epitaxial layers. Moreover, the laser beam of the scriber ought to be carefully aligned in the *a*-plane of the sapphire substrate (which corresponds to the m-plane of the epitaxial layers) to avoid zigzag breaking of the cleavage plane. The optimum laser ablation depth was found to be about 50um in the 160um thick wafer..

Conclusions

Due to the misalignment of the sapphire and AlGaN crystal planes, laser assisted scribing and cleaving is not a viable solution unless scribing parameters are carefully designed. By adjusting wafer thinning, scribing direction, scribing depth and distance, facet quality of AlGaN-based UV laser was improved and smooth sidewall was obtained. This approach simplifies facet fabrication and improve acceptable production yield for group-III nitride-based laser structure grown on sapphire substrate. It is particularly interesting for facet fabrication of self-standing nitride-based laser diodes.

Acknowledgments

This work was supported by the National Key Research and Development Project of China, under grant 2017YFB0404202.

References

1. Lochner, Z., et al., *Deep-ultraviolet lasing at 243 nm from photo-pumped AlGaN/AlN heterostructure on AlN substrate.* Applied Physics Letters, 2013. **102**(10): p. 4.

2. Kneissl, M., et al., *Dry-etching and characterization of mirrors on III-nitride laser diodes from chemically assisted ion beam etching.* Journal of Crystal Growth, 1998. **189**: p. 846-849.

3. van Look, J.-R., et al., *Laser Scribing for Facet Fabrication of InGaN MQW Diode Lasers on Sapphire Substrates.* Ieee Photonics Technology Letters, 2010. **22**(6): p. 416-418.

4. Tian, Y.D., et al., *Stimulated emission at 272 nm from an AlxGa1-xN-based multiple-quantum-well laser with two-step etched facets.* Rsc Advances, 2016. **6**(55): p. 50245-50249.

5. van Look, J.R., et al., *Laser Scribing for Facet Fabrication of InGaN MQW Diode Lasers on Sapphire Substrates.* Ieee Photonics Technology Letters, 2010. **22**(6): p. 416-418.

6. Miller, M.A., et al., *Smooth and Vertical Facet Formation for AlGaN-Based Deep-UV Laser Diodes.* Journal of Electronic Materials, 2009. **38**(4): p. 533-537.

7. Itoh, M., et al., *Straight and smooth etching of GaN (1100) plane by combination of reactive ion etching and KOH wet etching techniques.* Japanese Journal of Applied Physics Part 1-Regular Papers Brief Communications & Review Papers, 2006. **45**(5A): p. 3988-3991.

8. Zhuang, D. and J.H. Edgar, *Wet etching of GaN, AlN, and SiC: a review.* Materials Science & Engineering R-Reports, 2005. **48**(1): p. 1-46.

9. Scherer, M., et al., *Characterization of etched facets for GaN-based lasers.* Journal of Crystal Growth, 2001. **230**(3-4): p. 554-557.

10. Chen, C.H.S., et al., *Vertical high quality mirrorlike facet of GaN-based device by reactive ion etching.* Japanese Journal of Applied Physics Part 1-Regular Papers Short Notes & Review Papers, 2001. **40**(4B): p. 2762-2764.

11. Lee, W.J., et al., *Facet formation of a GaN-based device using chemically assisted ion beam etching with a photoresist mask.* Journal of Vacuum Science & Technology a-Vacuum Surfaces and Films, 1999. **17**(4): p. 1230-1234.

12. Mileham, J.R., et al., *WET CHEMICAL ETCHING OF ALN.* Applied Physics Letters, 1995. **67**(8): p. 1119-1121.

13. Okumura, H., *Fabrication of an AlN ridge structure using inductively coupled Cl-2/BCl3 plasma and a TMAH solution.* Japanese Journal of Applied Physics, 2019. **58**(2): p. 4.

14. Cimalla, I., et al., *Wet chemical etching of AlN in KOH solution,* in *Physica Status Solidi C - Current Topics in Solid State Physics, Vol 3, No 6,* S. Hildebrandt and M. Stutzmann, Editors. 2006, Wiley-Vch, Inc: New York. p. 1767-1770.

Studies on primary lens for LED light source to enhance lateral emission intensity

Wenting Tang[1,2], Baojin Chen[3], Rui Zhang[3], Baoxing Wang[4], Yunfei Sun[3], Jiahua Min[1], Shuqi Li[4**], Yong Cai[2*]

[1]School of Materials Science and Engineering, Shanghai University, Shanghai 200072, China

[2] Key Laboratory of Nanodevices and Applications, Suzhou Institute of Nano-tech and Nano-Bionics, Chinese Academy of Sciences, Suzhou 215123, China

[3]Suzhou University of Science and Technology, Suzhou 215009, China

[4]Ningbo Skytorch optoelectronics Co.Ltd, Ningbo, China

* Email: ycai2008@sinano.ac.cn, Tel: +86-0512-62872826

** Email: sqli@skytorch.cn, Tel: +86-0574-63223036

Abstract

Typical LED light sources have lower light utilization efficiency (LUE) in reflective directional illumination applications due to light pattern mismatch. Light patterns are generally controlled by secondary optical systems, which are heavy and large. Compared with the secondary optical system, the primary lens is smaller in size, lighter in weight, and more efficient in light extraction. This work is a preliminary study on the primary lens of LED in order to improve the LUE of LED in reflective directional illumination fields. Experimental result shows that the light extraction efficiency (LEE) of LED light source packaged by a primary optical lens is improved by a maximum proportion of 16.3%. The LUE of the LED packaged with the primary lens is up to 40.1% at 60°. After encapsulating the primary lens, the weighted mean angle of light intensity in the full space is at 57° which is bigger than that of LED without primary lens (46°).

Introduction

The ordinary LEDs with Lambertian radiation characteristics can't be directly applied in many applications. Chen et al. [1] designed a high-efficient LED headlamp lens, which was composed of Fresnel collimating lens and splicing free-form surface. Qin et al. [2] proposed a free-form lens in LED backlighting system, which could control light emitting from LED to form a specified pattern. Wang et al. [3] designed a compact free-form lens that formed a rectangular light pattern for road lighting. According to the illumination applications, the luminaire's radiation patterns are diverse. The secondary optical systems, which include lenses, reflectors, etc., are usually applied to adjust the Lambertian patterns of ordinary LEDs to the one the luminaire requires. Secondary optical systems design is very complicated. As the light-emitting area of the chip increases, the volume of the secondary optical luminaire needs to be multiplied, thereby increasing weight and cost.

It is desirable to replace the secondary optical system by a primary optical lens that is integrated with LED light source. Because compared with the secondary optical systems, the primary optical lens not only is more compact, i.e. smaller and lighter, but also improves the LEE of the LED light source.

This study focuses on the primary optical lens design to achieve such an ideal emission pattern for reflective applications.

Background

In reflective applications, such as reflective flashlights, reflective searchlights, head-mounted miner's lamps, reflective headlights, etc., the Lambertian pattern which has a strong forward light intensity and weak lateral light intensity is not an ideal emission pattern, as shown in Fig.1. In those applications, the ideal emission pattern requires suppressed forward emission intensity and enhanced lateral emission intensity, with which more lateral lights are collimated to forward direction by reflectors, as a result the light utilization efficiency (LUE) is improved. Light utilization efficiency (LUE) is defined as the ratio of the light collimated to forward direction by reflector to the light emitting from the LED light source. As shown in Fig.2 (a), light emitted from LED source in range of α to 90° can be collimated to forward direction by reflection of the reflector and light emitted from LED source in range of 0° to α is free to transmit. In reflective directional illumination applications, only light in range of α to 90° is effective. As can be seen from Fig.2 (b), LUEs of Lambertian light sources are 76.2%, 51.7% and 26.3% at 30°, 45° and 60°, respectively. The purpose of this paper is to enhance the lateral emission intensity by designing a suitable primary lens, thereby improving the LUE of the LED light source.

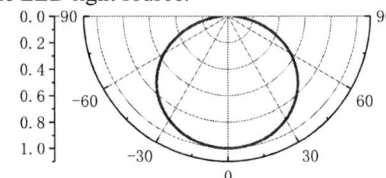

Fig.1. Lambertian pattern

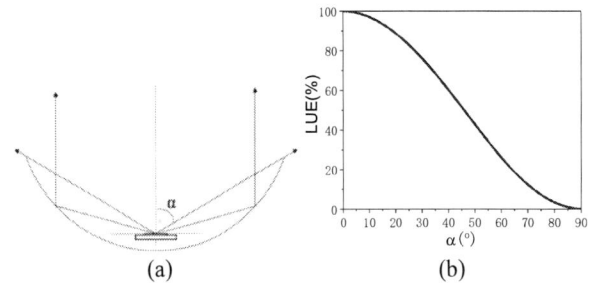

(a)   (b)

Fig.2. Light utilization diagram (a) and LUE of Lambertian light source (b)

Design of primary lens

Based on Snell' Law and Total internal reflection (TIR) principle, the designed primary lens consists of a top concave surface and a side curved surface. When building a lens model, primary optical lens is obtained by the Boolean operation of two balls, as shown in Fig.3(a). Six optical lenses were designed to reshape the light patterns. Three-dimensional shape and the structure parameters of lenses are given in Fig.3(b) and Table 1, respectively.

Table 1 Structure parameters of lenses

Sample Number	L1/mm	L2/mm	θ/°
S1	5	1.1	30
S2	5	1.1	45
S3	7	0.7	30
S4	7	0.7	45
S5	7	1.1	45
S6	9	0.7	45

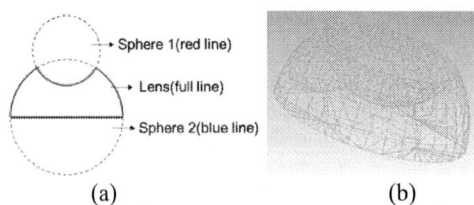

(a)   (b)

Fig.3. Boolean operation of two balls (a) and Three-dimensional lens (b)

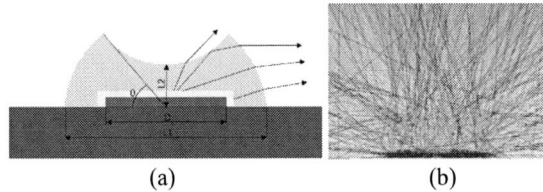

(a)   (b)

Fig.4. Geometry model (a) and Ray trace for an LED light source packaged with primary lens (b)

Table 2 The key parameters of the simulation models

Item	Specification
Chip size	3.3mm*3.3mm
Number of tracing rays	200 thousand
Radiation pattern	Uniform
Lens surface	Perfect transmission
Lens material	Silicone (n ~ 1.4)

The propagation of light rays in the lens is simulated by Ray-tracing software (Tracepro). The simulation model and light propagation in the lens are shown in the Fig.4(a). The simulation parameters of the geometry models are shown in Table 2. The diameter of the bottom surface of the lens is L1, the distance between the concave surface and the bottom surface is L2, θ is the angle between the bottom surface and the line connecting the intersection of two spherical surface and the center of the bottom surface, and D represents chip size. The ray trace for a LED light source packaged with a primary lens is shown in Fig.4(b). The light emits from the phosphor layer and propagates into the lens. The large-angle light is directly refracted by the lateral sphere surface. One part of small-angle light is refracted by the top sphere surface toward the sides. Another part of small-angle light is firstly reflected by the top sphere surface and then refracted by the lateral sphere surface to the sides. In this way，the lateral emission intensity

enhanced and LUE improved. The simulation results are shown with experiment results in "results and analysis" section.

Experiment

1) MI-LED

The blue LED chip used in this study is monolithically integrated light-emitting diode (MI-LED) with the dimensions of 3.3x3.3 mm². The MI-LED chip (named after its topological structure) is formed by interconnecting multiple small LED cells that are electrical isolated [4,5]. MI-LED chip features high current density and integration. The blue MI-LED chip was fabricated on the same LED epi-wafer with a peak wavelength of 450 nm. Other parameters of the MI-LED are shown in Table 3.

Table 3 Parameters of the MI-LED

Item	Specification
Material	GaN
Peak wavelength	450 nm
Dimensions	3.3×3.3 mm²
Power	10 W
Voltage	9 V
Drive current	1100 mA

2) White LED source

The blue MI-LED chip was mounted on a gold deposited copper-core PCB with silver-epoxy adhesive. After gold wire bonding, silicone resin mixed with a yellow phosphor crystal was dropped and flatten on the blue MI-LED chip in 90°C oven for 2h and 180°C oven for 1h, respectively.

3) Encapsulation of silicone lens

According to the simulation results, we fabricated a modular mold of lenses, as shown in Fig.5. E(i) in Fig.5(a) have the same dimensions as S(i) in Table 1, (i=1,2,3,4,5,6). As shown in Fig.6, the package process of the lenses mainly includes: a) applying some release agent to the mold and heating in 160°C hot plate for 3 min; b) injecting the silica gel (n~1.4) into the mold and white LED source after mold cooling; c) vacuuming to remove the bubbles; d) inverting white LED light sources on the mold; e) heating in 150°C hot plate for 10 min, curing the silicone and taking out the LED light sources with primary lenses after cooling. The completed samples after the package are shown in Fig.7.

To verify the optical properties of the LEDs packaged by primary lenses, luminous power and the light pattern were characterized using an integrating sphere and an optoelectronic test system, respectively. In the light pattern test, the distance between equivalent luminous surface and the measured target plane is 10 cm.

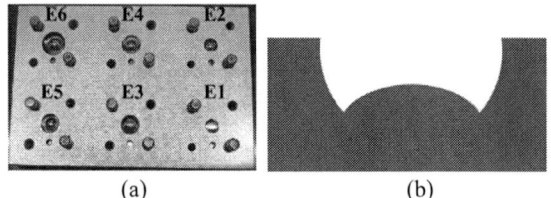

(a)   (b)

Fig.5. Top view (a) and side view (b) of the mold

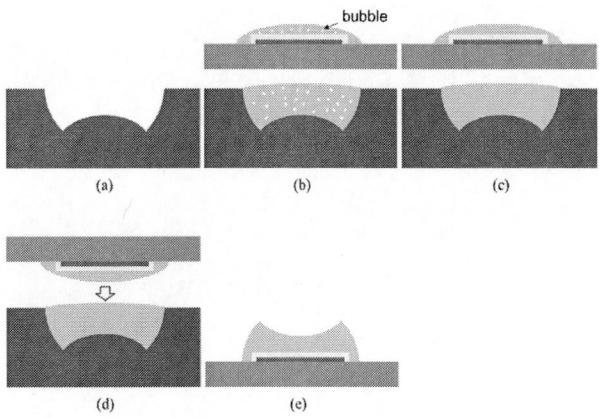

Fig.6. Package process

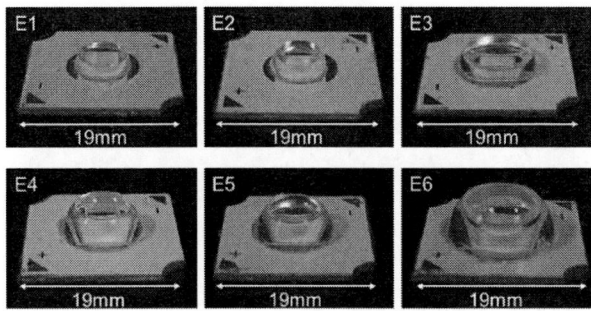

Fig.7. The prototype of LEDs

Results and Analysis

1) Light extraction efficiency (LEE)

The light extraction efficiency (LEE) is defined as the ratio light escaping from LED packaging module to the light emitting from the chips [6,7]. Due to the large refractive index difference between gallium nitride (n=2.35) and air(n=1), only a small fraction of the light emitted from gallium nitride can be incident into the air. In a general white LED source without primary lens, there is a layer mixed phosphor and silica gel between the gallium nitride chip and the air. Because the refractive index of silica gel is 1.4, more blue light can be extracted from the chip. But the interface between the phosphor layer and the air is a flat surface, much of the light is still reflected back into the chip due to the limitation of TIR principle. The designed primary lens can break the limits of TIR principle and improve LEE.

Table 4 LEE of LED sources

Sample Number	Initial luminous power (mW)	Cover lens Luminous power (mW)	Enhancement (%)
E1	2802	3089	10.2
E2	2752	3054	11.0
E3	2732	3051	11.7
E4	2795	3251	16.3
E5	2761	3207	16.1
E6	3329	3432	3.1

The measured light output powers of the light sources without/with primary lenses are listed in Table 4. It shows that luminous powers of all LED sources are improved after packaging the designed primary lenses. This indicates that the primary lens can improve the LEE. The maximum increase proportion 16.3% in E4 and the minimum increase proportion 3.1% in E6 indicate that the increase proportion of LEE is also related to the designed lens structure and its size.

2) Light pattern

To research the control function of the lens structure on light pattern, light intensity distribution of the LED is characterized by the photoelectric test system. As shown in Fig.8, imaginary black lines represent experimental normalized light patterns of general LEDs without primary lenses which are Lambertian patterns. Full red lines whose lateral light patterns are enhanced represent experimental normalized light patterns of LEDs with primary lenses. The simulated normalized light patterns are shown by imaginary blue lines. Although the full red lines do not perfectly match with the imaginary blue line, they show similar trends. There are some reasons for explaining the mismatch. a) The surface of the mold is not absolutely smooth. Therefore, there is a scattering effect on the surface of the lens; b) The number of tracing rays is not enough; c) The phosphor conversion mechanism between blue light and yellow light was not included in the simulation model. The light-emitting surfaces of blue light and yellow light are two independent surfaces. The actual situation is more complicated than the simulation.

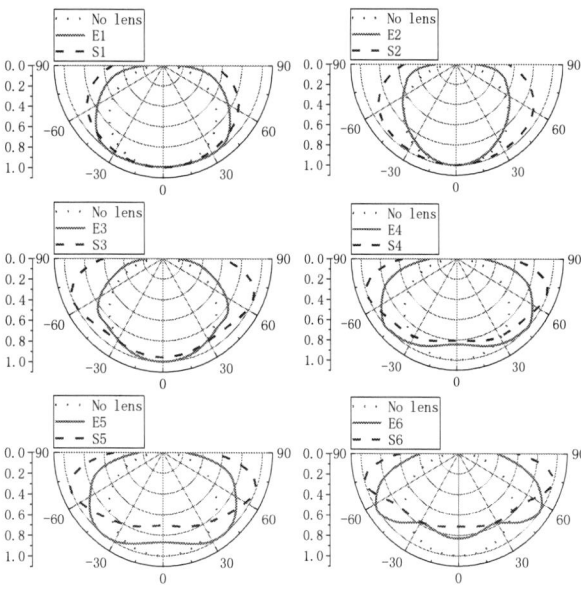

Fig.8. Normalized light patterns of simulation with lenses (imaginary blue lines), experiment with lenses (full red lines) and experiment without lenses (imaginary black lines)

In order to greatly quantify the effect of lateral emission intensity enhancement, the weighted mean angle of light intensity is defined as follows:

978-1-7281-5757-3/19 $31.00 © 2019 IEEE

$$M = \frac{\int_0^{90} \varphi 2\pi r^2 \sin\varphi P(\varphi)d\varphi}{\int_0^{90} 2\pi r^2 \sin\varphi P(\varphi)d\varphi},$$

where $P(\varphi)$ is light intensity, r is the distance between the receiving surface and the light emitting surface, and φ is the angle between the light and the optical axis.

The larger weighted mean angle is, the closer the light pattern is to the ideal light pattern in reflective applications. As shown in Table 5, sample E6 has the largest M ~57° in the designed lens. Compared to the E6, the E4 is smaller in volume and has a slightly smaller M ~54°. In the application, the advantage of E4 will be even greater.

Table 5 Center of gravity of light intensity

	No lens	E1	E2	E3	E4	E5	E6
M/°	46	51	52	51	54	53	57

3)   Light utilization efficiency (LUE)

The LUE as a very important performance index in reflective collimation application is shown in Fig.9 and Table 6. The imaginary black lines in Fig. 9 represent the LUE of the general LED without the primary lens, and the full red lines represent the LUEs of the LED light sources with the primary lenses. At the same angle, the full red lines are higher than the imaginary black lines, indicating that the designed primary lenses are advantageous for improving the LUEs of the LEDs. Comparing the LUEs in Table 6, it can be seen that the E6 with the largest volume has the highest LUE at the same angle. The smaller E4 has a slightly lower LUE than E6. To achieve the same LUE as the LED light source with a primary lens, the LED light source controlled by a secondary optical system requires a smaller angle α, which means that the volume and cost of the secondary optical system increase.

Table 6 LUE of LED sources

	30°	45°	60°
No lens	76.2%	51.7%	26.3%
E1	81.3%	59.7%	35.3%
E2	80.2%	59.9%	37.3%
E3	80.4%	59.3%	33.7%
E4	84.5%	65.0%	40.1%
E5	83.6%	63.1%	38.5%
E6	86.6%	69.1%	43.6%

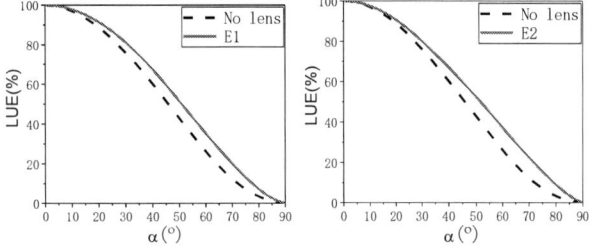

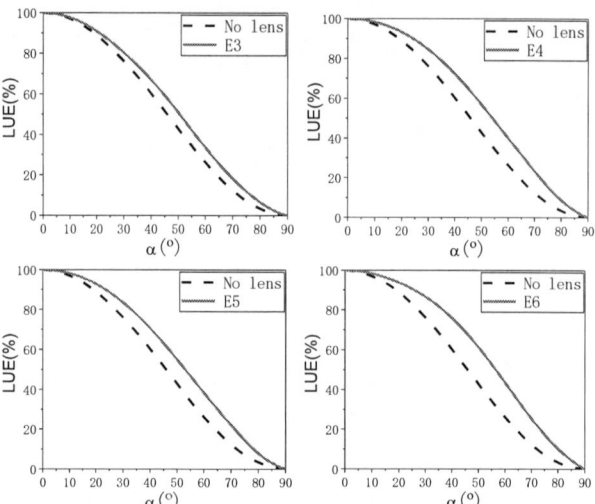

Fig.9. LUE of LED sources with lens (full red lines) and without lens (imaginary black lines)

Conclusions

Owing to the minimized package volume as well as lower cost, the designed primary optical lens shows a great potential in reflective lighting applications. This preliminary study also indicates that the primary lens can partly realize the functions of the secondary optical systems or ease the designs of secondary optical systems.

Acknowledgments

The authors would like to thanks for the support and help of Nanofabrication Facility, Nano-X, Suzhou Institute of Nano-Tech and Nano-Bionics (SINANO), Chinese Academy of Sciences (CAS).

References

1. F. Chen, K. Wang, Z. Qin, D. Wu, X. Luo, and S. Liu, "Design method of high-efficient LED headlamp lens," Opt. Express, vol. 18, no. 20, pp. 20926–20938, 2010.
2. Z. Qin, K. Wang, and S. Liu, "Energy-saving bottom-lit LED backlight with angle-control freeform lens," in Renew. Energy Environ. Tech. Dig.,2011, pp. 1–3.
3. K. Wang, F. Chen, Z. Liu, X. Luo, and S. Liu, "Design of compact freeform lens for application specific light-emitting diode packaging," Opt. Express, vol. 18, no. 2, pp. 413–425, 2010
4. Zhang, Yibin , et al. "Analysis and Modeling of Thermal-Electric Coupling Effect of High-Power Monolithically Integrated Light-Emitting Diode." IEEE Transactions on Electron Devices 65.2(2018):564-571.
5. Ding, Mingdi , et al. "High-power single-chip GaN-based white LED with 3058 lm." Electronics Letters (2016).
6. X. Luo, R. Hu, S. Liu, and K. Wang, "Heat and fluid flow in high-power LED packaging and applications," Prog. Energy Combustion Sci.,vol. 56, pp. 1–32, Sep. 2016.
7. Nian, Laixia, et al. "Review of Optical Designs for Light-Emitting Diodes Packaging." IEEE Transactions on Components Packaging and Manufacturing Technology (2019):1-1.

High color rendering index white LEDs fabricated using InP/ZnS green-emitting quantum dots and InP/ZnSe/ZnS red-emitting quantum dots

Doudou Zhang[1,2], Yuxian Yan[1,2], Fan Cao[2,5], Gongli Lin[4], Xuyong Yang[2], Wanwan Li[4], Luqiao Yin[2,3,*], Jianhua Zhang[2]

[1]School of Materials Science and Engineering, Shanghai University, Shanghai 200072, China
[2]Key Laboratory of Advanced Display and System Applications, Shanghai University, Ministry of Education, Shanghai, 200072, China
[3]School of Mechatronics and Automation, Shanghai University, Shanghai, 200072, China
[4]State Key Lab of Metal Matrix Composites, School of Materials Science and Engineering, Shanghai Jiao Tong University, 800 Dongchuan Road, Shanghai 200240, P. R. China
[5]School of Mechatronic Engineering and Automation, Shanghai University, Shanghai, 200072, China

Abstract

Quantum dots (QDs) have the advantages of spectrally adjustable, high quantum yield and narrow half-peak width, and are the most potential light-converting materials for white light that achieve high color rendering index and high saturation. The InGaN blue light chip currently used excites Ce-doped yttrium aluminum garnet (YAG: Ce^{3+}) yellow phosphor, and the white light-emitting diode (WLED) realized exhibits color rendering index (CRI) <80 and an unsatisfactory color coordinate because of lack of red spectral components. In order to improve the CRI of white LEDs, in this paper, we fabricated InP / ZnS green-emitting QDs and InP / ZnSe / ZnS red-emitting QDs, and excited YAG: Ce^{3+} yellow phosphor, InP / ZnS green-emitting QDs and InP / ZnSe / ZnS red-emitting QDs with blue chip , made WLED (Ⅰ), and compared with WLED (Ⅱ) generated by using blue LED chip to excite YAG: Ce^{3+} yellow phosphor. The result shows that the WLED (Ⅰ) achieves a CRI of 90.3 at a driving current of 5 mA. The Commission International de L'Eclairage (CIE) chromaticity coordinates are (0.3393, 0.3342), and the correlated color temperature (CCT) is 5186K, which is high quality white light close to sunlight.

1 Introduction

In the past few decades, white LEDs (WLEDs) have received great attention in the field of lighting due to their high efficiency and long life. For general illumination, WLEDs are usually formed in two ways. One is to use a blue InGaN LED chip to excite the yellow phosphor, and the blue and yellow light are mixed into white light; the other is a method of mixing various monochromatic lights, but the method of producing white light by various chips is complicated and the cost is too high to be suitable for making WLEDs. Therefore, the use of phosphor conversion materials to make white light is currently the most economical and effective method. [1] At present, the most widely used and well-developed conversion material is yttrium garnet yellow phosphor (YAG: Ce^{3+}). YAG: Ce^{3+} has many advantages, including high absorption blue light and extremely efficient emission processes with a quantum yield of about 90%. [2] The YAG: Ce^{3+} is excited by a blue InGaN LED chip, and the resulting white light spectrum usually has a color rendering index (CRI) of less than 80 due to the lack of a red spectrum component, and has no excellent color coordinates. One solution is to mix the red phosphor with the yellow phosphor to make WLEDs. However, because of the low efficiency of red phosphors, the development of high CRI WLEDs is limited. [3-5] The semiconductor quantum dots (QDs) have high Photoluminescence quantum yield (PLQY), wide absorption spectrum, size-adjustable emission, high resistance to photo-oxidation, low scattering effect and good color saturation. [6] Therefore, it is an effective solution to produce white light by replacing the phosphor with QDs material. CdSe QDs can emit a complete visible spectrum by adjusting the band gap. [1] In 2008, Wang et al. used a blue InGaN chip to excite three different sizes of CdSe/ZnS QDs to make white LEDs with a CRI of up to 76. [7] In 2011, Chandramohan, S and Ryu, Beo Deul et al. made hybrid WLEDs using core and core/shell CdSe NCs integrated on InGaN/GaN LEDs with a CIE chromaticity coordinates of (0.356, 0.330), CRI is 87.4. [8] In the same year, Wang, XB and Li, WW et al. synthesized CdSe/CdS/ZnS core/multi-shell QDs, which combined green, yellow and red emission QDs and epoxy composites with blue InGaN LEDs. The white LEDs produced has a CIE chromaticity coordinate of (0.35, 0.37) and a CRI of 88. [6]

Although cadmium (Cd)-containing QDs have excellent fluorescence characteristics and have good application prospects in white LEDs manufacturing. [6, 8-10] However, since the Cd element is toxic, the application is not environmentally friendly, so it is necessary to find a non-toxic environmental QDs that replaces the Cd element. It was found that WLEDs made by using InP QDs combined with phosphorescence can improve the defects of the red spectrum and improve the color rendering performance of WLEDs. [11-13] In this work, we synthesized InP/ZnS green-emitting QDs and InP/ZnSe/ZnS red-emitting QDs, and excited YAG: Ce^{3+}, InP/ZnS green-emitting QDs and InP/ZnSe/ZnS red-emitting QDs with blue LED chips，fabricated WLEDs with high CRI.

2 Experiment

2.1 Materials and equipment used in the experiment

In the experiment, blue LED chips, YAG: Ce^{3+}, silica gel (6550A and 6550B) and UV glue (U-613) were purchased from the market for LED packaging. The size and morphology of the QDs were measured using a field emission transmission electron microscope (JEM-2100F). The absorption characteristics and photoluminescence (PL) characteristics of the QDs were measured by a fluorescence spectrophotometer (HITACHI 4500) and a fluorescence spectrometer (U-3900H), respectively. The electroluminescent properties of the device were measured by an integrating sphere test system (HAAS-2000).

2.2 Synthesis of QDs

A Synthesis of InP/ZnS green-emitting QDs

Reagents used to synthesize InP/ZnS green-emitting QDs: Indium(III) chloride (99.999%), zinc(II) chloride ($\geq$ 98%), tris(diethylamino)phosphine (97%), zinc stearate (technical grade, 65%) were purchased from Sigma-Aldrich. sulfur powder were purchased from Strem Chemicals. Oleylamine (80~90%) was purchased from Acros Organics. Octadecene (technical 90%) was purchased from Alfa Aesar.

Reference synthesis scheme based on indium halide and aminophosphine precursors: 160 mg (0.45 mmol) of indium(III) chloride and 300 mg (2.2 mmol) of zinc(II) chloride were mixed in 5.0 mL (15 mmol) of technical oleylamine. The mixture was stirred and degassed at 120 °C for one hour and then heated to 180 °C under an inert atmosphere.A volume of 0.45 mL (1.6 mmol) of tris (diethylamino) phosphine (Phosphorus: Indium ratio = 3.6:1) was quickly injected into the above mixture. At 20 minutes, 1 mL of saturated TOP-S (2.2 M) was slowly injected. At 60 minutes: the temperature rose to 200 °C. At 120 minutes: 1 g of Zn(stearate)$_2$ was slowly injected into 4 mL of octadecene (ODE). The temperature rose to 220 °C. At 150 minutes: 0.7 mL stoichiometric TOP-S (2.2 M) was injected. The temperature rose to 240 °C. At 180 minutes: 0.5 g of Zn(stearate)2 was slowly injected into 2 mL of ODE. The temperature rose to 260 °C. At 210 minutes the reaction was over and the temperature was cooled. The InP / ZnS nanocrystals were then precipitated in ethanol and suspended in chloroform. The InP / ZnS QDs emission line width (fwhm) is 46nm-63 nm, and the PLQY is 20%-60%.[14]

B Synthesis of InP/ZnSe/ZnS red-emitting QDs

InP/ZnSe/ZnS red-emitting QDs were synthesized by a potentially efficient method previously reported. PLQY shows a maximum of 73% and has a narrow emission line width (up to 40 nm), wide spectral tunability and excellent stability. [15]

2.3 Production of WLED

WLED (Ⅰ) was fabricated using a 1*1 mm InGaN blue chip (peak wavelength of 450 nm). The WLED (Ⅰ) package is divided into three layers.

First layer: Weigh 0.5g 6550A and 0.5g 6550B for mixing, stir using a blender until the mixture is uniform, and weigh 0.2g YAG: Ce^{3+} and mix with the above colloid (YAG: Ce^{3+}: 6550A: 6550B=2:5:5), use a stirrer to stir a plurality of times until YAG: Ce^{3+} is uniformly dispersed in the silica gel, and the bubbles therein are removed. The obtained YAG: Ce^{3+} colloid was applied to the LED device in a spot coating manner, and the device was placed in an oven to carry out a curing process at 150 ° C for 5 min. The second layer: 500 ul of InP/ZnS green-emitting QDs solution having a concentration of 70 mg/ml and 2 g of UV glue were centrifugally stirred until uniformly mixed to form a QDs colloid. A small amount of InP/ZnS green-emitting QDs colloid was added over the cured yellow phosphor by spot coating, and then the second layer was cured by irradiation under an ultraviolet lamp. The third layer: 400 μl of InP/ZnSe/ZnS red-emitting QDs solution having a concentration of 30 mg/ml and 2 g of UV glue were

centrifugally stirred until uniformly mixed to form an InP/ZnSe/ZnS red-emitting QDs colloid. An appropriate amount of InP/ZnSe/ZnS red-emitting QDs colloid was applied over the InP/ZnS green-emitting QDs colloid by spot coating and cured by irradiation under an ultraviolet lamp. This completes the packaging of the WLED (Ⅰ).

A device having substantially the same performance as the above-mentioned blue LED chip was prepared, and a uniformly mixed YAG: Ce^{3+} colloid was applied and cured in an oven for 1 hour to form WLED (Ⅱ). Comparison test with WLED (Ⅰ).

3 Results and discussion

3.1 Morphology and size characterization of QDs

Fig. 1(a) shows a picture of InP/ZnS green-emitting QDs under transmission electron microscopy (TEM). It can be seen that the InP/ZnS green-emitting QDs are a multi-layered sheet-like particle structure with uniform dispersion. 100 QDs were randomly selected for size measurement, and the particle size distribution of InP/ZnS green-emitting QDs was obtained in fig. 1(b). It can be seen that the particle size distribution of the InP/ZnS green-emitting QDs is between 4.1 nm and 8.2 nm, and the average size is 6 nm. The inset of fig 1(a) is a high-resolution transmission electron microscopy (HRTEM) image of the InP/ZnS green-emitting QDs, and the lattice structure of InP/ZnS green-emitting QDs can be clearly seen.

Fig. 1(c) shows a TEM image of InP/ZnSe/ZnS red-emitting QDs. It can be seen that the InP/ZnSe/ZnS red-emitting QDs is a multi-layered sheet-like particle structure. 100 QDs were randomly selected for size measurement, and the particle size distribution of InP/ZnSe/ZnS red-emitting QDs was obtained as shown in fig. 1(d). It can be seen that the particle size distribution of the InP/ZnSe/ZnS red-emitting QDs is between 5nm and 15 nm, and the average size is 10.3 nm. The inset of fig 1(c) is an HRTEM image of the InP/ZnSe/ZnS red-emitting QDs, and the lattice structure of the red-emitting QDs can be clearly seen.

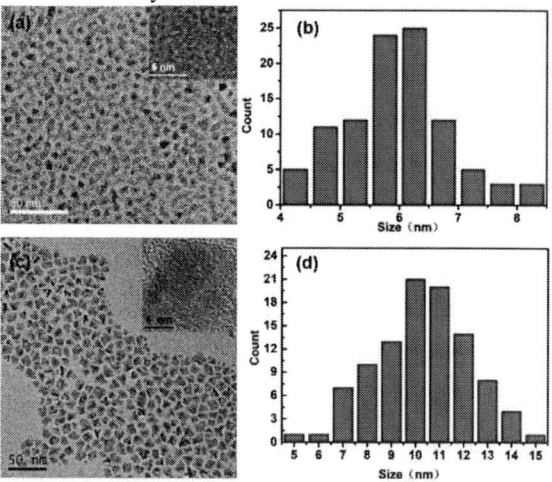

Fig. 1(a) TEM image of InP/ZnS green-emitting QDs, inset is HRTEM image (b) Size distribution of InP/ZnS green-emitting QDs (c) TEM image of InP/ZnSe/ZnS red-emitting QDs, illustration is HRTEM (d) Size distribution of InP/ZnSe/ZnS red-emitting QDs.

3.2 Photoluminescence properties of QDs

We also used a fluorescence spectrometer and a fluorescence spectrophotometer to test the PL spectrum and ultraviolet-visible absorption spectrum of InP/ZnS green-emitting QDs and InP/ZnSe/ZnS red-emitting QDs, respectively. Fig 2(a) and fig 2(b) are ultraviolet-visible absorption spectrum and PL spectrum of InP/ZnS green-emitting QDs /toluene solution and InP/ZnSe/ZnS red-emitting QDs /octane solution, respectively. Fig. 2(a) shows that the absorption spectrum of InP/ZnS green-emitting QDs has no obvious structure. It can be seen from the PL spectrum that InP/ZnS green-emitting QDs have the highest emission peaks at 467 nm and 520 nm, respectively. Fig. 2(b) shows that InP/ZnSe/ZnS red-emitting QDs have an absorption peak at 576 nm, and the emission peak is 606 nm.

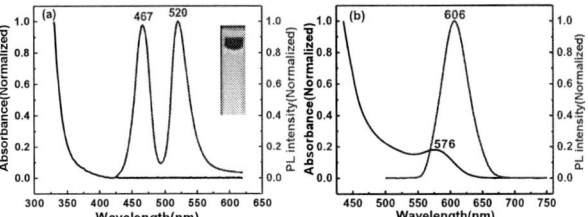

Fig 2 (a) UV-visible absorption / PL spectrum of InP / ZnS green-emitting QDs / toluene solution, inset of InP / ZnS green-emitting QDs / toluene solution under UV irradiation (b) UV-visible absorption / PL spectrum of / InP / ZnSe / ZnS red-emitting QDs octane solution.

3.3 Electroluminescent properties of QDs

We also tested the electroluminescence (EL) spectrum of the prepared WLED (I). As shown in fig. 3(a), at driving current of 5 mA, the WLED (I) with CRI of 90.3, CIE chromaticity coordinate of (0.3393, 0.3342) (fig. 3(b)) and CCT of 5186K is obtained. The illustration shows the luminescence of the WLED (I) device at 5 mA drive current.

At the same time, we also tested the EL spectrum of the prepared WLED (II). As shown in fig. 3(c), at driving current of 5 mA, the WLED (II) with CRI of 76.8, CIE chromaticity coordinate of (0.3311, 0.3480) (fig. 3(d)) and CCT of 5561K is obtained. The illumination of WLED (II) device at 5 mA can be seen in the inset.

As shown in fig 4, the CRI of WLED (I) device and WLED (II) device at driving current of 5 mA to 100 mA are also measured. It can be seen that the CRI of WLED (I) devices is always above 89, which is much higher than WLED (II) devices.

We also made a related study on the luminous efficiency of WLED (I) devices with the trend of current. Fig 5 shows the trend of luminous efficiency of WLED (I) devices under the driving current of 5mA~100mA. The overall trend of luminous efficiency can be found as follows: The luminous efficiency increases with the increase of the driving current, and the increasing trend becomes slow. The maximum luminous efficiency reaches 19.19 lm/w under the driving current of 100 mA.

In summary, it can be obtained that the YAG: Ce³⁺ yellow phosphor is excited by using a blue chip, adding InP / ZnS green-emitting QDs and InP / ZnSe / ZnS red-emitting QDs can make WLEDs with higher CRI and better CIE color coordinates.

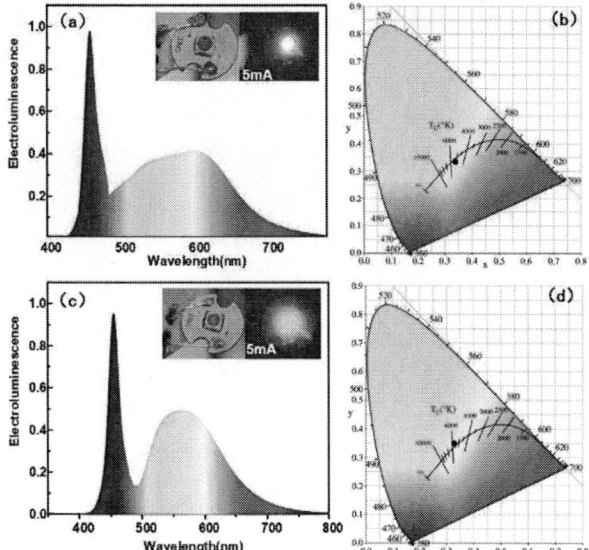

Fig 3 (a) The PL spectrum of the WLED (I) prepared, the illustration shows the state of the WLED (I) under natural light and the state of illumination in the dark. (b) shows the CIE color coordinates of the WLED (I) (c) the EL spectrum of the prepared WLED (II), the illustration shows the state of the WLED (II) under natural light and the state of illumination in the dark (d) showing the CIE color coordinates of the WLED (II).

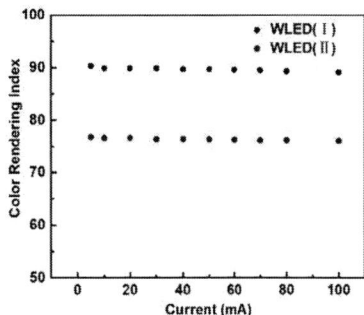

Fig 4 CRI comparison of the prepared WLED (I) device and WLED (II) device at 5mA~100mA drive current.

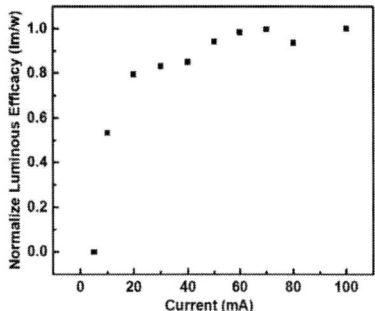

Fig 5 Trend of luminous efficiency of WLED (I) devices under 5mA~100mA driving current

4 Conclusions

In this paper, we made InP/ZnS green-emitting QDs and InP/ZnSe/ZnS red-emitting QDs, and combined with YAG:

Ce^{3+} yellow phosphors, we made high CRI WLED (Ⅰ). In comparison with WLED (Ⅱ) made using YAG: Ce^{3+} yellow phosphor, excellent luminescence properties were exhibited. Its CIE chromaticity coordinate, CRI and CCT are: (0.3393, 0.3342), 90.3, and 5186k, respectively.

5 Acknowledgments

This work was supported by the National Nature Science Foundation of China (NSFC) under the Grant Number (51605272) and Science and Technology Commission of Shanghai Municipality Program (19DZ2281000; 17DZ2281700).

References

1. W. Chung, K. Park, J. Y. Hong, J. Kim, B. H. Chun, and S. H. J. O. M. Kim, "White emission using mixtures of CdSe quantum dots and PMMA as a phosphor," vol. 32, no. 4, pp. 515-521, 2010.

2. K. A. Denault, A. A. Mikhailovsky, S. Brinkley, S. P. Denbaars, and R. J. J. o. M. C. C. Seshadri, "Improving color rendition in solid state white lighting through the use of quantum dots," vol. 1, no. 7, pp. 1461-1466, 2013.

3. E. F. Schubert and J. K. Kim, "Solid-State Light Sources Getting Smart," vol. 308, no. 5726, pp. 1274-1278, 2005.

4. E. F. Schubert, J. K. Kim, H. Luo, and J. Q. Xi, "Solid-state lighting—a benevolent technology," Reports on Progress in Physics, vol. 69, no. 12, pp. 3069-3099, 2006/11/02 2006.

5. Y. ZHU, B. XIE, and X. LUO, "Progress and expectation of quantum dots converted light emitting diode package," vol. 62, no. 0023-074X, p. 659, 2017.

6. X. Wang, W. Li, and K. Sun, "Stable efficient CdSe/CdS/ZnS core/multi-shell nanophosphors fabricated through a phosphine-free route for white light-emitting-diodes with high color rendering properties," Journal of Materials Chemistry, 10.1039/C1JM00061F vol. 21, no. 24, pp. 8558-8565, 2011.

7. H. Wang, K.-S. Lee, J.-H. Ryu, C.-H. Hong, and Y.-H. Cho, "White light emitting diodes realized by using an active packaging method with CdSe/ZnS quantum dots dispersed in photosensitive epoxy resins," Nanotechnology, vol. 19, no. 14, p. 145202, 2008/03/04 2008.

8. S. Chandramohan, B. D. Ryu, H. K. Kim, C.-H. Hong, and E.-K. Suh, "Trap-state-assisted white light emission from a CdSe nanocrystal integrated hybrid light-emitting diode," Optics Letters, vol. 36, no. 6, pp. 802-804, 2011/03/15 2011.

9. E. Jang, S. Jun, H. Jang, J. Lim, B. Kim, and Y. Kim, "White-Light-Emitting Diodes with Quantum Dot Color Converters for Display Backlights," vol. 22, no. 28, pp. 3076-3080, 2010.

10. S. Jun, J. Lee, and E. Jang, "Highly Luminescent and Photostable Quantum Dot–Silica Monolith and Its Application to Light-Emitting Diodes," ACS Nano, vol. 7, no. 2, pp. 1472-1477, 2013/02/26 2013.

11. S. Kim et al., "Highly Luminescent InP/GaP/ZnS Nanocrystals and Their Application to White Light-Emitting Diodes," Journal of the American Chemical Society, vol. 134, no. 8, pp. 3804-3809, 2012/02/29 2012.

12. W.-S. Song, S.-H. Lee, and H. Yang, "Fabrication of warm, high CRI white LED using non-cadmium quantum dots," Optical Materials Express, vol. 3, no. 9, pp. 1468-1473, 2013/09/01 2013.

13. J. Ziegler et al., "Silica-Coated InP/ZnS Nanocrystals as Converter Material in White LEDs," vol. 20, no. 21, pp. 4068-4073, 2008.

14. M. D. Tessier, D. Dupont, K. De Nolf, J. De Roo, and Z. Hens, "Economic and Size-Tunable Synthesis of InP/ZnE (E = S, Se) Colloidal Quantum Dots," Chemistry of Materials, vol. 27, no. 13, pp. 4893-4898, 2015/07/14 2015.

15. F. Cao, S. Wang, F. Wang, Q. Wu, D. Zhao, and X. Yang, "A Layer-by-Layer Growth Strategy for Large-Size InP/ZnSe/ZnS Core–Shell Quantum Dots Enabling High-Efficiency Light-Emitting Diodes," Chemistry of Materials, vol. 30, no. 21, pp. 8002-8007, 2018/11/13 2018.

Violet Chip Excited White LEDs for Sun-Like Lighting and Horticulture Lighting

Yue Zhuo, Hongyuan Zhu, Chongyu Shen, Guoxi Sun, Jay Guoxu Liu
ShineOn (Beijing) Technology Co. Ltd
58 Jinghai 5th Ave, Digital Factory,Building 3, 3rd floor, BDA, Beijing 10167

Abstract

With the latest progress in new generation of phosphor and packaging design, full spectrum, sun-like white LEDs become possible. We have developed such light sources by using a 410nm violet chip to excite a set of phosphors of blue, cyan, green and red, and then fine tune the spectrum power distribution (SPD). High color quality properties such as Ra of 98, Rg of 98 and Rf of 100 were achieved. With CCT of 5000K, it has an efficiency >105lm/W, a spectral continuity Cs >85 and low blue hazard ratio Br <40%. We also formulated spectra of LED by using a 410nm violet chip combined with a 450nm blue chip to excite cyan, green and red phosphors. High photosynthesis photo flux efficiency (PPE) of 2.35 umol/J was realized. These light sources are designed to mimic sun light to suit for both human centrical and horticulture lighting applications.

Keywords: White LEDs, full spectrum light, sun-like light, violet LED excitation, blue light hazard, horticulture lighting.

I. Introduction

Life on earth evolved in day-and-night cycles. Human, plants, and animals all have a biological clock that controls their circadian rhythms. Light is the strongest time cue to the circadian clock that keeps these rhythms. The rapid installation of artificial lightings such as LED is changing those rhythms, especially with high intensity blue wavelength that is mostly sensitive to human eye and circadian system[1]. Typical LED light is made with blue chip excited yellow phosphor to convert into white light. Therefore, LED lamp emits more blue light than a sunset or an incandescent lamp does. With more and more study showing the significant light impact on human life, higher quality, healthier and safer lighting are in demand. The concept of "lighting for health" or "human centric lighting" has emerged[2].

To make LED light more like a nature sunlight, the term "sun-like" or "full-spectrum" lighting is developed to imply that the spectral power distribution has some degree of uniformity throughout the visible spectrum and the violet or ultraviolet (UV) radiation. Their spectrum is continuous and includes most of the colors of the rainbow, much like a sunset. For a day light use, it would indicate the correlated color temperature (CCT) of 5000K or higher and a color rendering index (CRI) of 95 or higher.

A variety of benefits have been claimed for sun-like or full-spectrum lighting. In addition to help maintain the circadian rhythm, improve sleep quality and work efficiency, it provides high color rendering and is more natural to eye perception. With the reduced high energy blue light, it is safer to human eyes. In the meantime, the sun-like spectrum is also suitable for horticultural lighting, as it mimics the day light, and provides safe and pollution-free fruits and vegetables in agriculture facility. It fulfills the requirements of the light source for effective cultivation of all varieties of crops, flowers, and cannabis.

This paper attempted to use quantitative methods to describe color rendering index based on IES TM-30-15, spectral continuity Cs, blue light hazard ratio Br. These assessments provide a helpful method to guide the optimization of future sun-like light sources. The experiment will be centered on violet-excited white LEDs to achieve sun-like spectrum with target applications for human-centric lighting and horticultural lighting.

II. Design

Instead of using typical 450nm blue chip to excite the yellow phosphor, we took approach that uses 410nm violet chip to excite multiple phosphor mixture. The violet-excited full spectrum LED performs better in suppressing hazard blue light and complementing cyan. It can achieve a wide and continuous spectrum with violet spectrum that helps render more accurate and vivid color, especially on white objects. It also helps reduce the short-wavelength blue content while filling up the long-wavelength blue light. The short-wavelength blue between 415nm and 455nm has been observed to cause photochemical damage to retinal photoreceptors and pigment epithelium cells, leading to the retinal disorders such as age-related macular degeneration (AMD) and visual fatigue[3-4]. On the other hand, the long-wavelength blue light with a spectral range between 465nm and 495nm is most effective to affect circadian system. This wavelength of blue has an inhibitory effect on melatonin production, which keeps people awake, alert, and productive [5, 6]. The sun-like spectrum excited by violet light are engineered not only to reduce the hazardous blue light, but also to supplement the needed high wavelength blue, therefore help address the lack of daylight in indoor spaces.

From the perspective of horticultural lighting, the sun-like spectrum excited by violet light also has its advantages. Traditional horticultural illumination sources typically use 450 nm blue light and 660 nm red light as the overall spectrum power distribution, with the aim of matching the spectral response curves of chlorophyll A and chlorophyll B. The blue & red light are the strong absorption band. However, further study shows that these light spectra are not enough to enhance photosynthesis alone. Many secondary metabolites produced during the growth of plants can not solely rely on red and blue light to synthesize. The complete composition of plant nutrients, the appearance and taste may be affected [7]. According to McCree's research[8], different plants have different requirements for spectrum power distribution. One approach is to custom lighting spectrum for different plants. The other is to use the violet-excited full spectrum, which is very close to sunlight and suitable for majority plant's need.

978-1-7281-5757-3/19 $31.00 © 2019 IEEE

III. Quantitative Evaluation Methods

In the new color quality evaluation method IES TM-30 system [9], fidelity index Rf is used to characterize the degree of similarity of the Color Evaluation Samples (CES) under the illumination of the test source compared to the illumination under the reference source (sunlight in high CCT cases), 100 stands for the same, 0 stands for the biggest difference. This indicator is similar to the general color rendering index Ra when CIE evaluates CRI, but uses a new set of color samples and calculation method; It also introduces a color saturation Rg to characterize the saturation of each standard color compared to the reference source under the illumination of the test source, 100 means the same saturation, greater than 100 means that under this light source, the object looks more vivid and more vibrant. Below 100 means that the light source reduces the saturation of the color.

In order to quantify the fullness characteristics of a spectrum and to evaluate the ability of a light source to reveal color quality, we define spectral continuity Cs in the wavelength range of 400-700 nm, the difference between the reference source and the test source, which is then divided by the weighted value of the reference source's SPD. The value of spectral continuity Cs is expressed in percentage form [10], and defined as:

$$C_s = 1 - \frac{\left| \sum_{\lambda=400}^{700} (Y_R(\lambda) - Y_T(\lambda)) * \Delta\lambda \right|}{\sum_{\lambda=400}^{700} Y_R(\lambda) * \Delta\lambda} \times 100\%$$

Where

Cs is the spectral continuity;

$Y_T(\lambda)$ is the test source's SPD;

$Y_R(\lambda)$ is the reference source's SPD;

$\Delta\lambda$ is the bandwidth in nm

The greater the continuity Cs value is, the better fit it is to sunlight spectrum. The sunlight reference illuminant's continuity is 100%.

To quantify the impact of blue light on human eye, TÜV Rheinland has established a method to evaluate the blue light hazard ratio for a LED backlight TV as the ratio of light in the range from 415 nm - 455 nm compared to 400 nm - 500 nm. An eye safe TV is recommended with the ratio to be less than 50%. We define the blue light hazard ratio as Br by combining the methods used by IEC/EN 62471 and TÜV Rheinland, and the formula is as follows[11]:

$$B_r = \frac{\sum_{\lambda=415}^{455} Y_T(\lambda) * B(\lambda) * \Delta\lambda}{\sum_{\lambda=400}^{500} Y_T(\lambda) * \Delta\lambda} * 100\%$$

Where $Y_T(\lambda)$ is the spectral power distribution weighting function of a test source, $B(\lambda)$ is the blue light hazard weighting function, $\Delta\lambda$ is the bandwidth in nm.

With the above parameters established, the violet-excited white LED sun-like spectrum for general and healthy lighting can be evaluated.

For horticultural lighting, the parameters are different, and the complete spectrum power distribution, including violet light and far-red light is required [12]. There is no specific requirement for spectral continuity and blue light hazard concern. Instead, more attention is paid to photon flux efficiency (PFE) and wavelength specific for different plant applications. The photon flux efficiency (PFE) is defined by photon flux in the radiation range of 280 nm to 800 nm divided by the input power of the LED lamp, with the unit of micromolar per joule (μmol/J) [13].

IV. Experiment

The samples of LED are prepared by bonding a 410nm violet chip on to a 2.8mm x 3.5mm plastic lead-frame 2835 package with thermally conductive adhesive. The top electrode of chip is connected to the leadframe by a gold wire for electrical connection. Then a mixture of different phosphors and silicone gel is dispensed into the cavity of package to convert white light, followed with a curing process of the phosphor-silicone material. Design of experiment (DOE) was conducted to select suitable chip size that matches the phosphor mixing ratio of blue, cyan, green and red to optimize the targeted performance. Through repeated experiments, we have formulated violet-excited sun-like spectrum samples for CCT of 2700K, 5000K, and 6500K.

When preparing samples for horticulture lighting, we selected the leadframe 3030 package (3.0mm×3.0mm) for easy placement of dual-chip design. The 450nm blue chip and 410nm violet chip are connected in series. Blue chip has higher electro-optic conversion efficiency as well as higher phosphor excitation efficiency, while the violet chip is mainly used to supplement the violet spectrum power. A mixture of blue, green and red phosphors was selected and evaluated. By controlling the amount and types of phosphor and altering their adsorption, the violet peak can be adjusted to the right intensity. We also selected red phosphor that extend its peak to 660nm to achieve a balanced, sun-like full spectrum power distribution with high photon flux efficiency (PFE) at CCT of 3500K suitable for plant growth.

The photoluminescence measurements of above samples were carried in UV through VIS spectra region on the EVERFINE LED300E and EVERFINE HAAS-2000 instrument. The SPD, optical efficiency, and CRI of these samples were measured at driving power of 0.5W for 2835 packages and 0.2W for 3030 packages, respectively. The fidelity index (Rf), gamut index (Rg), the spectral continuity (Cs), blue hazard ratio (Br) and PFE were calculated accordingly from the SPD information.

V. Results and Discussion

After a series of experiment on the violet-excited 2835 package, two mixtures of phosphors with different emission wavelengths from different suppliers were settled. We denote them as Sample 1 and Sample 2. The design target for both samples is to optimize and balance various parameters including color rendering properties Ra, Rf, Rg, and luminous efficiency. Their spectral continuity Cs values are also calculated to evaluate how close their SPDs is to that of sunlight. Sample 2 is specifically formulated with eye safety in mind. The blue hazard ratio Br can be a good indicator for relative comparison of LED samples of similar power output after normalized their power intensity.

Table 1. Test parameters of light sources at 5000K CCT

Sample	Sample 1	Sample 2	Reference 1	Reference 2
Spectral				
Ra	98.9	97.8	97.5	100
Rf	98.1	95	94	100
Rg	100	99	101	100
Cs	88.7%	88.0%	80.6%	100%
Br	41%	39%	48.9%	37.80%
Eff (lm/W)	105	102	140	—

Table 1 shows the test parameters of the two different violet-excited white LED Sample 1 and Sample 2 along with blue-excited white LED sample (Reference 1) and sunlight (Reference 2). The comparison is made for the same CCT of 5000K using 0.5W 2835 packages. It was found that the violet-excited spectrum is better than the blue-excited spectrum in terms of Rf, Cs and Br. Sample 1 has Ra and Rf most close to the sunlight spectrum, and its Cs the highest among the three samples. Both violet-excited spectra met the targeted requirement for high CRI. Especially, the sample 1 achieved Ra close to 99, Rf above 98, indicating that the color of the object under illumination of this type of light source, can best rendering its true color. It is also confirmed that the better the continuity, the higher the Ra and Rf values. It is worthy noting that by varying the phosphor mixture, we can fine tune the spectrum band, especially between the 415-455nm, to further reduce blue hazard ratio. The Sample 2 is designed to minimize the short-wavelength blue light for healthy lighting application such as the class rooms and the elder centers. It has a Br value of only 39% well below the 50%-mark TUV recommended.

However, there is a downside for the violet-excited light. Its luminous efficiency is lower than that of blue-excited reference samples. It may be due to several reasons. The first is phosphor quantum loss, which is caused by limited quantum efficiency of a blue phosphor material in down-conversion emission. The second is the result of Stokes loss, which is caused by down conversion of the re-emitted photons excited by violet photons. Additionally, the quantum efficiency of the violet LED chip itself is slightly lower than that of blue LED chip. Further improvement of luminous efficiency of violet-excited samples is undertaken through improved blue phosphor material and packaging process that minimize the inter-absorption of different photons.

Table 2. Comparison the violet-excited samples under three different CCTs

Sample	2700K	5000K	6500K
Spectral			
Ra	97.2	98.9	97.8
Rf	94	98.1	97
Rg	102	100	102
Br	32%	41%	43.0%

Table 2 shows the key parameters of the violet-excited Sample 1 formulated for three different CCTs: 2700K, 5000K, and 6500K. As indicated in the table, all three CCT can achieve Ra > 97, with the blue hazard ratio Br the lowest at

2799K. It is clear that the blue hazard ratio reduces as the CCT decreases, due to the intensity of blue light drops.

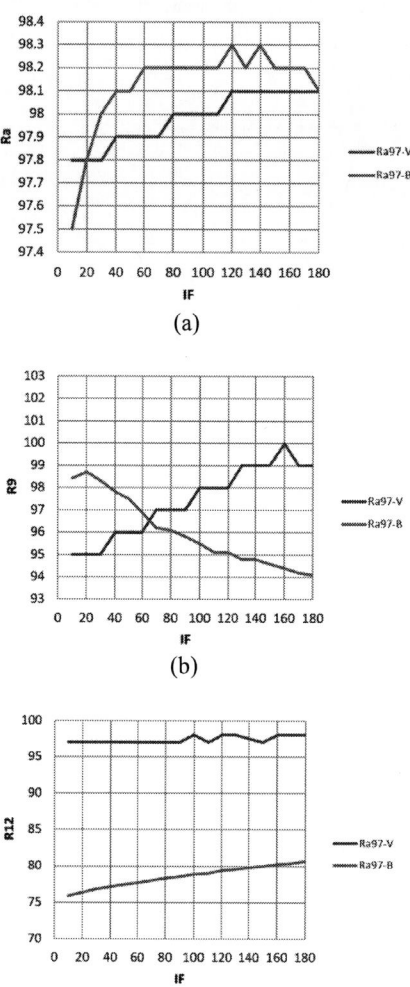

Fig. 1. Comparison of violet-excited and blue-excited samples for (a) Ra as a function of driving current, (b) R9 as a function of driving current, (c) R12 as a function of driving current.

In addition to Ra, R9 and R12 are most interesting color index as they represent for saturated red and saturated blue. Fig.1 shows compare these CRI property changes for violet-excited sample (Ra97-V) and blue-excited sample (Ra97-B) as a function of currents. The violet-excited samples didn't change much under current lower than 60 mA, while the Ra of the blue-excited sample drops rapidly as current reduces from 60mA to 10mA. CRI stability over driving current is important for warm dimming function of lamps. The violet-excited R9 increases with current, while blue-excited R9 decreases with current increase. The violet-excited R12 was also much higher than blue-excited R12, and does not change with the current.

As we have introduced some new phosphor materials in the study, the reliability of these samples needs to be tested and their stability for market application needs to be confirmed. The lumen maintenance under a high temperature operation life test at 105°C is shown in Fig. 2 for both Sample 1 and Sample 2 of Table 1. It can be found that Sample 1 is more stable than Sample 2. The difference is largely due to the different composition of phosphor from different vendor.

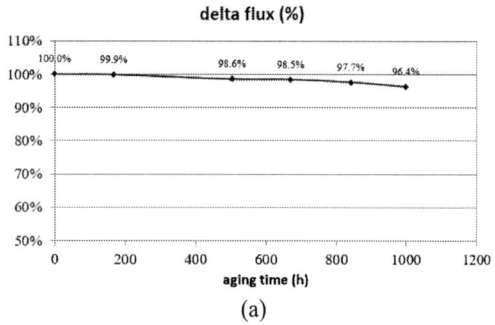

(a)

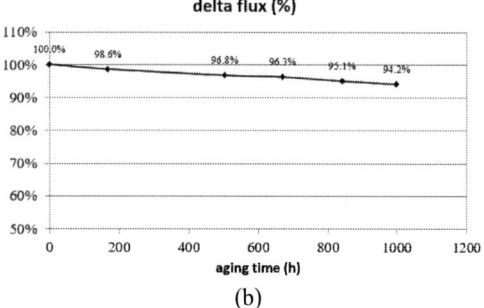

(b)

Fig. 2. Lumen maintenance of (a) Sample 1 and (b) Sample 2 under condition of Ts=105°C of high temperature life test.

In the study for horticulture lighting, blue and violet dual-chip excited sample is tested against single-violet excited sample. Both samples contain 380-410nm violet and 700-780nm far-red spectrum and can reach Ra greater than 95. The photo flux efficiency was measured for both samples at 0.2W and the results are shown in Table 3. The dual-chip excited sample has PFE > 2.35umol/J, while the single-violet excited sample can only reach 1.9umol/J under the same driving power. This is because the single-violet excited sample has the low electro-optical conversion efficiency of the violet chip and the low excitation efficiency of phosphor. The dual-chip excited improves the light extraction efficiency through a more efficient chip and phosphor combination.

Table 3. Comparison of PFE between dual-chip excited scheme and single-violet excited scheme

Scheme	Spectrum	PFE@0.2W(umol/J)
Single-violet excited		1.85
Dual-chip excited		2.35

The reliability of the dual-chip excited samples was also tested at Ts 105°C for high temperature operation life. The sample is very stable after 1000 hours of testing. Its reliability performance is close to those commonly used white LEDs installed in the horticulture market.

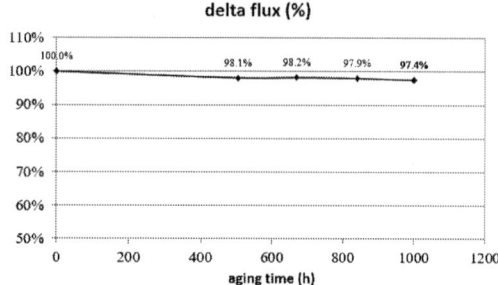

Fig. 3. The lumen maintenance of the dual-chip excited samples for horticulture lighting.

VI. Conclusions

This paper describes the different designs and advantages of violet-excited sun-like spectrum to meet healthy lighting and horticultural lighting application needs. White LED formulated by using a 410nm violet chip to excite a set of phosphors of blue, cyan, green and red demonstrated high color quality properties such as Ra of 99, Rg of 98 and Rf of 100 were achieved. With CCT of 5000K, it has an efficiency >105lm/W, a spectral continuity Cs >85 and low blue hazard ratio Br <40%, suitable for general lighting and education lighting where the color quality, eye safety, and human biorhythm are important. For horticultural lighting, a violet and blue light dual-chip excitation design showed CCT of 3500K and photon flux efficiency PFE > 2.35umol/J are achievable. The Ra>95 provides full spectrum power including violet and far-red for greenhouse and full artificial light plant factory application.

Acknowledgments

The authors would like to acknowledge the funding from The National Key Research and Development Program of China (No. 2017YFB0403902). Additionally, support from the colleagues in Shineon (Beijing) Technology Co., is also gratefully acknowledged.

References

1. Zielinska-Dabkowska, K M, "Make lighting healthier", Nature, Vol. 553, Januanury 18, 2018, 274-276.
2. Haitz R, Tsao J Y, "Solid- state lighting: 'The case' 10 years after and future prospects". *physica status solidi (a)*, 2011, 208(1): 17-29.
3. Kim H, Kim H, Jung C H, et al., "A Method for Reducing Blue Light Hazard from White Light-Emitting Diodes Using Colorimetric Characterization of the Display", *International Journal of Control and Automation*, 2015, 8(6): 9-18.
4. Sliney D H, Freasier B C, "Evaluation of optical radiation hazards", *Applied Optics*.1973; 12(1):1-24.

5. Hattar S, Lucas RJ, Mrosovsky N, et al., "Melanopsin and rod-cone photoreceptive systems account for all major accessory visual functions in mice", *Nature*. 2003 Jul 3; 424(6944):76-81. Epub 2003 Jun 15.

6. Berson D M, "Phototransduction in ganglion-cell photoreceptors", *Pflugers ArchEur J Physiol*. 2007; 454: 849-55.

7. Li Q, "Effects of quality on growth and phytochemicalaccumulation of lettuce and salvia miltiorrhiza bunce".Shanxi: Northwest A&F University 2010:6-12.

8. McCree K J, 1972. "The action spectrum, absorptance and quantum yield of photosynthesis in crop plants". *Agric. Meteorol*. 9, 191–216.

9. IES Technical Memorandum (TM) 30 30-15: IES Method for Evaluating Light Source Color Rendition.

10. Liu J G, Tang W, et al., "Quantitative Analysis of Full Spectrum LEDs for High Quality Lighting". *2018 15th China International Forum on Solid State Lighting: International Forum on Wide Bandgap Semiconductors China (SSLChina: IFWS)*.

11. Liu J G, Tang W, Shen C, "Blue light hazard optimization for high quality white LEDs". *IEEE Photonics Journal*, 2018,10(5).

12. Zou J, Zhang Y T, et al., "Morphological and physiological properties of indoor cultivated lettuce in response to additional far-red light". *Scientia Horticulturae* 257 (2019) 108725

13. Chinese Quality Certification Center (CQC), "Safety and performancer equirements for plant growth lighting p roducts". 2019-7-16.

High-efficiency GaN-based LED with patterned SiO₂ passivation layer and discontinuous current block layer

Jie Deng, Weiling Guo, Jianpeng Tai, Zehua Lu, Mengmei Li, Qinghua Yu

BeiJing University of technology

Beijing University of Technology, 100 Pingleyuan, Chaoyang District, Beijing

guoweiling@bjut.edu.cn

Abstract

GaN-based blue light-emitting diodes (LEDs) with patterned SiO₂ passivation layer and discontinuous current block layer (CBL) has been proposed and fabricated. The patterned SiO₂ passivation layer is inserted between Indium tin oxide（ITO）and P/N electrodes, and has a series of windows under the P and N electrodes, it can not only isolate the N electrode from the sidewall to prevent leakage, but also can re-contact the P electrode with ITO to form discontinuous P electrode structure and current injection. For the etching of active region, we use the method that partially etched to form a discontinuous N-GaN mesa, which can reduce the loss of the active area and improve the light emission intensity of the LEDs. It can be called discontinuous N-electrode structure. In addition, the discontinuous CBL deposited between P-GaN and ITO can increase the light extract efficiency from the active region under the P electrode compared to the conventional LED structure. As a result, at an injection current of 150mA, the light output power of the LED with patterned SiO₂ passivation layer and discontinuous CBL was 7.06% higher than that of conventional structure LED. What's more, at 50mA, the LED with the patterned SiO₂ passivation layer and discontinuous CBL structure shows the lower forward voltage of 3.29V compared with the conventional LED of 3.41V.

1. Introduction

In the passed few years, with the rapid development of semiconductor technology, LEDs, due to their outstanding advantages such as high efficiency, energy saving, environmental protection and long life, have gradually became the fourth generation of illumination sources to replace traditional light sources. they are widely used in general lighting, automotive lighting, and transportation lights and high resolution displays and other fields [1]. Improving the light extraction efficiency has always been the one of the most important path in improving efficiency of LEDs. As a result, some methods, such as optimizing electrode structure [2], photonic crystals [3], and patterned substrates [4], has been proposed and fabricated to improve the light extraction efficiency of LEDs and satisfy the demand of the entire market for LEDs [5].

Usually, the GaN-based lateral blue LED are grown on an insulated sapphire substrate, so the P\N electrode can only be on the same side. Since the sheet resistance of the indium antimony oxide (ITO) transparent electrode is much higher than that of the N-GaN layer, result in the P-electrode side current crowded [6] and reduce the luminous efficiency and reliability of LEDs, the traditional LED uses a current blocking layer to solve the problem, which is between the ITO transparent electrode and P-GaN layer. It can prevent the current from diffusing directly below the P electrode, and the current density distribution in the device is improved, thereby reducing the absorption of the light emitting by the metal electrode and alleviating the vicinity of the P-electrode current crowded. As a result, the light extraction efficiency of the LED can be increased. However, this will increase the operating voltage of the LED [7]. In addition, in order to deposit the N electrode, inductively coupled plasma (ICP) is used to etch the mesa to exposes the N-GaN layer of the LED, and there is a possibility of leakage current, thereby reducing the reliability of the LED [8], usually using SiO₂ passivation layer to solve it.

In this paper, we proposed the blue LED, which was partially etched the active region to increase the light-emitting area, as well as the discontinuous current blocking layer structure to reduce the forward voltage of the LED. And the SiO₂ passivation layer not only isolate the N electrode from the sidewall to prevent leakage, but also can re-contact the P electrode with ITO to form discontinuous P electrode structure and current injection.

Fabricated

The LED epitaxial wafer used in the experiment was grown by metal organic chemical vapor deposition (MOVCD) on a 430μm sapphire substrate. The epitaxial structure were consisted of a 15nm buffer layer, a 3.4μm intrinsic layer, a 2.3μm N-GaN layer, a 132nm InGaN/GaN layer, a 121nm MQW, a 18.5nm electron blocking layer, a 129nm P-GaN layer and so on.

In order to study the effects of partial etching the active region and discontinuous CBL structure on the optical and electrical properties of LEDs, two different structures of LED Ⅰ and LED Ⅱ were prepared. Among them, LED Ⅰ is a traditional structure LED, and LED Ⅱ has a partial etching the active region, a discontinuous CBL and patterned SiO₂ passivation structure. The detailed manufacturing process of LED Ⅱ is as follows: (1) The ICP partial etching was employed to define the mesa until the N-GaN layer is exposed to form a discontinuous mesa structure. (2) A 380nm SiO₂ current blocking layer, using plasma enhanced chemical vapor deposition (PECVD) was deposited on the P-GaN layer, then using photolithography and etching to form a discontinuous CBL structure. (3)An 110nm ITO was sputtered. (4) A 235nm SiO₂ passivation layer using PECVD was deposited on the ITO transparent electrode, then using photolithography and etching to form a lot of windows under the P and N electrodes. (5)Cr/Al/Ti/Pt/Ti/Pt/Ti/Pt/Au multi-layer metal stacks was deposited as P/N metal electrodes. (6) Finally, the LEDs were grinded and polished, and the size of the device is 210*630μm².

We used optical microscope to obtain the image of the device. I-V characteristic of the device was measured by the

Keysight B1500A. LED 201 was used to measure the light output power and the light efficiency of the LED.

Test result and discussion

Figure.1 (a-c) show the top-view of the optical microscope of fabricated LED Ⅰ and LED Ⅱ and the schematic of the LED Ⅱ respectively, and each of them has four samples. Fig.1(a) is the optical microscope image of the LED Ⅰ with traditional structure. Fig.1(b) is the optical microscope image of the LED Ⅱ showing the discontinuous CBL between the P-GaN layer and ITO transparent electrode, and SiO₂ passivation window-layer is inserted between ITO transparent electrode and P/N electrode. Fig.1(c) is the schematic diagram of LED Ⅱ. In the LED Ⅱ, patterned SiO₂ passivation window-layer allows interconnection between ITO and P-electrodes or N-GaN and N-electrodes and it can also isolate the N electrode from the sidewall to prevent leakage. On the one hand, the active region which is partially etched can increase the lighting-emitting area，but this will lead to a smaller contact area between the N-electrode and N-GaN, resulting in a slight increase in forward voltage of the device，and on the other hand, discontinuous CBL structure can reduce the series resistance of the device and current-spreading path, thereby reducing the operating voltage of the LED Ⅱ.

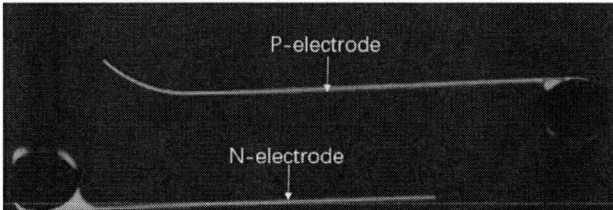

(a) The Top-view of the optical microscope diagram of LED Ⅰ

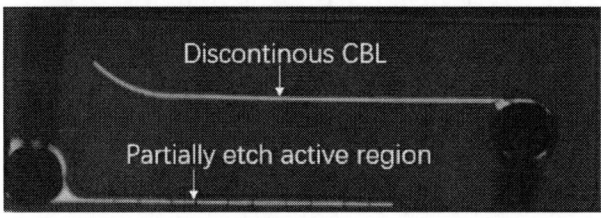

(b) The Top-view of the optical microscope diagram of LED Ⅱ

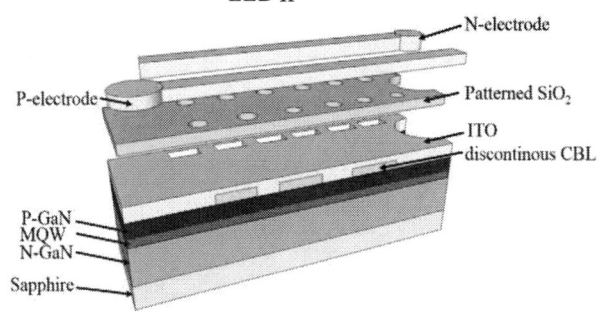

(c) Schematic diagram of LED Ⅱ.

Figure 1 Top-view of the optical microscope of (a) LED Ⅰ. (b) LED Ⅱ (c) Schematic diagram of LED Ⅱ.

In figure 2(a), we have measured the *I-V* characteristics of both LED Ⅰ and LED Ⅱ, each of them has four samples at different currents. At 50 mA, the forward voltages of LED Ⅰ and LED Ⅱ were respectively 3.41 V and 3.29 V. The forward voltage of LED Ⅱ were lower than that of LED Ⅰ, on the one hand, the patterned SiO₂ passivation window-layer and the active region which is partially etched will increase the series resistance of the device and cause a slightly higher forward voltage, on the other hand, the discontinuous CBL structure will shorten the current-spreading path and decrease the forward voltage of the device. The test result shows that the absolute value of the voltage reduced by discontinuous CBL is greater than the voltage increased by the patterned SiO₂ passivation window-layer, so the forward voltage of LED Ⅱ are lower.

Figure 2(b) shows the light output power of LED Ⅰ and LED Ⅱ. LED Ⅰ is less different from LED Ⅱ when the current is small, and the light output power of the two is gradually increased when the current increases. At 150mA, the light output powers of LED Ⅰ and LED Ⅱ is 155mW and 166mW respectively. The average light output power of LED Ⅱ is 7.06% higher than that of LED Ⅰ, the reason is that the area of the active region is increased due to the partially etching the active region, so that the light-emitting region of the LED Ⅱ is larger, what's more, since there is no current blocking layer under the SiO₂ passivation window-layer near P electrode of the LED Ⅱ compared with the LED Ⅰ, the injection current into the active region through the metal electrode is increased, especially in a big current injection.

Figure 2(c) shows light efficiency of LED Ⅰ and LED Ⅱ. At 50 mA of injection current, the light efficiency of LED Ⅰ and LED Ⅱ is 11.78 lm/W and 12.24 lm/W respectively, and at 150 mA of injection current, the light efficiency of LED Ⅰ and LED Ⅱ is 5.82 lm/W and 6.09lm/W respectively. As a result, the light efficiency of LED Ⅱ is higher than that of LED Ⅰ due to the increase of the light-emitting area of the active region and the decrease of the voltage of the discontinuous CBL structure.

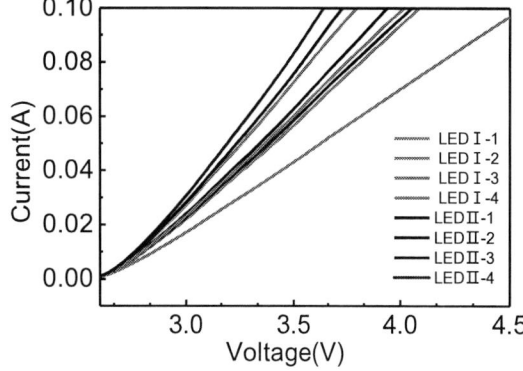

(a) *I-V* characteristics of LED Ⅰ and LED Ⅱ

978-1-7281-5757-3/19 $31.00 © 2019 IEEE

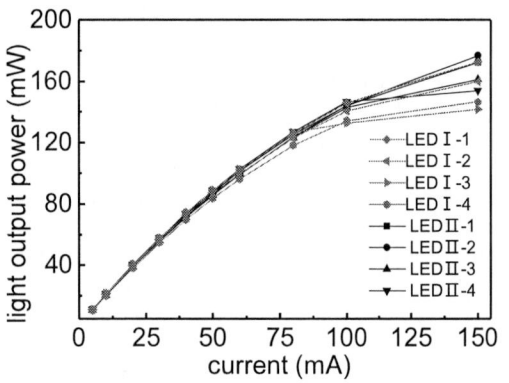

(b) Comparison of light output power of LED I and LED II

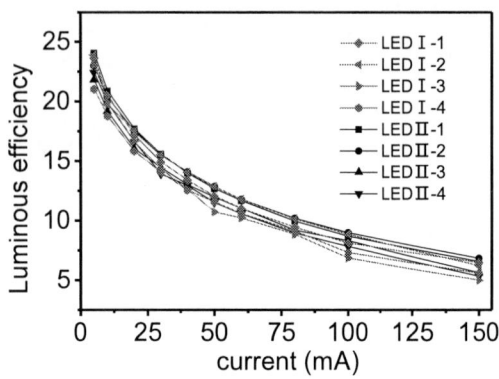

(c) Luminous efficiency of LED I and LED II

Figure 2 (a) *I-V* characteristics of LED I and LED II.

(b) Comparison of light output power of LED I and LED II.

(c) Luminance efficiency of LED I and LED II

Conclusions

In conclusion, we investigated the patterned SiO_2 passivation window-layer and discontinuous CBL structure on the electrical and optical properties of blue LEDs. The active region which is partially etched has increase the light-emitting area and discontinuous CBL structure has decreased the forward voltage of the device. The patterned SiO_2 passivation layer can not only isolate the N electrode from the sidewall to prevent leakage, but also can re-contact the P electrode with ITO to form discontinuous P electrode structure and the current injection of the metal electrode is enhanced. As a result, at 50 mA, the LED with the patterned SiO_2 passivation layer and discontinuous CBL structure shows the lower forward voltage of 3.29 V compared to the conventional LED of 3.41 V. At 150mA, the light output power of the experimental LED is 166mW compared to the conventional LED of 155mW; what's more, the luminance efficiency is also higher than that of the conventional LED.

Acknowledgments

This work was supported by National key R&D program of China (Grant No.2017YFB0403100, 2017YFB04 03102)

References

1. Dupuis R D , Krames M R . "History, Development, and Applications of High-Brightness Visible Light-Emitting Diodes," *Journal of Lightwave Technology*, Vol.26,No.9 (2008), pp.1154-1171.

2. Lee Y C.,*et al*, "Experimental and Numerical Analysis of P-Electrode Patterns on the Lateral GaN-Based LEDs," *Journal of Lightwave Technology*, Vol.32, No.15(2014), pp.2643-2648.

3. Lin C H, *et al*, "Enhancement of InGaN-GaN indium-tin-oxide flip-chip light-emitting diodes with TiO_2-SiO_2 multilayer stack omnidirectional reflector," *IEEE Photonics Technology Letters*,Vol.18, No.19(2006), pp.2050-2052.

4. Tang Y.,*et al*, "Research on Luminous Efficiency Improvement of GaN-based High Power LED Chips," *The 7th China International Semiconductor Lighting Forum*, october.2010,pp.198-203.

5. Mo H P. "The National Strategy of Solid State Lighting Development in the United States and Its Enlightenment to China," *China Illuminating Engineering Journal*, Vol.23, No.04(2012), pp.7-17.

6. Liu Meng-Ling.,*et al*, "Effect of interdigitated SiO2 current blocking layer on external quantum efficiency of high power LED," *Chinese Journal of Luminescence*, Vol.38, No.06(2017), pp. 786-792.

7. Chen Jia-Cai., *et al,* "Study on Saturation Characteristics and Lifetime of Novel AlGaInP Light Emitting Diodes" *Chin. Phys. Soc*, Vol.63, No.3(2014), pp. 370-375.

8. Yang C M., *et al*, "Improvement in Electrical and Optical Performances of GaN-Based LED With Double Dielectric Stack Layer," *IEEE Electron Device Letters,* Vol.33, No.4(2014), pp. 565-566.

Thermal Simulations of a UV LED module with nanosilver sintered die attach process on graphene-coated copper substrates

Pan Liu[1], Yong Li[1], Xiaobin Jian[1], Chen Jing[1], Min Li[1], Shurong Ding[1#], Guoqi Zhang[2#]

[1]Fudan University, Handan Road 220, Yangpu District, Shanghai, China
[2]Delft University of Technology, Feldmannweg 17, 2628CT, Delft, the Netherlands
panliu@fudan.edu.cn
[#]These authors are both corresponding authors to this work.

Abstract

The industrial market has been growing for high-power ultraviolet (UV) LEDs in curing, water purifying, and other applications that required high light output. As a rule of thumb, LED lumen output usually drops 0.3-0.5% for each 1°C increase in temperature while operating within the typical working temperature range. Such requirements for high-power UV LED modules indicate that innovative materials and processes for module packaging are needed to reduce thermal conductivity and to ensure high reliability.

In this work, UVA (wavelengths between 365-405nm) LEDs were chosen, since the increasing usage in curing equipment for drying paints, adhesives, and other curable materials. In order to reduce the thermal conductivity of UVA LED packages, a high power UV Chip on Board (COB) based LED module was simulated. Compared with traditional modules using silver-filled adhesive and metal-core printed circuit boards(MCPCB) substrate, the novel high power UV COB based module was using nano-silver material for die attach process. Such silver sintering method is emerging for high power electronics applications. Such substrate is then sintered on copper heatsink which is covered by a thin layer of graphene, for better thermal management. Such novel structure was investigated using ABAQUS software and the Heat Transfer Module. The simulation is intended to first check how much thermal conductivity reduce caused by nanosilver sinter die attach process. Secondly, the simulation will also investigate the graphene influence on the copper heatsink. Results from the simulation will show the typical structural function of a 200W LED module with reduced thermal resistance up to 7%. The module temperature with edge-to-center temperature difference will also be checked as a standard to compare the heat dissipation capacity. Besides, the simulation will provide temperatures in various parts of the LED module in steady-state conditions, as an indication for reliability.

1. Introduction

Ultraviolet (UV) LEDs are electronic devices with light-emitting in the wavelength range of 200-300 nm (deep-ultraviolet) and 300-400 nm (near-ultraviolet) range. UVA LEDs, with the wavelength range of 315-400 nm, have been available since the late 1990s. [1] These LEDs have been traditionally used in applications such as counterfeit detection or validation. Since the past several years, the majority of UVA LED applications were extended to UV curing for both commercial and industrial materials such as adhesives, coatings, and inks. Considering the advantages of increased efficiency, lower cost, and system miniaturization, traditional mercury or fluorescent lamps are replacing by UVA LEDs.

However, UV LEDs are facing significant challenges in terms of thermal management, light output, and efficiency. Most UVA light output around 385nm wavelength is only up to 15% for efficiency. Therefore, one of the major challenges for UVA LED modules is thermal management and reliability. [2]

Similar to power electronics, UVA LED modules are with high power densities. This results in the transfer of power electronics packaging technologies to UV LED packaging applications. In order to optimize the thermal paths and improve the optical output power, new materials which are commonly used in power electronics are applied to UV LED modules, such as ceramic substrates such as aluminum nitride (AlN), silver sintering technology, and graphene-coated copper substrates. [2, 3] Therefore, the aim of this paper is to check through simulation how to improve thermal management of UVA LED module through advanced packaging materials/processes.

2. Module Design and Approach

In this work, a module layout as Figure 1 was designed. The module consists of 108 LED dies, with a wavelength of 380nm, provided by LUMILEDS (Luxeon UV FC Line). Such flip chip UVA LED was chosen to test silver sintering technology. Compared with other UVA LEDs, such layout eliminates wire bonds completely in the system. For better analyze the influence of sintered silver, solder (SAC305) was chosen and compared in the simulation work.

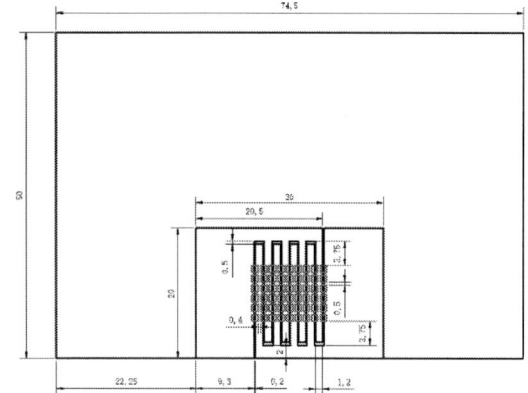

Fig.1 Top view of schematic drawing of half UV LED module (LED dies are marked in red.)

The AlN substrate was designed with dimensions of 30 mm by 40 mm, with the thickness of copper of 70 μm. The copper heatsink (74.5 mm by 100 mm) was coated with monolayer graphene. The detailed layout of this module was listed in Table 1. The chips can be sintered/soldered onto AlN substrate, while the AlN substrate can also be sintered/soldered onto the copper heatsink. To better analyze the graphene influence for the module in terms of heat transfer, copper heatsink without graphene coating was also compared.

Table 1 Components with the detailed layout and physical parameters

Component	Length (mm)	Width (mm)	Height (mm)	Conductivity (W/m/K)	Density (g/cm3)	Specific heat capacity (J/kg/K)
LED die	1	1	0.120	130	6.15	490
Solder on DPC	0.91	0.375	0.150	34	7.40	180
Silver on DPC	0.91	0.375	0.150	100	10.50	240
Cu of DPC	40	30	0.070	390	8.95	390
AlN of DPC	40	30	0.635	150	3.00	750
Solder on heatsink	40	30	0.150	34	7.40	180
Silver on heatsink	40	30	0.150	100	10.50	240
Graphene on heatsink	100	74.5	4.E-07	5300	7.7E-10	1500
Cu heatsink	100	74.5	38.500	390	8.95	390

To analyze the heat transfer in the UVA LED module, the heat sources must be identified in advance. The total heat is regarded as the sum of heat from the LED junction (UVA LED). The total generated heat is mainly dissipated in three ways: conduction, convection, and radiation. [4, 5]

Conduction of heat transfer through solid mediums can be described by Fourier's law, and calculated as:

$$\frac{\partial}{\partial x}\left(k\frac{\partial T}{\partial x}\right) + \frac{\partial}{\partial y}\left(k\frac{\partial T}{\partial y}\right) + \frac{\partial}{\partial z}\left(k\frac{\partial T}{\partial z}\right) + \dot{q} = \rho c_p \frac{\partial T}{\partial t}$$

where $\frac{\partial}{\partial x}\left(k\frac{\partial T}{\partial x}\right) + \frac{\partial}{\partial y}\left(k\frac{\partial T}{\partial y}\right) + \frac{\partial}{\partial z}\left(k\frac{\partial T}{\partial z}\right)$ is the net transfer of thermal energy into the control volume, $\dot{q}$ is the thermal energy generation, and $\rho c_p \frac{\partial T}{\partial t}$ is the change in thermal energy storage. k is thermal conductivity(which might be a function of temperature), ρ is density, c_p is heat capacity. [6]

For one-dimensional heat conduction, thermal resistance of solids is introduced. Therefore, heat transfer through conduction can be written as:

$$q_{cond} = \frac{\Delta T}{R}; R = \frac{d}{kA}$$

where A is the cross-section for the transferred amount of heat q_{cond}, d is the thickness, ΔT is the temperature difference. [7]

Convection of heat transfer can be regarded as:

$$q_{conv} = hA(T_s - T_\infty)$$

where h is convective heat transfer coefficient, A is surface area implied in the heat transfer process, T_s is the system temperature and T_∞ is the reference temperature. Parameter h is not a constant, but depends on the type of convection, surface structure and other geometrical parameters. [6, 8]

Natural convection is a rather complex mechanism. Thermal expansion causes density differences, and thus hot air rises driven by force of density differences. The velocity of air is usually dependent on the geometric environment.

Radiation of heat transfer between surfaces is usually explained as:

$$q_{rad} = \varepsilon\sigma AT^4$$

where σ is Boltzmann constant (5.6704×10-8 W/m2K4), A is the radiating surface area, T is the temperature, ε is the emissivity. [9]

The Optical energy loss is partially converted into heat by the efficiency of phosphor. The energy of lumen output was calculated based on the following equation:

$$\varepsilon_{e,white}\left[\frac{lm}{W}\right] = WPE\,(T,I) \cdot \varepsilon_{o,ph}\left[\frac{lm}{W}\right] \cdot \eta_{QD} \cdot \eta_{ph}(T) \cdot \eta_{pkg}$$

where $WPE\,(T,I)$ is the output optical power at junction temperature T and forward current I. It is also called wall-plug-efficiency which equals to internal quantum efficiency multiply extraction efficiency and then multiply electrical efficiency. $\varepsilon_{o,ph}$ is the luminous efficacy of phosphor; η_{QD} is quantum deficit in pumping phosphor; η_{ph} is phosphor quantum efficiency; and η_{pkg} is package efficiency. [10]

3. Simulation

After fully understand the generated heat source and heat dissipation methods, software ABAQUS® 6.14 was chosen as a finite element method (FEM) tool for simulation. The heat transfer module was used to calculate the temperature distribution.

In order to develop a UVA LED module, we have chosen UVA LEDs flip chip mounted on the ceramic metal substrate, then mounted on the copper heatsink. An equivalent geometry model was designed. Due to symmetries, we utilize only 1/2 of the real sample thus reducing the FEM calculation effort.

The 3D thermal model of the UVA LED package has been created and the partial front view of module with graphene clad layer and nano-silver material for die attach process is shown in Fig.2. Seven layers of different materials are characterized in the model from top to bottom. It is assumed that all the layers in the model are homogenous and the properties of the materials are isotropic. Interface heat resistance is neglected, since it is quite small compared with the whole thermal resistance of the whole module.

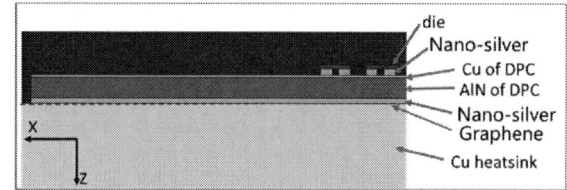

Fig. 2 FEM crosssection view (partial)

Different meshing size was applied considered the different scale of the components. The structured meshing technique is adopted to create a mesh for the UVA LED module, in which every domain is divided into a set of hexahedron elements. Different sizes of elements were chosen to save the time cost of simulation and ensure the accuracy. Fig.3 shows the meshed model of UV-LED module. As is shown, the meshed elements of dies regions are refined and the elements of Cu heatsink regions far from dies are coarse. A special meshed component was created to simulate the graphene coating layer, since the thickness of this layer is rather small compared with other components. The type of this single unit is DC3D8 (An 8-node line heat transfer brick). In total, 140637 single units generated for this special meshed layer.

The starting temperature, as well as the environment temperature was set to 298.15K (25° C). No heat was tranfered or gained through the symmetry face. Convection heat transfer with water is through the bottom surface of copper heasink. Different coefficient heat transfer with water was simulated. All the rest surfaces were set with radiation and convection heat transfer with air.

Fig. 3 Meshed view of half module

It is important to define the boundary conditions in thermal simulations. In this work, the boundary condition should provide sufficient cooling to make LED module properly transferred through heatsink. In this work, water cooling was expected for the lower surface of heatsink. Therefore, different heat transfer coefficient (h=0.8, 0.5, 0.3, 0.2)of water was chosen to check. (Fig. 4-7)

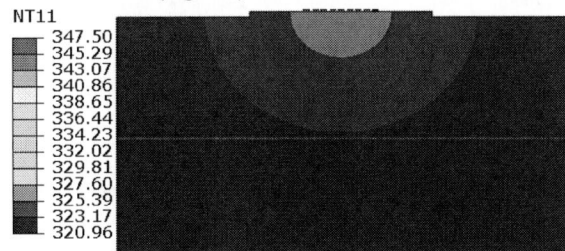

Fig. 4 Crosssection simulation results when h=0.8

Fig. 5 Crosssection simulation results when h=0.5

Fig. 5 Crosssection simulation results when h=0.3

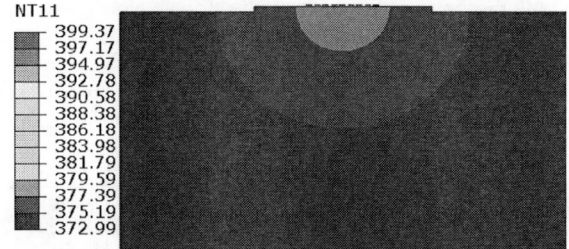

Fig. 5 Crosssection simulation results when h=0.2

The junction temperature was greatly influenced by the convection heat transfer coefficient with water. As calculated through the equation marked in Fig. 8. Taken into consideration of LED chip proper working temperature marked in datasheet, the junction temperature is preferably controlled under 120°C, thus h=0.3 was finally chosen as boundary condition in the following discussion.

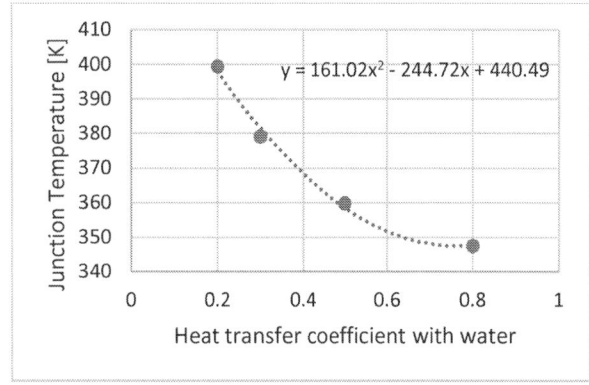

Fig 8. The relationship of heat transfer coeffieicnt with water with junction temperature

4. Result Discussion

When h=0.3, junction temperature of LED chip is around 371-379K. When juntion temperature is at room temperature (Tj=298.15K), LED die is powered by 500mA, 3.2V, while the radiometric power is 0.45W. As shown in the datasheet (Fig. 9), the higher the junctin temperature, the less optical power generated. When Tj=60, the optical power drops to 0.34W.

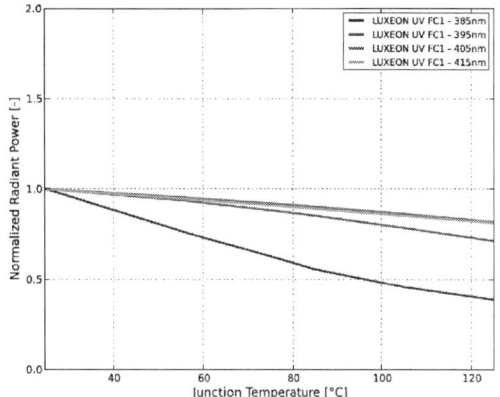

Fig. 9 Typical normalized radiant power vs junction temperature for 500mA [11]

978-1-7281-5757-3/19 $31.00 © 2019 IEEE

With this optical power output, the heat generated was then calculated. Based on the heat generated by each LED, the thermal behavior is then simulated. As seen from the crosssection view of temperature distribution, the highest temperature is the middle junction temperature of the LED die. The lowest temperature is the edge of the copper heatsink.

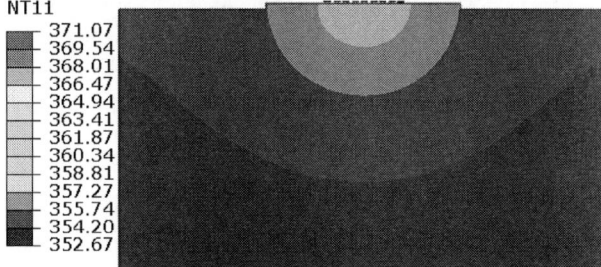

Fig. 10 Crosssection temperature distribution of sintered silver module with copper heat sink

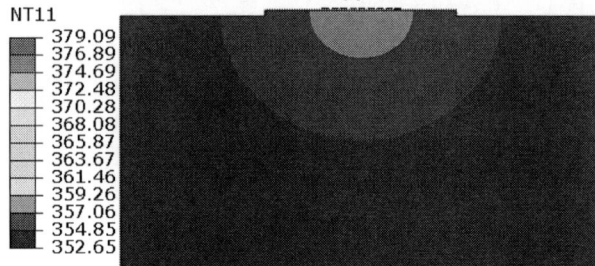

Fig. 11 Crosssection temperature distribution of solder module with copper heat sink

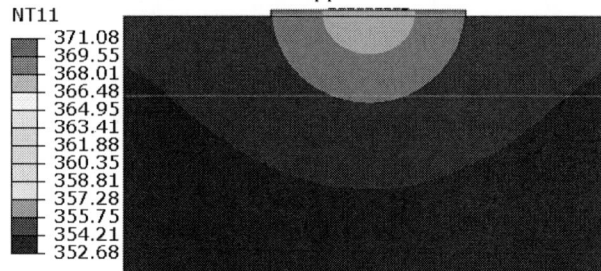

Fig. 12 Crosssection temperature distribution of sintered silver module with graphene-coated copper heat sink

Fig. 13 Crosssection temperature distribution of soldered module with graphene-coated copper heat sink

The thermal resistance R_{th} is then calculated through the equation:

$$R_{th} = \frac{T_j - T_{amb}}{P}$$

where P is the heat dissipated from the heat source, T_j is junction temperature, T_{amb} is the ambient temperature. Therefore, based on the results from Fig. 10-13, the results of junction temperature and thermal resistance were compared in Table 2.

Table 2 Junction temperature and thermal resistance comparison with different packaging process/materials

	Copper Heatsink	Graphene-coated copper heatsink
Junction temperature with Sintered silver process [K]	371.07	371.08
Junction temperature with soldering process [K]	379.09	377.84
Thermal resistance with sintered silver process[K/W]	0.72	0.72
Thermal resistance with soldering process[K/W]	0.78	0.77

It is clear through Fig. 10-13, the sintered silver enhanced heat dissipation of flip chip joined LEDs. Thermal resistance dropped 7% compared to the traditional soldering process using SAC 305. Moreover, the sintered layer can be adjusted to only several micros to enable lower shrinkage and a better adhesion for the chip joinings. In Table 3, the trend of silver sintering layer thickness with thermal resistance was listed. In this calculation, the thickness of silver sintering layer of LED chip to DPC was kept the same with the thicknes of this layer inbetween DPC to copper heatsink. It is clear from the table that thermal resistance drops futher when reducing the thickness of silver sintering layer. Through simulation results, it is clear that silver sintering technology is promising in UV LED packaging. However, the electromigration of silver and the voids percentage should be taken into account in real applications.

Table 3 Silver sintering layer thickness vs Junction temperature

Thickness of silver sintering layer [mm]	Junction temperature [K]	Thermal resistance [K/W]
0.15	371.07	0.72
0.12	370.33	0.72
0.10	369.78	0.72
0.08	369.23	0.72
0.04	368.19	0.71
0.02	367.72	0.71
0.01	367.52	0.71

As seen from Table 2, the influence of graphene can be neglected. A mono layer graphene brings only limited reduction to the whole thermal resistance. However, the process brings the extra cost up to a 20% increase in terms of the whole LED module packaging. Therefore, mono layer graphene coated heatsink is not recommended for such module design.

For better analyze the influence of graphene coating layer, another simulation was carried out. Graphene-coated AlN copper substrate was simulated for sintered silver and soldering technology.

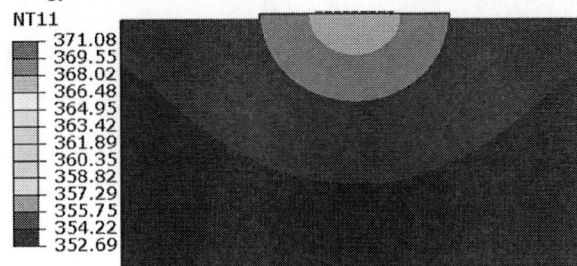

Fig. 14 Crosssection temperature distribution of sintered silver module with graphene-coated AlN copper substrate

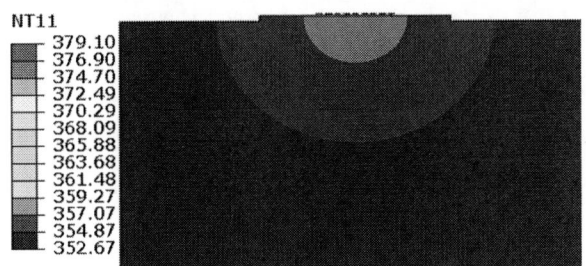

Fig. 15 Crosssection temperature distribution of soldered module with graphene-coated AlN copper substrate

It is clear from the results that graphene-coating greatly improved the thermal conductivity for soldering, but not as much as for silver sintering. This is due to the high thermal resistance of soldering is comparable high with graphene material. Therefore, it is clear that graphene as a high performance thermal interface material is more promising for modules using soldering technology.

5. Conclusions

An essential challenge in LED packaging is the thermal management, especially for UV LEDs due to the low optical power efficiency. The technology gap for UV LED manufactures is to develop proper process/material combination that enables heat to be conducted efficiently away to the environment. Similar to power electronics, UV LED packages are with high power densities. Therefore, in this work, power electronics packaging materials/processes were studied through simulation.

In this report, different heat transfer coefficient of the bottom surface of the heatsink was first calculated. Through this data, the junction temperature was calibrated, thus lead to the calculation of optical power. Based on the heat generated, thermal analysis of UVA LED module based on different packaging materials was conducted. It has been demonstrated that silver sintering process can significantly influence the junction temperature. Through thermal simulation results, it is clear that sintering process enables high thermal conductivity.

Thermal resistance dropped up to 7% compared with the traditional soldering process. However, the graphene-coated copper heatsink brings no clear benefits in the terms of thermal management. It is also checked through simulation, soldered module gains more benefits of graphene coating when coated on AlN copper substrate, instead of copper heatsink.

References

1. Ploch, N.L., et al., *Effective Thermal Management in Ultraviolet Light-Emitting Diodes With Micro-LED Arrays.* IEEE Transactions on Electron Devices, 2013. 60(2): p. 782-786.
2. Albar, I. and N. Donmezer. *Phonon Mean Free Path - Thermal Conductivity Relation in AlN.* in *2019 18th IEEE Intersociety Conference on Thermal and Thermomechanical Phenomena in Electronic Systems (ITherm).* 2019.
3. Broughton, J., et al., *Review of Thermal Packaging Technologies for Automotive Power Electronics for Traction Purposes.* Journal of Electronic Packaging, 2018. 140.
4. Jou, R.-Y., *Heat Transfer Measurements of Compact High Power LED Illumination Cooled by Different Fluids.* ASME Conference Proceedings, 2010. 2010(49163): p. 477-486.
5. Chang, C.-C., et al., *Novel heat dissipation design for light emitting diode applications.* Microsystem Technologies, 2010. 16: p. 519-526.
6. Chang, C.-C., et al., *Novel heat dissipation design for light emitting diode applications.* Microsystem Technologies, 2010. 16(4): p. 519-526.
7. Schubert, E.F., T. Gessmann, and J.K. Kim, *Light Emitting Diodes,* in *Kirk-Othmer Encyclopedia of Chemical Technology.* 2000, John Wiley & Sons, Inc.
8. Jou, R.-Y. *Heat Transfer Measurements of Compact High Power LED Illumination Cooled by Different Fluids.* in *ASME 2010 10th Biennial Conference on Engineering Systems Design and Analysis.* 2010.
9. Ye, H., et al. *Numerical modeling of thermal performance: Natural convection and radiation of solid state lighting.* in *Thermal, Mechanical and Multi-Physics Simulation and Experiments in Microelectronics and Microsystems (EuroSimE), 2011 12th International Conference on.* 2011.
10. Martin, P.S. *Light from Silicon Valley.* 2003.
11. *DS185 LUXEON UV FC line Product Datasheet,* L.H. B.V., Editor.

The contrast ratio improvement of perovskite nanocrystals LEDs devices based on carbon nanotubes

Caiman Yan[1], Hanguang Lu[1], Zongtao Li[1*], Jiexin Li[1], Jiasheng Li[1], Yong Tang[1], Binhai Yu[1]

[1]National & local joint engineering research center of semiconductor display and optical communication devices, South China University of Technology

Guangzhou 510641, Guangdong Province, China.

*Corresponding Author: Zongtao Li (e-mail: meztli@scut.edu.cn)

Abstract

Light emitting diodes (LEDs) are showing explosive growth in the display filed. To obtain high quality color performance display devices, ultraviolet excitation of quantum dots has become an effective solution. However, its contrast ratio (CR) performance limits its further development. Herein, for attaining high CR and color purity display performance, we combined the $CsPbBr_3/Cs_4PbBr_6$ perovskite nanocrystals (Pero NCs) and carbon nanotubes (CNTs) for LED package. Although the light output combined with CNTs is decreasing because of the CNTs light absorption property, the CR is improved due to the reduced reflectance. Therefore, the green LED with the appropriate CNT concentration could balance the light emission and CR performance. Result shows that with 0.2% CNTs, the green Pero NCs LED boosts the contrast by 18.02% under 50 lx ambient illuminance while the luminous flux drops by 25.0% under 50 mA. And the color purity could be guaranteed over 84% at this CNT concentration. This study provides a better understanding of perovskite NCs LED and guidance on the contrast ratio improvement.

1 Introduction

Light emitting diodes (LEDs) have attracted increasing attention in the field of backlighting and direct display, due to its high brightness, good stability, long life and environmental protection[1]. For display, red, green and blue (RGB) are three primary colors. To attain these RGB colors, there are two main technical routes. The first one is that the RGB light are obtained from three LED chips, respectively. But the red and green LED chips are not as mature as the blue LED chip [2], which increases the device cost. Besides, three LED chips would give rise to the mismatched voltage problem[3],increasing difficulty in drive design. Therefore, there is another route with only one LED chip. This LED chip could be blue or ultraviolet (UV) chip with short wavelength. Then the longer emission like green light could be attained by exciting the fluorescent material with the short wavelength[4]. This single chip solution demands the fluorescent materials with superior color performance for high quality display.

Because of the narrow emission full width at half maximum (FWHM) and wide spectrum range, nanocrystals (NCs) has been widely studied for display and backlight[5]. Nowadays, the display based on CdSe NCs can be found in the market, leading to an extensive attention. Different from the CdSe NCs, the lead halide perovskite NCs have presented a more significant potential due to their narrower FWHM and higher color purity[6]. The perovskite NCs are mainly composed of ABX_3 structure. The A could be Cs^+, MA^+, FA^+

or their mixed cation; the B could be Pb^{2+}, Sn^{2+}; and X would be Cl^-, Br^-, I^-. The luminescent color of the perovskite NCs can be easily tunable by adjusting the proportion of the halogen element[7]. In our previous study, the transformation from $CsPbBr_3$ NCs to $CsPbBr_3/Cs_4PbBr_6$ NCs in a short time was feasible. And by taking advantage of this high color purity $CsPbBr_3/Cs_4PbBr_6$ perovskite NCs, a wide color gamut of 122.8% NTSC was attained[8].

In addition to the color purity, the contrast ratio (CR) is also an important parameter for the display devices[9]. The CR is used to evaluate the difference between the light on and off state of LED devices. High CR is conductive to improve the resolution of the displayed image. In organic light emitting diodes (OLED), CR enhancement has been a hot research. Mingxiao Zhang at al[10] reported that replacing the high reflectivity Ag anode with $Ag:MoO_3$ anode. The CR of OLED was enhanced, resulting in a better color quality. Feng Xu at al[11] introduced the carbon nanotubes as the anodes of OLED. With the considerably reduced spectral reflectance of anode, the enhanced CR was attained. Yet there are few reports about the CR improvement of LED.

Herein, to improve the CR of the LED devices, we introduced the carbon nanotubes (CNTs) as packaging additives. The green LED is chosen to verify the function of CNTs. To guarantee the superior color performance of the LED, the green emission is attained by using the UV 385 nm LED chip to excite the $CsPbBr_3/Cs_4PbBr_6$ perovskite NCs (Pero NCs). The green LEDs with different CNT concentration were fabricated and their optical performances were compared. The intrinsic reason for adding CNTs to improve CR was also analyzed.

2 Experiment

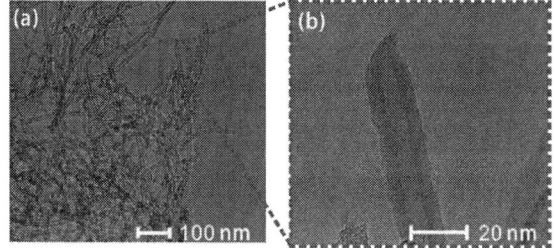

Figure 1. The TEM graph of carbon nanotubes (CNTs): magnification (a) 74 Kx; (b) 630 Kx

From the transmission electron microscope (TEM) graph shown in Figure 1, the CNTs are exhibiting the tube shape in nanoscale. Single CNT is about 18 nm of width and its wall is made up of multiple layers of C atoms. Further, as shown in Figure 2, the X-Ray Diffraction (XRD) confirms the CNTs

978-1-7281-5757-3/19 $31.00 © 2019 IEEE

owns two main diffraction peaks: 25.6° and 42.9°, which is corresponding to (002) and (100) crystal orientations. This is consistent with the reported carbon nanotubes[12]. The narrow diffraction peak also supports that these CNTs material are with high crystallinity.

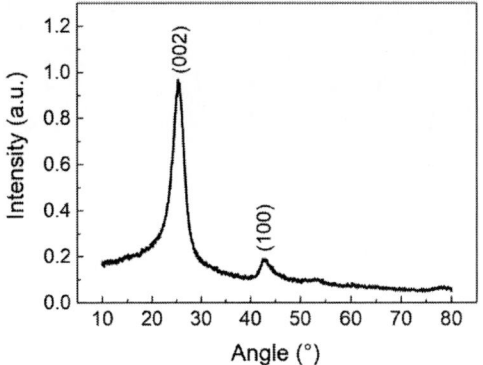

Figure 2. XRD pattern of CNTs

To obtain LED devices with different concentrations of CNTs, firstly, the UV chip (385 nm) was fixed on the surface of the framework. Then the LED chip was connected to the circuit board using a gold wire connection. For convenience of testing, these samples were welded to a plum blossom board. The $CsPbBr_3/Cs_4PbBr_6$ Pero NCs were synthesized by a microchannel reactor method as we had reported before[8]. Then the CNTs and Pero NCs were mixed into the PDMS silicone. After defoaming, the mixed silicone was injected into the cavity as package material. After 60 minutes of curing at 80℃, the LED sample preparation was completed. To find out the effect of CNTs, the concentration of Pero NCs was fixed at 0.6%, and the CNT concentration was set from 0.0% to 1.2% gradually. Figure 3 shows three LED devices with different CNT concentrations: 0.0%, 0.6% and 1.2%. From left to right, as the CNT concentration increases, the blackness of whole LED becomes more apparent.

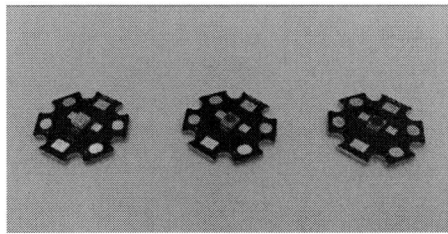

Figure 3. The physical picture of LED samples. The CNT concentrations from left to right were 0.0%, 0.6% and 1.2%, respectively.

The CNTs was purchased from National Technology Co., Ltd. company. The PDMS was from Dow Corning Corporation, USA. The TEM picture was captured by a TEM electron microscope operating at 200 KV (JEM-2100F, JEOL, Japan). The XRD was tested by an X-ray diffractometer from 5 degrees to 80 degrees at low speed scanning (D8-Advance, Bruker, Germany). The absorption and reflectance spectrum was tested by a UV-Vis spectrophotometer (TU-1901, Persee, Beijing, China). The EL spectra and optical performance were measured by an optical measurement system with an integrating sphere.

3 Discussion

The green LED devices are obtained by exciting the Pero NCs with ultraviolet light (385 nm) here. Then the luminous flux of green LEDs with different CNT concentration were tested. As the current increases, the luminous flux shows an increasing trend. The best luminous flux is the reference LED without any CNTs. After introducing the CNTs into LED package, the luminous flux decreases gradually. For example, the luminous flux of LED with 0.0%, 0.2%, 0.4%, 0.8%, 1.0% and 1.2% is 2.0, 1.50, 1.04, 0.63, 0.36, 0.24 and 0.14 lm under 50 mA, respectively. So the luminous flux decreases by 25.0% with 0.2% CNT concentration. And the luminous efficiency is also calculated to evaluate the electro-optical conversion efficiency. As shown in Figure 4(b), under large current, the luminous efficiency of all LED samples shows a downward trend, which means the LED device is suffering the droop effect. Moreover, the LED with CNTs has a lower luminous efficiency than without CNTs. The lowest luminous efficiency is presented on the highest concentration of CNTs (1.2%).

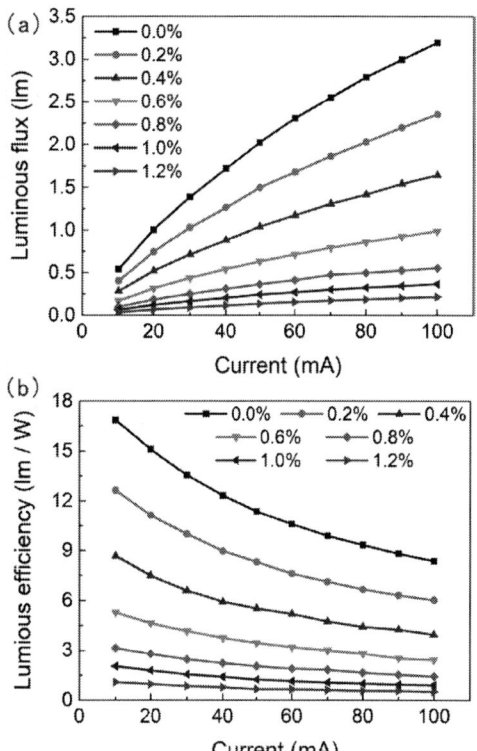

Figure 4. (a) Luminous flux and (b) luminous efficiency of LED with different CNT concentration

To find out the reason responsible for the light output decline, the visible electroluminescence (EL) spectra were tested at 50 mA. As shown in Figure 5, all the LED samples emit 522 nm green light with 21 nm FWHM. The green light is obtained by ultraviolet light exciting Pero NCs. The narrow FWHM shows the Pero NCs have great potential in the future display field. This also confirms that the added CNTs have no effect on the original excellent Pero NCs properties. However, as the concentration of CNTs increases, the intensity of green emission gradually decreases, which partly explains the decrease in luminous flux with the addition of CNTs.

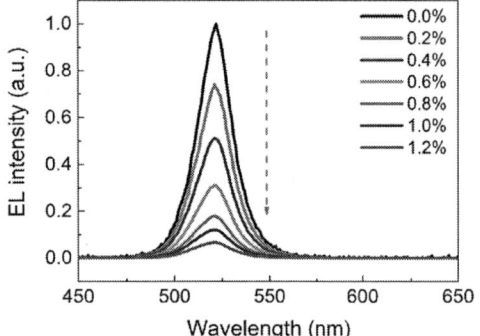

Figure 5. EL spectra of LED with different CNT concentration

As the CNTs is black materials under daylight, it is speculated that the green peak is lowered due to the light absorption of CNTs. To confirm this hypothesis, the absorption spectrum of CNTs after dispersing in DMF solution was tested. From Figure 6, the CNTs have the capacity to absorb the light from 300 to 800 nm. What is more, this absorbing ability of CNTs increases as the wavelength of light becomes shorter. In other words, the absorption capacity of CNTs for ultraviolet light (385 nm) is stronger than that of green light (522 nm). So the CNTs could absorb both the UV light and green light, resulting in the decrease the light output.

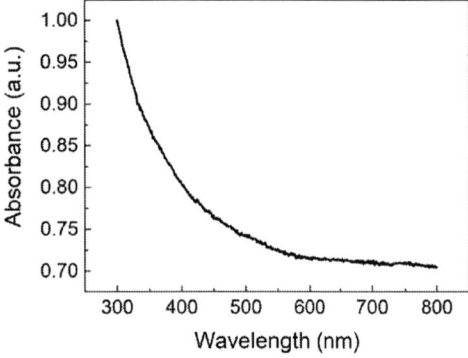

Figure 6. The absorption spectrum of CNTs

Meanwhile, we found that the reflectivity of LED devices surface is decreased. As shown in Figure 7, without any CNTs, the reflectivity of surface is quite high up to over 70% in long wavelength. Although it drops below 40% at wavelengths shorter than 500 nm, the reflectance is still as high as 66.8% near 522 nm. High device surface reflectance is not conducive to CR. After adding the CNTs into the green Pero NCs LED, the reflectively could be decreased effectively. The reflectivity of LED with 0.2%, 0.4%, 0.6%, 0.8%, 1.0% and 1.2% is 46.08%, 38.98%, 35.71%, 32.80%, 28.83% and 27.76% under 522 nm. So only with 0.2% CNTs, it could decrease the 20.72% reflectance.

To obtain the CR of these LED devices with different CNT concentration, the CR of the LED is calculated as formula (1):

$$CR = \frac{L_{on} + R_l L_{ambient}}{L_{off} + R_l L_{ambient}} \qquad (1)$$

In which, L_{on} and L_{off} are the luminance when LED is at "on" and "off" state. $L_{ambient}$ is the ambient illumination, and the R_l is the luminous reflectance, which could be defined as formula (2):

$$R_l = \frac{\int_{\lambda 1}^{\lambda 2} V(\lambda)S(\lambda)R(\lambda)d(\lambda)}{\int_{\lambda 1}^{\lambda 2} V(\lambda)S(\lambda)d(\lambda)} \qquad (2)$$

In which, the $V(\lambda)$ is the standard photonic curve, $S(\lambda)$ is the spectrum of green light emission in our experiment, and $R(\lambda)$ is the spectral reflectance of the device. The measurement was carried out from wavelength $\lambda_1 = 300$ nm to $\lambda_2 = 800$ nm to cover the whole visible spectrum.

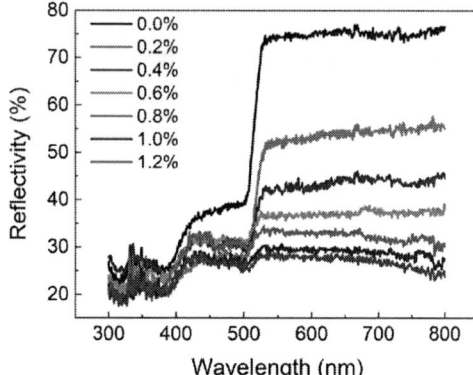

Figure 7. The reflectance spectrum of LED device with different CNT concentration

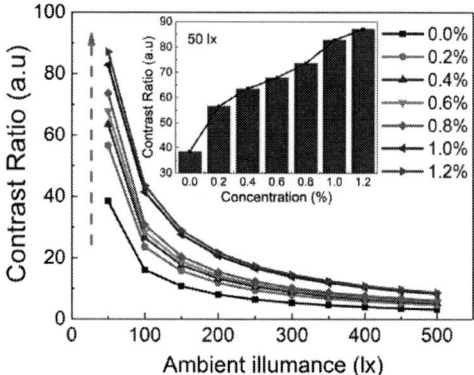

Figure 8. the contrast ratio of green LED with different CNT concentration versus the ambient illuminance from 50 to 500 lx.

For calculation, the L_{on} and L_{off} state is set as 1000 and 0 cd/m^2, which is similar to the lighting and black state of LED device. Then the CR versus the ambient illuminance is plotted as the graph, as shown in Figure 8. The CR of the LED device and the ambient illuminance is negatively correlated, which means that the high ambient illumination would lower the device contrast. As the CNT concentration increases, the CR is improved effectively. Specifically, under 50 lx, the CR of the LED with 0.0%, 0.2%, 0.4%, 0.6%, 0.8%, 1.0%, 1.2% is 38.53, 56.55, 63.59, 67.92, 73.60, 82.82, 87.13. So the CR of

LED device could be improved 18.02% with 0.2% CNT concentration.

Finally, the color purity of the LED devices with different CNT concentrations was also compared. As the current increases, the color purity shows a decreasingly trend. It should be noted that the color purity of all the LED samples is over 70% under 10 mA. This is the unique advantage of UV excitation, because the UV light (385 nm) almost cannot be perceived by the human eye. But too much CNTs addition would lower the color purity of LED device. This is due to the ability of CNTs to absorb ultraviolet and green light, which reduces energy utilization. However, the color purity of LED with 0.2% CNT concentration is still over 84%, only slightly lower than the reference. Therefore, the color purity with high quality at this low CNT concentration (0.2%) still could be guaranteed.

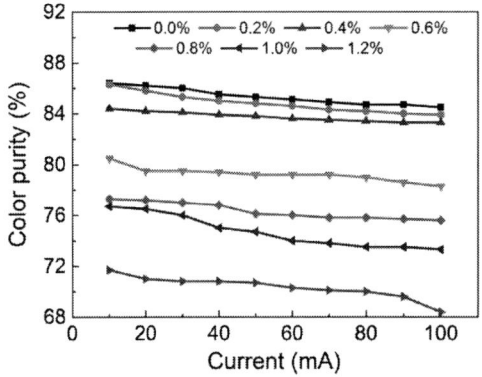

Figure 8. the color purity of green LED with different CNT concentration

Through above experiment, although the CNTs are negative in luminous flux because of its strong light absorption, they are beneficial for the improvement in CR due to the reduced reflectance. For display, the brightness requirement of the device is not as strict as the illumination. What's more, the excessively bright light is harmful for the human eye. In this case, it is acceptable to sacrifice little brightness of the device in exchange for high CR. But too much CNTs addition results in weak brightness, which is also not recommended. To balance the brightness and CR performance, the recommended solution is 0.2% CNT concentration.

4 Conclusions

To attain high performance green LED for RGB display, the Pero NCs LEDs devices excited by UV light with different CNT concentration were fabricated. After the CNT is added to the LED package, the light output of the LED is reduced due to the light absorbing ability of the CNTs. Meanwhile, the CR has an effective improvement due to the reduced reflectivity of the LED. In order to balance light and contrast, proper CNT concentration is the key and 0.2% concentration is the recommended value after optimization. At this CNT concentration, although the luminous flux drops by 25.0% under 50 mA, the CR increases by 18.02% under 50 lx ambient illuminance and the color purity of the LED is still over 84%. This study provides a solution to improve contrast

with CNTs, which is beneficial to display quality improvement.

Acknowledgments

This work was supported by the Science & Technology Program of Guangdong Province (No. 2019B010130001), the National Natural Science Foundation of China (No.51735004), the Science & Technology Program of Guangdong Province (No. 2017B010115001), and National Natural Science Foundation of China (No. 51775199).

References

1. S. Pimputkar, J. S. Speck, S. P. Denbaars *et al.*, "Prospects for LED lighting," *Nature Photonics,* vol. 3, No. 4 (2009), pp. 180-182.
2. Z. Li, K. Cao, J. Li *et al.*, "Investigation of Light-Extraction Mechanisms of Multiscale Patterned Arrays With Rough Morphology for GaN-Based Thin-Film LEDs," *IEEE Access,* vol. 7 (2019), pp. 73890-73898.
3. J. Hasan, and S. S. Ang, "A RGB-driver for LED display panels." pp. 750-754.
4. C.-J. Chen, J.-Y. Lien, S.-L. Wang *et al.*, "P- 91: Highly - Efficient LEDs with On- Chip Quantum- Dot Package for Wide Color Gamut LCD Display." pp. 1465-1468.
5. Z. Luo, Y. Chen, and S.-T. Wu, "Wide color gamut LCD with a quantum dot backlight," *Optics express,* vol. 21, No. 22 (2013), pp. 26269-26284.
6. X. Li, F. Cao, D. Yu *et al.*, "All Inorganic Halide Perovskites Nanosystem: Synthesis, Structural Features, Optical Properties and Optoelectronic Applications," *Small,* vol. 13, No. 9 (2017), pp. 1603996.
7. X. M. Li, Y. Wu, S. L. Zhang *et al.*, "CsPbX3 Quantum Dots for Lighting and Displays: Room-Temperature Synthesis, Photoluminescence Superiorities, Underlying Origins and White Light-Emitting Diodes," *Advanced Functional Materials,* vol. 26, No. 15 (Apr, 2016), pp. 2435-2445.
8. H. Lu, Y. Tang, L. Rao *et al.*, "Investigating the transformation of CsPbBr3 nanocrystals into highly stable CsPbBr3/Cs4PbBr6 nanocrystals using ethyl acetate in a microchannel reactor," *Nanotechnology,* vol. 30, No. 29 (2019), pp. 295603.
9. H. H. Wang, J. W. Pan, Y. C. Huang *et al.*, "56.1: Distinguished Paper: A Higher-Contrast, Ghost-Ray-Deflecting, Total-Internal-Reflection Light Separator for LED DLP Projectors," *Sid Symposium Digest of Technical Papers,* vol. 45, No. 1 (2015), pp. 813-816.
10. M. Zhang, Z. Chen, L. Xiao *et al.*, "High-Color-Quality Blue Top-Emitting Organic Light-Emitting Diodes with Enhanced Contrast Ratio," *Japanese Journal of Applied Physics,* vol. 52, No. 5 (2013), pp. 492-494.
11. X. Feng, Q. Z. Wen, Y. Long *et al.*, "Single walled carbon nanotube anodes based high performance organic light-emitting diodes with enhanced contrast ratio," *Organic Electronics,* vol. 13, No. 2 (2012), pp. 302-308.

12. Ş. S. Bayazit, and Ö. Kerkez, "Hexavalent chromium adsorption on superparamagnetic multi-wall carbon nanotubes and activated carbon composites," *Chemical Engineering Research & Design,* vol. 92, No. 11 (2014), pp. 2725-2733.

Preparation of Translucent Al₂O₃ Ceramic Substrates for LED Filament Bulb

Yizheng Zhang[a], Ling Gao[a], Fengpo Yuan[b], Yaguang Wu[a], Dongsheng Wang[a], Caihua Ren[a], Hongbo Bai[a]

a. Hebei Semiconductor Research Institute
b. Hebei Key Lab of New Semiconductor Optoelectronic Devices
Shijiazhuang, Hebei
Email: zhangyizheng117@126.com

Abstract

As the most possible substitutes for traditional incandescent bulbs, LED filament bulbs are becoming a research hotspot all over the world in recent years. The LED filament bulb has many advantages, including its similar appearance to incandescent bulbs, 360° lighting angle, high brightness, and low energy consumption. However, its price is so expensive that it cannot occupy the lighting market completely. So a big challenge of LED filament bulbs is how to reduce the cost of production. For the actual low-cost application requirements of a LED filament bulbs manufacturer, who is our customer, a translucent Al₂O₃ ceramic substrate has been prepared by tape casting and sintering process. We optimize the formulation of tape casting and technological process of preparation continually, in order to satisfy the needs of the customer as well as reduce the cost, finally 2# formula (1.2wt.%Ca-3wt.%Si-0.8wt.%Mg sintering additives with 95wt.% α-Al₂O₃ powder) is selected and used for preparing the ceramic substrate. The substrate can dramatically reduce the cost of LED filament bulbs from two aspects: materials and process. In terms of materials, the ceramic substrate is sintered by low-cost α-Al₂O₃ green tape at 1600°C in the air furnace, and the performance fully reaches the targets of the customer: the flexure strength is 470MPa, the light transmittance is 22%, and the fluctuation of shrinking percentage is ±0.14%. From the aspect of process, the ceramic substrate is divided into strips efficiently and conveniently, through scribing before sintering with hot knives on the flexible green tape substrate and breaking after sintering with automatic machines on the ceramic substrate. The dimensional precision of ceramic strips is very high, so the customer can employ automatic equipment for fabricating LED filament strips, not only reducing the process cost, but also increasing the production efficiency.

1. Introduction

The USHIO INC. in Japan developed the first LED filament bulb in 2008. The appearance of LED filament bulbs is similar to traditional incandescent bulbs, which arouses public interest extremely. On the other hand, the filament is made up of several LEDs connected in series, so it is low energy consumption compared with the incandescent bulb. In addition, the LED filament bulb makes the 360° lighting angle come true, and its high brightness brings an unprecedented experience to people.[1]

In recent years, LED filament bulbs are becoming a research hot spot all over the world as the most possible substitutes for traditional incandescent bulbs. In 2015, there are more than 600 LED filament bulb manufacturers in China. However, the price of the LED filament bulb is higher than incandescent bulbs, which impedes its popularization and application. So the challenge is how to reduce the cost.

The LED filament bulbs include LED filament strips, transformers, stents, bulbs, bulb holders, etc. The cost of LED filament strips is the highest in the series. Moreover, a filament strip consists many LEDs, a substrate (cut into strips), phosphor and so on (see in Fig. 1.1).[2,3] According to the report, the cost of LEDs and substrates occupies above 50% in the LED filament bulb. For example, sapphire is the most common substrate materials, but its preparation and process costs are very high: sapphire adopts seed-crystal induction method to grow into monocrystal, in the course of preparation, the temperature of equipment need to be kept above 2045°C (melting point of sapphires).[4] For another, sapphire substrates need to be ground, polished, and cut into narrow strips, the rate of finished products is low in the process because of its natural brittleness. So exploiting a low-cost substrate materials and simplifying the technological process are extremely urgent.

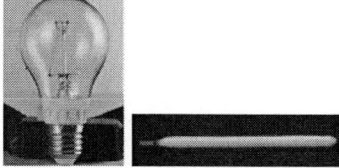

Fig. 1.1 The LED filament bulb and filament strip

A LED filament bulbs manufacturer, who is our customer, calls for low-cost translucent substrates to expand the lighting market. The specific targets of the customer are listed in Table 1.1.

Table 1.1 The specific targets of our customer

Items	Targets
Substrate materials	Ceramics
Preparation method	Tape casting+sintering
Cutting process from substrate to strips	Scribing and breaking
The quantity of strips in a substrate	20
Dimension of a strip (mm)	$30.0^{\pm0.10}*3.0^{-0.03}*0.4^{-0.03}$
Fluctuation of shrinking percentage	$<\pm0.3\%$
Light transmittance	$\geq20\%$
Flexure strength (MPa)	≥400

2. Formulation design of translucent Al₂O₃ ceramic substrates

According to the targets of Table 1.1, translucent Al₂O₃ ceramic substrates can satisfy the requirements of the customer basically. Firstly, the ceramics are sintered by α-Al₂O₃ powder, which is the most stable phase in all kinds of

978-1-7281-5757-3/19 $31.00 © 2019 IEEE

Al$_2$O$_3$ crystal forms, and have a property of high strength, hardness and chemical stability.[5]　Secondly, the Al$_2$O$_3$ ceramic substrate can be easily prepared with low-cost tape casting and sintering in the air furnace, and the operating temperature of which is below 1700℃.　Finally, the flexible Al$_2$O$_3$ green tape substrate obtained by tape casting can be scribed with hot knives before sintering, and broke into strips with automatic machines after sintering, which improves the automation level and the rate of finished products.　Fig. 2.1 shows the detailed preparation process mentioned above.

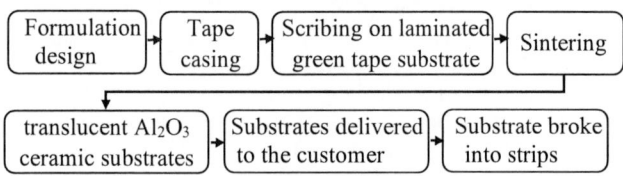

Fig. 2.1 The preparation process of translucent Al$_2$O$_3$ ceramic substrates and strips

However, there are unexpected difficulties in the actual preparation process.　On the one hand, it is difficult to simultaneously accomplish the light transmittance target and flexure strength target, especially the ceramics have to be sintered fully below 1700℃.　Therefore trace amounts of sintering additives should be added in the formulation, in order to bring down the sintering temperature, and improve the performance of the Al$_2$O$_3$ ceramics.　On the base of our experience, Ca-Si-Mg system is well suited as the sintering additives and the additive amount in ceramics may as well control within 5%.　On the other hand, the dimensional precision target and the shrinking percentage fluctuation target of ceramic strips are very strict, for purpose of automatic production of the customer.　It is well known that the boundary dimension of ceramics will shrink a lot after sintering, so dimensional precision of strips is closely linked to the fluctuation of shrinking percentage.　According to the dimension data of strips provided by the customer, the fluctuation of shrinking percentage should be under ±0.3%.　The fluctuation of shrinking percentage is affected by the particle size distribution (D10～D90), dispersibility and solid load of α-Al$_2$O$_3$ powder in solvent.

In conclusion, we design the formulation as follows: prepare 95wt.% α-Al$_2$O$_3$ ceramic substrate accompanied with 5wt.% Ca-Si-Mg sintering additives through tape casting and sintering in the air furnace, enhance the light transmittance and flexure strength, and decrease the fluctuation of shrinking percentage.

3. Formulation experiments of translucent Al$_2$O$_3$ ceramic substrates

We design four sintering additives formulas to conduct the experiments: 1# (1.2wt.%Ca-2.5wt.%Si-1.3wt.%Mg), 2# (1.2wt.%Ca-3wt.%Si-0.8wt.%Mg), 3# (1.2wt.%Ca-3.3wt.%Si-0.5wt.%Mg) and 4# (1.2wt.%Ca-3.5wt.%Si-0.3%wt.Mg).　The total quantity of every sintering additive takes up 5wt.% of Al$_2$O$_3$ ceramics.　Except for the solids including sintering additives and α-Al$_2$O$_3$ powder (D10～D90=0.3～6μm) which occupy 65wt.% (solid load), the constituents of tape casting slurry contain liquids, that is, 24wt.% alcohol, 1wt.% menhaden fish oil, 7wt.% PVB

(polyvinyl butyral), and 3% wt.DBP (dibutyl phthalate).　All of them are mixed in ball mill tanks for three days, turning into the slurry fit for tape casting.　Then transform the slurry into green tapes through tape casting machine, and the thickness of green tapes includes 0.13mm and 0.26mm.

Vernier caliper is used to measure the dimension for calculating the shrinking percentage.　Density balance (METTLER TOLEDO, ME204TE) is used to measure the density of ceramics.　The flexure strength is measured by universal materials testing machine (CMT 6303).　The light transmittance is measured by spectrophotometer (CARY 500).　Scanning electron microscopy (SEM, ZEISS EVO 18) is used to observe the size and microtopography of ceramics.

1) Confirm the sintering temperature of four formulas, and test the shrinking percentage and density of ceramics

According to the sintering mechanisms, the density of ceramics increases with the furnace temperature rising until it comes up to maximum and the temperature at this moment is the sintering temperature.　If the temperature continues rising, the density no more increases or even decreases.

For testing the shrinking percentage and density, 5 green tape substrate samples of every formula are made with boundary dimension of 35mm (X direction)* 20mm (Y direction)* 0.52mm (Z direction) and laminated with 500psi. Of which, X, Y, Z stand for the direction parallel to tape casting, the direction perpendicular to tape casting, and the thickness direction, respectively.　After sintering, the dimension shrinks and becomes a (X direction) mm* b (Y direction) mm* c (Z direction) mm.　So we can figure up the shrinking percentage: 100%-a/35 (X direction), 100%-b/20 (Y direction), 100%-c/0.52 (Z direction).　Before testing the shrinking percentage, we should confirm the sintering temperature of four formulas in advance.

We test the density of the substrate samples mentioned above with drainage method.　The relationship between density and furnace temperature of four formulas is drawn into Fig. 3.1, and we can confirm that the sintering temperature of 2# formula, 3# formula and 4# formula is 1600℃，1550℃ and 1500℃, respectively.　The density of 1# formula keeps increasing until 1650℃, but we don't know if it will increase because the furnace temperature cannot reach 1700℃.　So the 1650℃ is selected as the sintering temperature of 1# formula .

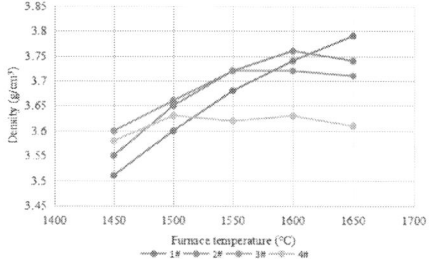

Fig. 3.1 The relationship between density and furnace temperature of four formulas

We measure and calculate the shrinking percentage of four formulas on their own sintering temperature in Table 3.1.　The shrinking percentage data is important and will be used to calculate the dimension of green tape substrate according to the dimension of final product.　The data from the table shows the fluctuation of shrinking percentage (X and Y directions) is ±0.3%, at the critical position of the target. So the fluctuation of shrinking percentage need to be decreased none the less.

Table 3.1 The shrinking percentage of four formulas

Formula	Shrinking percentage (n=5)		
	X direction	Y direction	Z direction
1#	15.5%±0.3%	15.8%±0.3%	27%±0.2%
2#	15.0%±0.3%	15.3%±0.3%	25%±0.2%
3#	14.8%±0.3%	15.1%±0.3%	25%±0.2%
4#	14.1%±0.3%	14.4%±0.3%	24%±0.2%

2) Test the flexure strength, light transmittance and grain size

On the basis of GB/T 4741-1999, the dimension of flexure strength test samples is 120mm*4mm*4mm. According to the shrinking percentage, we get the dimension of green tape samples. The green tape samples are machined with hot knives and sintered in the air furnace. The flexure strength of four formulas is tested by universal materials testing machine and the average results (n=12) are summarized in Table 3.2.

According to the dimension of strips of LED filament and shrinking percentage, we calculate the dimension of green tape strips. The green tape strips are machined with hot knives and sintered in the air furnace. The light transmittance of four formulas is tested by spectrophotometer and the average results (n=3) are also summarized in Table 3.2. Fig. 3.3 shows the translucent Al$_2$O$_3$ ceramic substrate (the thickness is 0.4mm).

The microtopography of ceramic strips mentioned above are observed by SEM, and the SEM images of four formulas are arranged in Fig. 3.2. The maximum grain size is listed into Table 3.2.

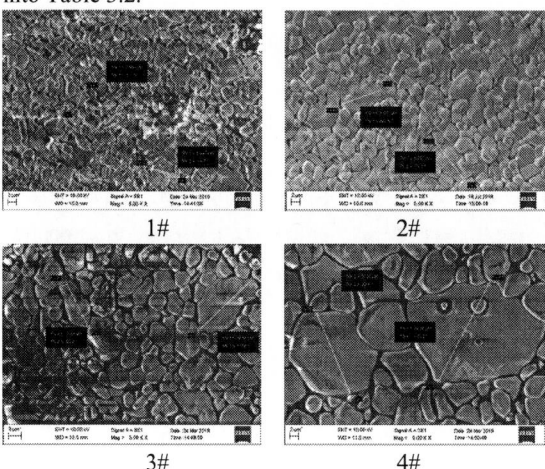

1#   2#

3#   4#

Fig. 3.2 SEM images of four formulas

Observing Table 3.2, the flexure strength has an inversely proportional relation with grain size, while the light transmittance has a proportional relation with grain size. Analyzing this phenomenon: the smaller the grain sizes are, the more the grain boundaries are. More grain boundaries are beneficial to increasing the strength through impeding the micro crack extending, yet are bad for light transmittance because of the boundary scattering. Contrasted with the targets of the customer, 1# and 4# formulas are not qualified. Although 3# formula reaches the target, its flexure strength is

the critical value. Finally, 2# formula is selected and used for preparing the translucent Al$_2$O$_3$ ceramic substrate.

Table 3.2 The test results of flexure strength, light transmittance and maximum grain size

Formula	1#	2#	3#	4#
Flexure strength (MPa)	504	470	405	363
Light transmittance	14.4%	22.0%	25.5%	30.3%
Maximum grain size (μm)	6	10	17	25

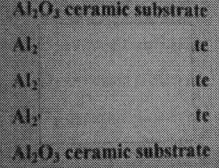

Fig. 3.3 The translucent Al$_2$O$_3$ ceramic substrate

4. Optimization of formulation and process

1) Optimization of formulation

As mentioned above, the fluctuation of shrinking percentage closely related to boundary dimension of strips is ±0.3% (X and Y directions), at the critical position of the target, and needs to be decreased through optimization of formulation. Base on the previous experiments, the fluctuation of shrinking percentage is related to the particle size distribution (D10～D90, see in Fig.4.1), dispersibility and solid load of α-Al$_2$O$_3$ powder in solvent. So we design a simple orthogonal experiment: the influence factors and their variate levels are listed in Table 3.2 and the experimental designs are in Table 3.3 including 4 trials. Prepare the slurry of tape casting of 4 trials with 2# formula and the parameters in the tables, then make the samples of shrinking percentage. After sintering, the data of shrinking percentage fluctuation (X and Y directions) is also summarized in Table 3.3.

Table 3.2 Orthogonal factor level table

Level Factor	A Size distribution	B Additive amount of the dispersant	C Solid load
1	0.3～4μm	1wt.%	63wt.%
2	0.3～6μm	1.5wt.%	65wt.%

Table 3.3 L$_4$ (2^3) Orthogonal experimental table

No. Col.	A	B	C	Shrinking percentage fluctuation (X/Y)
Trial 1	1 (0.3～4μm)	1 (1wt.%)	1 (63wt.%)	±0.18%
Trial 2	1 (0.3～4μm)	2 (1.5wt.%)	2 (65wt.%)	±0.14%
Trial 3	2 (0.3～6μm)	1 (1wt.%)	2 (65wt.%)	±0.30%
Trial 4	2 (0.3～6μm)	2 (1.5wt.%)	1 (63wt.%)	±0.27%

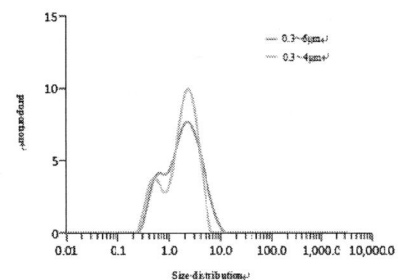

Fig. 4.1 Size distribution of two α-Al$_2$O$_3$ powders

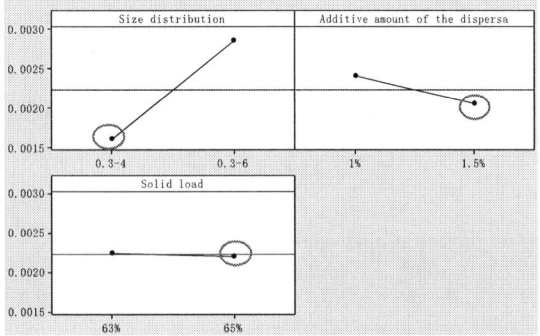

Fig. 4.2 Impact factor trend graph

Utilizing range analysis, we get the impact factor trend graph as shown in Fig.4.2. The importance ranking of impact factors is: Size distribution>Additive amount of the dispersant>Solid load, and the optimal level group is: A1B2C2. That is to say, when the size distribution is 0.3～4µm, the additive amount of the dispersant is 1.5wt.%, and the Solid load is 65wt.%, the fluctuation of shrinking percentage can reach minimum, namely ±0.14%.

2) Optimization of scribing and breaking process

The green tape substrate is laminated by flexible green tapes, so it is also flexible and easily scribed with hot knives. For the scribing and breaking process, the deeper the depth of scribing on the green tape substrate is, the better the effect of breaking process of ceramic substrate is. However, sometimes the micro crack tips formed by scribing expand easily, leading to breakage of green tape substrate. It goes against the automatic production of our customer. Through experimenting constantly, we control the green density within 2.5～2.6 g/cm^3 and the depth of scribing at the point of 2/3 of the green tape substrate thickness approximately (see in Fig.4.3). As a result of process optimization above and tiny fluctuation of shrinking percentage, the dimension precision of ceramic strips after breaking is very high, which satisfies fully the dimension target of the customer.

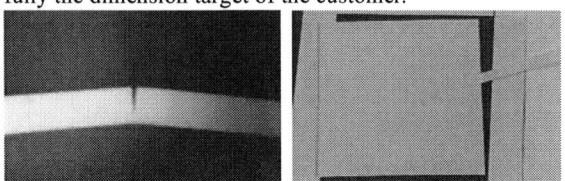

Fig. 4.3 Scribing process on flexible green tape substrate

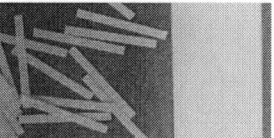

Fig. 4.4 Breaking process on ceramic substrate

5. Conclusions

For reducing the cost of substrate of LED filament damps, we prepare a low-cost translucent Al$_2$O$_3$ ceramic substrate, reaching the targets of the customer completely:

1) Through formulation designs and experiments, employ 2# sintering additives formula (1.2wt.%Ca-3wt.%Si-0.8wt.%Mg) with 95wt.% α-Al$_2$O$_3$ powder for preparing the translucent Al$_2$O$_3$ ceramic substrate: the sintering temperature is 1600℃, the flexure strength is 470MPa, and the light transmittance is 22%.

2) Through orthogonal design and formulation optimization, confirm the importance ranking of impact factors in connection with the fluctuation of shrinking percentage: Size distribution>Additive amount of the dispersant>Solid load, and the optimization parameters: size distribution (0.3～4µm), additive amount of the dispersant (1.5wt.%) and solid load(65wt.%), reducing the fluctuation of shrinking percentage (X and Y directions) to ±0.14%.

3) Through optimization of scribing and breaking process, control the green density within 2.5～2.6 g/cm^3 and the depth of scribing at the point of 2/3 of the green tape substrate thickness approximately. As a result of process optimization above and tiny fluctuation of shrinking percentage, the dimension precision of ceramic strips after breaking the ceramic substrate is very high, which is beneficial to automatic production, not only reducing the process cost, but also increasing the production efficiency.

References

1. Huai Z., *et al*, "A Novel Cooling Method for LED Filament Bulb Using Ionic Wind", *Thermal and Thermomechanical Phenomena in Electronic Systems, 16th IEEE Intersociety Conf*, May. 2017, pp. 998-1003.

2. Wilson F., *et al*, "Simulation and Optimization on Thermal Performance of LED Filament Light Bulb", *Solid State Lighting, 12th China International Forum*, November. 2015, pp. 88-92.

3. Chongguang Y., "Technological Innovation of LED Filament bulb", *China Light and Lighting*, No. 2 (2014), pp. 26-28.

4. Jie L., "Preparation Methods for Sapphire Crystal and Their Features", *Mining and Metallurgical Engineering*, Vol. 31, No. 5 (2011), pp. 102-106.

5. Yi W.,*et al*, "Influencing factors of alpha alumina's crystal structure", *China Nonferrous Metallurgy*, No. 6 (2017), pp. 72-74.

Study on high power density light-emting diodes light source

Huaiwen Zheng[1], Zhonghua Deng[2], Zhuguang Liu[2], Fei Yu[1], Yan Li[1], Yuming Yang[1], Qiao Liang[3], Hua Yang[1*](corresponding author), Xiaoyan Yi[1], Junxi Wang[1], Jinmin Li[1]

1. Institute of Semiconductors, Chinese Academy of Sciences
No.A35, QingHua East Road, Haidian District, Beijing P R China

2. Key Laboratory of Optoelectronic Materials Chemistry and Physics, Fujian Institute of Research on the Structure of Matter, Chinese Academy of Sciences, Fuzhou, China 350002, China

3. Department of Computer，North China Institute of Science and technology

huayang@semi.ac.cn

Abstract

This paper introduces three kinds of high power density light sources which are independently designed. Two of them use ceramic substrate and flip chip. Another one use metal substrate and lateral structure chip. The tests of their light efficiency, color temperature and color rendering index are conducted. After that, the three kinds of light sources are lighten for 1500 hours at the environment temperature of 55 ℃. After the aging test, the parameters such as light efficiency, color temperature and color rendering index are recorded. All the data are compared at the end of the paper.

Introduction

COB package has many advantages such as high luminous flux density, uniform luminescence, high power and integration. COB package has become a trend of LED package technology. In order to promote the development of the industrialization and application of high-power LED, the research on the characteristics of high-power LED has gradually become one of the focuses of the semiconductor industry. With the increasing of LED size and power of COB light source, heat dissipation of COB LED is particularly important [1-4]. Due to the defect of semiconductor crystal, the photoelectric conversion efficiency of LED chip is not high, and a large part of electric energy is converted into heat [5]. With the increase of the power density and package density of LED products, it will cause heat accumulation inside the chip, which will lead to a series of problems, such as the drift of luminous wavelength, the decrease of light output efficiency, the accelerated aging of fluorescent powder and the shortening of service life. In the electronics industry, failure rates tend to increase by an order of magnitude for every 10℃ increasing of the device's ambient temperature. At present, most of the methods adopted refer to the materials of the circuit board, some materials with high thermal conductivity and stability are selected, such as copper, aluminum, ceramics and so on. However, it is not enough to improve the heat dissipation problem only through the circuit board. It is also necessary to improve the heat dissipation performance of LED through other thermal design methods [6-9].

With the development of LED, especially high-brightness LED, the requirements on LED chip packaging technology are getting higher, and the technical difficulty is also getting larger. Packaging is an important part of LED, although the function of LED package is roughly the same as that of other semiconductor components, the function is protecting the chip against the impact of the outside environment and improve the ability of heat dissipation. However, the packaging of LED is also special, it is more important to improve the efficiency of light, and achieve a specific optical distribution, output visible light [9]. In order to reduce junction temperature of LED chip, currently, we mainly try to improve the heat loss ability of products in three aspects. On one hand, we improve the luminous efficiency of LED chip and reduce its thermal power by improving technology and adopting new materials. On the other hand, it is looking for new packaging methods and materials to export the heat generated by LED chips more effectively. Moreover, an efficient external heat dissipation method is adopted to quickly conduct heat from the LED heat sink to the environment [10-12].

Flip –chip high power light source on ceramic substrate

As shown in figure 1, we designs a light source, 200 W high power density on ceramic substrate, the ceramic substrate size is 75 mm * 75 mm * 1 mm, the material of the substrate is aluminum nitride ceramic, thermal conductivity is 280W/m•K, light emitting area is circular, light emitting area has the diameter of 60 mm, chip is blue flip chip of the size 40mil * 40mil, the working voltage of the chip is from 2.7 V to 3.2 V, working current is 350 mA, the dominant wavelength is from 450 nm to 460 nm, flip chip is bonded to the ceramic substrate by solder paste, The equipment used for the die-bonding is AD860M produced by ASM. After the die bonding, the luminescent area is coated with a layer of silica gel and phosphor powder mixture, and the phosphor powder is deposited on the surface of chip and substrate by centrifugal equipment.

Fig.1: The multi-chip light source 1

As shown in figure 2, the light source is installed on the heat sink with the structure of heat pipe and fin. Four fans are installed around the fins of the heatsink, and the heat will be timely exported to the air. The silicone grease with thermal

978-1-7281-5757-3/19 $31.00 © 2019 IEEE

conductivity of 12W/m·K is used between the light source and the heatsink.

Fig.2: The light source is installed on the heat sink.

After the installation of the light source, HAAS-2000 is used for the test. The ambient temperature is 25.2 ℃, ambient humidity is 55%RH, atmospheric pressure is 101.5KPa. The initial test parameters are as follows: test voltage 134.4V, test current 1.485A, test power 199.6W, luminous flux 32616lm, luminous efficiency 163.41lm/W, color temperature 5765K, color rendering index 71.7.

Later, the sample is lighten for 1500 hours under the condition of ambient temperature 55°C. The final parameters are as follows: test voltage 134.1V, test current 1.484A, test power 199W, luminous flux 33519lm, luminous efficiency 168.48lm/W, color temperature 5826K, color rendering index 71.9

Figure 3 is the test spectrum of the light source. The dominate wavelength is 520.2nm and the peak wavelength is 453nm

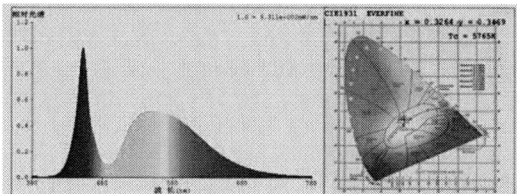

Figure 3: The test spectrum of the light source 1.

As shown in figure 4, we have designed another light source, 200W high power density on ceramic substrate, ceramic substrate has thickness of 1 mm and diameter of 70 mm, the materials is aluminum nitride, thermal conductivity is 280 W/m·K, the light emitting zone is circular, light emitting zone has diameter of 62 mm, the chip is blue flip chip with size of 40 mil * 40 mil, chip working voltage is from 2.7 V to 3.2 V, working current is 350 mA, the dominate wavelength is from 450 nm to 460 nm, flip chip is bonded to the substrate by using the ASM AD860M machine. After the bonding is finished, the light emitting zone is coated with a layer of mixture with silica gel and phosphor powder, and the phosphor powder is deposited on the surface of chip and substrate by centrifugal equipment.

The light source is mounted on the heatsink as the light source 2 is, the heat sink is the same as the heat sink used in the first light source, and the thermal conductivity of the silicone grease is the same. After the light source is installed, relevant tests are carried out. The HaasSuite (EVERFINE)

HASS-2000 is conducted with ambient temperature of 25 ℃, ambient humidity of 40%RH, atmospheric pressure of 101.5KPa. The initial test parameters are as follows: test voltage 133.9V, test current 1.501A, test power 200.9W, luminous flux 33042lm, luminous efficiency 164.46lm/W, color temperature 6599K, color rendering index 73.7.

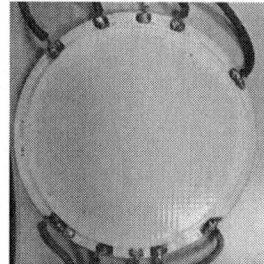

Figure 4: The multi-chip light source 2.

The sample is lighten for 1500 hours under the condition of ambient temperature 55℃. The final parameters are as follows: test voltage 133.9V, test current 1.5A, test power 200.8W, luminous flux 32310lm, luminous efficiency 160.93lm/W, color temperature 6669K, color rendering index 73.9.

Figure 4 is the test spectrum of the light source 2. The dominate wavelength is 487.3nm and the peak wavelength is 452nm.

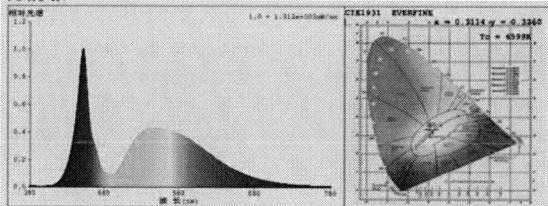

Figure 5: The test spectrum of the light source 2.

Lateral structure chip high-power light source on metal substrate

Figure 6 is light source 3, 200 W high power density on the copper substrate, copper size is 84 mm * 84 mm * 2 mm, materials is copper, thermal conductivity of 350 W/m·K, light emitting area is 65 mm * 65 mm square, chip is lateral structure blue chip with the size of 45 mil * 45 mil, chip working voltage is from 2.7 V to 3.2 V, working current is 350 mA, the chip is bonded to the copper substrate by using ASM AD860M machine. After completion of the die bonding, the light emitting zone is coated with a mixture with phosphor powder and silicone gel, the phosphor powder is deposited onto the surface of the chip and substrate using a centrifugal device.

The installation and silicone grease used are the same with the first two light sources. After the light source is installed, the HaasSuite (EVERFINE) HASS-2000 test is conducted with ambient temperature of 25.2 ℃, ambient humidity of 55%RH, atmospheric pressure of 101.5KPa. The initial test parameters are as follows: test voltage of 81.96V, test current of 2.436A, test power is 199.7W, luminous flux of 32245lm,

978-1-7281-5757-3/19 $31.00 © 2019 IEEE

luminous efficiency of 161.47lm/W, color temperature of 4766K, color rendering index is 82.5.

Figure 6: The multi-chip light source 3.

Under the condition of ambient temperature 55 ℃, the sample is continuously lighten for 1500 hours. The final parameters are as follows: test voltage is 82V, test current is 2.435A, test power is 199.7W, luminous flux is 31938lm, luminous efficiency is 159.93lm/W, color temperature is 4872K, color rendering index 82.9.

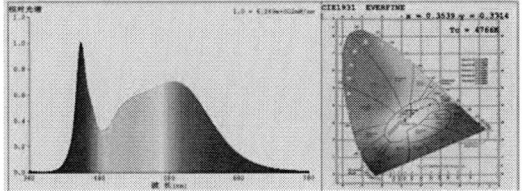

Figure 7: The test spectrum of the light source 3.

Figure 7 is the test spectrum of the light source, the dominate wavelength is 570.8nm and the peak wavelength is 453nm

Conclusions

1) The initial test parameters of light source 1 are as follows: test voltage 134.4V, test current 1.485A, test power 199.6W, luminous flux 32616lm, luminous efficiency 163.41lm/W, color temperature 5765K, color rendering index 71.7. The initial test parameters of light source 2 are as follows: test voltage 133.9V, test current 1.501A, test power 200.9W, luminous flux 33042lm, luminous efficiency 164.46lm/W, color temperature 6599K, color rendering index 73.7.The initial test parameters of light source 3 are as follows: test voltage of 81.96V, test current of 2.436A, test power is 199.7W, luminous flux of 32245lm, luminous efficiency of 161.47lm/W, color temperature of 4766K, color rendering index is 82.5.

The initial luminous efficiency of the three light sources are very close; The color temperature of the light source 2 is the highest, that of the light source 1 is lower, and that of the light source 3 is the lowest. The color rendering index of the light source 3 is higher than that of the light source 2 and 1, and the color rendering index of the light source 1 is close to that of the light source 2.

2) After aging test, the final parameters of light source 1 are as follows: test voltage 134.1V, test current 1.484A, test power 199W, luminous flux 33519lm, luminous efficiency 168.48lm/W, color temperature 5826K, color rendering index 71.9. The final parameters of light source 2 are as follows: test voltage 133.9V, test current 1.5A, test power 200.8W,

luminous flux 32310lm, luminous efficiency 160.93lm/W, color temperature 6669K, color rendering index 73.9.The final parameters of light source 3 are as follows: test voltage is 82V, test current is 2.435A, test power is 199.7W, luminous flux is 31938lm, luminous efficiency is 159.93lm/W, color temperature is 4872K, color rendering index 82.9.

After 1500 hours high-temperature aging experiment, the light efficiency of the light source 1 increases slightly, the light efficiency of the light source 2 decreases slightly, and the light efficiency of the light source 3 decreases slightly.

Acknowledgments

1. Supported by the National Key R&D Plan,

No. 2017YFB0403201

2. Supported by Project of science and technology research and development of Langfang, No.2019011032

References

[1]Zhang Jianping, Wen Shangsheng, Chen Yu. COB LED Thermal Performance Analysis Using Metal Substrate or Ceramic Substrate [J]. China light & lighting, 2016(10).

[2] liu songhao, li chunfei. Photonics technology and application [M].Part ii. Guangzhou: guangdong science and technology press, 2006.

[3] Ge baogui. Fluorescent Lighting Has a Brilliant Future in the Field of Industrial Lighting. Shanghai Energy Conservation, 2015(3) : 140-144.

[4] FANG Fu-bo, WANG Yao-hao, SONG Dai-hui, YU Bin-hai. Spectroscopic Analysis of White LED Attenuation. Chinese Journal of Luminescence, 2008, 29(2): 353-357.

[5] Wei yuefeng, Xu Minwei, Shen Yuanyuan. Study on the thermal reliability mechanism and evaluation method of a semiconductor light-emitting diode [J].

[5] Wei yuefeng, Xu Minwei, Shen Yuanyuan. Study on the thermal reliability mechanism and evaluation method of a semiconductor light-emitting diode [J]. Lamps and Lighting, 2016(1):4-6.

[6] QI Shu-qi, DING Shen-dong, QIN Hui-bin, ZHENG Peng, YU Yang, YU Dong. Thermal Performance Analysis of LED Based on the COB Technology [J]. Electronics and Packaging,2012,12(08):36-39. [2017-08-29].

[7] HU Ming-yu, WU Yi-ping, YANG Zhuo-ran. Thermal Design of High Power Plate LED COB Lighting Source [J]. Electronics Process Technology,2015,36(02):63-68. [2017-08-29].

[8]Wu Huiying , Qian Keyuan, Hu Fei, Luo Yi. study on thermal performances of flip-chip high-power white leds [J] .Journal of Optoelectronic • Laser, 2005 , 16(5) :511 - 514 .

[9] A rik M , Petro ski J , Weave r S .T her mal challenges in the futur e generatio n so lid state lig hting application[C] / / Pr oc . IEEE Intersocie ty Co nf . , 2002 : 113 -120 .

978-1-7281-5757-3/19 $31.00 © 2019 IEEE   109

[10] QI Shu-qi, DING Shen-dong, ZHENG Peng, QIN Hui-bin. Research on Optical Performance of LED by COB Packaging [J]. ELECTRONICS & PACKAGING, 2012, 12(3):6-9.

[11] MA Lu , LIU Jing. Latest Advancement of Thermal Management for High Power LED [J]. SEMICONDUCTOROPTOELECTRONICS,2010,31(01): 8-15. [2017-08-29].

[12] Richard K.Ulrich, William D.Brown. Advanced ElectronicPackaging（2nd Edition）[M]. New York: John

Wiley & Sons, 2010.

[13] Ganasan, J.R. Chip on chip（COC）and chip on board（COB）assembly on flex rigid printed circuit assemblies[J].IEEE transactions onelectronics packaging manufacturing.2006, 23（1）: 28-31.

Review of High Power Phosphor-Converted Light-Emitting Diodes

Yan Li[1,2], Yuming Yang[1,2], Huaiwen Zheng[1], Fei Yu[1], Qiao Liang[3], Hua Yang[1,2*], Xiaoyan Yi[1,2], Junxi Wang[1,2], Jinmin Li[1,2]

1. Center for Semiconductor Lighting, Institute of Semiconductors, Chinese Academy of Sciences, No.A35, QingHua East Road, Haidian District, Beijing P R China
2. Center of Materials Science and Optoelectronics Engineering，University of Chinese Academy of Sciences, Beijing 100049, China
3. Department of Computer, North China Institude of Science and technology
* Corresponding auther: huayang@semi.ac.cn

Abstract

With the increasing demand for high-brightness white light illumination, Researchers work on phosphor-converted light-emitting diodes (PC-LED) solid-state lighting from the begining. A variety of parameters play a significant role in determining the quality of PC-LED, which can meet the requirements of high power density, lightweight and small volume. Therefore, the encapsulant in PC-LED package , phosphors must have good thermal stability and optical properties. In order to comply with this demand, the effects of the encapsulant material, phosphor layer thickness, concentration, particle size, distance between the phosphor layer and the chip on the optical properties and thermal properties of the LED were summarized. It will be useful for future research on PC-LED.

1. Background

In recent years, the development of solid-state lighting has provided great potential for high-power white light-emitting diodes (WLED) to replace ordinary lighting and lighting equipment [1,2]. Most white LEDs are produced by mixing the blue light emitted by the LED chips with yellow light from yellow phosphor, called phosphor-converted white light-emitting diode (PC-LED) [3,4]. In such a pc-LED, the colour conversion element is usually composed of phosphor particles embedded in silica gel. There are three main forms of existence: dispersed in the encapsulant surrounding the chip, conformally coated on the chip, or away from the chip (remote phosphor) [5,6]. However, blue-light-excited WLED still has some problems, such as low color rendering index, uneven light distribution, low light extraction efficiency, the color shift caused by current and temperature [7,8].

Studies have shown that the optical properties of phosphor-converted LEDs are heavily dependent on the design of phosphor conversion components and are strongly influenced by the shape, material and arrangement within the LED package [9-12]. In addition to the direct influence on optical properties, phosphor conversion materials and encapsulants have a great influence on the thermal performance of LEDs during operation [13,14]. The matrix and encapsulant of phosphor conversion elements commonly used in LEDs are mainly epoxy resin and silicone, because the thermal conductivity of silicone is very low, the heat generated by the phosphor layer and temperature increasement is very high, and the silicone layer even appears to be carbonized [14]. In addition, due to the demand for high-brightness white light illumination, lighting equipment is developing in the direction of high current, high power density, lightweight and small volume to cope with market competitiveness [15]. A multi-chip LED package with chip-on-board structure (COB) and laser-driven white illumination were developed to achieve the desired lumen output, but then generate large thermal stress [16,17]. An increase in the temperature of the phosphor layer is usually accompanied by a decrease in the intensity of the luminescence, which means a reduction in the emission of the yellow light, and it is necessary to ensure that the temperature change in the LED package is as low as possible.

This paper mainly summarizes the influence of pc-WLED components (phosphor, encapsulant) on the optical properties and thermal properties of LEDs, including the type of materials, material parameters, package shape and other factors. Classification according to the way the phosphors are distributed, detailing the advantages and disadvantages of each category, and how to improve them.

2. Phosphor-converted LED

Representative research progress of white light illumination was introduced in this chapter. In addition, latest technology and research results of white light components (encapsulant materials and phosphors) were summarized.

2.1 Encapsulant material

The encapsulant material is a medium filled around the LED chip, as shown in Fig 1(a). On the one hand, the optically transparent encapsulant material can provide the necessary protection for the chip. On the other hand, due to the difference in refractive index between the chip and the air, total reflection is one of the factors that affect optimal light extraction. The encapsulant material can better increase light extraction by creating a refractive index gradient between the chip material and the air. Wiesmann et al. [18] studied the effects of three different phosphor mixed matrices on device temperature and light output: phosphor particles in silicone (refractive index of 1.41, thermal conductivity of 0.15W/m·K), particles in glass (refractive index of 1.52, thermal conductivity of 1.1W/m·K), and particles in alumina (refractive index of 1.77 ,thermal conductivity of 30W/m·K). As the thermal conductivity (alumina > glass > silica gel) increases, the phosphor layer temperature decreases and the temperature distribution are more uniform, but the light output decreases (refractive index alumina > glass > silica gel).

978-1-7281-5757-3/19 $31.00 © 2019 IEEE

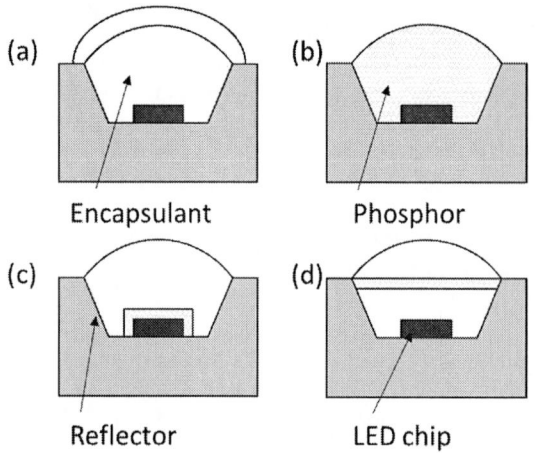

(a) Encapsulant (b) Phosphor (c) Reflector (d) LED chip

Fig.1 Scheme of different phosphor configurations: (a) no coating, (b) conventional coating, (c) conformal coating, (d) remote phosphor

Arik et al. demonstrated by simulation that the high thermal conductivity silicone can reduce the temperature of the phosphor layer by about 150 ℃ compared with ordinary silicone [19]. Subsequent studies have shown that an increase in the thermal conductivity of the mixed medium with the phosphor reduces the temperature of the phosphor layer [18,20,21]. This is because the thermal resistance of the phosphor layer is lowered, providing a better heat dissipation path for the heat generated by the phosphor layer. In addition to enhancing heat dissipation, Chen et al. [22] added silicone mixed with ZrO_2 nano-particles on the surface of the phosphor layer. Compared to conventional remote phosphor structures, the luminous flux of the ZrO_2-remote phosphor structure increases by 2.25% at a drive current of 120 mA. Rao et al. [23] integrated node-like, sheet-like and rod-like ZnO nanostructures into pc-LEDs. The results show that the scattering effect of ZnO nano-particles can effectively improve the uniformity of scattering energy. However, strong scattering is usually accompanied by strong absorption, which reduces the luminous efficiency. Li et al. [24] incorporated TiO_2 nanoparticles (TiO_2NPs) into silicone. By controlling the concentration of TiO_2NPs, a tunable emission spectrum between 430-600 nm and a wide wavelength range can be obtained. When the concentration of TiO_2NPs is 0.05 wt%, the luminescence intensity can be increased by 31%. This is because the light scattering of TiO_2 increases the likelihood of light reabsorption. Zheng et al. [25] proposed a double encapsulation layer structure in which TiO2 nanoparticles and an auxiliary encapsulation layer were added, which improved the light extraction efficiency of the COB package by 65%.

In order to develop environmentally benign materials with improved properties, a deep eutectic solvent is prepared by selecting oxalic acid dihydrate and choline chloride to remove lignin from balsa wood [26]. At the same time, multiple-color-emission carbon dots (CD) were synthesized by controlling the molar ratio and the reaction temperature of citric acid to urea. Poly (acrylic acid) (PAA) was chosen as the filled polymer because the index of refraction between delignified wood (n = 1.54) and PAA (n = 1.51) matched throughout the visible spectrum. The CD and PAA were filled into the delignified wood by in-situ polymerization to prepare

a transparent wood film embedded with a multicolor CD having a thickness of 550 μm and a transmittance of up to 85% in the range of 400-800nm. The encapsulated white LED has excellent optical characteristics and good luminescence stability. When the CIE color coordinates is (0.33, 0.32), the CRI value is 83 and the CCT value is 5237K.

2.2. Parameters of the phosphor

In PC-LEDs, there are three different distributions of phosphor: conventional coating, conformal coating and remote phosphor (Fig.1(b-d)). Phosphor layer thickness [13,14,27-29], concentration [13,29,30-32], size [33,24], the distance between the phosphor layer and the chip [5,19,27,30,35-37] will affect the optical performance and thermal performance of the white LED. This section reviews and discusses these parameters of the phosphor.

2.2.1. Thickness of the phosphor layer

As the thickness of the phosphor layer increases, the total radiation power decreases [32], Hu et al. [28] used Monte Carlo ray tracing simulation to study the light extraction efficiency (LEE), correlated color temperature (CCT) and angular color uniformity (ACU) of conformal phosphor coating. The LEE of the blue light is reduced, and the LEE of the yellow light is increased because when the thickness of the phosphor layer is increased, more blue light is absorbed and more yellow light is re-emitted. At the same time, it is observed that as the thickness increases, the CCT decreases and the ACU is greatly reduced. The thickness can be optimized to achieve good optical properties.

Yan et al. [27] assumed that the amount of phosphor was fixed to produce white light with a correlated color temperature of 4600 K, so that as the thickness of the phosphor layer increased, the concentration of the phosphor decreased, and eventually the temperature of the phosphor layer increases as the thickness increases. Later studies have also shown that as the thickness of the phosphor layer increases, the phosphor layer temperature increases [14]. This is because as the thickness of the phosphor layer increases, the average distance of the heat conduction increases and the phosphor concentration decreases, resulting in a decrease in the effective thermal conductivity of the phosphor layer, and thus the phosphor temperature increases.

2.2.2. phosphor concentration

Hu et al. [30] simulated remote phosphors and directly coated phosphors. The optical simulation was first performed to calculate the heat flux accumulated by the chip and phosphor layer light absorption, and then the FEM model was built to simulate the temperature field of the LED package. The results show that as the concentration of the phosphor layer increases, the heat of the two phosphor coating packages is greatly increased. The increase in heat is mainly attributed to the fact that high-concentration phosphors generate more heat, while the thermal conductivity of silicone is very low, and the heat generated in the phosphor is difficult to transfer to the surrounding air. Yan et al. [27] believed that as the concentration of phosphor increases, the temperature of the phosphor decreases first and then increases slightly. It can be explained in two aspects: on the one hand, the increase in concentration means that the rate of heat generation increases

978-1-7281-5757-3/19 $31.00 © 2019 IEEE   112

and tends to increasing the temperature of the phosphor; on the other hand, the phosphor concentration is increased, so that the phosphor particles are arranged more densely and the thermal conductivity is improved.

You et al. [31] considered the phosphor concentration-dependent characteristics of white LEDs under different current regulation conditions. The conversion efficiency of the phosphor of the white LED under constant current driving is lower than that of the phosphor driven by the pulse current. As the phosphor concentration increases, the LED exhibits relatively stable optical characteristics at higher drive currents. Nguyen T found that the CCT and output power of LEDs in planar and convex lens packages decrease with increasing in phosphor concentration, and that increasing the concentration of phosphors increases the possibility that blue light will be absorbed and converted to yellow light [32].

2.2.3 Phosphor particle size

The light loss in the traditional white LED is mainly caused by the scattering and reflection of the excitation light back into the LED chip, about 40% of the light passes through the phosphor layer, and about 60% of the light is reflected back [6]. The photoluminescence (PL) of YAG:Ce phosphor particles can be improved by optimizing the phosphor particle size [38], and their results indicate that the smaller average phosphor size has a lower PL intensity.

Nguyen The Tran et al. [55] studied the effect of YAG:Ce phosphor particle size on the lumen output and the conversion efficiency of conventional phosphors and remote phosphor packages. As the phosphor particle size increases from nanometer size to submicron size, the lumen output decreases and then increases as the particle size continues to increase to the micron size. Sommer et al. [39] examined the effect of phosphor particle size on angular color uniformity. They found that smaller particles were suitable for higher color temperatures and larger particles were suitable for lower color temperatures. Shi et al. [40] considered the effect of phosphor particle size on light extraction, indicating that large particles (20 μm) and nanoparticles (0.1 μm) can enhance light extraction.

Liu et al. [34] constructed a conformal phosphor-coated CREE white LED as an optical model, and analyzed the effects of phosphor particle size on the luminous flux and angular color uniformity of white LEDs. The simulation results indicate that the 2 μm particles show the lowest luminous flux and the best angular color uniformity, and for particle sizes larger than 2 μm, the luminous flux can be enhanced but the angular color uniformity is deteriorated.

2.2.4. Distance between the phosphor layer and the chip

Arik et al. [19] found that the increase in the distance between the phosphor layer and the LED chip resulted in an increase in the temperature of the phosphor layer, while the thermal load of the LED chip was reduced owing to the increased distance. Huang et al. [35] proposed that the temperature distribution of the distance between the phosphor layer and the chip has little effect without considering the heat generated by the phosphor layer, and the maximum temperature appears in the phosphor layer in consideration of the heat generated in the phosphor layer. Studies have further shown that as the distance between the LED chip and the phosphor layer increases, the temperature of the phosphor layer also increases [36]. In this case, the phosphor layer is far away from the chip, and it is difficult for heat to dissipate through the chip from the heat sink, and it can only be dissipated through a path with a large thermal resistance [5].

Yan et al. [27] showed that the remote phosphor package structure reduces the temperature of the LED chip as the phosphor layer moves away from the chip (ie, from the conformal phosphor model to the remote phosphor model). However, compared with the conformal phosphor package structure, the operating temperature of the phosphor layer in the remote phosphor package structure is higher, and the light output power is increased. The reason is that when the phosphor is directly coated on the chip in the conformal phosphor model, a large amount of backscattered light from the phosphor particles is absorbed by the chip, so light extraction from the phosphor coating and the chip array is significantly reduced. In addition, the lowest temperature of the phosphor layer in the conformal phosphor model can be attributed to two reasons: ① the heat source (phosphor particles) is closer to the chip, and can be better cooled by the heat sink; ② the effective thermal conductivity of the phosphor layer in the conformal phosphor model is higher, because the phosphor particles are arranged more densely and the thermal conductivity is improved [41]. Fan et al. [37] considered that the thermal isolation package has a lower surface temperature than the phosphor layer of the conventional phosphor package because the thermal isolation package prevents the heat generated by the chip from being transferred to the phosphor layer.

3. Summary and Outlook

In this review, the latest technologies and research results of encapsulant materials and phosphor conversion elements in PC-LED packages are introduced to improve thermal and optical performance in white light illumination. Despite the above results, there are still some unresolved problems in white lighting:

(1) The doping of ZrO_2, ZnO, TiO_2 and other nanoparticles can not only increase the thermal conductivity of the phosphor mixed medium, but also the scattering effect of the nanoparticles can effectively improve the uniformity of the scattering energy and increase the light extraction. However, the matrix is limited to materials such as epoxy resin and silicone, and the encapsulant material may change color under high power density exposure, resulting in reduced optical performance [8,42]. Improved encapsulants should be developed to address these issues. The improved encapsulant should have good thermal stability, high transmission, high refractive index, and a coefficient of thermal expansion that matches other packaging materials.

(2) For different phosphor parameters, such as phosphor layer thickness, concentration, size, phosphor and chip distance, etc., an optimized design should be performed to achieve optimal optical and thermal performance.

(3) For the conformal phosphor package structure, a good ACU can be achieved, but the reflection from the phosphor increases the absorption of light by the chip, and some optimized conformal phosphor package structures can improve the luminous efficiency of the LED [43,44]. The

optimization of the phosphor deposition structure has a certain auxiliary effect on the package temperature, ACU, and luminous efficiency. The remote phosphor reduces the absorption of the down-converted phosphor emission by the chip, which helps to increase photon extraction [45]. However, the operating temperature is higher [36,44,59,78], new phosphor conversion elements such as PiG and CPP have been proposed, in which PiG is easy to manufacture, the refractive index is adjustable, and the chromaticity coordinates can be controlled by mixing different phosphors [47-49]. Therefore, the use of a single-layer graphene-coated PiG can effectively act as a heat dissipation medium on the surface of the phosphor layer and reduce thermal quenching [50]; introduced phosphor-aluminum composite with high thermal conductivity of 31.6 $Wm^{-1}K^{-1}$ and excellent luminous flux and chromatic stability at 4W laser [51].

Much progress has been made in optimizing encapsulant materials and phosphor conversion elements in PC-LED packages, However, in order to meet the demand for high-brightness white light illumination to adapt to market competitiveness, the development of new encapsulant materials and thermally stable phosphor conversion components is required to a large extent. At the same time, there is still a need to further improve color quality and reduce costs.

Acknowledgments

Thanks are due to the financial support provided for this study by the (1) National Key R&D Plan (No. 2017YFB0403201) and (2) Project of science and technology research for development of Langfang (No. 2019011032).

References

1. Krames M R. et al, Status and Future of High-Power Light-Emitting Diodes for Solid-State Lighting[J]. Journal of Display Technology, 2007, 3(2), pp. 160-175.
2. Li S. et al, Color Conversion Materials for High-Brightness Laser-Driven Solid-State Lighting[J]. Laser & Photonics Reviews, 2018.
3. Narendran N, Improved performance white LED[J]. Proceedings of SPIE - The International Society for Optical Engineering, 2005,vol. 5941, pp. 594108.
4. Li J S. et al, ACU Optimization of pcLEDs by Combining the Pulsed Spray And Feedback Method[J]. Journal of Display Technology, 2016, 12(10):1-1.
5. Yan B. et al, Influence of phosphor configuration on thermal performance of high power white LED array[J]. 2013.
6. Narendran N. et al, Extracting phosphor-scattered photons to improve white LED efficiency[J]. Physica Status Solidi (A) Applications and Materials, 2005, 202(6), pp. R60-62.
7. Ueda J. et al, Insight into the Thermal Quenching Mechanism for $Y_3Al_5O_{12}:Ce^{3+}$ through Thermoluminescence Excitation Spectroscopy[J]. Journal of Physical Chemistry C, 2015, 119.
8. Meneghini M. et al, Extensive analysis of the degradation of phosphor-converted LEDs[C] Proceedings of SPIE - The International Society for Optical Engineering. 2009.
9. Xiao H. et al, Improvements on Remote Diffuser-Phosphor-Packaged Light-Emitting Diode Systems[J]. IEEE Photonics Journal, 2014, 6(2):1-8.
10. Shuai Y. et al, Angular CCT Uniformity of Phosphor Converted White LEDs: Effects of Phosphor Materials and Packaging Structures[J]. IEEE Photonics Technology Letters, 2011, 23(3):137-139.
11. Xie B. et al, Effect of packaging method on performance of light-emitting diodes with quantum dot phosphor [J]. IEEE Photonics Technology Letters, 2016: 1-1.
12. You J P. et al, Light extraction enhanced white light-emitting diodes with multi-layered phosphor configuration. [J]. Optics Express, 2010, 18(5): 5055-5060.
13. Fulmek P. et al, On the Thermal Load of the Color-Conversion Elements in Phosphor-Based White Light-Emitting Diodes [J]. Advanced Optical Materials, 2013, 1(10): 753-762.
14. Luo X. et al, Phosphor self-heating in phosphor converted light emitting diode packaging [J]. International Journal of Heat and Mass Transfer, 2013, 58(1-2): 276-281.
15. Moon S H. et al, A single unit cooling fins aluminum flat heat pipe for 100W socket type COB LED lamp[J]. Applied Thermal Engineering, 2016, pp. 1164-1169.
16. Ying S P , Shen W B , Thermal Analysis of High-Power Multichip COB Light-Emitting Diodes With Different Chip Sizes[J]. IEEE Transactions on Electron Devices, 2015, 62(3):896-901.
17. Tsai PY. et al, High-power LED Chip-on-Board Packages with Diamond-like Carbon Heat-spreading Layers[J]. Journal of Display Technology, 2016, 12(4):357-361 .
18. Wiesmann C. et al, Estimating the performance of remote phosphor SSL devices by simulations[J]. Proceedings of SPIE - The International Society for Optical Engineering, 2012, 8550(1):85502G-85502G-5.
19. Arik M. et al, Effects of Localized Heat Generations Due to the Color Conversion in Phosphor Particles and Layers of High Brightness Light Emitting Diodes[C], Asme International Electronic Packaging Technical Conference & Exhibition. 2003.
20. Shih B J. et al, Study of temperature distributions in pc-WLEDs with different phosphor packages[J]. Optics Express, 2015, 23(26):33861-33869.
21. Wang P C . Improving Performance and Reducing Amount of Phosphor Required in Packaging of White LEDs With TiO2-Doped Silicone[J]. IEEE Electron Device Letters, 2014, 35(6):657-659.
22. Chen H C. et al, Improvement in uniformity of emission by ZrO_2 nano-particles for white LEDs[J]. Nanotechnology, 2012, 23(26):265201.
23. Rao L. et al, Effect of ZnO nanostructures on the optical properties of white light-emitting diodes[J]. Optics Express, 2017, 25(8):A432.
24. Li J S. et al, Investigation of the Emission Spectral Properties of Carbon Dots in Packaged LEDs using TiO_2 Nanoparticles[J]. IEEE Journal of Selected Topics in Quantum Electronics, 2017,23(5).
25. Zheng H. et al, Optical Performance Enhancement for Chip-on-Board Packaging LEDs by Adding TiO2/Silicone Encapsulation Layer[J]. IEEE Electron Device Letters, 2014, 35(10):1046-1048.
26. Zhihao B. et al, Transparent Wood Film Incorporating Carbon Dots as Encapsulating Material for White Light-Emitting Diodes[J]. ACS Sustainable Chemistry &

Engineering, 2018, pp. 9314-9323.

27. Yan B. et al, Can Junction Temperature Alone Characterize Thermal Performance of White LED Emitters?[J]. IEEE Photonics Technology Letters, 2011, 23(9):555-557.

28. Hu R , Luo X , Liu S . Study on the Optical Properties of Conformal Coating Light-Emitting Diode by Monte Carlo Simulation[J]. IEEE Photonics Technology Letters, 2011, 23(22):1673-1675.

29. Mou X , Narendran N. et al, Optical and thermal performance of a remote phosphor plate[J]. Proceedings of SPIE - The International Society for Optical Engineering, 2014, 9190:91900Q-91900Q-6.

30. Hu R , Luo X , Zheng H . Hotspot Location Shift in the High-Power Phosphor-Converted White Light-Emitting Diode Packages[J]. Japanese Journal of Applied Physics, 2012, 51(9):09MK05.

31. You J P. et al, Phosphor-Concentration-Dependent Characteristics of White LEDs in Different Current Regulation Modes[J]. Journal of Electronic Materials, 2009, 38(6):761-766.

32. Tran N T , Shi F G . Studies of Phosphor Concentration and Thickness for Phosphor-Based White Light-Emitting-Diodes[J]. Journal of Lightwave Technology, 2008, 26(21):3556-3559.

33. Tran N T. et al, Effect of Phosphor Particle Size on Luminous Efficacy of Phosphor-Converted White LED[J]. Journal of Lightwave Technology, 2009, 27(22):5145-5150.

34. Liu ZY. et al, Effects of YAG: Ce Phosphor Particle Size on Luminous Flux and Angular Color Uniformity of Phosphor-Converted White LEDs[J]. Journal of Display Technology, 2012, 8(6): 329 -335.

35. Hwang J H. et al, Study on the effect of the relative position of the phosphor layer in the LED package on the high power LED lifetime[J]. Physica Status Solidi, 2011, 7(7-8):2157-2161.

36. Dong M. et al, Thermal analysis of remote phosphor in LED modules[J]. Journal of Semiconductors, 2013, 34(5):053007.

37. Fan B F B. et al, Study of Phosphor Thermal-Isolated Packaging Technologies for High-Power White Light-Emitting Diodes [J]. IEEE Photonics Technology Letters, 2007, 19(15): 1121-1123.

38. Yuan F , Ryu H . Ce-doped YAG phosphor powders prepared by co-precipitation and heterogeneous precipitation[J]. Materials Science & Engineering B (Solid-State Materials for, Advanced Technology), 2004, 107(1):14-18.

39. Sommer C. et al, The Effect of the Phosphor Particle Sizes on the Angular Homogeneity of Phosphor-Converted High-Power White LED Light Sources[J]. IEEE Journal of Selected Topics in Quantum Electronics, 2009, 15(4):1181-1188.

40. Shuai Y , Tran N T , Shi F G . Nonmonotonic Phosphor Size Dependence of Luminous Efficacy for Typical White LED Emitters[J]. IEEE Photonics Technology Letters, 2011, 23(9):552-554.

41. Furgel I A. et al, Thermal conductivity of polymer composites with a disperse filler[J]. Journal of Engineering Physics and Thermophysics, 1992, 62(3):335-340.

42. Yang S C. et al, Failure and degradation mechanisms of high-power white light emitting diodes[J]. Microelectronics Reliability, 2010, 50(7):959-964.

43. Meng Q. et al, Optimized Self-Adaptive Phosphor Coating Structure of White LEDs by Conventional Dispensing Method[J]. IEEE Transactions on Components, Packaging and Manufacturing Technology, 2017:1-4.

44. Wu J. et al, Realization of Conformal Phosphor Coating by Ionic Wind Patterning for Phosphor-Converted White LEDs[J]. IEEE Photonics Technology Letters, 2017, 29(3):299-301.

45. Kim J K , Luo H , Schubert E F , et al. Strongly Enhanced Phosphor Efficiency in GaInN White Light-Emitting Diodes Using Remote Phosphor Configuration and Diffuse Reflector Cup[J]. Japanese Journal of Applied Physics, 2005, 44(21), pp. L649-L651.

46. Ma Y. et al, A modified bidirectional thermal resistance model for junction and phosphor temperature estimation in phosphor-converted light-emitting diodes [J]. International Journal of Heat and Mass Transfer, 2017, 106:1- 6.

47. Zhang X. et al, All-Inorganic Light Convertor Based on Phosphor-in-Glass Engineering for Next-Generation Modular High-Brightness White LEDs/LDs[J]. ACS Photonics, 2017, 4(4):986-995.

48. Zheng P. et al, A unique color converter architecture enabling phosphor-in-glass (PiG) films suitable for high-power and high-luminance laser-driven white lighting[J]. ACS Applied Materials & Interfaces, 2018, 10:14930-14940.

49. Zhang X. et al, Facile Preparation and Ultra-stable Performance of Single-Component White-Light-Emitting Phosphor-in-Glass used for High-Power Warm White LEDs[J]. ACS Applied Materials & Interfaces, 2015, 7(51).

50. Kim E. et al, Effective Heat Dissipation from ColorConverting Plates in High-Power White Light Emitting Diodes by Transparent Graphene Wrapping[J], ACS Nano, 2016, 10(1):238-245.

51. Park J , Kim J , Kwon H , Phosphor-Aluminum Composite for Energy Recycling with High-Power White Lighting[J]. Advanced Optical Materials, 2017:1700347.

Effects of Different Ratiosof Red and Blue Light on the Morphology and Photosynthetic Characteristics of *Anoectochilus roxburghii*

Rui Li[1], Yinghui Mu[2],Hongyu Wei[1], Lixue Zhu[1], Wenqi Tang[1], Zhiyu MA[1]*

1. Zhongkai University of Agriculture and Engineering,No. 500, Zhongkai Road, Haizhu District, Guangzhou;
2. South China Agricultural University, No. 483, Wushan, Tianhe District, Guangzhou,
Email 931291088@qq.com

Abstract

This paper studied the effects of different light qualities on the growth and photosynthesis of *Anoectochilusroxburghii*. Under uniform temperature, humidity, CO_2 concentration, and photosynthetic photon flux density conditions, the morphological indexes, chlorophyll,photosynthetic rate,stomatal conductance, and transpiration rate of nine kinds of red and blue LED combination light sources were analyzed.The results showed that the 1:1 blue to red light treatment was more conducive to the accumulation of fresh weight of*A.roxburghii* than other light; these conditions also caused an increase in the overall plant height and stem diameter of*A.roxburghii*. White light treatment had a significant effect on the accumulation of abovegroundorgansof *A.roxburghii*. The net light and rate were the highest under the white light treatment, and were negatively correlated with intercellular CO_2 concentration. Blue light treatment promoted the increase of stomatal conductance and the transpiration rate of *A. roxburghii*, but the net photosynthetic rate wasthe lowest compared with other light treatments. The greater the proportion of red light, the lower the chlorophyll content, and the difference was significant. The blue chlorophyll content has a significantly higher chlorophyll content.

Keywords: *Anoectochilusroxburghii*, light quality, photosynthesis, growth characteristics

0 Introduction

*Anoectochilus*roxburghii is a famous medicinal plant known as the "King of Medicine" and "Golden Grass" in tropical and subtropical regions of China. As aperennial herb [1], *A.roxburghii*is widely distributedin China, Japan, Nepal, Sri Lanka, India, and other regions [2], mainly distributed in Taiwan [3], Fujian [4] waiting for the southeastern region. Due to its remarkable medicinal effects, it has attracted the attention of researchers. *A.roxburghii*is primarily used to treat hypertension, diabetes, rheumatism, dysplasia, and other symptoms, and the effect is remarkable [5].*A. roxburghii*is ashade plantthat has strict requirements on the growth environment, especially the temperature and illumination requirements. Because of habitat destruction, pollution, overharvesting, and climate change, wild populations of *A.roxburghii* are decreasing. This species also has weak breeding abilities, which also impacts the wild population. In recent years, artificial reproduction has been focused on, but related researchhas primarily addressed rapid propagation techniques, as well as the scale breeding of different growth

environments.Fewer studies have been conducted on the growth of *A.roxburghii*under different light qualities.

Light is used as energy for plant growthand as a light signal to regulate its morphogenesis [6]. In this paper, we studied the effects of different light qualities on the growth of *A.roxburghii* in order to increase the yield and quality of *A.roxburghii* and provide a reference for actual production and yield improvement.

An LED is a kind of semiconductor electronic component that can convert electrical energy into light energy. It is a high-efficiency energy-saving light source with a single wavelength and simple combination of light quality. At present, it is considered to be the ideal light source for plant factories. LEDs have excellent characteristics such as energy saving, easy adjustment, beam concentration and easy maintenance[7]. They are widely used in light energy experiments in plant factories,and with the development of high brightness and multi-color LEDs,their application is expanding. The use of LEDs is an important way to achieve plant factory production by artificial light cultivation. As the core technology of artificial light in plant factories, the identification and optimization of LED light formula parameters is to improve plant productivity and reduce labor is key [8]. Therefore, this paper uses an intelligent LED combined light source system in a closed plant factory to explore the effects of different combinations of spectral qualities on the growth and physiological characteristics of *A.roxburghii*, in order to provide a reference for the artificial cultivation of *A.roxburghii*.

1Materials and methods
1.1 Materials

This study was conducted in the closed artificial light-type cultivation room of the Precision Agricultural Intelligent Equipment Technology Innovation Team of Zhongkai Agricultural Engineering College of Guangdong Province. The ambient temperature was set to 24 °C during the day and 20 °C at night, and the air humidity was maintained at 60%. The material used in the test was the *A. roxburghii* from Fujian, which was transplanted by Geo (Xiamen Biotechnology Co., Ltd. *A.roxburghii*is bottled seedlings.Two weeks after refining, the plants were carefully removed from the tissue culture bottles and cleaned.They were then transplantedin plastic trays (50 cm X 30 cm X 5 cm) on September 6, 2018. A total of 30 plants were planted in each tray, and the culture medium was perlite: peat: bark = 1:2:4 mixed.Plants were irrigated using a chassis sink.

1.2 Experimental design

The experiment used a combined light source (self-made) based on red (658 nm), blue (456 nm), and white LEDs. The light quality treatments were blue (B), red (R), blue: red 9:1 (B9R1), blue:red 6:1 (B6R1), blue:red 3:1 (B3R1), blue: red 1:1 (B1R1), blue:red 1:3 (B1R3), blue:red 1:6 (B1R6), blue:red 1:9 (B1R9), and white light (WL). Two trays of seedlings (60 in total) were placed under each treatment, the photoperiod was set to 12h, and the light intensity was 40 ± 2 $\mu mol/m^2 \cdot s$. Figure 1 shows the spectra of the combined light sources for each process.

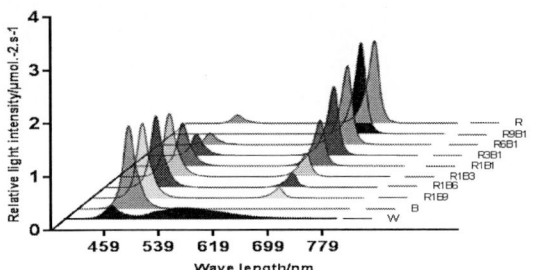

W=white light,R=red light,B=bluelight,
R9B1=red:blue9:1,R6B1=red:blue6:1,R3B1=red:blue3:1,
R1b1=red:blue1:1,R1B3=red:blue1:3,R1B6=red:blue1:6,R1B9=red:blue1:9

Fig. 1 Spectral composition of each processed LED light source

1.3 Experiment methods

1.3.1 Growth indicators

After treatment with different light qualities for 30 days, five plants were randomly collected under each treatment, washed with water, and the surface moisture was removed. Plant height was measured with a ruler and the base of the stem was measured with a Vernier caliper. After measuring the fresh weight with an electronic balance, it was dried in an oven at 105 °C for 5 min, then at 65 °C until a constant weight was reached to determine the dry weight.The water contentwas calculated as water content = (fresh weight - dry weight) / fresh weight X 100%. Leaf area was obtained using scanner (HP Laser JetVI1005 MFP) and PhotoshopCS6 software. The leaf image of the second fully expanded leaf of *A.roxburghii* was obtained. The image was analyzed based on Image-Pro Plus 7.0 image analysis software. The leaf area of *A.roxburghii* was calculated [9].

1.3.2 Photosynthetic characteristics

According to the method of DIAO Kai [10], the photosynthesis rate (Pn), stomatal conductance (Gs), and transpiration rate of the photosynthetic parameters of *A.roxburghii* were measured using a Li-6400 photosynthetic apparatus every 7 days after the start of the test treatment.Photosynthetic indicators, such as intercellular CO_2 concentration (Ci), were measured and compared. Three*A. roxburghii* were randomly selected in each treatment and measurements were repeated in triplicate.

1.3.3 Data Processing

The data were analyzed by ANOVA using SPSS Statistical software v 20. Duncan's method was used to determine whether differences were significantusing Excel mapping, with a significant level of 5% (P < 0.05).

2 Results and analysis

2.1 The effect of LED light quality on the growth of *A.roxburghii*

2.1.1The effect of LED light quality on the stem diameter of *A.roxburghii*

There were significant differences in plant height between *A.roxburghii* under different light quality treatments (Fig. 2-a). In theB1R1 treatment, *A. roxburghii* was significantly taller than *A. roxburghii* in other treatments; while in blue light (B), blue red 9:1 (B9R1), blue red 3:1 (B3R1) and blue red,The plant height in the B1R1 treatment showed a significantly decreasing trend, indicating that increasing blue light inhibited the elongation of *A. roxburghii*internodes,and affecting the overall height. In B1R3, B1R6,B1R9, and Rand the difference in plant height of *A.roxburghii* under treatment with a large proportion of red light was not significant. The plant stem diameter showed a different overall trend(Fig. 2-b). As the proportion of red light in the light source increased, the stem size of*A. roxburghii*decreased.In theB1R1 treatment, the stems of *A.roxburghii* were the largest, and the stems of *A. roxburghii* plants under the R treatment were the smallest. This indicates that under controlled environmental conditions, the cultivation of *A. roxburghii*was better than under natural conditions.Increasing the proportion of blue light in the cultivation light source was conducive to dwarf *A.roxburghii*with robust growth.

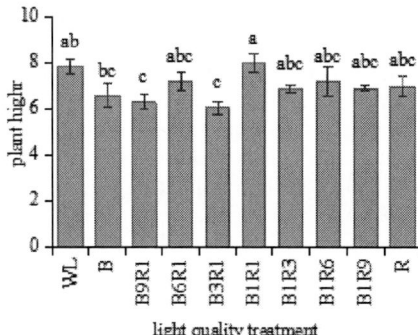

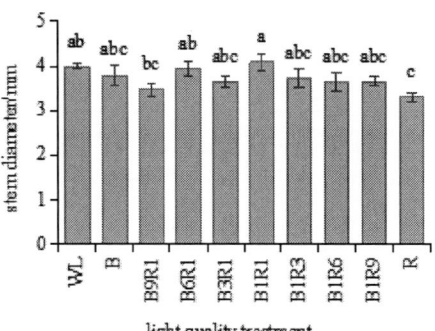

Fig. 2 Effect of LED light quality on plant height and stemdiameter of *A. roxburghii*

978-1-7281-5757-3/19 $31.00 © 2019 IEEE

2.12 The effect of LED light quality on the biomass of *A. roxburghii*

As shown in Table 1,different light quality treatments have significant effects on the dry weight of the shoots, the dry weight of the roots, and the total fresh weight. In the WL and B1R1 treatments, the fresh weight of the whole plant was significantly greaterthan other treatments, whereas the B9R1 treatment had the lowest fresh weight.Compared with the WL treatment, the B1R1 treatment, it was reduced by 37.94% and 36.75%, respectively. These results indicate that the B9R1 treatment is not conducive to the accumulation of biomass of *A. roxburghii*. The maximum dry and fresh weight of the aboveground organsoccurredin theWL treatment, followed by the B1R1 treatment, but under the white light (WL) and red and blue 1:1 (B9R1) treatment. The difference between dry weight and fresh weight was not significant; the minimum value of dry and fresh weight of the aboveground organsoccurred in theB9R1 treatment.Compared with WL and B1R1 treatments, there was a decrease of 34.75% and 31.14% in the dry weight of the aboveground organsdecreased by 53.3% and 34.38%, respectively. The maximum dry and fresh weight of the underground organs was observed in the B1R1 treatment, and the minimum dry and fresh weight occurred in theB9R1 treatment, compared with the underground organsof the B1R1 treatment. The fresh weight was reduced by 49.23% and 46.15%, respectively. The dry weight of the aboveground and underground parts of the pure red R treatment was lower, indicating that red light is not conducive to the biomass accumulation of Anoectochilusroxburghii.

2.1.3 The effect of LED light quality on chlorophyll content of *A.roxburghii*

Under the WL treatment, the chlorophyll content of *A. roxburghii*was the highest, followed by the B treatment (Fig. 3).Then, as the proportion of red light added to the combined light source increased, the chlorophyll content of *A. roxburghii* decreased. However, when the proportion of red light increased to a certain value (B1R3), the effect of increasing the proportion of red light on chlorophyll content was not significant, indicating that red light is not conducive to the accumulation of chlorophyll content and has an inhibitory effect on the increase of chlorophyll content in*A. roxburghii*.

Table 1 Effect of LED light quality on the fresh weight and dry weight of *Anoectochilusroxburghii* organs

Light Quality	Total fresh weight	Shoot fresh weight	Root fresh weight	Shoot dry weight	Root dry weight
WL	4.27±0.42a	3.05±0.28a	1.22±0.16a	0.45±0.09a	0.12±0.02a
B	3.31±0.51ab	2.21±0.37ab	1.11±0.15a	0.23±0.04b	0.09±0.01ab
B9R1	2.65±0.22b	1.99±0.06b	0.66±0.16b	0.21±0.01b	0.07±0.02b
B6R1	3.98±0.25ab	2.83±0.18ab	1.15±0.07a	0.31±0.03b	0.09±0.01ab
B3R1	3.32±0.56ab	2.32±0.40ab	1.01±0.17ab	0.27±0.06b	0.10±0.02ab
B1R1	4.19±0.58a	2.89±0.39ab	1.30±0.20a	0.32±0.04b	0.13±0.02a
B1R3	2.95±0.36ab	2.06±0.30ab	0.89±0.07ab	0.24±0.04b	0.09±0.01ab
B1R6	3.38±0.49ab	2.15±0.45ab	1.23±0.09a	0.24±0.04b	0.07±0.01b
B1R9	3.51±0.27ab	2.53±0.22ab	0.98±0.05ab	0.27±0.02b	0.08±0.00b
R	3.02±0.20ab	2.08±0.17ab	0.94±0.06ab	0.20±0.02b	0.11±0.01ab

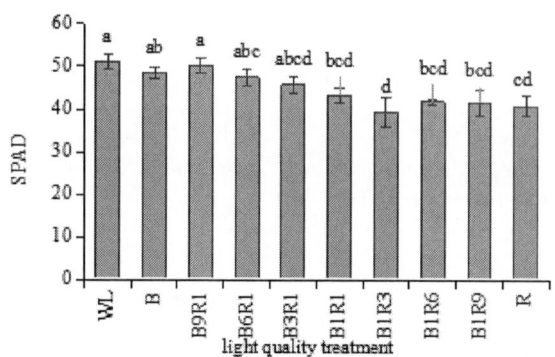

Fig. 3 Effect of LED light quality on chlorophyll content of *A. roxburghii*

2.2 Influence of LED light quality on photosynthetic characteristics of *A. roxburghii*

As shown in Table 2,the net photosynthetic rate (Pn) is the greatest in the WL treatment, which is significantly higher than other treatments. In the combined light source treatments, the net Pn was the lowest under B treatment.The net photosynthetic rate of *A. roxburghii* increased with the increase of the proportion of red light in the light source, and the net photosynthetic rate when the ratio of red and blue reached 1:1. When the maximum is 2.5 μmol/(m·s), the red light ratio was further increased and the net photosynthetic rate decreased. The stomatal conductance (Gs) reached the highest value in the B treatment, but when the proportion of red light was increased in the light source, the Gsshowed an upward trend, and rose to 0.035 mmol/m^2.When the red to blue ratio was 3:1, the Gsdecreased again as the proportion of red light increased. The intercellular CO_2 concentration (Ci) was the highest in the B1R3 treatment, and the intercellular CO_2 concentration showed a significant downward trend with the increase of the red light ratio. The intercellular CO_2 concentration was the lowest in the WL treatment. The Pnwas negatively correlated. The transpiration rate reached the highest value in the blue light treatment, but as the proportion of blue light decreased, the transpiration rate continued to increase. When the proportion of red and blue was 1:1, the transpiration rateincreased to 0.41mmol/m^2 · s, and then increased with the proportion of red light.At the same time, the transpiration rate gradually decreased, and the lowest transpiration rate occurred when the ratio of blue to red was 1:9.

Table 2 Effect of LED light quality on photosynthetic characteristics of *A. roxburghii*

Light Quality	Pnμmol/(m·s)	Gs mmol/(m^2·s)	Ci (μmol/mol)	Trmmol/(mol·s)
W	2.99±0.21b	0.020±0.0004c	311.15±15.50b	0.32±0.007cd
B	1.34±0.03a	0.045±0.0013de	400.14±2.52c	0.53±0.021d

B9R1	1.61±0.04b	0.012±0.0002f	322.78±2.55bc	0.20±0.002f
B6R1	1.39±0.11b	0.020±0.0018de	408.19±3.14a	0.33±0.026d
B3R1	1.79±0.31b	0.035±0.0038cd	341.13±19.85a	0.46±0.046bcd
B1R1	2.50±0.27b	0.027±0.0019b	347.82±23.39bc	0.41±0.030ab
B1R3	1.69±0.15b	0.021±0.0016d	428.72±10.58a	0.33±0.025d
B1R6	1.43±0.17b	0.025±0.0018ef	402.84±7.12a	0.39±0.027ef
B1R9	1.36±0.21a	0.016±0.0008c	407.68±14.38bc	0.25±0.011bc
R	1.60±0.19b	0.027±0.0019d	358.51±14.15bc	0.36±0.023de

3 Discussion

Light is essential for plant growth [11]. Different plants have different requirements for light quality and light intensity. A large number of studies have shown that red and blue light can effectively promote the accumulation of various growth factors in plants. Different from monochromatic light conditions, the wavelengths of the combined light red and blue light was more suitable form for plant growth. In this paper, by comparing the cultivation A. roxburghiigrown under different light qualities, we found a significant difference in the overall growth. Under the same growth conditions and light intensity, the plants in the B1R1 treatment were the tallest and had the greatest stem diameter,indicating that a certain proportion of blue to red light is beneficial to the internode elongation and stem width of A. roxburghii. In plants grown in the B1R1 treatment, the stem wasthick and sturdy, which is more conducive to the growth of the seedlings of A. roxburghii. This is consistent with the research results of Kim [12] and Tang Dawei[13] on chrysanthemum and radix puerariae,Adjusting the ratio of blue to red light can effectively increase the number of chrysanthemums and the stem length of the plant.

Chlorophyll in leaves is an important indicator of plant growth and development, and light quality can regulate the photosynthetic pigment content, morphology, and photosynthesis of A. roxburghii leaves[14-15]. This study showed that the B1R1and W treatments were more conducive to the increase of chlorophyll content during the planting process of A. roxburghii. However, one study has found that blue light increases the chlorophyll content during the growth of pea seedlings [16]. Monochromatic red light promotes the growth of strawberry seedlings but reduces the chlorophyll content [17], while the ratio of blue light in the red:blue combined light source.The chlorophyll content of the leaves of 30% of the strawberry seedlings was significantly improved, and the chlorophyll content of blue and red light treated cucumbers was higher than that of white light treatment [18], indicating that different red and blue combinations of light sources have different effects on chlorophyll synthesis in different species. The results of this study showed that the combined light quality of red light and blue light can promote the growth and photosynthesis of A. roxburghii, and the red-blue combined light source will increase the photosynthetic pigment content of A. roxburghii.

Photosynthesis is a biophysical and chemical process in which plants convert light energy into chemical energy used in life activities and synthesize organic matter. Intercellular CO_2 concentration is one of the main factors affecting photosynthesis. Photosynthetic rate affects plant assimilation ability and actual production. Some researchers have studied the effects of different ratios of red and blue on the leaf structure and photosynthetic characteristics of tomato seedlings, and found that the net photosynthetic rate was the highest under white light treatment [19-20]. In our study, it was found that white light effectively promoted the transformation of plant photosynthetic characteristics. The intercellular CO_2 concentration was the lowest, and it had a negative correlation with the photosynthetic rate. This is consistent with the effect of different light qualitieson the growth and secondary metabolite content of A. roxburghii and the lower the intercellular CO_2 concentration[21]. The higher the photosynthetic rate. In this study, it was also found that under the blue light treatment, the net light and rate showed a negative correlation with stomatal conductance and the transpiration rate, which is similar tothe photosynthetic performance of potato buds and potato in studiespublished by GuoJinting [22-25]. The results were positively correlated with the rate, which had a great influence on the environmental stress in the experiment. It has been reported that excessive water stress [26], light stress [27], fertilization level [28], and other conditions may affect plant light, causing a change in function.

4 Conclusions

In this study, plant height, stem diameter, and fresh weight of A. roxburghiiwere highest under the 1:1 blue to red treatment, followed by white light and 6:1 red to blue light. As the main photosynthetic pigment, chlorophyll plays a central role in light absorption [29-30]. In this study, the results show that the greater the proportion of red light, the lower the chlorophyll content, and the difference is significant. The chlorophyll content is significantly increased under the blue light treatment. Plants treated with white light had a significantly higher plant chlorophyll synthesis and plant photosynthetic characteristics than other treatments. Blue light also had a positive effect on plant photosynthesis. Based on the above results, 1:1 blue to red light was an ideal artificial light source for cultivating A. roxburghii. The results of this study also showed that varying differences in water and light intensity affect the morphology and photosynthetic characteristics of A. roxburghii. Further studies should be conducted on the combination of different ratios of red and blue light, and the growth and photosynthesis of A. roxburghiiin different light intensity and moisture conditions.

References

1. ZHANG Yi-zhong, DENG Lin-qiong,HUANG Li-hua,CHEN Kun-hao. STATUS AND PROSPECT OF RESEARCH ON ANOECTOCHILUS ROXBURGHII[J]. GUIZHOU SCIENCE,2007,(02):81-84.
2. CAI Wen-yan,XIAOHua-shan, FAN Xiu-zhen. A review of research on Anoectochilusroxburghii[J]. SUBTROPICAL PLANT SCIENCE,2003,(03):68-72.
3. WEI Cui-hua,ZHOUHui-jun,XIE Yu, QING Jian-bin,CHEN Fang-lan. Growth Characteristics and Yield of Anoectochilusroxburghii Cultivars[J]. Fujian Journal of Agricultural Sciences,2018,(05):491-494.

4. PANG Jing,LIYa,JINGYue-bo,ZHAO Yong-hong,SHIChun-juan. Comparison of Active Components and Yield of Different Varieties Anoectochilusroxburghii(Wall.)Lindl[J].Journal of Anhui Agricultural Sciences,2018,(10):104-105+144.[16].

5. SHAO Qing-song, HUANG Yu-qiu,HU Run-huai, HU Bing-kang,LI Yan, LI Ming-yan. Multiple analysis of relationship between morphologic traits and yield formation of Anoectochilusroxburghii[J]. China Journal of Chinese Materia Medica,2014,(13):2456-2459.

6. SHIMAZAKIK, DOI M, ASSMANN S M, ctal, Light regulation of stomatal movement[J].Annual Review of Plant Biology, 2017, 267（1）：27-33.

7. Thao L. Nguyen,Mahmoud A. Saleh. Effect of exposure to light emitted diode (LED) lights on essential oil composition of sweet mint plants[J]. Journal of Environmental Science and Health, Part A,2019,54(5).

8. Science; Studies from Department of Biology Update Current Data on Science (Enhanced growth and cardenolides production in Digitalis purpurea under the influence of different LED exposures in the plant factory)[J]. Science Letter,2019.

9. Mandarim-de-Lacerda C.A., Fernandes-Santos C., Aguila M.B. (2010) Image Analysis and Quantitative Morphology. In: Hewitson T., Darby I. (eds) Histology Protocols. Methods in Molecular Biology (Methods and Protocols), vol 611. Humana Press, Totowa, NJ.

10. DIAO Kai.Study on Photosynthetic Characteristics of Different Jujube Varieties[D].Xinjiang Agricultural University,2016.

11. Kowallik W Bule light effects on respiration}J}. Annu Rev Plant Physiology,1998,33:51-72.

12. Kim S J,Hahn E J, Heo J W,et al. Effects of LEDs on net photosynthetic rate,growth and leaf stomata of Chrysanthemum plantlets invitro[J].Sci Hortic,2004,101:143-15.

13. Tang DW, Zhang G B}Zhang F}Pan X M}Yu J H. 2011. Effects of different LED light qualities on growth and bio-chemical characteristics of cucumber seedlings }J}.Journal of Gansu Agricultural University } 46 <1):44-48.

14. TERASHIMA I, HANBA Y T, THOLEN D, et al.Leaf functional anatomy in relation to photosynthesis[J].Plant physiology, 2011,155（1）:108-116.

15. XU Kai,GUO Yan-ping,ZHANG Shang-long. Effect of Light Quality on Photosynthesis and Chlorophyll Fluorescence in Strawberry Leaves,SCIENTIA AGRICULTURA SINICA,2006,38(2):369-375.

16. Wu M C, Hou C Y , Jiang C M , et al. A novel approach of LED light radiation improves the antioxidant activity of pea seedlings[J]. Food Chemistry, 2007, 101 (4):1753 -1758.

17. NHUTD T,TAK AMURA T,WATANABE H,etal. Responses of strawberry plantlets cultured in vitro undersuperbright red and bluelight-emitting diodes (LEDs) [J]. PlantCell Tissue& Organ Culture, 2003, 73 (1): 43-52.

18. WANG Hong, JIANG Yu-ping, SHI Kai, ZHOU Yan-hong, YU Jing-quanEffects of Light Quality on Leaf Senescence and Activities of Antioxidant Enzymes in Cucumber Plants, SCIENTIA AGRICULTURA SINICA, 2010,43 (3) :529 534.

19. MA Zhao, WANG Li-juan, Kiriiwa Yoshikazu, LYU Si-yu.Study on Photosynthesis Characteristics and Fruit Quality of Different Varieties of Strawberry[J].Tianjin Agricultural Sciences,2019(10):637-646.04.

20. YANG Junwei, BAO Encai, ZHANG Kejia, PAN Tonghua, CAO Yanfei, ZHANG Jing, ZOU Zhirong.Effects of Different Ratios of Red and Blue Light on Anatomic Structure and Photosynthetic Characteristics of Tomato Leaf[J], ActaAgriculturaeBoreali-occidentalis Sinica,2018,27(05):716-726.

21. WangW,SuMH,LiHHetal. Effectsof supplemental lighting with different light qualities on growth and secondary metabolite content of Anoectochilusroxburghii[J]. PeerJ,2018. 6:p. e5274.

22. GUOJin-ting,TENGYue,GAO Yu-liang,ZHANGYan,LIKui-hua.Effects of Different Light Qualities on Potato Characters and Photosynthetic Characteristics of Potato Sprouts[J/OL].Crops:1-7[2019-10-22].

23. LI Huimin, LU Xiaomin.Growth and Physiological Characteristics of Rapeseed Seedlings under Different Light Quality. ActaBotanicaBoreali-OccidentaliaSinica, 2015, 35(11): 2251-2257.

24. JIANG Xiaojun,JIANGYing.Effects of Different LED Light Quality on Chlorophyll Fluorescence, Photosynthetic Parameters and SPAD in Cucumber Leaves[J].Tianjin Agricultural Sciences,2019,25(09):7-9.

25. Tholen D, Pons T L, Voesenek L A．Ethylene insensitivity results in down － regulation of rubisco expression and photosynthetic capacity in tobacco［J］．Plant Physiology, 2007, 144(3):1305 － 1315.

26. TouzyGaëtan,RincentRenaud,BogardMatthieu,LafargeStephane, DubreuilPierre,MiniAgathe,Deswarte Jean-Charles,BeauchêneKatia,LeGouisJacques,PraudSébastien. Using environmental clustering to identify specific drought tolerance QTLs in bread wheat (T. aestivum L.).[J]. TAG. Theoretical and applied genetics. Theoretische und angewandte Genetik,2019,132(10).

27. ZhengYoufei, Li Jian, Wu Rongjun, Mai Boru,XuJingxin,SunJian, Wu Fei. Effects of Solar Radiation Weakening on Morphology and Photosynthesis Characteristics of Winter Wheat. Crops, 2012(4):69-74.

28. Annick M, Elisabeth P, Gerard T. Osmotic adjustment, gas exchanges and chlorophyll fluorescence of a hexaploid triticale and its parental species under salt stress. Journal of Plant Physiology, 2004, 161(1):25-33.

29. WANG Pei-ling, XU Yu-bin, SONG Shu-ying. Effects of Doubled Atmospheric CO2 Concentration and Nitrogen Application on Photosynthetic Characteristics and Chlorophyll Fluorescence Characteristics of Winter Wheat. ActaBotanicaBoreali-OccidentaliaSinica, 2011, 31(1):144-151.

30. Annick M, Elisabeth P, Gerard T. Osmotic adjustment, gas exchanges and chlorophyll fluorescence of a hexaploid triticale and its parental species under salt stress. Journal of Plant Physiology, 2004, 161(1):25-33.

Effect of Different LED Light Sources on Growth and Development of Cherry Radish

Zhipeng Wen, Shaoming Luo, Hongyu Wei, Xiaomin Li, Jiawei Liu, Zhiyu Ma*
Zhongkai University of Agriculture and Engineering
No. 500, Zhongkai Road, Haizhu District, Guangzhou
Email:846882689@qq.com, phone: 15820286024

Abstract

To explore suitable LED light sources for the growth and development of cherry radish, two kinds of LED light sources designed to cover the full spectrum with either staggered or linear arrangement were used as the experimental groups. Traditional white light LEDs were used as the control group to study the uniformity of the three kinds of light source and their influence on the growth of cherry radish. The results showed that the uniformity of the photosynthetic photon flux density of staggered and linear light sources was more than 86% compared to the conventional LED white light, which had a uniformity of 69.55%. Cherry radishes grown under staggered and linear light sources had significantly higher aboveground and belowground biomass compared to LED white light, showing a significant difference. No significant difference was found between staggered and linear groups.

1. Introduction

Regulation of the light environment is an effective way to increase yield and quality of crops being grown under artificial light. [1] At present, leafy vegetables, such as lettuce and spinach, have been successfully grown in plant factories [2, 3]. In contrast, fewer studies have reported the influence of light environments on root vegetables. In the case of cultivated crops, small leafy vegetables, root vegetables, medicinal plants, etc., are more suitable for cultivation in plant factories than larger plants. Cherry radish is an economically important root vegetable crop with a short growth cycle and plant habit; both the aboveground and belowground parts nutritious. Thus, cherry radish may be very suitable for cultivation in plant factories. [1]

Light quality, intensity, and photoperiod are key components of the light environment that have a significant impact on the growth of root crops. [4] A large number of studies have focused on the effect of light quality, light intensity, and photoperiod on plant growth. [4-8] Some studies have shown that red and blue light were more effective than other light wavelengths to promote photosynthesis in plants. [9-10] Red light can effectively promote photosynthetic organ development, plant photomorphogenesis and photochemical synthesis; blue light plays an indispensable role in chloroplast development and stomatal opening. [11-14] It was also found that a 1:1 ratio of red to blue light can increase the dry and fresh weight of plants [15-17] and the photosynthetic rate of leaves [18]. In addition, some plants responded to green light [19-20] and ultraviolet light [21]. However, little research had been done on the effect of different arrangements of lamp beads composing light sources on the growth and development of cherry radish. Based on the existing research, we studied the uniformity of different light treatment sources and their influence on the growth of cherry radish. Specifically, two kinds of full-spectrum LED light sources that have different arrangements (staggered and linear) of lamp beads were used as the experimental groups, and the control consisted of the traditional LED white light.

2. Materials and methods

2.1. Construction of test platform

1) Design of plant growth light source

In this study, high-power (single bead power 1 W) red (peak 660 nm), blue (peak 455 nm), and white LED lamp beads were used to design two kinds of full-spectrum LED strip light sources for plant growth with different arrangements (staggered and linear) as shown in Fig. 1. The total length of the light source was 1.2 m, the highest rated voltage was 48 V, and the highest rated power was 50.4 W. The total number of lamp beads was 13 with a 1:1:1 ratio of red, blue, and white. The same color lights were controlled separately. In the linear arrangement, the three kinds of lamp beads were alternately arranged evenly with an interval of 30 mm; in the staggered arrangement, the red and blue lamp beads were alternately distributed with the arrangement of the symmetric dual arithmetic progression (First Term 40 mm, common difference 0.5 mm) as a line, and the white lamp beads were evenly spaced 100 mm apart in a line, the two lines were separated by 15 mm, and a white lamp bead was placed at the center of the light source. Both sets of LED light sources were powered by a switching power supply (ERPF-400-48, Ming Wei, Taiwan, China). The control group used the LED white lights (KES-GL-002, Keisue, Shenzhen, China) that commonly used in plant lighting. The light source had a length of 1.2 m, a rated voltage of 220 V, and a rated power of 36 W. The spectra of the three sources were measured using a fiber optic spectrometer (FLA5000+, Jingfei Technology, Hangzhou, China) (Fig. 2).

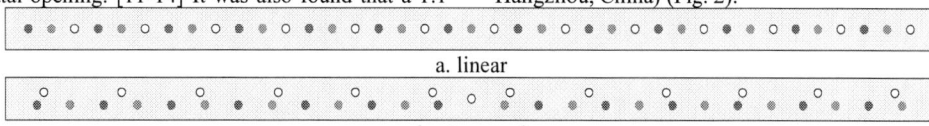

a. linear

b. staggered

Fig. 1 Schematic diagram of two full-spectrum LED plant growth sources

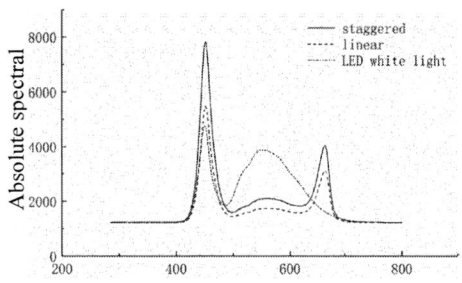

Fig. 2 Spectral distribution of three different light source treatments

2) Construction of the experimental platform

This experiments were carried out in the closed plant factory of the Key Laboratory of Intelligent Equipment Technology Innovation of Guangdong Province, Zhongkai University of Agriculture and Engineering. The plant factory covers an area of 30 m² and has a computer-controlled automatic system inside, which can monitor and automatically regulate the ambient temperature and humidity. There were two rows of cultivation racks in the cultivation room. The total height of the cultivation rack was 2,000 mm, the width was 600 mm, the length was 1,300 mm, and it was divided into two layers. Each layer of the cultivation frame had six light sources arranged side by side, and the distance between the light sources was 65 mm, 101 mm, 108 mm, 101 mm, and 65 mm. The average light intensity of the canopy (M) was 215±1 μmol·m⁻²·s⁻¹.

2.2. Test design

The test material was cherry radish (Red Angel, Beijing Juhong Seedling Technology Co., Ltd), the daytime temperature in the plant factory was 24 °C, the nighttime temperature was 20 °C, the relative humidity was 60%, the light intensity was 215±1 μmol·m⁻²·s⁻¹, and the photoperiod was 12 h/d. From the day of transplanting, random sampling was carried out every 4 days, with 18, 18, 10, 10, and 10 samples collected during each random sampling period. After sampling, the remaining samples were evenly aligned.

2.3. Measurement indicators and methods

The light source intensity (Photosynthetic Photon Flux Density) was measured with a photosynthetically active radiometer (GLZ-B, Top, Zhejiang, China). The light intensity was evaluated by the maximum value (E_{max}) and the average value (M) of 60 measurement points in the target light receiving surface. The uniformity of illumination was the ratio of the mean to the maximum [22] and was calculated as follows:

$$\text{Uniformity:} \quad U = \frac{M}{E_{max}} \times 100\%$$

Fresh weights of aboveground and belowground organs were weighed using an electronic balance (AUY220, SHIMADZU, Japan). Then, after determining the fresh weight, the material was heated in an electric oven (DHG-9420A, Yiheng, Shanghai, China) at 105 °C with 30 min, and dried at 60 °C until a constant weight was reached. The dry weight of the material was then recorded.

The volume of all the cherry radishes collected during the harvest was measured by using a measuring cylinder containing a moderate amount of water. All the belowground organs were immersed in water for measurement, then the volume of the water was read before and after immersion. The scale difference was the total volume.

2.4. Statistical analysis

The data was statically analyzed using Duncan's multiple range test to determine the whether differences in the data were significant at 5% (P ≤ 0.05) using SPSS 20 (IBM SPSS Statistics Version 20, USA) software.

3. Results and discussion

3.1. Uniformity of different light sources

The color change at the top of Fig. 3 a–c represents the change of light intensity. The arrangement and combination of LED lamp beads had a significant effect on the uniformity of the light intensity at receiving surface with a distance of 15 cm. The overall color distribution under the staggered and linear processing was relatively uniform. Under the traditional white LED light treatment, the overall color distribution in the light receiving surface had a large step difference, with the middle concentrated with red, rapidly turning yellow to green on both sides, with blue at the four corners, indicating that the light intensity varies greatly. For the traditional white LED light treatment, the uniformity of light intensity distribution was poor. The order of uniformity of light intensity distribution in the light receiving surface was staggered (86.72%)>linear (86.54%)> LED white light (69.55%). The uniformity of illumination was best in the staggered treatment. Some research results also showed that by optimizing the permutation of the LED lamp beads to obtain non-uniform arrangement, the illumination uniformity was increased by approximately 60% compared to before optimization. [23] The above shows that the non-uniform arrangement of LED lamp beads can effectively reduce the concentration of the central light intensity effect, providing more uniform light for plants, which improves the uniformity of plant growth.

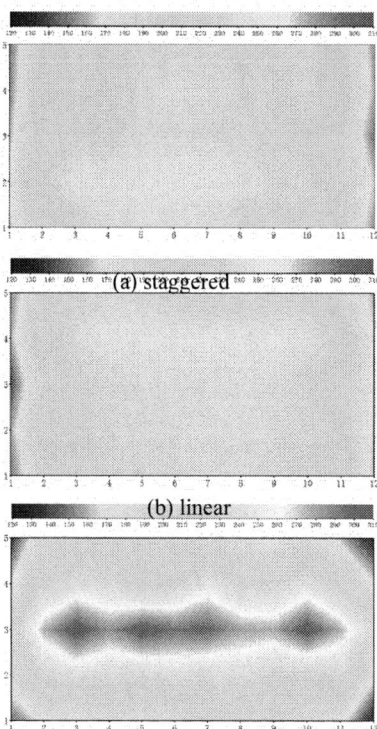

(a) staggered

(b) linear

(c) LED white light

Fig. 3 The intensity distribution map of staggered, linear, and white LED light sources

3.2. Effect of different light sources on fresh and dry weight

For the aboveground organs, the dry weight of the three treatments began to increase rapidly eight days after transplanting, and the gain in dry and fresh weight under the staggered and linear treatments was primarily concentrated between DAY8 and DAY20, which was better than the white LED light treatment (Fig. 4; Fig. 5). On DAY12 and DAY16, the dry and fresh weights of cherry radishes in both the staggered and linear treatments were significantly different from the white LED light treatment (Fig. 4; Fig. 5). For the belowground organs, except for the fresh weight growth slows down after entering DAY20 under the linear treatment, with the increase of the planting time, the dry and fresh weight grows faster and faster; before DAY20, the dry weight growth under staggered and linear treatments were significantly faster than that of LED white light, but after DAY20, the opposite was true; except for the dry weight of DAY4, the dry and fresh weight of both the staggered and linear treatments were significantly different from the LED white light treatment (Fig. 5).

In general, staggered and linear LED treatments were more conducive to the growth of the aboveground and belowground organs of cherry radish and there were no significant differences between them, which may be due to the addition of red and blue light in the staggered and linear light sources. Moreover, the optimization of the non-uniform arrangement of lamp beads may have improved the uniformity of the light source and illumination efficiency, promoting the belowground biomass accumulation of cherry radish. Some previous studies have also showed that red and blue light quality had a significant effect on the growth and morphology of radish. [24,25]

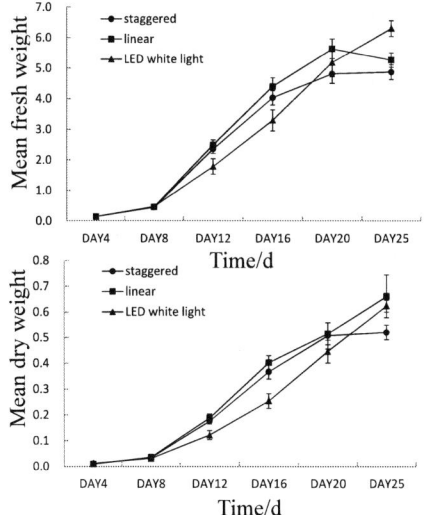

Fig. 4 Variation of dry and fresh weight of aboveground organs with time

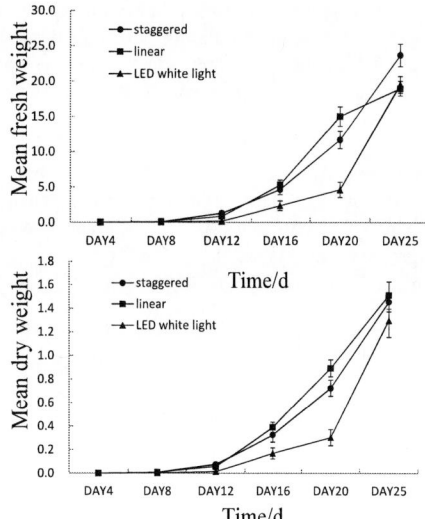

Fig. 5 Variation of dry and fresh weight of the belowground organs with time

3.3. Effect of different light sources on the volume of the belowground organs of cherry radishes

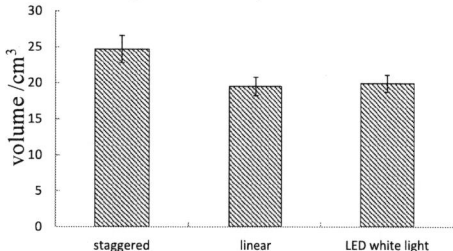

Fig.6 The volume of the belowground cherry radishes in three different treatments at DAY25

The volume of the fresh belowground sample in the three treatments showed the following order: staggered (24.7 cm^3)>LED white light (19.9 cm^3)>linear (19.5 cm^3), and the order of fresh weight was: staggered (23.70 g)>LED white light (19.54 g)>linear (18.99 g) (Fig. 6). The density of the belowground fresh samples was: staggered (0.96 g/cm^3), linear (0.97 g/cm^3), and LED white light (0.98 g/cm^3), meaning that the density of the radishes in the different treatments was nearly the same and slightly less than the density of water (1 g/cm^3). The dry weight of the belowground organs was the greatest under the linear treatment at DAY25, but the fresh weight was the smallest, which indicated that the moisture in the belowground organs was the lowest and the volume was the smallest. These findings might be explained by the rapid growth of the cherry radish in the linear treatment, which exceeded the growth period of radish on DAY25, leading to the radish hollowing and the reduction of water content and fresh weight.

4. Conclusions

In view of the low uniformity and poor illumination effect of the current white LED light source on growth and development of plants, two types of full-spectrum LED light sources were designed (staggered and linear arrangement) to study the uniformity of the three treatment sources and their influence on the growth and development of cherry radish. The conclusions were as follows.

1) The order of light source uniformity was staggered (86.72%) > linear (86.54%) > LED white light (69.55%);

2) The staggered and linear light sources were more conducive to the accumulation of aboveground and belowground biomass of cherry radishes than the traditional white LEDs, but no significant difference was found between the treatments.

References

1. ZHA Ling-yan，LIU Wen-ke. Effects of Red/Blue Light Ratio with Different Light Intensity on Growth and Yield of Cherry Radish[J]. Chinese Journal of Agrometeorology，2018.

2. Shiina T , Hosokawa D , Roy P , et al. LIFE CYCLE INVENTORY ANALYSIS OF LEAFY VEGETABLES GROWN IN TWO TYPES OF PLANT FACTORIES[J]. Acta Horticulturae, 2011(919):115-122.

3. Kang, J.H., Gyeongsang National University, Jinju, Republic of Korea. Light intensity and photoperiod influence the growth and development of hydroponically grown leaf lettuce in a closed-type plant factory system[J]. Horticulture Environment & Biotechnology, 2013, 54(6):501-509.

4. ZHA Ling-yan，LIU Wen-ke. Effects of Light Quality and Photoperiod on Growth and Yield of Cherry Radish Grown under Red Plus Blue LEDs [J]. China Illuminating Engineering Journal ， 2017, v.28(06):105-108+113.

5. Lee S H , Tewari R K , Hahn E J , et al. Photon flux density and light quality induce changes in growth, stomatal development, photosynthesis and transpiration ofWithania Somnifera(L.) Dunal. plantlets[J]. Plant Cell Tissue and Organ Culture, 2007, 90(2):141-151.

6. XIE Bao-xing,WEI Jing-jing,ZHANG Yi-ting,SONG Shi-wei,SU Wei,SUN Guang-wen,HAO Yan-wei,LIU Hou-cheng.Supplemental blue and red light promote lycopene synthesis in tomato fruits[J].Journal of Integrative Agriculture,2019,18(03):590-598.

7. Kim S. Effects of LEDs on net photosynthetic rate, growth and leaf stomata of chrysanthemum plantlets in vitro[J]. Scientia Horticulturae, 2004, 101(1):143-151.

8. Bian Z H , Yang Q C , Liu W K . Effects of light quality on the accumulation of phytochemicals in vegetables produced in controlled environments: a review[J]. Journal of the Science of Food and Agriculture, 2015, 95(5):869-877.

9. Pfündel E, Baake E. A quantitative description of fluorescence excitation spectra in intact bean leaves greened under intermittent light[J]. Photosynthesis Research, 1990, 26: 19－28.

10. Kim H. Green-light supplementation for enhanced lettuce growth under red- and blue-light-emitting diodes[J]. Hortscience A Publication of the American Society for Horticultural Science, 2004, 39(7):1617.

11. Saebo A . Light quality affects photosynthesis and leaf anatomy of birch plantlets in vitro[J]. Plant Cell Tissue and Organ Culture, 1995, 41.

12. Choi H G , Moon B Y , Kang N J . Effects of LED light on the production of strawberry during cultivation in a plastic greenhouse and in a growth chamber[J]. Scientia Horticulturae, 2015, 189:22-31.

13. Terfa M T , Solhaug K A , Gisler?D H R , et al. A high proportion of blue light increases the photosynthesis capacity and leaf formation rate of\r, Rosa\r, ×\r, hybrida\r, but does not affect time to flower opening[J]. Physiologia Plantarum, 2013, 148(1):146-159.

14. Senger H . Effect of blue light on plants and microorganisms[J]. Photochemistry & Photobiology, 2010, 35(6):911-920.

15. Liu X Y, Guo S R, Xu Z G, et al. Regulation of Chloroplast Ultrastructure, Cross-section Anatomy of Leaves, and Morphology of Stomata of Cherry Tomato by Different Light Irradiations of Light-emitting Diodes[J]. Journal of Biotechnology, 2011, 24(2):129-39.

16. KIM, HAHN, EunJoo, et al. Effects of LEDs on net photosynthetic rate, growth and leaf stomata of chrysanthemum plantlets in vitro[J]. Scientia Horticulturae, 2004, 101(1):143-151.

17. Lian M L, Murthy H N, Paek K Y. Effects of light emitting diodes (LEDs) on the in vitro induction and growth of bulblets of Lilium oriental hybrid 'Pesaro'[J]. Scientia Horticulturae, 2002, 94(3):365-370.

18. Lee S H , Tewari R K , Hahn E J , et al. Photon flux density and light quality induce changes in growth, stomatal development, photosynthesis and transpiration ofWithania Somnifera(L.) Dunal. plantlets[J]. Plant Cell Tissue and Organ Culture, 2007, 90(2):141-151.

19. Johkan M , Shoji K , Goto F , et al. Effect of green light wavelength and intensity on photomorphogenesis and photosynthesis in Lactuca sativa[J]. Environmental & Experimental Botany, 2012, 75(none):128-133.

20. Kim H H , Wheeler R M , Sager J C , et al. LIGHT-EMITTING DIODES AS AN ILLUMINATION SOURCE FOR PLANTS: A REVIEW OF RESEARCH AT KENNEDY SPACE CENTER[J]. Habitation, 2005, 10(2):71-78.

21. DENG Yifang, XU Xiaohui, SU Yanmang, et al. Research on Design and Uniformity of LED Array for Plant Tissue Culture[J]. Jiangsu Agricultural Sciences, 2017, 45(24):225-228.

22. WEN Zhipeng, MA Zhiyu, WEI Hongyu, et al. Optimized Design of Combined Spectrum LED Light Source for Plant Growth[J]. China Illuminating Engineering Journal, 2018, v.29(04):39-44.

23. FANG Ying, JI Hangfeng, JI Huihua, et al. Design of High Illumination Uniformity LED Luminaires for Plant Tissue Culture[J]. China Light & Lighting，2014(2).

24. Samuolienė G , Ramūnas Sirtautas, Brazaitytė A , et al. The impact of red and blue light-emitting diode illumination on radish physiological indices[J]. Central European Journal of Biology, 2011, 6(5):821-828.

25. Kwack Y , Kim K K , Hwang H , et al. Growth and quality of sprouts of six vegetables cultivated under different light intensity and quality[J]. Horticulture, Environment, and Biotechnology, 2015, 56(4):437-443.

Calculation Method for Chicken Perceived Light Intensity

Zhichao Li, Xiaocui Wang, Baoming Li, Weichao Zheng,

Zhengxiang Shi, Qin Tong*

College of Water Resources & Civil Engineering, China Agricultural University, Beijing 100083, China

17 Qinghua East Road, Xueyuan Road, Haidian District, Beijing

Email: 15350519016@163.com, phone: 15350519016

* Corresponding author. Email: tongqin@cau.edu.cn

Abstract

Poultry has superior visual function. Its visible spectrum is wider than humans (380-760 nm) and its sensitivity is high. To study chicken perceived light intensity, seven light intensities from full-spectrum white light, monochromatic blue light and monochromatic green light were set-up. Measurement and comparison of differences between chicken perceived light and human perceived light were conducted. Results showed that the difference between them was significant ($P<0.01$). Both wavelength band and light intensity had effect on the ratio of chicken perceived light and human perceived light ($P<0.01$). Therefore, lx-meter and photometric unit lx used in poultry production and research were not appropriate. In order to scientifically set poultry house lighting conductions and reasonably evaluate the illumination environment, the actual chicken perceived light intensity should be considered and applied.

1. Introduction

The visual function of birds is well developed. Optical information mainly acts on animals through photoreceptors (Sharp et al., 1979). There are two kinds of photoreceptors on the avian retina: rod and cone cells. Rod cells are sensitive to deem light (< 0.4 lx) and have the greatest sensitivity at 507 nm (blue-green light). Cone cells are sensitive to bright light (> 0.4 lx) and can distinguish color, which gives chickens amazing color perception (Hart et al., 1999; Kram et al., 2010). Human cone cells are sensitive to light between 400 and 730 nm, with the most sensitive at 555 nm (Sagawa et al., 1986). There are only three types of cone cells in humans. But poultry has one more type of cone cells than humans, with a peak sensitivity at 415 nm (Govardovski et al., 1977; Hart et al., 1999). The maximum sensitive spectra of birds and humans are similar and at 545-575 nm. However, the spectral sensitivity of poultry at 400-480nm and 580-700 nm are stronger than that of humans (Prescott et al., 1999; Wortel et al., 1987). Lux (lx) is the international unit of illumination and usually used to measure the intensity of the light environment for human beings. There were a large number of researches which indicated that light played an important role in poultry. Appropriate light intensity, color and photoperiod can promote the growth and development of chicken embryos during incubation (Duncan et al., 1978; Shafey et al., 2002) and affect the post-hatching activities, reproduction, and growth for chicken (Phillips, 1992). Especially in the modern poultry house where the artificial light is fully applied for the illumination. Therefore, the quality and quantity if poultry house lighting are important.

Illumination Ev, commonly known as Lux (lx), is a photometric unit to measure the human perceived light intensity of the light environment. It is calculated from the spectral power of the light source and the spectral sensitivity of human.

$$Ev=683.002 lm/W \cdot \int_0^\infty \overline{V}(\lambda)\emptyset_{e\lambda}(\lambda)d\lambda/(4\pi r^2) \qquad (1)$$

where, $V(\lambda)$ is the human photometric function (CIE, 1983); $\emptyset_{e\lambda}$ is the spectral radiation power of each nm, also known as spectral radiation flux. The maximum luminous efficacy for human photopic vision is 683.002 lm w^{-1}, and r is the distance (m) to the light source.

For illumination level calculation, it refers to the radiation value of the lamp, such as the spectral power distribution or the spectral irradiance, and calculates the perceived illumination of the species through the photometric function of the species. Some researchers have validated the photometric functions of chicken (Prescott et al., 1999), turkey (Barber et al., 2006), duck (Barber et al., 2006), rat (Jacobs et al., 2001), mouse (Jacobs et al., 2004) and cat (Brown et al., 1964; Pasternak et al., 1981). The results showed that the visual sensitivity coefficients of different species were obviously different. Therefore, human light intensity (lx) cannot reflect the actual chicken perceived light intensity (Chicken lx, Clx) (Nuboer et al., 1992), nor can it accurately evaluate the light environment of poultry house. According to the photometric function and illumination calculation formula of human (Sagawa et al., 1986) and poultry (Prescott et al., 1999), a method for measuring the chicken perceived light intensity was established and integrated into a conventional spectrometer. Furthermore, human perceived light intensity (lx) and chicken perceived light intensity (Clx) under three different light sources were compared using the calculation data of integral sphere as a reference. The reliability of the spectrometer measurement was proved by comparing the date between these two methods. Human perceived light intensity (lx) and chicken perceived light intensity (Clx) were measured using spectrometer. The ratios of them were calculated and the differences of the ratios for different wave bands and illumination levels were compared. At present, normally a lx-meter was used to measure the light intensity in livestock and poultry houses. There are not many instruments on the market which measures chicken perceived light intensity. Therefore, the ratio for certain spectrum can be used to convert human perceived light intensity (lx) to chicken perceived light intensity (Clx), if only lx can be measured. It helps for light evaluation and production management.

2. Materials and Methods

The illumination and chicken perceived light intensity obtained by integral sphere calculation and directly measured by spectrometer were compared. Furthermore, the difference between illumination and chicken perceived light intensity was compared under the same intensity level of the same LED.

978-1-7281-5757-3/19 $31.00 © 2019 IEEE

2.1. Light source

Three wave bands of LEDs (Yimeixinguang Technology Co., Ltd., Beijing, China), which were white light (380-780 nm), blue light (447.5-462.5 nm) and green light (515-535 nm), were controlled using iLMS lighting management software (Aviation Energy Technology Nanjing Co., Ltd.,) for reference setting, light attenuation compensation and real-time monitoring.

2.2. Formula for calculating chicken perceived light intensity

According to the formula of human perceived light, the formula of chicken perceived light intensity at a fixed distance from the light source is as follows:

$$Es = 683.002 lm/W \cdot \int_0^\infty \overline{S}(\lambda) \emptyset_{e\lambda}(\lambda) d\lambda / (4\pi r^2) \qquad (2)$$

where, Es is the chicken perceived light intensity; $\emptyset_{e\lambda}$ is the spectral radiation power of each nm, also known as spectral radiation flux; $\overline{S}(\lambda)$ is the photometric function of poultry (Prescott et al., 1999). It is assumed that the optimal spectral luminescence efficiency of chicken photopic vision is the same as that of human beings (683.002 lm w[-1]); r is the distance of light source (m).

The maximum spectral luminous efficacy of radiation for photopic vision for humans was used for both calculations because no poultry data are available. However, if the figure for birds differs from 683lumens/W, then the calculation will be different.

2.3. Calculation and measurement of human perceived light and chicken perceived light

2.3.1. Calculation method using integral sphere

The white, blue and green LED were placed in the integral sphere with a diameter of 2.0 m. The intensity was adjusted to 7 levels using iLMS lighting management software and spectral radiation power was measured using a LED Spectral analyzer (HAAS-1200, Hangzhou Yuanfang Photoelectric Information Co., Ltd., Hangzhou, China). Then, according to formulas (1) and (2), human perceived light intensity (lx) and chicken perceived light intensity (Clx) from three LEDs were calculated.

2.3.2. Spectrometer measurement method

In collaboration with Shangze Photoelectric Co., Ltd., a software was developed and integrated into a conventional spectrometer (SRI-LM-2000, Shangze Photoelectric Co., Ltd., Taiwan, China). It can simultaneously measure the human perceived light and chicken perceived light. The LEDs were fixed on the shelf and adjusted to seven illumination using the control software in the dark condition of the Animal Behavior and Environmental Physiology Lab (China Agricultural University). The illumination level (lx & Clx) at 16 cm vertical distance to the LEDs were measured using SRI-LM-2000 spectrometer.

2.4. Statistical analysis

SPSS (IBM statistics 25) software was used to analyze the data and results were expressed as mean ± SD. Independent sample T test was performed for human perceived light and chicken perceived light under the same intensity. A general linear model was used to analyze the effects of spectrum (LED) and intensity level on the ratio: Y = μ + LED + intensity + LED * intensity + ε, μ is the overall average value, ε is the residual error term, LED and intensity were fixed effects, and LED * intensity was the cross-influence factor. When the effect was significantly different ($p<0.05$), multiple comparisons were further conducted.

3. Results and Discussion

Table 1 shows that the average values of the three repeated measurements of the spectrometer are in good agreement with the integral sphere calculated data as a reference, especially the full spectrum white light. The difference between the two methods was small. Meanwhile, the output values of SRI-LM-2000 spectrometer at different LEDs and intensity levels were also stable, and the difference among repeat measurement data was small. This indicated the accuracy and reliability of the data measured using SRI-LM-2000 spectrometer. The method of measuring the chicken perceived light intensity established in this study could be used to measure the illumination in livestock and poultry house or for scientific research.

The difference between human perceived light and chicken perceived light under the same intensity level of the same LED is shown in Table 2. The difference between them at the seven intensity levels of white LED was significant (P < 0.01); and the latter was about 1.5 times of the former. At the seven intensity levels of blue LED, the difference of perceived light intensity between human and chicken was significant (P < 0.01). Compared with human, the intensity change of blue LED caused a big increase in chicken perceived light intensity. This indicated that poultry was extremely sensitive to blue light. It was corresponded to the known spectral sensitivity curve of poultry and human, and bird was more sensitive to blue light than human beings. The difference of perceived light intensity between human and chicken under green LED was also significant (P < 0.01), although the ratio of the latter to the former was about 1.0. The results showed that the perceived light intensity of human and chicken had significant difference under different intensity levels of different LEDs. Current illumination measurement tools and unit could not reflect the actual chicken perceived light intensity, especially for some monochrome light.

Table 1. Human and chicken perceived light intensity obtained using spectrometer and integrating sphere and the ratio between them.

LED	Intensity level	Human (lx)		Chicken (Clx)		Ratio	
		Spectro-meter	Integral sphere	Spectro-meter	Integral sphere	Spectro-meter	Integral sphere
White	1	136.82	120.43	204.30	182.20	1.49	1.51
	2	261.05	248.40	388.45	374.22	1.49	1.51
	3	924.63	945.28	1359.05	1407.35	1.47	1.49
	4	1040.53	1084.62	1526.87	1612.06	1.47	1.49
	5	1181.19	1223.83	1730.60	1816.32	1.46	1.48
	6	1312.89	1362.36	1921.35	2019.78	1.46	1.48
	7	1439.51	1500.32	2102.44	2221.38	1.46	1.48
Blue	1	23.37	24.15	205.10	200.11	8.78	8.29
	2	44.78	49.91	401.30	421.78	8.96	8.45
	3	152.42	186.48	1446.14	1656.32	9.49	8.88
	4	172.83	213.37	1647.87	1904.72	9.54	8.93
	5	194.94	237.40	1865.53	2147.33	9.57	9.05
	6	210.88	264.62	2025.35	2396.00	9.60	9.05
	7	232.09	266.33	2237.15	2410.78	9.64	9.05
Green	1	191.61	230.93	194.09	234.18	1.01	1.01
	2	192.34	231.76	195.06	234.98	1.01	1.01
	3	675.52	760.04	690.17	773.43	1.02	1.02
	4	1164.54	1298.11	1192.80	1325.48	1.02	1.02
	5	1640.70	1812.84	1688.16	1856.87	1.03	1.02
	6	2095.86	2312.21	2164.70	2374.59	1.03	1.03
	7	2522.07	2787.40	2613.32	2869.65	1.04	1.03

Note: Spectrometer data was the average value of three repeat measurements and integrating sphere data was the calculated value using the spectral radiation power measured in integrating sphere.

From the results of the general linear model analysis, the influence of LED on the ratio was significant ($P < 0.01$) (Table 2). The ratio of perceived light intensity of chicken to human was blue LED > white LED > green LED. Furthermore, the intensity level also had significant effects on the ratio ($P < 0.01$). Multiple comparisons of ration at different intensity levels under the same LED had done and the ratio in different LEDs showed a certain change trend with the intensity level. The ratio of white LED decreased with the increase of light intensity. However, the ratio of blue LED and green LED increased with the increase of light intensity.

The data showed that the ratio was relatively stable in the white and green LEDs. Lewis and Morris had proved that both poultry and human are sensitive to green light (Prescott et al., 1999). Lewis showed that the ratio in green light was close to 1 (Lewis et al., 2000). So birds would perceive the intensity of green light similarly to humans. The ratio of blue light fluctuated from 3.24 to 13.3 in previous reports. However, the results in this study showed that the ratio was around 8.78 to 9.64 for blue LED. It was possible that the spectrum and intensity of blue-light sources were different between this study and the previous ones. The illumination conditions in this study were 447.5–462.5 nm blue light (23.37 lx–232.09 lx) and in Lewis experiment were 440-500 nm spectrum (3.0 lx–16.4 lx). For full spectrum white light, the ratio was also stable at 1.5. The relative visual sensitivity of poultry and human were known and the spectral range was determined. Therefore, the corresponding chicken perceived light intensity could be obtained by multiplying the value of illumination by 1.5 in practical application.

Table 2. Comparisons of perceived light intensity between human and chicken measured using spectrometer under different light intensity of three LEDs.

LED	Intensity level	Human (lx)	Chicken (Clx)	Ratio
White	1	136.82±0.25[A]	204.30±0.34[B]	1.493±0.000[T]
	2	261.05±0.87[A]	388.45±1.41[B]	1.488±0.001[U]
	3	924.63±2.71[A]	1,359.05±3.61[B]	1.470±0.000[V]
	4	1,040.53±3.05[A]	1,526.87±4.32[B]	1.467±0.001[W]
	5	1,181.19±2.36[A]	1,730.60±3.64[B]	1.465±0.001[X]
	6	1,312.89±2.07[A]	1,921.35±2.61[B]	1.463±0.001[Y]
	7	1,439.51±6.79[A]	2,102.44±10.81[B]	1.460±0.001[Z]
Mean				1.473±0.121
Blue	1	23.37±0.10[A]	205.10±0.82[B]	8.777±0.011[Y]
	2	44.78±0.23[A]	401.30±0.74[B]	8.962±0.031[X]
	3	152.42±0.68[A]	1,446.14±2.84[B]	9.488±0.024[W]
	4	172.83±1.38[A]	1,647.87±11.61[B]	9.535±0.027[V]
	5	194.94±0.33[A]	1,865.53±1.53[B]	9.570±0.010[UV]
	6	210.88±0.76[A]	2,025.35±4.38[B]	9.604±0.014[TU]
	7	232.09±0.45[A]	2,237.15±3.44[B]	9.640±0.020[T]
Mean				9.368±0.330
Green	1	191.61±0.35[A]	194.09±0.31[B]	1.013±0.002[Y]
	2	192.34±0.23[A]	195.06±0.24[B]	1.014±0.000[Y]
	3	675.52±1.72[A]	690.17±2.04[B]	1.022±0.001[X]
	4	1,164.54±0.98[A]	1,192.80±0.99[B]	1.024±0.001[W]
	5	1,640.70±3.42[A]	1,688.16±3.82[B]	1.029±0.000[V]
	6	2,095.86±1.77[A]	2,164.70±1.77[B]	1.033±0.000[U]
	7	2,522.07±6.23[A]	2,613.32±7.12[B]	1.036±0.000[T]
Mean				1.024±0.008

Note: [A][B] values with different superscripts in the same row means significant difference ($P < 0.01$); [T-Z] values with different superscripts within one LED in the same column means significant difference ($P < 0.01$).

4. Conclusions

The chicken perceived light intensity (Clx) measured by updated spectrometer were accurate and reliable in this study. The perceived light intensities between human and chicken under the same intensity level of the same LED were significantly different. The ratios between them were relatively stable in the white and green LEDs and were about 1.5 and 1, respectively. However, the ratio of blue LED varied from 8.8 to 9.6. Therefore, it was inaccurate to evaluate the chicken perceived light intensity (Clx) using lx-meter for various monochrome light application.

Acknowledgements

This study was supported by the National Key R&D Program of China (grant numbers 2017YFE0122200 and 2017YFB0404000).

References

Barber, C.L., N.B. Prescott, J.R. Jarvis, C.L. Sueur, G.C. Perry, C.M. Wathes, 2006. Comparative study of the photopic spectral sensitivity of domestic ducks (Anas platyrhynchos domesticus), turkeys (Meleagris gallopavo gallopavo) and humans. British Poultry Science, 47(3), 365-374.

Brown, J.L., F.D. Shively, R.H. Lamotte, J.A. Sechzer, 1964. Color discrimination in the cat. Journal of Comparative & Physiological Psychology, 144(3617), 427-429.

CIE, 1983. The Basis of Physical Photometry. Commission Internationale de L'Eclairage (CIE) Publication 18.2, Paris, France.

Duncan, I.J.H., C.J. Savory, D.G.M. Wood-Gush, 1978. Observations on the reproductive behaviour of domestic fowl in the wild. Applied Animal Ethology, 4(1), 29-42.

Govardovski, V.I., L.V. Zueva, 1977. Visual pigments of chicken and pigeon. Vision Research, 17(4), 537-543.

Hart, N.S., J.C. Partridge, I.C. Cuthill, 1999. Visual pigments, cone oil droplets, ocular media and predicted spectral sensitivity in the domestic turkey (Meleagris gallopavo). Vision Research, 39(20), 3321.

Jacobs, G.H., G.A. Williams, J.A. Fenwick, 2001. Cone-based vision of rats for ultraviolet and visible lights. Journal of Experimental Biology, 204(14), 2439-2446.

Jacobs, G.H., G.A. Williams, J.A. Fenwick, 2004. Influence of cone pigment coexpression on spectral sensitivity and color vision in the mouse. Vision Research, 44(14), 1615-1622.

Kram, Y.A., M. Stephanie, J.C. Corbo, 2010. Avian cone photoreceptors tile the retina as five independent, self-organizing mosaics. Plos One, 5(2), e8992.

Lewis P.D., M.T. Morris, 2000. Poultry and coloured light. Worlds Poultry Science Journal, 56(3), 189-207.

Nuboer, J.F.W., M.A.J.M. Coemans, J.J. Vos, 1992. Artificial lighting in poultry houses: Are photometric units appropriate for describing illumination intensities? British Poultry Science, 33(1), 135-140.

Pasternak, T., W.H. Merigan, 1981. The luminance dependence of spatial vision in the cat. Vision Research, 21(9), 1333-1339.

Phillips, C.J.C., 1992. Environmental factors influencing the production and welfare of farm animals: Photoperiod. Pages 49–65 in Farm Animals and the Environment. C.J.C. Phillips and D. Piggins, ed. CAB International, Oxford, UK

Prescott, N.B., C.M. Wathes, 1999. Spectral sensitivity of the domestic fowl (Gallus g. domesticus). British Poultry Science, 40(3), 332-339.

Sagawa, K., K. Takeichi, 1986. Spectral luminous efficiency functions in the mesopic range. Journal of the Optical Society of America A Optics & Image Science, 3(1), 71.

Shafey, T.M., T.H. Al-Mohsen, 2002. Embryonic Growth, Hatching Time and Hatchability Performance of Meat Breeder Eggs Incubated under Continuous Green Light. Asian Australasian Journal of Animal Sciences, 15(12), 1702-1707.

Sharp, P.J., C.G. Scanes, J.B. Williams, S. Harvey, A. Chadwick, 1979. Variations in concentrations of prolactin, luteinizing hormone, growth hormone and progesterone in the plasma of broody bantams (Gallus domesticus). Journal of Endocrinology, 80(1), 51.

Wortel, J.F., H. Rugenbrink, J.F.W. Nuboer, 1987. The photopic spectral sensitivity of the dorsal and ventral retinae of the chicken. Journal of Comparative Physiology A, 160(2), 151-154.

Effects of LED Light Color and Intensity on Feather Pecking and Fear Responses of Layer Breeders in Natural Mating Colony Cages

Haipeng Shi [1,2,3], Baoming Li [1,2,3], Qin Tong [1,2,3], Weichao Zheng [1,2,3,*], and Dan Zeng [4] and Guobin Feng [4]

[1]Department of Agricultural Structure and Bioenvironmental Engineering, College of Water Resources & Civil Engineering, China Agricultural University, Beijing 100083, China

[2]Key Laboratory of Agricultural Engineering in Structure and Environment, Ministry of Agriculture and Rural affairs, Beijing 100083, China

[3]Beijing Engineering Research Center on Animal Healthy Environment, Beijing 100083, China

[4]Hebei Industrial Technology Research Institute of Layers, Handan 056800, China

*Correspondence: weichaozheng@cau.edu.cn, 17 Qinghua East Road, Beijing 100083, 010-62736181

Abstract

Natural mating colony cages for layer breeders have become commonplace for layer breeders in China. However, feather pecking (FP) and cannibalism are prominent in this system. The objective of this study was to investigate the effects of 4 LED light colors (white: WL, red: RL, yellow-orange: YO, blue-green: BG) with 2 light intensities for each color on FP, plumage condition, cannibalism, fear, and stress. A total of 32 identical cages were used for the 8 treatments (4 replicates for each treatment). For both light intensities, hens in RL had a lowest frequency of severe FP, whereas hens in WL had a highest frequency of severe FP. Hens in RL and BG had a better plumage condition than in WL and YO. Compared with RL and BG, hens in WL and YO had a significant longer TI duration. Hens in RL had a higher concentration of 5-HT, a lower concentration of CORT, and a lower heterophil to lymphocyte ratio than WL and YO. Furthermore, RL could significantly reduce mortality from cannibalism. Overall, hens treated with RL and low light intensity showed a lower frequency of severe FP, less damaged plumage, were less fearful, and lower physiological indicators of stress, and had reduced mortality from cannibalism. Transforming the light color to red or dimming the light could be regarded as an effectively method to reduce the risk of FP and alleviate the fear responses of layer breeder.

1. Introduction

The rising public concern for poultry welfare and increasing labor costs have resulted in in stacked natural mating colony cages becoming a trend in housing systems for commercial layer breeders in China. Layer breeders are the parent-stock of laying hens and in colony cages are confined together with roosters. The ratio of roosters and hens is generally kept between 1:10 and 1:8 and the flock size is usually maintained between 40 and 100 per individual cage. Compared with the cage system using artificial insemination, the natural mating behavior of breeding hens can be expressed in the natural mating colony cage system, taking into account animal welfare, high efficiency, energy savings, and clean production characteristics [1]. However, this housing system is still in the stage of exploration and optimization. Behavioral issues such as feather pecking (FP) and cannibalism are prominent in this system, contributing to economic losses and diminished health and welfare of hens. Currently, limited systematic research on FP and cannibalism in natural mating colony cages can be found. Available and efficient management measures are urgently required to ease the

negative effects caused by FP and cannibalism in this colony cage system.

Feather pecking and cannibalism can occur as a result of numerous factors including genetic background [2], hormones [3], nutrition [4], group size and stocking density [5], and environmental enrichment [4]. Indeed, light management is a crucial eliciting factor of the incidence and severity of FP and cannibalism of hens [6]. Measures such as keeping the hens under a reduced light intensity or altered light color are usually adopted to alleviate FP and cannibalism when necessary [6,7]. The objective of dimming the light or altering the light color is to diminish the birds' perception of colors and visual detection among them [8].

Excessive light is a vital factor initiating and favoring FP and cannibalism [9]. It was reported by Blokhuis and Arkes [10] that higher light intensity strongly impacts the occurrence and severity of FP in hens, resulting in more pecking damage. Reduced feather pecking behaviors and incidence of aggressive behaviors were observed by lowering the light intensity according to the results of Braastaad [6]. Hens confined close to light sources at an intensity level of 11–44 lux were more likely to perform FP than those further away where the light intensity ranged from 1 to 11 lux [11]. However, Kjaer and Sørensen [12] found that light intensity had no impact on the frequency of FP in any of the tested genotypes. Experimental results on the effects of light color on FP or aggression behavior are contradictory [7,13,14]. Due to other environmental effects and the strain differences between hens, it is difficult to draw any firm conclusions from these experiments.

Light-emitting diodes (LEDs) are a special kind of semiconductor diode which can give monochromatic light. Compared to incandescent light and fluorescent light, LED light has a marked longer life, specific spectrum, lower thermal output, higher energy efficiency, and higher reliability and frequency, as well as lower maintenance costs [7,13,15]. Knowledge about the influence of light condition on FP behavior is well documented for laying hens in other housing systems and strains, such as Oakham Blue [8], White Leghorn [12], and Brown Nick laying hens [16] in free-range systems, ISA Brown [9] and Lohmann Brown [17] hens in deep litter systems, Dekalb white breed hens in aviary systems [18], White Leghorns hens in battery cages [19], and so on. However, effects of LED light wavelength and intensity on FP and cannibalism have rarely been investigated in natural mating colony cages. The results of both wavelength and light intensity on the behaviors of laying hens in other housing

978-1-7281-5757-3/19 $31.00 © 2019 IEEE

systems may not be applicable to this colony cage system. Therefore, it is crucial to explore the effects of LED light color and intensity on FP and cannibalism in order to provide a basis for the regulation of light environment for layer breeders in natural mating colony cages. The objectives of this study were to investigate the effects of four LED light colors (white, red, yellow-orange, blue-green) with two light intensities in each color on FP, plumage condition, mortality from cannibalism, fear, and stress hormones for layer breeders in natural mating colony cages.

2. Materials and Methods

Eight treatments were offered in this study with four LED light colors, each at two light intensities, and giving four replicate cages for each light treatment. As shown in Figure 1, the four LED light colors were (1) red LED light (RL), at a peak wavelength (λp) of 660 nm and a dominant wavelength (λd) of 641 nm, half band width ($\Delta \lambda d$) of 20 nm; (2) yellow-orange LED light (YO), $\lambda p = 616$ nm, $\lambda d = 600$ nm, $\Delta \lambda d = 38$ nm; (3) blue-green LED light (BG), $\lambda p = 445$ nm, $\lambda d = 479$ nm, $\Delta \lambda d = 21$ nm; and (4) white LED light (WL), $\lambda p = 449$ nm, $\lambda d = 491$ nm, $\Delta \lambda d = 23$ nm. All LED light lamps (Huazhaohong Optoelectronic Technology Co. Ltd., Wuxi, China) were installed at the upper-tier cages, which were attached to the two sides of the cage celling. For all rows, starting from one end of the house, the cages were lit with red, yellow-orange, blue-green, and white LED light, respectively (Figure 2). Voltage for red, yellow-orange, blue-green, and white LEDs was tuned based on the relative spectral sensitivity curve indicated by Prescott and Wathes [20], so that the four lightings appeared iso-illuminant to hens. Light intensity was measured at the level of birds' heads using a precision luminometer (SRI-PL-6000, Shang Ze Photoelectric Co. Ltd., Taiwan, China) with a resolution of 0.01 lux according to human spectral sensitivity. The light intensity of the upper tier was 25 lux (high light intensity: HLI), and the lower tier was 10 lux (low light intensity: LLI). Experimental cages of different light colors were separated by an empty colony cage to avoid light pollution between different light colors. During the experiment period, the lighting rhythm was adjusted based on the different age phase, with a starter 8-h light at the age of 16 and 17 weeks and 10-h light at the age of 18 weeks, and then increased stepwise each week to reach 16-h light at the age of 30 weeks.

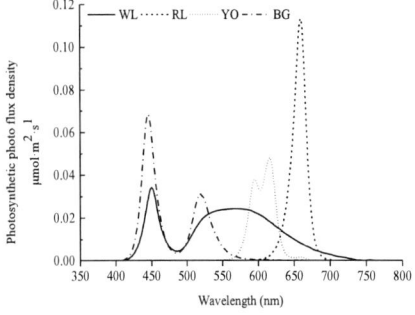

Fig. 1. Light spectral distribution of four light-emitting diode (LED) lights (WL: white, RL: red, YO: yellow-orange, BG: blue-green).

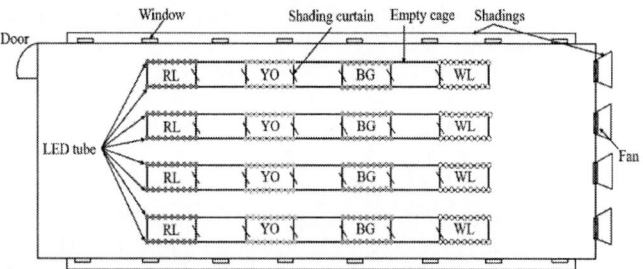

Fig. 2. Schematic diagram of the layout of the four LED tubes (WL: white, RL: red, YO: yellow-orange, BG: blue-green).

2.2. Behavioral Observations

The pecking behavior of the birds in each experimental cage was recorded by direct behavior sampling for 1 h periods. For each cage, 12 focal hens from tag numbers 1 to 12 were separately observed, lasting 5 min for each of them. Observations were made by two trained people over 4 days during 34 weeks. Hens from four cages were observed by each observer in 1 day: two cages in the morning, two in the afternoon. The order of observing each treatment cage, and time of day (am and pm) were balanced in a Latin square design to guarantee inter-observer agreement on behavior recording, the two observers developed proficiency in use of the ethogram before commencing formal data collection. Observation principles were brought into correspondence with each other and frequent checks were made for the consistency of inter-observer reliability during data collection. Frequencies of severe FP (SFP, forceful pecks, sometimes with feathers being pulled out, with the recipient bird moving away), gentle FP (GFP, slow and calm pecks, not resulting in feathers being pulled out, usually without reaction from the recipient bird), aggressive pecking (SP, fast and singular pecks, mainly directed at the head or other parts of the facial region), environmental pecking (ENP, pecks at the floor and other objects in the cage), and food pecking (FOP, pecks at the feeder and drinker) were recorded on a prepared check-sheet. A new bout of pecking behavior was recorded when there had been an interval 4 s or more between two feather pecks. Throughout this experiment, hens with bleeding wounds caused by injurious pecks by other conspecifics in all experimental cages were recorded. Injurious pecks targeted to cloacal, cannibalism of feathered body parts were separated. Number of dead birds led by cannibalism and other casualties was also recorded.

2.3. Fear Tests

2.3.1. Open Field Test

Six focal hens from each experimental cage were tested individually for their responses to an open field (OF) test for 10 min at the age of 35 weeks using similar method to Rodenburg et al. [21]. All tested hens were carried to an adjacent separate room, containing a 1.5 × 1.5 m test arena. The testing room was equipped with four LED light lamps, which could be switched according to the hens from different experimental treatments. The arena consisted of four walls (0.8 m high) and a floor made by galvanized iron sheets. In order to prevent unnecessary stress of an individual before the test, all hens were transported from the home cage to the testing arena in a cardboard box and were placed in the middle of the testing arena in darkness. The light was then turned on and the testing person left the room. The experimenter stood behind the door with a viewing window and was not visible to the bird. Measurements taken were the latencies to peep, defecation, walk, the duration of freezing, the number of vocalization and jump.

2.3.2. Avoidance Distance Test, Novel Object Test, Tonic Immobility Test

The procedure for the avoidance distance (AD) test, the novel object (NO) test, and the tonic immobility (TI) test was derived from the Welfare Quality protocol [22] and modified by the previous study by Shi et al. [5]. Fear tests (except for the OF test), were performed at the age of 36 weeks. The AD test was first, and then the NO test. Afterwards, 12 focal hens were selected from each experimental cage for the TI test. For the AD test, the distance from the experimenter's hand to the front wire mesh of the experimental cage was measured. Six hens were selected from each side of the cage, giving a total of 12 hens per cage. For the NO test, the selected novel object was a plastic stick measuring 60 cm in length with a 3-cm diameter. It was covered with five different colored bands of approximately 2 cm width. The number of hens within 30 cm of the NO was counted every 10 s for a duration of 2 min. Two positions on each side of the treatment cages were chosen (four positions per cage). The number of hens in the four positions in each cage was averaged. The TI test was conducted at the end of the house. For the TI test, the number of inductions and head movements needed and TI duration and latency were recorded for each hen. If the hens were not put into TI after 5 inductions, scores of 0 s for the duration and latency were given to hens, whereas a maximum of 5 was given for the number of inductions. If a hen remained in TI for the maximum testing period of 5 min, a score of 600 s was given for the duration of TI.

2.4. Plumage Scores

At 36 weeks of age, after TI test, the plumage coverage condition of 12 focal hens in each experimental cage was individually determined using the three-point scale method described in the Welfare Quality protocol [20] as follows—score 3: no or slight wear, (nearly) complete feathering (only single feathers lacking); score 2: moderate wear, i.e., damaged feathers (worn, deformed) or one or more featherless areas < 5 cm in diameter at the largest extent; score 1: at least one featherless area ⩾ 5 cm in diameter at the largest extent. The back, rump, tail, and belly regions of the hen were evaluated. A single score for overall plumage condition was also calculated.

2.5. Blood Measurements

All brachial blood samples were collected after fear tests and plumage coverage evaluation over 2 days during week 56. Six focal hens were randomly selected from the marked hens in each experimental cage giving a total number 192 birds. The samples were taken between 14:00 and 17:00 h each day. Blood samples were collected into 2-mL EDTA tubes within

2 min from bird handling to being stored on ice immediately after collection. Then blood samples were sent to Beijing Sino-uk Institute of Biological Technology for basal plasma corticosterone (CORT), thyroxine (T4), and triiodothyronine (T3) analysis, and for whole blood 5-hydroxytryptamine (5-HT) analysis.

2.6. Statistical Analysis

Data for each of the individual cage were averaged before analysis, as each experimental cage was treated as a statistical unit. Data were first checked for normality and heterogeneity of variance with and without transformations. Then the statistical analysis was performed using the linear mixed models procedure of SPSS software (IBM SPSS Statistics 22.0, Armonk, NY, USA). Fixed effects included light wavelength and light intensity, while the cage was considered as a random effect. The common model for each parameter contained the two qualitative factors as well as their interactions. Each model was reduced in a stepwise fashion, removing the least significant, highest order interaction in turn until only significant risk factors and interactions remained in the model. Post hoc analyses included pair-wise comparisons between significant factors in order to determine the nature of the significant effects (p < 0.05). Pecking frequency and plumage score showed non-normal distributions that were not suitable for transformation, so the Mann–Whitney U test was applied for post hoc group comparisons. Mean comparisons were evaluated on fear responses, cannibalistic injuries and blood parameters by Duncan's Multiple Range test. Statistical significance was determined at p < 0.05 unless otherwise stated.

3. Results

3.1. Behavioral Observations

The influence of light colors, light intensities and their interaction on the pecking behaviors are shown in Table 1. In comparison with BG, hens in the WL and RL groups had lower frequency of GFP (p ⩽ 0.004), and no significant difference was found between YO group and other groups for the frequency of GFP. Hens in RL group had the lowest frequency of SFP (p ⩽ 0.003), whereas hens in the WL group had the highest frequency of SFP (p ⩽ 0.025), and intermediate frequency of SFP for YO and BG. No significant difference was observed on the SP frequency for all light colors and both light intensities. In addition, hens in the RL and BG groups showed higher ENP activity (p ⩽ 0.025) than other groups. A significant effect of light intensity on GEP, SFP, and FOP was found. Compared with LLI, HLI showed a higher frequency of GFP (p ⩽ 0.037), SFP (p ⩽ 0.023), and FOP (p ⩽ 0.016). A significant intensity × color interaction was noted for GEP (p ⩽ 0.001), SFP (p ⩽ 0.017), ENP (p ⩽ 0.036), and FOP (p ⩽ 0.044).

Table 1. Means (± SE) of pecking behaviors of hens in response to light colors and light intensities *.

Item	Behaviors [3]				
	GFP	SFP	SP	ENP	FOP
Light intensity [1]					
HLI	7.83 ± 0.54 [a]	4.02 ± 0.28 [a]	0.56 ± 0.08	11.24 ± 0.92	18.49 ± 2.13 [a]
LLI	5.87 ± 0.61 [b]	2.91 ± 0.32 [b]	0.49 ± 0.07	11.01 ± 1.03	16.76 ± 2.05 [b]

		Light color [2]			
WL	5.78 ± 0.62 [b]	3.78 ± 0.33 [a]	0.51 ± 0.03	8.93 ± 0.88 [c]	20.17 ± 1.87 [a]
RL	6.22 ± 0.55 [b]	0.88 ± 0.12 [c]	0.58 ± 0.08	14.32 ± 1.75 [a]	17.20 ± 1.56 [bc]
YO	7.25 ± 0.63 [ab]	2.87 ± 0.35 [ab]	0.56 ± 0.09	9.84 ± 1.02 [c]	17.07 ± 1.88 [bc]
BG	7.38 ± 0.68 [a]	2.31 ± 0.04 [b]	0.46 ± 0.05	11.41 ± 1.33 [b]	16.03 ± 1.96 [c]
		Intensity-Color			
WL–HLI	6.94 ± 0.63 [b]	4.63 ± 0.66 [a]	0.49 ± 0.05	8.68 ± 1.13 [c]	22.83 ± 3.01 [a]
WL–LLI	4.62 ± 0.42 [c]	2.92 ± 0.18 [b]	0.52 ± 0.03	9.18 ± 1.02 [c]	17.51 ± 1.92 [bc]
RL–HLI	7.35 ± 0.56 [b]	0.83 ± 0.07 [c]	0.68 ± 0.05	15.41 ± 1.75 [a]	18.35 ± 1.87 [b]
RL–LLI	5.08 ± 0.37 [c]	0.92 ± 0.09 [c]	0.47 ± 0.06	13.22 ± 1.24 [b]	16.04 ± 1.44 [c]
YO–HLI	8.87 ± 0.89 [ab]	3.41 ± 0.42 [b]	0.58 ± 0.06	9.33 ± 0.88 [c]	16.96 ± 2.11 [c]
YO–LLI	5.62 ± 0.38 [c]	2.33 ± 0.21 [b]	0.54 ± 0.07	10.35 ± 1.65 [c]	17.17 ± 1.65 [bc]
BG–HLI	8.14 ± 0.76 [a]	2.75 ± 0.03 [b]	0.49 ± 0.04	11.54 ± 1.59 [bc]	15.81 ± 1.66 [c]
BG–LLI	6.62 ± 0.64 [bc]	1.87 ± 0.03 [bc]	0.42 ± 0.03	11.27 ± 1.73 [bc]	16.32 ± 2.25 [c]
		Source of variation			
Light intensity	0.004	0.003	0.234	0.316	0.037
Light color	0.001	0.001	0.104	0.006	0.005
Intensity × Color	0.001	0.017	0.742	0.036	0.044

[a–c] Means within a column and effects that lack common superscripts differ significantly ($p \leq 0.05$); * Values shown are the pecking frequency (number of pecks per bird/5 min) of four replicate cages with 12 hens per cage; [1] Light intensity: HLI = high light intensity, LLI = low light intensity; [2] Light colors: WL = white light, RL = red light, YO = yellow-orange light, BG = blue green light; [3] Behaviors: GFP = gentle feather pecking, SFP = severe feather pecking, SP = aggressive pecking, ENP = environmental pecking, FOP = food pecking.

3.2. Fear Responses

Table 2 presents the effects of light colors, light intensities and their interaction on the responses of hens to OF tests. No significant differences for light colors were found in the latency to first peep or number of jumps. Hens in the YO group had a significantly longer duration of freezing ($p \leq 0.023$) than other groups. WL and YO groups had a shorter latency to first defecation ($p \leq 0.033$) and more vocalizations ($p \leq 0.005$) than the RL and BG groups. Compared with other groups, hens in the RL group showed a shorter time to first pacing ($p \leq 0.044$). Hens under HLI had a longer latency to first peep ($p \leq 0.003$) and a shorter latency to first defecation ($p \leq 0.034$). A significant intensity × color interaction was noted for the duration of freezing ($p \leq 0.042$), the latency to first defecation ($p \leq 0.022$), the latency to first pacing ($p \leq 0.034$) and the number of vocalizations ($p \leq 0.028$).

Table 2. Means (± SE) of responses of hens to the OF test in response to light colors and light intensities *

Item	OF Tests [3]					
	Duration of Freezing (s)	Latency to First Peep (s)	Latency to First Defecation (s)	Latency to First Pacing (s)	Number of Vocalizations	Number of Jumps
			Light intensity [1]			
HLI	240.52 ± 21.25	17.44 ± 2.04 [a]	264.45 ± 34.52 [b]	6.88 ± 1.87	54.55 ± 7.62	5.78 ± 0.85
LLI	211.23 ± 22.33	13.92 ± 2.04 [b]	310.17 ± 41.56 [a]	5.62 ± 1.65	59.68 ± 7.56	5.86 ± 1.02
			Light color [2]			
WL	229.92 ± 25.23 [b]	14.56 ± 1.95	241.26 ± 35.23 [b]	8.29 ± 1.15 [a]	65.67 ± 8.55 [a]	7.89 ± 1.45
RL	216.57 ± 24.38 [b]	17.43 ± 1.34	365.64 ± 26.58 [a]	3.77 ± 1.45 [b]	34.62 ± 9.67 [b]	4.37 ± 0.98
YO	254.62 ± 21.16 [a]	17.84 ± 2.04	211.48 ± 23.14 [b]	7.96 ± 1.36 [a]	90.06 ± 11.22 [a]	7.06 ± 1.04
BG	211.05 ± 27.45 [b]	12.80 ± 2.15	332.55 ± 35.26 [a]	6.99 ± 1.47 [a]	39.81 ± 8.26 [b]	3.95 ± 0.55
			Intensity–Color			
WL–HLI	236.44 ± 27.14 [b]	17.24 ± 2.25	205.15 ± 43.88 [b]	10.15 ± 2.33 [a]	56.24 ± 12.56 [bc]	6.88 ± 1.62
WL–LLI	209.40 ± 23.44 [b]	11.88 ± 2.14	277.37 ± 48.36 [ab]	6.42 ± 1.78 [bc]	75.10 ± 18.36 [ab]	8.90 ± 1.56
RL–HLI	227.80 ± 21.71 [b]	20.18 ± 2.36	357.45 ± 38.74 [a]	3.25 ± 1.78 [c]	44.78 ± 8.95 [cd]	5.10 ± 1.56
RL–LLI	205.33 ± 24.56 [b]	14.68 ± 2.14	373.83 ± 32.35 [a]	4.28 ± 1.78 [c]	24.45 ± 6.25 [d]	3.64 ± 1.33
YO–HLI	276.23 ± 27.14 [a]	18.94 ± 3.24	166.60 ± 63.24 [c]	8.36 ± 2.33 [ab]	83.33 ± 19.33 [ab]	6.74 ± 1.33
YO–LLI	233.64 ± 26.38 [b]	16.74 ± 2.25	256.35 ± 33.64 [b]	7.55 ± 2.02 [ab]	96.80 ± 20.14 [a]	7.38 ± 2.02
BG–HLI	223.35 ± 28.24 [b]	13.40 ± 2.36	328.37 ± 35.56 [a]	5.76 ± 1.13 [bc]	36.62 ± 6.55 [cd]	4.40 ± 0.64
BG–LLI	198.83 ± 30.15 [b]	12.36 ± 1.98	336.46 ± 41.37 [a]	4.23 ± 2.14 [c]	43.28 ± 7.69 [cd]	3.50 ± 0.53
			Source of variation			
Light intensity	0.164	0.003	0.034	0.416	0.057	0.753
Light color	0.021	0.501	0.004	0.016	0.005	0.175
Intensity–Color	0.042	0.717	0.022	0.034	0.028	0.864

[a-c] Means within a column and effects that lack common superscripts differ significantly ($p \leq 0.05$); * Values shown are the responses of hens to OF test of four replicate cages with six hens per cage; [1] Light intensity: 1: HLI = high light intensity, LLI = low light intensity; [2] Light colors: WL = white light, RL = red light, YO = yellow-orange light, BG = blue green light; [3] OF test = open field test.

Table 3 shows the effects of light colors, light intensities and their interaction on the responses of hens to TI tests, NO test, and AD test. Compared with RL and BG, hens in the WL and YO groups had a significantly longer TI duration ($p \leq 0.042$). More hens in the RL went significantly closer ($p \leq 0.026$) to the novel object and within a shorter distance ($p \leq 0.026$) to human compared with WL, YO and BG. Compared with LLI, hens under HLI showed a significant longer TI duration ($p \leq 0.011$), and within a shorter distance to the human ($p \leq 0.042$) in AD test. In addition, there was a significant intensity × color interaction for the duration ($p \leq 0.041$) and latency ($p \leq 0.035$) of the TI test, and the responses to the NO test ($p \leq 0.034$), and to the human ($p \leq 0.022$).

Table 3. Means (± SE) of responses of hens to the TI test, NO test, and AD test in response to light colors and light intensities *.

Item	TI Tests [3]				NO Test [4]	AD Test [5]
	Duration (s)	Latency (s)	Induction (no)	HM (no)	Number of hens	Distance (cm)
			Light intensity [1]			
HLI	109.77 ± 3.43 [a]	20.67 ± 1.02	2.65 ± 0.08	4.72 ± 0.23	12.78 ± 0.37	23.92 ± 0.42 [a]
LLI	95.23 ± 4.86 [b]	19.31 ± 0.88	2.52 ± 0.11	4.80 ± 0.23	13.66 ± 0.28	18.38 ± 0.61 [b]
			Light color [2]			
WL	118.86 ± 3.44 [a]	19.21 ± 1.45	2.54 ± 0.11	5.43 ± 0.22	11.66 ± 0.23 [b]	25.20 ± 0.47 [a]
RL	85.43 ± 4.01 [b]	17.67 ± 2.01	2.47 ± 0.17	4.21 ± 0.24	14.85 ± 0.23 [a]	22.27 ± 0.56 [ab]
YO	110.39 ± 3.05 [a]	21.93 ± 1.88	2.79 ± 0.15	4.31 ± 0.31	12.96 ± 0.32 [b]	24.15 ± 0.81 [a]
BG	93.11 ± 4.64 [b]	21.17 ± 2.75	2.54 ± 0.15	5.09 ± 0.24	13.42 ± 0.31 [b]	18.08 ± 0.33 [b]
			Intensity–Color			
WL–HLI	125.36 ± 3.87 [a]	20.35 ± 1.05 [ab]	2.65 ± 0.14	5.02 ± 0.41	12.24 ± 0.31 [b]	28.83 ± 0.56 [a]
WL–LLI	112.35 ± 3.73 [ab]	18.07 ± 1.05 [bc]	2.42 ± 0.11	5.83 ± 0.34	11.07 ± 0.29 [b]	21.57 ± 0.59 [b]
RL–HLI	90.27 ± 5.45 [cd]	16.17 ± 1.88 [c]	2.75 ± 0.13	4.25 ± 0.48	13.88 ± 0.35 [ab]	29.35 ± 0.81 [a]
RL–LLI	85.58 ± 4.85 [d]	19.16 ± 2.02 [abc]	2.18 ± 0.15	4.17 ± 0.35	15.82 ± 0.28 [a]	15.18 ± 0.62 [c]
YO–HLI	118.33 ± 5.75 [a]	23.67 ± 2.04 [a]	2.75 ± 0.15	4.68 ± 0.29	13.75 ± 0.27 [ab]	27.77 ± 0.67 [a]
YO–LLI	102.44 ± 5.32 [bc]	20.18 ± 2.46 [ab]	2.83 ± 0.12	3.93 ± 0.52	12.16 ± 0.31 [b]	20.52 ± 0.73 [b]
BG–HLI	105.13 ± 3.24 [b]	22.50 ± 3.48 [a]	2.44 ± 0.15	4.92 ± 0.32	11.25 ± 0.30 [b]	19.92 ± 0.66 [bc]
BG–LLI	81.08 ± 3.09 [d]	19.83 ± 1.73 [ab]	2.64 ± 0.15	5.25 ± 0.28	15.58 ± 0.33 [a]	16.23 ± 0.72 [bc]
			Source of variation			
Light intensity	0.003	0.244	0.739	0.243	0.684	0.004
Light color	0.034	0.612	0.832	0.252	0.029	0.006
Intensity–Color	0.041	0.035	0.466	0.715	0.034	0.022

[a–d] Means within a column and effects that lack common superscripts differ significantly ($p \leq 0.05$); * Values shown are the responses of hens to the TI test, AD test of four replicate cages with 12 hens per cage, and the responses to the NO test of four replicate cages; [1] Light intensity: HLI = high light intensity, LLI = low light intensity; [2] Light colors: WL = white light, RL = red light, YO = yellow-orange light, BG = blue green light; [3] TI test = tonic immobility test, HM = head movement; [4] NO test = novel object test; [5] AD test = avoidance distance test.

3.3. Plumage Evaluation

Table 4 shows the effects of light colors, light intensities and their interaction on the plumage condition of four specific body regions and the overall score of the plumage evaluation. Hens in WL and YO had a lower score for back ($p \leq 0.036$) compared with RL and BG. Hens in YO had the lowest score for rump ($p \leq 0.046$) in comparison with other groups. There were no significant differences for the score of tail between light color treatments. For belly region, hens caged in RL and BG groups had a higher score ($p \leq 0.037$) than WL and YO groups. For overall score, hens in RL and BG were highest ($p \leq 0.003$), whereas hens in the YO were lowest ($p \leq 0.025$), and intermediate for WL. Compared with HLI, hens under LLI had a higher score for back region ($p \leq 0.022$) and a higher overall plumage score ($p \leq 0.006$). In addition, significant intensity × color interactions were noted for plumage score of all body parts and the overall score ($p \leq 0.05$).

Table 4. Means (± SE) of plumage score of hens in response to light colors and light intensities *.

Item	Body Part				
	Back	Rump	Tail	Belly	Overall
		Light intensity [1]			
HLI	2.54 ± 0.17 [b]	2.32 ± 0.22	2.67 ± 0.15	2.76 ± 0.22	2.52 ± 0.13 [b]
LLI	2.72 ± 0.17 [a]	2.45 ± 0.21	2.87 ± 0.18	2.71 ± 0.23	2.76 ± 0.15 [a]
		Light color [2]			
WL	2.58 ± 0.15 [b]	2.41 ± 0.21 [a]	2.71 ± 0.21	2.57 ± 0.22 [b]	2.45 ± 0.20 [b]
RL	2.85 ± 0.15 [a]	2.50 ± 0.21 [a]	2.85 ± 0.24	2.94 ± 0.14 [a]	2.81 ± 0.23 [a]
YO	2.52 ± 0.17 [b]	2.17 ± 0.17 [b]	2.73 ± 0.24	2.54 ± 0.19 [b]	2.20 ± 0.23 [c]
BG	2.72 ± 0.13 [a]	2.47 ± 0.18 [a]	2.80 ± 0.19	2.91 ± 0.21 [a]	2.71 ± 0.18 [a]
		Intensity–Color			
WL–HLI	2.63 ± 0.17 [a]	2.37 ± 0.24 [b]	2.58 ± 0.23 [b]	2.58 ± 0.18 [b]	2.43 ± 0.18 [b]
WL–LLI	2.52 ± 0.15 [b]	2.44 ± 0.21 [ab]	2.83 ± 0.24 [a]	2.55 ± 0.19 [b]	2.47 ± 0.15 [b]
RL–HLI	2.81 ± 0.15 [a]	2.42 ± 0.19 [ab]	2.77 ± 0.19 [ab]	2.92 ± 0.20 [a]	2.77 ± 0.16 [a]
RL–LLI	2.89 ± 0.14 [a]	2.58 ± 0.24 [a]	2.92 ± 0.26 [a]	2.96 ± 0.18 [a]	2.84 ± 0.14 [a]
YO–HLI	2.44 ± 0.18 [b]	2.08 ± 0.23 [c]	2.59 ± 0.23 [b]	2.58 ± 0.24 [b]	2.17 ± 0.16 [b]
YO–LLI	2.73 ± 0.15 [a]	2.25 ± 0.24 [bc]	2.88 ± 0.24 [a]	2.49 ± 0.22 [b]	2.42 ± 0.15 [b]
BG–HLI	2.67 ± 0.14 [a]	2.41 ± 0.20 [ab]	2.75 ± 0.22 [ab]	2.96 ± 0.24 [a]	2.70 ± 0.20 [a]

BG–LLI	2.73 ± 0.13 [a]	2.52 ± 0.23 [ab]	2.85 ± 0.24 [a]	2.85 ± 0.14 [a]	2.72 ± 0.16 [a]
		Source of variation			
Light intensity	0.001	0.474	0.524	0.175	0.024
Light color	0.01	0.012	0.132	0.014	0.019
Intensity–Color	0.01	0.005	0.024	0.036	0.004

[a-c] Means within a column and effects that lack common superscripts differ significantly ($p \leq 0.05$); * Values shown are the plumage score of four replicate cages with 12 hens per cage; [1] Light intensity: 1: HLI = high light intensity, LLI = low light intensity; [2] Light colors: WL = white light, RL = red light, YO = yellow-orange light, BG = blue green light.

3.4. Blood Parameters, Mortality, and Cannibalistic Injuries

Table 5 shows the effects of light colors, light intensities and their interaction on blood parameters, mortality, and cannibalistic injuries. Hens in RL had a higher concentration of 5-HT ($p \leq 0.05$) and a lower concentration of CORT ($p \leq 0.007$) than WL and YO. There was a significant difference between groups for the heterophil to lymphocyte ratio (H/L ratio), with the H/L ratio of hens in YO being the highest ($p \leq 0.013$), and that of hens in RL being the lowest ($p \leq 0.028$). The H/L ratio of hens in WL was higher compared with BG ($p \leq 0.044$). Compared with HLI, hens under LLI

had a higher 5-HT concentration ($p \leq 0.042$) and a lower CORT concentration ($p \leq 0.006$). The heterophil to lymphocyte ratio ($p \leq 0.024$) was significantly higher under HLI than LLI. No significant differences were found in the concentration of T3 and T4 between light treatments. Mortality from cannibalism for RL was significantly lower ($p \leq 0.026$) compared with other groups. Cannibalistic injuries for the light treatments presented a similar trend toward to the mortality from cannibalism. In comparison with HLI, hens under LLI had less cannibalistic injuries ($p \leq 0.010$), and a lower rate of mortality from cannibalism ($p \leq 0.026$).

Table 5. Means (± SE) of blood parameters, mortality and cannibalistic injuries of hens in response to light colors and light intensities *.

Item	Blood Parameters [3]					Mortality and Injuries [4]	
	T3 (ng/mL)	T4 (ng/mL)	5-HT (ng/mL)	CORT (ng/mL)	H/L	Cannibalism (%)	Injuries (n)
			Light intensity [1]				
HLI	0.55 ± 0.01	15.10 ± 0.32	28.11 ± 1.22 [b]	5.19 ± 0.15 [a]	0.50 ± 0.01 [a]	5.88 ± 0.32 [a]	0.48 ± 0.44 [a]
LLI	0.60 ± 0.01	15.75 ± 0.24	30.64 ± 0.93 [a]	4.66 ± 0.31 [b]	0.41 ± 0.01 [b]	4.44 ± 0.58 [b]	0.28 ± 0.36 [b]
			Light color [2]				
WL	0.65 ± 0.01	16.76 ± 0.28	27.04 ± 1.22 [b]	4.90 ± 0.55 [a]	0.49 ± 0.01 [b]	5.31 ± 0.57 [a]	0.40 ± 0.02 [a]
RL	0.49 ± 0.02	14.04 ± 0.44	30.73 ± 0.89 [a]	4.06 ± 0.25 [b]	0.32 ± 0.01 [d]	2.65 ± 0.33 [c]	0.18 ± 0.01 [c]
YO	0.56 ± 0.01	15.05 ± 0.32	28.14 ± 2.13 [b]	5.08 ± 0.27 [a]	0.60 ± 0.01 [a]	6.06 ± 0.46 [a]	0.46 ± 0.04 [a]
BG	0.60 ± 0.02	15.85 ± 0.17	30.61 ± 1.07 [a]	4.52 ± 0.33 [ab]	0.41 ± 0.01 [c]	4.63 ± 0.32 [b]	0.29 ± 0.04 [b]
			Intensity–Color				
WL–HLI	0.62 ± 0.01	16.48 ± 0.40	26.10 ± 4.61 [c]	4.96 ± 0.62 [a]	0.53 ± 0.01 [b]	5.93 ± 0.72 [a]	0.51 ± 0.04 [a]
WL–LLI	0.67 ± 0.01	17.03 ± 0.29	27.98 ± 1.77 [bc]	4.84 ± 0.18a	0.44 ± 0.01 [c]	4.69 ± 0.65 [b]	0.29 ± 0.04 [b]
RL–HLI	0.45 ± 0.01	13.71 ± 0.44	28.81 ± 0.84 [b]	5.63 ± 0.30 [a]	0.35 ± 0.01 [d]	2.46 ± 0.30 [c]	0.19 ± 0.03 [c]
RL–LLI	0.53 ± 0.02	14.36 ± 0.38	32.64 ± 0.88 [a]	4.08 ± 0.33 [b]	0.28 ± 0.01 [e]	2.83 ± 0.33 [c]	0.17 ± 0.02 [c]
YO–HLI	0.52 ± 0.02	14.66 ± 0.22	28.51 ± 0.89 [b]	5.04 ± 0.21 [a]	0.65 ± 0.01 [a]	6.30 ± 0.85 [a]	0.49 ± 0.04 [a]
YO–LLI	0.59 ± 0.02	15.44 ± 0.21	31.76 ± 0.83 [a]	5.11 ± 0.39 [a]	0.55 ± 0.01 [b]	5.81 ± 0.72 [a]	0.42 ± 0.04 [a]
BG–HLI	0.59 ± 0.02	15.53 ± 0.17	29.03 ± 1.48 [b]	5.11 ± 0.52 [a]	0.45 ± 0.01 [c]	4.81 ± 0.60 [b]	0.32 ± 0.03 [b]
BG–LLI	0.61 ± 0.02	16.16 ± 0.20	30.19 ± 0.91 [b]	4.60 ± 0.39 [b]	0.37 ± 0.01 [d]	4.44 ± 0.54 [b]	0.25 ± 0.02 [b]
			Source of variation				
Light intensity	0.431	0.579	0.042	0.001	0.041	0.010	0.005
Light color	0.643	0.412	0.021	0.003	0.037	0.010	0.007
Intensity–Color	0.233	0.283	0.018	0.001	0.008	0.001	0.003

[a-e] Means within a column and effects that lack common superscripts differ significantly ($p \leq 0.05$);* Values shown are the plumage score of four replicate cages with 12 hens per cage and the mortality and cannibalistic injuries of four replicate cages; [1] Light intensity: 1: HLI = high light intensity, LLI = low light intensity; [2] Light colors: WL = white light, RL = red light, YO = yellow-orange light, BG = blue green light; [3] Blood parameters: T3 = triiodothyronine, T4 = thyroxine, CORT = corticosterone, 5-HT = serotonin; H/L = the ratio of heterophil to lymphocyte. [4] Cannibalism = mortality from cannibalism, Injuries = cannibalistic injuries.

4. Discussion

In this experiment, hens under RL and BG tended to express more frequent GFP and ENP than birds exposed to the other two lighting colors, but a lower SFP frequency under RL (especially compared with WL and YO, with BG being intermediate). These results suggest that hens under RL and BG were more engaged in explorative behavior. We also noted that the pecking activities were promoted by high light intensity and hens under high light intensity were more vulnerable to suffering from SFP. Clearly, pecking behavior

may be affected by the wavelength of light as well as by light intensity. Huber-Eicher et al. [16] investigated the effects of colored LED illumination on behavior of laying hens. Hens under green light spent more time on pecking at objects and had more frequent pecking at conspecifics compared with red and white light. Hens under red lighting showed less often severe pecks or distress calls than hens under white light, with green light being intermediate. Mohammed et al. [17] looked at the behavior of laying hens under four different light sources. Higher frequency of GFP and aggressive behavior

978-1-7281-5757-3/19 $31.00 © 2019 IEEE

were increased by blue light and high light intensity. This current study confirm these findings that red light alleviates SFP. The higher contribution of longer wavelengths contained in red light may have reduced SFP behavior, although this needs confirmation. This effect was due to the wavelength and should not be confused with eventual effects of intensity. There is now a general agreement that a particular causative factor that is positively correlated with FP is the inhibition of foraging or dust bating behaviors, such as ground pecking or ENP [23]. It has been suggested that FP is a redirection of oral behavior toward conspecific under barren conditions [24]. In our study, the hens under RL and BG spent more time in their explorative pecking behaviors (GFP and ENP) compared to hens of the other treatments; therefore, attention and severe pecks of the hens shifted from conspecifics towards the surroundings. In other studies, Sultana et al. [7] studied the effect of various LED light color on the behavior of laying hens and indicated that hens in red light were more active and expressed more feather pecking than those of hens in blue light. Prayitno et al. [13] suggested that broilers illuminated with red light showed more aggression and did more floor pecking than birds under white, green, or blue light. Similar increases in aggressive behaviors were recorded in a separate investigation of broilers maintained under red, compared with blue, lighting through to 8 weeks [25]. These results are likely a consequence of the perceived increased intensity, as broilers are more sensitive to this range of the spectrum than that measured by lux [26], and birds have greater visual acuity in red light, while higher light intensity increases aggression. Long wavelengths may alter the reflectance of both the plumage of hens and the appearance of the experimental houses [20,27]. This may well make plumage and objects within the environment more attractive for the birds to peck at and explore. However, Leighton et al. [28] suggested that light sources do not affect these behaviors. Lewis and Morris [29] also mentioned that light color appears to have minimal influence on FP, as red light would reach the hypothalamus more rapidly than blue light. In the above studies that differ from the results in the present experiment, only Sultana et al. used LED light. It is difficult to reach a consistent conclusion from previous studies about wavelength effects upon FP. The discrepancy between the results may be caused by the differences of spectral sensitivity of the fowl, the spectral output of the light sources, the adaptability of birds to particular light environment over time, the housing system, stocking density, group size, and so on. Those aspects complicate direct comparisons of the data. The reduction in SFP under red light needs further evaluation because it could be of interest in commercial production situations.

The results in the present experiment indicated that hens caged in RL had effectively reduced mortality from cannibalism and cannibalistic injuries, in accordance with the finding of Wells [14] who found that the employment of red filters or red paint to light sources may be a simple and effective method in alleviating SFP and cannibalism. However, it may be surprising that in spite of the probable differences in the intensity perceived by hens, even where the light had been adjusted being equated for irradiance, wavelength generally did not significantly affect mortality rates in broilers [30]. The parent-stock hens in colony cages were confined together with roosters. The frequent mounting behavior may generate inferior back and rump plumage conditions, which resulted in hens suffering from injuries or scratches on the back and rump. There is a risk of severe feather pecking and cannibalism, especially if there is hemorrhage, broken skin, and fresh wounds. Therefore, the explanation of the red light reducing mortality from cannibalism and cannibalistic injuries may be that the birds cannot easily see red blood or fresh wounds in red light [31]. The elevated mortality and cannibalistic injuries under high light intensity noted in the present study was in accordance with Kjaer and Vestergaard [9], who suggested that high light intensity in both rearing and laying periods tended to increase mortality during laying, especially due to cannibalism.

Light sources have influences on plumage condition of hens through the influences on FP, as described by Long et al. [18], who showed that different light sources might affect plumage condition as judged by the incidence of feather pecking. In the current study, back, rump, belly, and overall plumage condition of the hens under RL and BG tended to be superior to those under WL and YO. Also, the increased plumage damage under high light intensity found in the current experiment confirms previous findings by Hughes and Duncan [11], Hughes and Black [32], and Allen and Perry [33], who indicated that high light intensity strongly affects the occurrence and severity of FP in laying hens with higher light intensity resulting in more damage. According to Bilcík and Keeling [34], GFP does not contribute to feather damage, while SFP is identified as the major cause of feather pulling, damaging, and plucking. Huber-Eicher and Sebö [35] suggested that at an early-age GFP is prevalent, whereas more SFP can develop later, resulting in more deteriorated plumage in older hens, consistent with our observations. Therefore, it could be speculated that hens under RL and BG had a better plumage condition which may attribute to being engaged less in SFP.

Reactions to humans or a new environment are widely employed to estimate the fearfulness of hens [36]. The ability to deal with this situation reflects the stability of the nervous system and the degree of individual excitability [3]. In this experiment, it seems that the likelihood that hens under RL and BG and caged in low light intensity approaching the NO was higher, the duration of the TI test was shorter, and the distance of the AD test was closer than hens in WL and YO and high light intensity. In addition, WL and YO tended to cause longer freezing time, longer latencies, and more distress calls. This indicated that hens caged in WL and YO were more fearful and susceptive to fear tests. However, the results of the study were in disagreement with those of Scott and Siopes [36], who found that no behavioral indications of stress were observed when mature turkey hens were exposed to blue, green, red, or white illumination of the same photon flux from commercial lamps between 30 and 53 weeks. One possibility for the discrepancy may be that the different breeds of hens may respond to light conditions differently. Studies have shown that the fearfulness of hens was associated with feather damage in commercial breeding [37]. The results are in accordance with those of the study by Johnsen et al. [38], which reported that severely feather-pecked birds tended to have an inferior feather coverage condition and were more

978-1-7281-5757-3/19 $31.00 © 2019 IEEE

fearful than birds with minor pecking damage. Hughes and Duncan [11] also found that fearful behavior was associated with greater feather loss. Other studies suggested that on an individual and flock level, having high levels of fear at a young age can become a risk factor for developing feather pecking as adult [39]. Therefore, the effects of light condition on behavioral response to fear tests of hens might through the effects on FP.

According to previous studies, thyroidal hormones are considered to be physiological indicators of various forms of stress in fowl [40]. Triiodothyronine (T3) regulates the metabolic rate and T4 is considered to be inducing molting of laying hens [41]. However, in this experiment, T3 and T4 concentrations were not affected by the light treatments. The hormones may be correlated with the quality of feather coverage. In addition, the CORT and 5-HT levels have been proven to be associated with fearfulness and feather pecking [42]. Hens caged under RL and low light intensity tended to have a higher concentration of 5-HT, a lower CORT concentration, and a lower ratio of heterophils to lymphocytes than WL and YO, which suggested that hens treated with RL and low light intensity showed a lower stress response. The results of the study were in disagreement with those of Olanrewaju et al. [43]. who found that there were no effects of light sources on plasma CORT concentrations. Scott and Siopes [36] also indicated that blue, green, red, and white lights were not stressful to the birds. However, sampling data for the 45 and 53 week showed that the birds exposed to red light had the lowest proportion of heterophils and the narrowest H/L ratio [36]. However, the effect of light color on the significant effect on CORT concentration and H/L ratio in this study was not clear; this effect might be caused by the effect of light condition on hens' behavior. As previous studies regarding the effect of light color on fear response of layer hens are scarce, a direct comparison is difficult. However, these results showed a consistent tendency towards greater CORT concentration [3], lower levels of whole blood 5-HT [44], and a higher H/L ratio [44] in highly fearful hens, which showed long tonic immobility durations, a far avoidance distance, and particular fearfulness of novel objects in this study. Cockrem [3] found that corticosterone responses and fearfulness were linked and indicated that greater fearfulness was accompanied by larger corticosterone responses to potentially threatening stimuli. Bolhuis et al. [42] suggested that hens from the generation of the low mortality line showed less fear-related behavior and displayed higher whole-blood 5-HT concentrations. José et al. [44] indicated that hens suffering from cloacal cannibalism were more asymmetrical, stressed, and fearful than non-vent pecked birds, with increased heterophil to lymphocyte ratio and tonic immobility duration. However, in the present study, the differences of the level of fearfulness were not reflected precisely in the concentration of thyroidal hormones. Therefore, measuring the thyroid hormone may not be a particularly appropriate method for evaluating stress in hens, because some factors related to welfare appear to lead to a rise, whereas others result in a fall [40]. Under closely controlled conditions, circulating stress hormones can be a measure of the hen's reaction to its environment. The condition is apparently not so straightforward in actual operations. The only safe conclusion seems to be that for stress hormones too many uncontrolled factors exert an effect to permit these indicators to be employed as simple and practical assessment of welfare.

Conclusions

The results of this study illustrate that different light color and light intensity influenced the behavior and fear response of laying hens. Hens treated with RL and low light intensity in natural mating colony cages during the laying period showed a lower frequency of SFP, less damaged plumage, were less fearful, and had lower physiological indicators of stress. In addition, RL could reduce mortality from cannibalism and cannibalistic injuries. Transforming the light color to red or dimming the light could be regarded as an effective method to reduce the risk of FP and cannibalism and alleviate the fear responses of layer breeders in natural mating colony cages. Such knowledge might help to understand FP behavior and stress susceptibility of hens in natural mating colony cages and will provide a basis for the development and optimization of cage equipment and regulation of the light environment.

Acknowledgments

The study was supported by the National Key R&D Program of China (2017YFB0404000). We acknowledge the manger and staff of Hebei Huayu Poultry Breeding Co. Ltd., Handan, Hebei, China. Help and support from colleagues at the department during the project are also appreciated.

References

1. Shi, H.P.; *et al,* In fluence of nest boxes and claw abrasive devices on feather pecking and the fear responses of layer breeders in natural mating colony cages. *Appl. Anim. Behav. Sci.* **2019**, 104842, doi:10.1016/j.applanim.2019.104842.

2. Rodenburg, T.B.; *et al,* Selection method and early-Life history affect behavioural development, feather pecking and cannibalism in laying hens: A review. *Appl. Anim. Behav. Sci.* **2008**, *110*, 217–228.

3. Cockrem, J.F. Stress, corticosterone responses and avian personalities. *J. Ornithol.* **2007**, *148*, 169–178.

4. Savory, C.J.; *et al,* Incidence of pecking damage in growing bantams in relation to food form, group size, stocking density, dietary tryptophan concentration and dietary protein source. *Br. Poult. Sci.* **2000**, *40*, 579–584.

5. Shi, H.P.; *et al,* Effects of different claw-Shortening devices on claw condition, fear, stress and feather coverage of layer breeders. *Poult. Sci.* **2019**, *98*, 3103–3113.

6. Braastaad, B.O. Rearing pullets in cages: High crowding has unfortunate effects. *Poult..* **1986**, 2, 38–41.

7. Sultana, S.; *et al,* Effect of various LED light color on the behavior and stress response of laying hens. *Indian. J. Anim. Sci.* **2013**, *83*, 829–833.

8. Bright, A. Plumage colour and feather pecking in laying hens, a chicken perspective? *Br. Poult. Sci.* **2007**, *48*, 253–263.

9. Kjaer, J.B.; *et al,* Development of feather pecking in relation to light intensity. *Appl. Anim. Behav. Sci.* **1999**, *62*, 243–254.

10. Blokhuis, H.J.; *et al,* Some observation on the development of feather pecking in poultry. *Appl. Anim. Behav. Sci.* **1984**, *12*, 145–157.

11. Hughes, B.O.; *et al,* The influence of strain and environmental factors upon feather pecking and cannibalism in the fowl. *Br. Poult. Sci.* **1972**, *13*, 525–547.

12. Kjaer, J.B.; *et al,* Feather pecking and cannibalism in free-range laying hens as affected by genotype, dietary level of methionine + cystine, light intensity during rearing and age at first access to the range area. *Appl. Anim. Behav. Sci.* **2002**, *76*, 21–39.

13. Prayitno, D.S.; *et al,* The effects of color of lighting on the behavior and production of meat chickens. *Poult. Sci.* **1997**, *76*, 452–457.

14. Wells, R.G. A comparison of red and white light and high and low protein regimes for growing pullets. *Br. Poult. Sci.* **1971**, *12*, 313–325.

15. Yang, Y.; *et al,* A new method to manipulate broiler chicken growth and metabolism: Response to mixed LED light system. *Sci. Rep.* **2016**, *6*, 25972.

16. Huber-Eicher, B.; *et al,* Effects of colored light-emitting diode illumination on behavior and performance of laying hens. *Poult. Sci.* **2013**, *92*, 869–873.

17. Mohammed, H.H.; *et al,* The effects of lighting conditions on the behavior of laying hens. *Eur. Poult. Sci.* **2010**, *74*, 197–202.

18. Long, H.; *et al,* Effect of light-emitting diode vs. fluorescent lighting on laying hens in aviary hen houses: Part 1-Operational characteristics of lights and production traits of hens. *Poult. Sci.* **2015**, *95*, 1–11.

19. Shinmura, T.; *et al,* Effects of light intensity and beak trimming on preventing aggression in laying hens. *Anim. Sci. J.* **2010**, *77*, 447–453.

20. Prescott, N.B.; *et al,* Spectral sensitivity of domestic fowl (*Gallus g. domesticus*). *Br. Poult. Sci.* **1999**, *40*, 332–339.

21. Rodenburg, T.B.; *et al,* Comparison of individual and social feather pecking tests in two lines of laying hens at ten different ages. *Appl. Anim. Behav. Sci.* **2003**, *81*, 133–148.

22. Butterworth, A.; *et al,* Welfare Quality^R Assessment Protocol for Poultry, 1st ed.; Welfare Quality^R Consortium: Lelystad, Netherlands, 2009; pp. 60-81

23. Gilani, A.M.; *et al,* The effect of rearing environment on feather pecking in young and adult laying hens. *Appl. Anim. Behav. Sci.* **2013**, *148*, 54–63.

24. Hartcher, K.M.; *et al,* The effects of environmental enrichment and beak-Trimming during the rearing period on subsequent feather damage due to feather-Pecking in laying hens. *Poult. Sci.* **2015**, *94*, 852–859.

25. Prayitno, D.S.; *et al,* The effects of color and intensity of lighting on behavior and leg disorders in broiler chickens. *Poult. Sci.* **1997**, *76*, 1674–1681.

26. Prescott, N.B.; *et al,* Reflective properties of domestic fowl (*Gallus g. domesticus*), the fabric of their housing and the characteristics of the light environment in environmentally controlled poultry houses. *Br. Poult. Sci.* **1999**, *40*, 185–193.

27. Maddocks, S.A.; *et al,* Rapid behavioural adjustments to unfavourable light condition in European starlings (*Sturnus vulgaris*). *Anim. Welf.* **2002**, 11, 95–101.

28. Leighton, A.T.; *et al,* Effect of light sources and light intensity on growth performance and behavior of male turkeys. *Br. Poult. Sci.* **1990**, *30*, 563–574.

29. Lewis, P.D.; *et al,* Poultry and colored light. Worlds. *Poult. Sci. J.* **2000**, *56*, 189–207.

30. Wabeck, C.J.; *et al,* Influence of radiant energy from fluorescent light sources on growth, mortality and feed conversion of broilers. *Poult. Sci.* **1974**, *53*, 2055–2059.

31. Savory, C.J. Feather pecking and cannibalism. *Worlds. Poult. Sci. J.* **1995**, *51*, 215–219.

32. Hughes, B.O.; *et al,* The effect of environmental factors on activity, selected behaviour patterns and 'fear' of fowls in cages and pens. *Br. Poult. Sci.* **1974**, *15*, 375–380.

33. Allen, J.; *et al,* Feather pecking and cannibalism in a caged layer flock. *Br. Poult. Sci.* **1975**, *16*, 441–451.

34. Bilcík, B.; *et al,* Changes in feather condition in relation to feather pecking and aggressive behaviour in laying hens. *Br. Poult. Sci.* **1999**, *40*, 444–451.

35. Huber-Eicher, B.; *et al,* Reducing feather pecking when raising laying hen chicks in aviary systems. *Appl. Anim. Behav. Sci.* **2001**, *73*, 59–68.

36. Scott, R.P.; *et al,* Light color: Effect on blood cells, immune function and stress status in turkey hens. Comp. *Biochem. Physiol.* **1994**, *108*, 161–168.

37. Hrabcakova, P.; *et al,* Evaluation of tonic immobility in common pheasant hens kept in different housing systems during laying period. *Arch. Tierz.* **2012**, *55*, 626–632.

38. Johnsen, P.F.; *et al,* Influence of early rearing conditions on the development of feather pecking and cannibalism in domestic fowl. *Appl. Anim. Behav. Sci.* **1998**, *60*, 25–41.

39. de Haas, E.N.; *et al,* Parents and early life environment affect behavioral development of laying hen chickens. *PLoS ONE* **2014**, *9*, e90577.

40. Gibson, S.W.; *et al,* Plasma concentrations of corticosterone and thyroid hormones in laying fowls from different housing systems. *Br. Poult. Sci.* **1986**, *27*, 621–628.

41. Siegel, H.S. Physiological stress in birds. *Bioscience* **1980**, *30*, 529–534.

42. Bolhuis, J.E.; *et al,* Effects of genetic group selection against mortality on behavior and peripheral serotonin in domestic laying hens with trimmed and intact beaks. *Physiol. Behav.* **2009**, *97*, 470–475.

43. Olanrewaju, H.A.; *et al,* Interactive effects of photoperiod and light intensity on blood physiological and biochemical reactions of broilers grown to heavy weights. *Poult. Sci.* **2013**, *92*, 1029–1039.

44. José, L.C.; *et al,* Association between vent pecking and fluctuating asymmetry, heterophil to lymphocyte ratio, and tonic immobility duration in chickens. *Appl. Anim. Behav. Sci.* **2008**, *113*, 87–97.

Effects of Illumination and Color Temperature Distribution on Subjective Perception

Dandan Hou[1,2], Congshan Dai[1], Yan Lu[1,2], Yandan Lin[2,3*]

1Institute for Electric Light Sources, Fudan University, Shanghai, China
2Engineering Research Center of Advanced Lighting Technology, Ministry of Education, Shanghai, China
3 Institute of Engineering and Applied Technology, Fudan University, Shanghai, China
ydlin@fudan.edu.cn, +86-13501903746

Abstract

There is ample evidence to support the hypothesis that indoor lighting environment has an impact on human affective and cognitive processes. Light distribution is one of the important aspects of the indoor lighting environment. With the development of various pendant luminaires and LED technology, both photometric and colorimetric distribution in vertical can be adjustable thus different lighting schemes can be achieved to satisfy the need for lighting design.

The present study of light distribution in a real lit office conducted two parts of experiments. The aim is to discuss the impact of proportions of direct/indirect lighting on both subjective evaluation and work performance, also include the discussion of the distribution of correlated color temperature (CCT) of direct /indirect lighting in an office space.

1.Introduction

There is ample evidence to support the hypothesis that indoor lighting environment has an impact on human affective and cognitive processes. Light distribution is one of the important aspects of the indoor lighting environment. With the development of various pendant luminaires and LED technology, both photometric and colorimetric distribution in vertical can be adjustable thus different lighting schemes can be achieved to satisfy the need for lighting design.

Laboratory studies have shown that people tend to prefer lighting systems with indirect lighting components to systems providing just direct lighting and the ratio of direct/indirect lighting can modulate spacious perception [1][2]. Yu-bin Shin's study that indirect lighting makes people feel happy and awake through physiological parameter analysis and subjective evaluation analysis [3]. However, there are some inconsistent conclusions from previous studies and not all of them support mixed lighting. A field study by Collins et al. further shows that indirect lighting system was less preferred to other mixed lighting system [4]. A Norwegian study proposed that the light distribution ratio of 75% indirect lighting and 25% direct lighting could improve the lighting environment and visual conditions sustainably [5]. In contrast, Wolska revealed increased eye fatigue with mixed lighting [6]. KI Fostervolda believes that the proportion of light distribution has no obvious influence on people's work performance in a one-year intervention experiment [7]. Besides, since these researches focused on the illuminance distribution of direct/indirect lighting system, there is a lack of study considering the colorimetric distribution of mixed lighting. Will different CCT of uplight and downlight affect

occupants' assessment and behavior to the lighting environment?

As an important occasion of work, offices of different scale have got the most attention from different researches. For the lighting environment in office space, both occupants' feeling and evaluation and visual behavior matter. The present study of light distribution in a real lit office conducted two parts of experiments. The aim is to discuss the impact of proportions of direct/indirect lighting on both subjective evaluation and work performance, and also include the discussion of the distribution of correlated color temperature (CCT) of direct/indirect light in an office space.

2.Methods

15 Chinese participants aged from 21 to 30 years (mean age=23.4 years, SD=2.77 years) were recruited for this experiment, including 7 males and 8 females. The Ishihara test was used to examine vision and all participants had normal vision.

The experiments were carried out to study the distribution of illuminance and color temperature in a real lit office. The room had a size of 3.2m x 3m x 2.8m. Six pendant fixtures were installed and the brightness and CCT of direct/indirect lighting could be controlled respectively. The experiments were separated into two parts. In the first part, the proportions of direct/indirect lighting changed. Lighting parameters including illuminance and CCT on working plane were held constantly ($6000\pm100K$, $300\pm5lx$). In the second part, the CCT of direct/indirect lighting changed and the CCT and illuminance on working plane were also kept in constant, which are $4800\pm100K$ and$300\pm10lx$ for all these three light settings. Figure 1 shows the experimental environment.

Table 1. Lighting parameters of each light setting

Light Setting		1	2	3	4	5
Illuminance proportions (%)	indirect	100	75	50	25	0
	direct	0	25	50	75	100
CCT (K)				6000		
Light Setting		6	7	8		
Illuminance proportions (%)	indirect		50			
	direct		50			
CCT (K)	indirect	6000	3800	3000		
	direct	4000	6500	8600		

Figure 1. Illumination distribution diagram - five illuminance ratios

978-1-7281-5757-3/19 $31.00 © 2019 IEEE

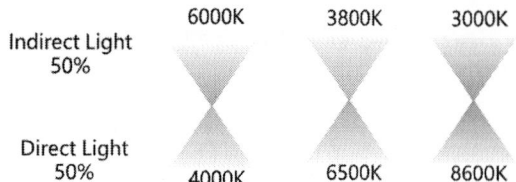

Figure 2. Color temperature distribution diagram - three color temperature ratios

In the first part of the experiment, each observer viewed the test lighting conditions following a random order. In the second part of experiment, the three light settings were viewed in three sessions. Each session was separated by 2 days due to the time setting up the experimental light. Observers were asked to assess the same direct lighting condition (6000K, 300lx on work plane) in every session to test the baseline of each observer in different experimental days.

Figure 3. Photographs showing lighting settings of proportions (top row) and CCT distributions (bottom row) of direct/indirect light

For each light setting, observers were asked to assess the lighting environment in terms of subjective evaluation, work performance, and fatigue. Subjective impressions were evaluated using 7-point categorical judgment methods. The subjective evaluation questionnaire was designed based on KW Houser's experiment [8], which were modified considering 7 parts of content: subjective brightness, objective evaluation, visual comfort, uniformity, spaciousness, readability, and preference. For work performance assessment in office, the d2 Test of Attention (D2) [9], a cancellation test, and the Paced Visual Serial Addition Task (PVSAT) [10] are used as visual tasks. Besides, critical fusion frequency (CFF) [11] and questionnaires were used to reflect fatigue in different light settings. CFF values were measured before and after each task session while questionnaires were used after visual tasks.

Table 2. The subjective evaluated scales studied in seven categories

subjective brightness	Bight (ceiling, desktop, walls, floor, room overall)
objective evaluation	Harsh/mild shadow, dull/sharp edges, attractive
visual comfort	Comfort, glare, light quality
uniformity	Uniform (ceiling, desktop, walls, floor, room overall)
spaciousness	Large, spacious
readability	Clear, recognizable (while reading),
preference	Like, pleasant, satisfied, energetic

3. Results

3.1 Statistical analysis

Data were analyzed in the Statistical Package for Social Science version 22.0 (SPSS Inc., Chicago, IL). The alpha level was set at 0.05 for all statistical tests, and all p-values were two-tailed. Analysis of variance (ANOVA) was made to analysis people's assessments of the distribution of illuminance and color temperature of office lighting. Multiple comparison groupings based on LSD method were also used to isolate which of the differences between sets of means contributes to the overall difference.

3.2 Subjective Evaluation

For the first part of the experiment, the results of subjective brightness show a significant downward trend on the perceived brightness of the ceiling when the proportions of indirect decrease (p=0.000). The brightness of the wall shows a small downward trend when the room is lit more direct but the noticeable difference only shows between light setting 1&4 and 1&5, which means noticeable wall's brightness change needs large difference. Bivariate correlation analysis shows there are significant linear correlations between overall brightness and brightness of the wall (Pearson's r=0.822), floor (Pearson's r=0.711), desk (Pearson's r=0.711), and ceiling (Pearson's r=0.475). The perceived overall brightness of the experimental office room can be described as the following equation:

$$B_{overal} = 0.25 + 0.357B_{wall} + 0.241B_{floor} + 0.282B_{desk} + 0.103B_{ceiling}$$

The overall brightness trend shows similar results with the previous study but the correlation results between different scale are different, which indicates brightness perception is related to the specific office space.

According to the subjective ratings, light settings with more indirect light and less direct light also lead to milder shadows of objects (p=0.000), more eye's comfort (p=0.046), more feelings of spaciousness (p=0.011) and pleasure (p=0.008). Observers also express more preference when more light was upward (p=0.002). However, people tend to feel tiredness easily in a lighting environment contains more indirect component (p=0.028).

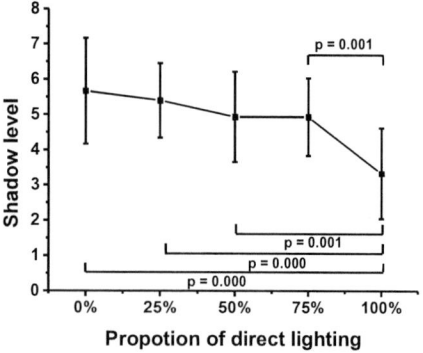

Figure 4. Ratings for shadow level

Under different illuminance distribution conditions, there were no significant differences in the evaluation of edge sharpness (P=0.404>0.05) and attractiveness of object (P=0.425>0.05), while the perception of shadow (P=0.000<0.05) showed significant differences. Figure 4 showed that the environment containing indirect light will deepen the shadow of the object.

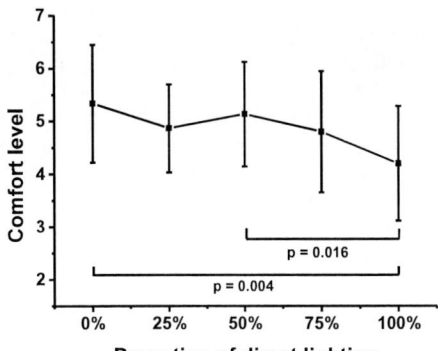

Figure 5. Ratings for comfort level

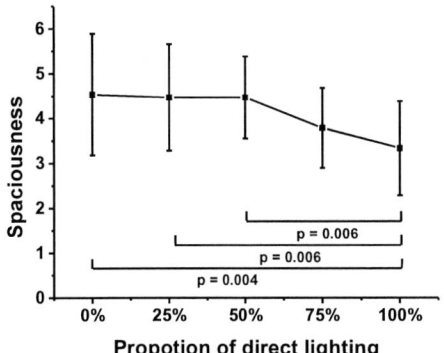

Figure 6. Ratings for spaciousness

As we can see in figure 6, suggesting environment with more than 50% indirect lighting appeared more capacious, and environment with 100% indirect lighting looked most spacious.

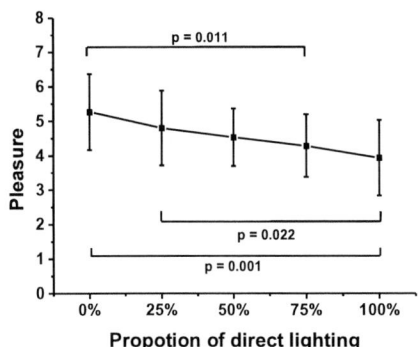

Figure 7. Ratings for pleasure

Under different illumination distribution conditions, a significant difference appeared in subjects' preference (P=0.002<0.05), pleasure (P=0.008<0.05). It could be seen from figure 7 and figure 8 that the changing trend of preference and pleasure were obvious and similar. The higher the indirect light proportion, the higher the preference and pleasure degree. The environment with 100% indirect light obtained the highest preference and pleasure rating.

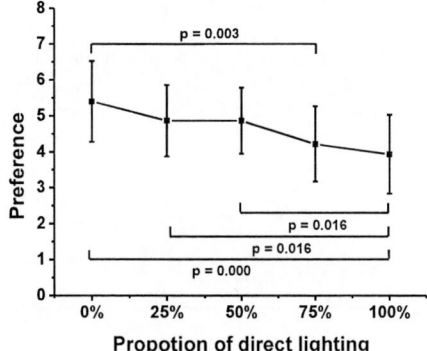

Figure 8. Ratings for preference

The uniformity of ceiling, wall, floor, and the overall room changes significantly when light changes and the uniformity except for the floor reaches the highest in the light setting 3 (direct/indirect light). The overall room uniformity decreases significantly (p=0.000) when more direct light is supplied.

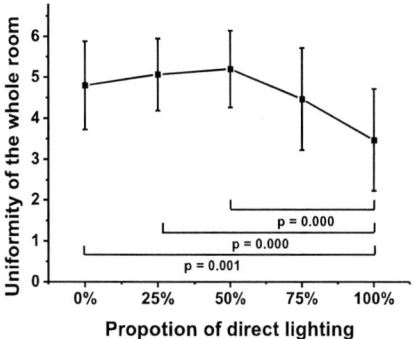

Figure 9. Ratings for uniformity of the whole room

For the CCT distribution of direct/indirect light, there is a significant decrease in perceived walls' uniformity when the difference of CCT between uplight and downlight becomes larger (p=0.031).

The CCT distribution seems do not affect the perceived brightness of different part of the office. Bivariate correlation analysis also shows significant linear correlations between overall brightness and brightness of the wall (Pearson's r=0.706), floor (Pearson's r=0.620), desk (Pearson's r=0.696), and ceiling (Pearson's r=0.398). The correlation between brightness perception results here and the linear regression model in the first part was tested and the r²=0.80. This shows that the linear regression model has a relatively good fitting degree when light settings changes, which may offer some solutions to modify the space brightness with light distribution.

For other scales used in the experiment, the different observer seems to have different preference or feeling about these three light settings. No obvious relationships are found between CCT distribution and other subjective evaluation results.

Compared all the light settings in view of the same baseline, conditions with indirect components improved the spaciousness of the room and preference for the environment. Overall, direct lighting supplemented by indirect lighting with consistent or inconsistent CCT is superior to direct lighting in the office in terms of subjective evaluation.

3.2 Cognitive performance and fatigue

Table 3 and Table 4 show the results of the D2 test and the PVSAT test. One-way ANOVA for repeated measurements was conducted. The results for two cognitive tests showed no significant difference between the five light settings (1-5) with different proportions of direct/indirect light and between the three light settings (6-8) with different CCT distributions as well. No effect of light settings on fatigue was found either.

Table 3. Test results of the D2 test (Mean and standard deviation)

Mea sure	1	2	3	4	5	6	7	8
TN	607.5 (40.0)	612.9 (39.8)	592.4 (58.0)	608.5 (46.6)	603.9 (45.4)	618.0 (40.1)	629.5 (37.7)	627.0 (35.5)
E1	12.5(16.8)	11.9(9.4)	12.6(12.9)	15.6(20.5)	12.9(13.1)	9.0(1 0.6)	4.9(6. 3)	7.5(9. 0)
E2	1.4(2. 9)	1.1(1. 3)	1.6(1. 6)	1.1(1. 6)	1.7(1. 8)	2.2(2. 3)	1.8(2. 6)	1.9(2. 1)
E	13.9(19.7)	12.9(10.6)	14.2(14.5)	16.7(22.1)	14.7(14.9)	11.2(12.9)	6.7(8. 9)	9.5(1 1.1)
TN-E	593.5 (47.0)	599.9 (43.0)	578.2 (63.7)	591.9 (56.9)	589.3 (49.7)	606.8 (47.8)	622.8 (42.3)	617.5 (38.2)
CP	304.2 (4.9)	305.4 (3.87)	305.5 (3.9)	306.1 (2.5)	305.1 (3.23)	305.8 (2.3)	306.2 (2.6)	306.1 (2.1)
FR	6.1(3. 3)	6.7(5. 3)	6.4(4. 7)	5.5(3. 4)	6.1(4. 0)	5.9(4. 2)	5.3(5. 6)	4.1(4. 0)

Notes: TN = total number; E1 = Omissions; E2 = Commissions; E=errors; N-E = total-errors; CP = concentrations performance; FR = fluctuation rate.

Table 4. Test results of the PVSAT test (Mean and standard deviation)

Measur e	1	2	3	4	5	6	7	8
Time/s	142. 6(12. 9)	143. 3(17. 9)	151. 7(19. 5)	144. 5(14. 9)	142. 1(11. 5)	140. 9(15. 1)	131. 8(7.5)	134. 5(9. 9)
Accura cy/%	98.9(2.7)	98.0(3.73)	96.7(5.77)	98.0(5.2)	98.0(3.3)	98.9(2.1)	98.7(2.5)	98.9 (1.6)
Time/A ccuracy /10^-3	7.0(0 .7)	7.0(1 .0)	6.5(1 .0)	6.9(1 .0)	7.0(0 .7)	7.1(0 .7)	7.5(0 .5)	7.4(0.6)

These results are partly consistent with KI Fostervold's study which revealed that change of the proportions of indirect and direct light in a long time intervention (half or one year) didn't affect health, well-being and cognitive performance. Our experiments indicate that as long as the illuminance and CCT at work plane are held constantly, neither the illuminance nor CCT distribution of the space can influence the work performance or fatigue of office workers'.

4. Conclusions

Experiments were conducted to assess perception and cognitive performance under a series of light settings with different proportions and CCT of direct/indirect lighting in the office. From the results above, the following conclusions can be drawn:

➢ It is worth considering the photometric and colorimetric distribution of office lighting because occupants are able to discriminate the differences.

➢ Ambient light in an indoor environment can significantly influence subject evaluation in many aspects especially the brightness and uniformity perception of the different parts of the room. Supplementing indirect lighting will improve the spaciousness of the room and preference for the environment.

➢ Neither the illuminance nor CCT distribution of the space influences work performance or fatigue of office workers as long as the illuminance and CCT at work plane are held constantly, which may offer more freedom to office lighting designers.

Acknowledgments

The authors disclosed receipt of the following financial support for the research, authorship, and/or publication of this article: This research was supported by National Key R&D Program of China (Project No.2017YFB0403700).

References

[1] HARVEY, L. O. D. D. (1984), "Quantifying Reactions of Visual Display Operators to Indirect Lighting ", *Journal of the Illuminating Engineering Society,* Vol. 14 No. 1, pp. 515-546.

[2] ROBERT YEAROUT, S. K. (1989), "Visual display unit workstation lighting", *International Journal of Industrial Ergonomics,* Vol. 3 No. 3, pp. 265-273.

[3] SHIN, Y., WOO, S., KIM, D., KIM, J., KIM, J. & PARK, J. Y. (2015), "The effect on emotions and brain activity by the direct/indirect lighting in the residential environment", *Neuroscience Letters,* Vol. 58428-32.

[4] BELINDA L. COLLINS, W. F. G. G. (1990), "Second-Level Post-Occupancy Evaluation Analysis", *Journal of the Illuminating Engineering Society,* Vol. 19 No. 2, pp. 21-44.

[5] ARNE AARÅS, G. H. H. B. (2001), "Musculoskeletal, visual and psychosocial stress in VDU operators before and after multidisciplinary ergonomic interventions. A 6 years prospective study F Part II", *Applied Ergonomics,* Vol. 32 No. 6, pp. 559-571.

[6] WOLSKA, A. (2003), "Visual Strain and Lighting Preferences of VDT Users Under Different Lighting Systems", *International Journal of Occupational Safety and Ergonomics,* Vol. 9 No. 4, pp. 431-440.

[7] FOSTERVOLD, K. I. & NERSVEEN, J. (2008), "Proportions of direct and indirect indoor lighting — The effect on health, well-being and cognitive performance of office workers", *Lighting Research & Technology,* Vol. 40 No. 3, pp. 175-200.

[8] HOUSER, K. W., TILLER, D. K., BERNECKER, C. A. & MISTRICK, R. G. (2002), "The subjective response to linear fluorescent direct/indirect lighting systems", *Lighting Research & Technology,* Vol. 34 No. 3, pp. 243-260.

[9] Zillmer, E. A., & Kennedy, C. H. (1999). Construct validity for the d2 test of attention. *Archives of Clinical Neuropsychology*, 14(8), 728-728.

[10] Feinstein A, Brown R, Ron M. (1994). Effects of practice of serial tests of attention in healthy subjects. *J Clin Exp Neuropsychol* ,16: 436–447.

[11] Wang, Y., Zhong, X., Tu, Y., Wang, L., Zhang, Y., & Wang, T., et al. (2017). A model for evaluating visual fatigue under led light sources based on long-term visual display terminal work. *Lighting Research & Technology*, 147715351769001.

Research on train light environment evaluation method

Sijie He[1,3], Jinrong Liu[2], Shuo Jing[3,4], Yandan Lin[1,4,*]

[1]Institute for Electric Light Sources, Fudan University

[2] Changchun Railway Vehicles Co., LTD.(CRC)

No 220 Handan Road, Shanghai, China

[3]Engineering Research Center of Advanced Lighting Technology, Ministry of Education

[4]Academy for ENGINEERING&TECHNOLOGY, Fudan University

No 220 Handan Road, Shanghai, China

ydlin@fudan.edu.cn

Abstract

With the continuous development of EMU technology, the role of light rail trains in cities is becoming more and more important. A good overall light environment is of great significance for the safe and efficient operation of light rail trains. This experiment is based on the purpose of exploring the evaluating the train light environment method. In this study, we adopted the method of first using optical software DIALux and TracePro simulation, then conducting physical test and subsequent ergonomics experimental verification. By comparing the simulation results with the experimental results, we gave the suggestion about the final evaluation of the train light environment. It is hoped that the results of this study will provide a scientific method and basis for the design of the light environment of the train cab and passenger compartment in the future.

Introduction

With the rapid development of high-speed rail technology, the safety of driving trains and the comfort of taking trains need to be paid more attention. Improving the light environment of the train cab and passenger cars has become an effective means. The effects of light on the human body are complex and not limited to the visual pathway. The parameters of the light environment can affect the fatigue, comfort and alertness of the human body, such as uncomfortable glare, light distribution, contrast, and the like.

In previous studies, the cockpit glare was focused, the effects of glare on the deterioration of visual function, like Ranney's [1] and Osterhaus's [2]. Meanwhile, in more extensive areas such as offices, bedrooms, classrooms, etc. The display also works in a time-varying light environment. The spatial light distribution also has a great influence on work performance and mood. Harvey [3], Hedge [4], Shin [5] found that people prefer an environment containing indirect lighting to only direct lighting. Besides, a series of studies have been conducted on the effects of monitors on sleep, alertness and EEG power in recent years, like Yosuke's [6], Brittany's [7], and Mariana's [8]. People pay more and more attention to the evaluation of the train's light environment to improve the safety and the comfort of cabs.

1.Simulation process

The simulation part is mainly for the two light environments of the driver's cab and the passenger compartment. The driver's cab and passenger compartment of the light rail are used as different missions. The light environment, internal light source distribution, visual mission requirements, and internal space environment are different. Therefore, in the light analysis of the driver's cab and the passenger compartment, different typical working conditions and evaluation indicators were selected. In this part, we use Tracepro and DIALux to simulate the light environment of two situation.

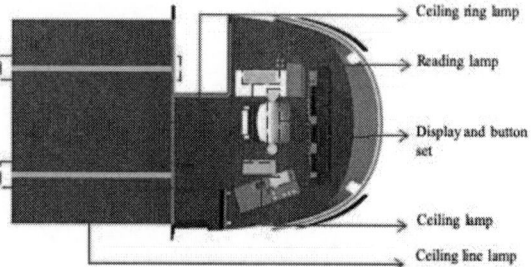

Figure 1 Tracepro simulation light source in driver's cab and passenger compartment

According to the analysis of simulation results, the main problems in the driver's cab include serious reflection glare interference and high direct glare. It is suggested to reduce the power of the overhead light and reading light in the driver's cab. The front windscreen should adopt a higher transmittance, and the windscreen should be equipped with an adjustable length shade. For the passenger compartment, the illumination of typical working face is on the high side, and the UGR glare value of typical eye position is between 22 and 28. It can be concluded that the lighting power of the coach room is too large for the environment.

2.Physical test

According to different situations, physical tests have been carried out on the driver's cab and passenger compartments. The light environment of driver's cab and passenger compartment is described in the Table 1 and Table 2.

Table 1 Description of driver's cab light environment

Code of the lighting scene	Description
A	Inner light off. HID power 100%.
B	Inner light off. HID power 67%.
C	No outer light, Ceiling lamp 100%, Reading lamp open.
D	No outer light, Ceiling lamp 50%, Reading lamp open.
E	No outer light, Ceiling lamp closed, Reading lamp open.

Table 2 Description of passenger compartments light environment

Code of the lighting scene	Description
A	Ceiling Line Lamp power 100%，Ring lamp power 100%.
B	Ceiling Line Lamp power 30%，Ring lamp power 30%.
C	Ceiling Line Lamp power 100%，Ring lamp closed.
D	Ceiling Line Lamp power 30%，Ring lamp closed.

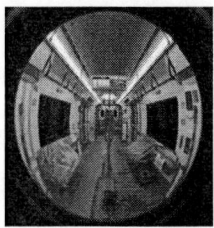
F compartment

MC compartment

TP Compartment

Figure 2 Sketch of the passenger compartment

Figure 3 Sketch of the driver's cab

According to different situations, two different eye positions are chosen to perform the analysis of the UGR. The sitting position is 1.2m above the ground, and the standing position is 1.5m above the ground. And the testing procedure including four steps:

1) Confirm the situation, and set the light-emitting device according to the Table 1 and Table 2.

2) Confirm the adjustable part of this situation.

3) Test the parameters of the environment in the sub situation.

4）Test the parameters using devices according to the analysis needed.

The tested illuminance and illuminance uniformity results are shown in Table 3 and Table 4.

Table 3 Illuminance of eye positions in driver's cab

| Mode | Eye position | | | | | | | |
	Front-standing	Left-standing	Right-standing	Rear-standing	Front	Left	Right	Rear
A	21.7	858.6	31.1	38.7	21.7	804.6	33.9	54.6
B	16.9	512.2	27.9	67.4	19.8	522.3	42.6	7.7
C	12.8	262.4	157.6	296.7	24.9	186.3	191.0	56.3
D	9.2	148.8	68.3	163.8	18.8	151.9	204.6	72.4
E	3.6	35.1	6.1	6.4	7.5	94.0	39.9	25.0

Table 4 Illuminance uniformity in driver's cab

| Illuminance uniformity | | | | | |
Mode	Work panel	Working plane	Ground	Monitor-left	Monitor-right
A	0.87	0.78	0.82	0.89	0.98
B	0.88	0.96	0.69	0.6	0.96
C	0.54	0.94	0.92	0.92	0.99
D	0.5	0.97	0.73	0.92	0.96
E	0.26	0.85	0.8	0.78	0.86

According to the testing results, the illumination uniformity of working face, right display and ground is better. They are over 0.7. In the case of simulated cloudy day, the illumination uniformity of left display and ground decreases obviously. According to the actual situation on the spot, the reason is that the simulated sunlight shines of metal halogen lamps rays directly into a part of this area, which leads to the decrease of illumination uniformity. With the decrease of light flux, the uniformity of illumination in the driver's cab area of night mode decreases, but the decrease is not obvious except for the worktable.

The glare test results are expressed by Unified Glare Index (UGR). CIE defines UGR (Unified Glare Rating) for glare assessment that covers general parameters of various models. It can precisely describe and calculate the glare of the simulation, so it is widely used for all kinds of discomfort glare assessment. Calculated by the unified glare formula: [9]

$$UGR = 8\lg \frac{0.25}{L_b} \square \frac{L_a^2 \leftarrow \omega}{p^2}$$

With: L_b—background luminance（cd/m2）;

L_a—luminaire luminance in observer direction;

ω —solid anger formed on the eye by the light emitting part（sr）;

P —position index of the luminaire;

r —the distance between the center of luminaire and the observer's eyes.

Glare test includes two types of eye positions, one is standing, that is, 1.5m high eye position, the other is sitting, that is, 1.2m high eye position, all located in the driver's seat, height from the ground of the train. The results of eye glare test are shown in Table 5.

Table 5 UGR values of the driver's cab

Mode	Front	Left	Front-standing	Left-standing
A	11.8	45.8	12.0	45.3
B	12.0	44.7	13.1	47.5
C	13.0	15.5	13.4	23.6
D	14.6	11.7	13.3	23.7
E	15.6	16.6	17.3	29.1

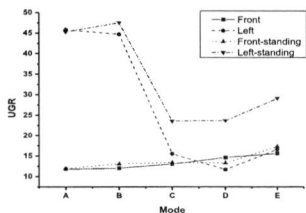

Figure 4 UGR values of the driver's cab in different modes

As can be seen from Figure 4 and Table 5, the left UGR index is higher in daytime, reaching about 45. And the left-standing UGR value is higher as a whole, over 20. In other cases, UGR values are all below 19, which meets the requirements for glare. The front glare value increases with the decrease of luminous flux at night conditions. The reason is that the decrease of luminous flux at night leads to the decrease of background luminance and the increase of UGR value.

Table 6 Illuminance and Illuminance uniformity of passenger compartments

Coach	Mode	Front	Seat	Center	Standing side
MC	A	20.5	19.5	21	-
	B	18.5	17.2	19	-
	C	21.2	25.2	21	-
	D	17.5	19.4	18.7	-
F	A	21.4	23	19.9	21.3
	B	17.6	17.1	15.9	16.9
	C	20.6	25.2	18.8	20.8
	D	16.4	18.4	14.1	17
TP	A	20	20.6	20.1	-
	B	17.2	18.6	18.8	-
	C	21	21	20.5	-
	D	17.9	17.7	18.1	-

Table 6 is the illuminance and illuminance uniformity of passenger compartments. The working plane is 0.8m high from the ground. It can be seen that in the D mode with the lowest light flux, the illumination of F-coach working plane is lower than the requirement of 150lx, while the illumination of other coaches in each case meets the requirement of 150lx. Combining with the analysis of the illuminance and the working mode, we can see that when the power of the line lamp drops, the illumination value of the whole coach decreases seriously, so the line lamp has a great influence on the illumination value of the coach.

The physical tests showed the luminous flux of compartments' interior environment can be adjusted appropriately to reduce the UGR value on the premise of ensuring certain working face illumination. In addition, for the increase of UGR caused by the change of night light flux in the driver's cab, it is suggested that the cab should be kept in a brighter state, which can reduce the UGR value.

2.Ergonomics experiment

The whole ergonomics experiment aims to take subjective evaluation to validate simulation results. Sixteen subjects are recruited for this experiment. 8 female and 8 females, and average age is 25. we measure visual comfort and working performance under real conditions in driver's cab and passenger compartment. And we also analyze influences of lighting factors on driver's visual comfort, work performance and the readability of passengers' indoor LCD screens. Three outdoor lighting scene modes are set in the driver's cab, including two static modes: daylight mode and nighttime mode. And one dynamic mode to simulate the situation when the external tunnel light environment changes abruptly from dark to bright. Four kinds of light environment modes P1-P4 are set in passenger's cab. As Table 7 showed, two typical passenger positions (typical eye height) are set in 4 scene modes: sitting position 1.2m and standing position 1.5.m, i.e. 8 experimental conditions.

Table 7 Device settings under all conditions in driver's cab

Device / Condition	Out environment	Sunlight (MH)	Top light	Ring light	Typical eye height	Instruction light	Button	Driver's cab light
P1	Night	/	100%	100%	1.2m	○	○	○
	Night	/			1.5m	○	○	○
P2	Night	/	35%	35%	1.2m	○	○	○
	Night	/			1.5m	○	○	○
P3	Night	/	100%	0%	1.2m	○	○	○
	Night	/			1.5m	○	○	○
P4	Night	/	35%	0%	1.2m	○	○	○
	Night	/			1.5m	○	○	○

Table 8 Device settings in MC passenger compartment

Device / Condition	Out environment	Sunlight (MH)	Plant light	Top light	Reading light	VMS	Instruction light	Button	Pointer table	Passenger compartment light
N1	Night	/	/	0%	100%	○	○	/	○	○
N2	Night	/	/	50%	100%	○	○	/	○	○
N3	Night	/	/	100%	100%	○	○	/	○	○
D1	Daytime	100%	/	/	/	○	○	/	○	○
D2	Daytime	67%	/	/	/	○	○	/	○	○
OUT	Night	off-on	/	100%	100%	○	○	/	○	○

Night mode   Daylight mode

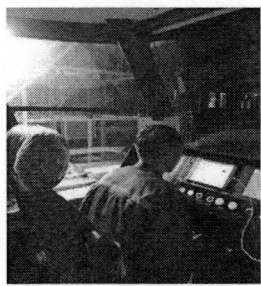

Dynamic mode

Figure 5 Different pictures of modes in driver's cab

Stance experiment   sitting experiment

Figure 6 Passenger compartment – sitting experiment

Figure 5-6 shows the ergonomic experiments in different light modes. The subjects have different experimental tasks when conducting experiments. In the nighttime mode of the driver's cab scene, the subjects were required to observe the surrounding environment in each lighting condition, fill in the nighttime mode light quality satisfaction evaluation scale, the discomfort glare and uniformity evaluation scale, and then complete the Landolt-C ring task outside the window and the Ann fermo J identify tasks on the console display device(Figure 8). In the daytime mode, the participants were required to fill in the daytime mode light quality satisfaction evaluation scale, the discomfort glare and uniformity evaluation scale, and then complete the Ann fermo J identify tasks. In the dynamic mode, the subjects need to complete the visual adaptation evaluation scale and the discomfort glare and uniformity evaluation scale after observing the dynamic changes of the light environment.

In the nighttime mode of the passenger compartment, the subjects need to observe the surrounding environment in each lighting condition, fill in the night mode light quality satisfaction evaluation scale, the discomfort glare and uniformity evaluation scale, and then complete the d2 correct word task.

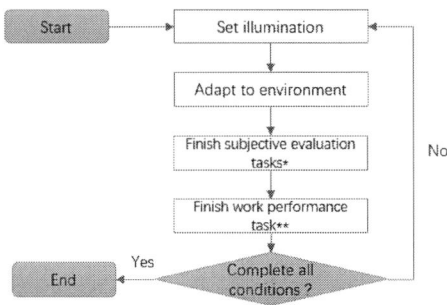

Figure 7 Experimental process

*: subjective evaluation tasks vary for different modes;
**: dynamic mode has no work performance tasks.

Figure 8 Landolt rings and Ann fermo J task

In the night scene in the driver's cab, the experimental data indicates that the luminance of the ceiling light is increased, the display screen information and the outdoor target are disturbed, the ceiling light luminance can be reduced to improve the impact on the driver's work. And the imaging problem of the windshield glass will affect the driver's observation on the external environment. We suggest to adjust the windshield transmittance to improve the problem. And when the luminance of the ceiling light is 100%, there will be an unbearable glare experience, indicating that the glare index of a typical eye position in such a scene may be too high. Compared to the night mode, more discomfort glare and lower uniformity caused by direct sunlight in the day mode. We suggest to consider adding a certain shading device to improve the driver's visual comfort. In the dynamic tunneling scene (Figure 9), the discomfort glare is more intense due to the sudden illumination of the front light, which may affect the driver's observation of the environment outside the cabin.

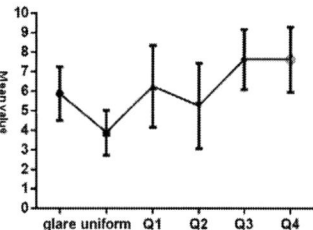

Figure 9 Mean value of the subjective evaluation of out-of-tunnel model

Table 9 DeBoer rating scale [10]

Rank	Description
1	Unbearable
2	Between 1 and 3
3	Disturbing
4	Between 3 and 5
5	Just admissible
6	Between 5 and 7
7	Acceptable
8	Between 7 and 9
9	Unnoticeable

After repeated testing of multi-factor analysis of variance, the data of the passenger compartment showed the luminance of the ceiling light has a significant effect on the feeling of discomfort glare in the sitting position experiment (F=10.182,

P=0.015). When the luminance of the ceiling light is reduced, the passenger's discomfort glare will be reduced. At the same time, it was found that after the luminance of the ceiling light reduced, the subject did not interfere with the reading of the LCD screen. In the standing position experiment, it was found that the imaging problem of the night glass was weakened after the ring light was turned off, and the feeling of discomfort glare of the subject was smaller. Since the typical eye height of a passenger who is experimenting in a standing position is 1.5 m, the ceiling light strip near the passenger will have a significant impact on each light quality perception of the passenger. When the luminance of the dome light is reduced, the problem of glass imaging at night will be weakened, and the glare of the strip and ring light will be improved, and the interference when viewing the LCD screen will be reduced. For the visual performance task, after the luminance of the ring light and the ceiling light is reduced, the score of the subject will increase. Which may be because when the luminance of the ring light and the ceiling light completely open, it will give the subject a discomfort glare feeling, affecting the subject's work performance.

Conclusion

In this study, we used physical tests combined with ergonomic experiments to verify the simulation results for the train light environment, and worked to form a more complete and effective system to evaluate the train's light environment. After the whole evaluation, the suggestion for driver's cab is to reduce the transmittance of the side window and turn on the interior lighting when exiting the tunnel. For the passenger compartment, we suggest to reduce the top light flux appropriately to improve the glare problem, and the luminance of the LCD display can be appropriately increased to reduce the interference of the indoor light source on reading the information.

We found that the results of using only the simulation will be somewhat singular, and the results of the evaluation cannot be obtained from the human body. On-site physical testing and subsequent ergonomic experimental verification can provide constructive advice for the improvement of the light environment. In the future, we will add more experimental scenarios and obtain more test data so that the evaluation system of the train light environment can form an evaluation model.

Acknowledgments

This research is in National Key R&D Program of China (Project No. 2017YFB0403700).

References

[1] "Flight vehicle integration panel working group 21 on glass cockpit operational effectiveness," pp. 189, 1996.

[2] Osterhaus, W., K. E., Bailey, & I.L., "Large area glare sources and their effect on discomfort and visual performance at computer workstations", 1992.

[3] Harvey L O , Dilaura D L , Mistrick R G . Quantifying Reactions of Visual Display Operators to Indirect Lighting[J]. Journal of the Illuminating Engineering Society, 1984, 14(1):515-546.

[4] Hedge A , Sims W , Becker F . Lighting the Computerized Office: A Comparative Study of Parabolic and Lensed-Indirect Office Lighting Systems[J]. Proceedings of the Human Factors and Ergonomics Society Annual Meeting, 1989, 33(8):521-525.

[5] Shin Y B, Woo S H, Kim D H, et al. The effect on emotions and brain activity by the direct/indirect lighting in the residential environment[J]. Neuroscience Letters, 2015, 584:28-32.

[6] Okamoto, Y., Rea, M. S., & Figueiro, M. G., "Temporal dynamics of eeg activity during short- and long-wavelength light exposures in the early morning," Bmc Research Notes, 2017, 7(1), 113-113

[7] Wood, B., Rea, M. S., Plitnick, B., & Figueiro, M. G., "Light level and duration of exposure determine the impact of self-luminous tablets on melatonin suppression", Applied Ergonomics, 2013, 44(2), 237-240.

[8] Figueiro M G, Wood B, Plitnick B, et al. "The impact of watching television on evening melatonin levels", Journal of the Society for Information Display, 2014, 21(10):417-421.

[9] Commission Internationale de l'Eclairage.Discomfort Glare in Interior Lighting CIE Publication 117-1995[M]. Vienna: CIE, 1995.

[10] De Boer J B , Schreuder D A . Glare as a Criterion for Quality in Street Lighting[J]. Lighting Research and Technology, 1967, 32(2 IEStrans):117-135.

[11] The IESNA "Lighting Handbook" Reference& Application

[12] GB/T 6769-2016 "Regulations Governing the Layout of Driver's Cabs of Locomotives"

[13] TB/T 2011-1987 "Lighting Measurement Method for Locomotive Cab"

[14] MIL-STD-1472G "Department of Defense Design Criteria Standard Human Engineering" 11 January 2012

Effect of Illuminance and Light Strobe on Attention and Visual Fatigue in Indoor Lighting

Jin Yang[1,2], Tianchi Zhang[1,2], Yandan Lin[1,3], Wei Xu[2,3*]

[1]Institute for Electric Light Sources, Fudan University, Shanghai, China
[2]Engineering Research Center of Advanced Lighting Technology, Ministry of Education, Shanghai, China
[3]Academy for Engineering & Technology, Fudan University, Shanghai, China
weixu@fudan.edu.cn, +86 18602105795

Abstract

For indoor lighting, illuminance is one of the important factors which can indicate the quality of lighting condition. However, with the popularity of LEDs, temporal light artefacts (TLA) caused by LED drivers or PWM technique has also become another consideration for evaluating the indoor lighting. One major factor that affects TLA is the frequency of light strobe. Even frequency is too high to be perceived, it will also have a negative impact on people's attention and visual perception to a certain extent. Previous studies seldom studied the combination effect of illuminance and frequency. Therefore, this study combined these two factors to investigate the change of various indicators that reflect people's attention and visual fatigue under different indoor lighting conditions. Three illuminance levels, 50lx, 150lx and 500lx and three frequency levels, 100Hz, 300Hz and 1000Hz were analyzed. A within-subject design was adopted. Twelve healthy students participated in the experiment. Subjective evaluation scales and the D2 task were used as the experimental methods. Factor analysis and MANOVAs were used for the data analysis. It was found out that illuminance significantly affected subjects' attention and alertness ($p < 0.05$). Meanwhile, visual fatigue such as ghosting and blurring mainly caused by frequency was affected by illuminance ($p = 0.038 < 0.05$), which proved our hypothesis.

Introduction

With the development of solid state lighting, LEDs are widely used in some common indoor lighting environments, such as office room, restaurant and museum because of its tunable illumination characteristics. However, such functions often realized by pulse width modulation (PWM) driven methods with different frequencies, which will introduce unsteadiness into the light output. This phenomenon is actually called light strobe or temporal modulated light.

According to the official technical report of International Commission on Illumination (CIE), light modulations can cause temporal light artefacts (TLAs), defined as "change in visual perception, induced by a light stimulus the luminance or spectral distribution of which fluctuates with time, for a human observer in a specified environment"[1]. Flicker (or strobe), which is the focus of this study, is considered to be one type of TLAs. It is known that low-frequency (below 60-90 Hz) flicker can be perceived by human eyes while high-frequency (above 100Hz) flicker cannot be perceived directly, which is also called non-visible flicker. However, whether being directly perceived or not, light strobe does have a significant physiological and psychological effect on human.

Studies have showed that workers under flickering lights may suffer from headaches and eye strains[2], or in some more extreme cases epileptic seizure[3-5], while may also find it harder to focus or have a worse mood [6]. Generally speaking, the research of flicker mainly focuses on two aspects. One is about the parameters affecting flicker, the other cares more about the physiological and psychological effect of flicker. Bullough et al. [7] found out that frequency, modulation depth and duty cycle were the most significant parameters affecting the perception of stroboscopic effect and discomfort. Other studies also drew the same conclusion [8]. Apart from that, the physiological and psychological effect of flicker also had been well studied for decades. Wilkins et al. [2] first reported more incidents of headaches and eye strains in office workers when their rooms were lit by conventional low frequency lighting compared to new high frequency lighting. Veitch et al. [9] concluded that subjects performed poorer under low frequency lighting. In addition, flicker also influence cognition and mood. Knez [6] conducted a series of experiments including short and long term memory, problem solving and subjective assessment and found out that subjects felt more activated and pleasant and performed better under high frequency lighting. Thus, the effect of the flicker parameters itself on human has already been well-discussed.

However, since flicker is just one aspect of light quality, more attention should be paid to the combined effect of flicker together with other lighting parameters, such as illuminance. There were already few studies involved in this field. Wang et al. [10] studied the effect of illuminance on the visibility threshold of flicker frequency. It was found out that for a LED light source, when driven at 100Hz, illuminance had a significant effect on visibility threshold. Such threshold being higher at lower (5lx) and higher (500lx) illuminance while being lower in the middle (50lx). Similar result was also obtained by Perz et al. [11], visibility threshold was highest at 5lx, and decreased up to illuminance level at 50lx, where it reached the lowest value. While for higher illuminance level such threshold increased again. It is reasonable to believe that since illuminance affects the perception of TLAs, it may also affect how subjects react to flicker in different lighting conditions

Thus, this study aims to investigate the combined effect of illuminance and frequency on subjects' attention and visual perception under some typical indoor lighting environments. Subjective evaluation methods and non-subjective task were used to measure such effect. It is expected to find the relationship between illuminance and frequency of how they affect people's feelings and visual perception, and provide effective guidance for indoor lighting design work.

Methods

The experiment was conducted in a simulated indoor environment (3.10m×1.95m×2.00m) with walls painted off-white as shown in Figure 1. A LED panel light (600mm×

600mm) was installed in the ceiling in order to create the required experimental conditions. There were table and chair in the middle of the room just under the LED panel. Subjects can adjust the height of chair and rotate freely when seated. To generate the required lighting parameters, a controllable LED driven system was designed in this experiment, which included a waveform signal generator and a power amplifier.

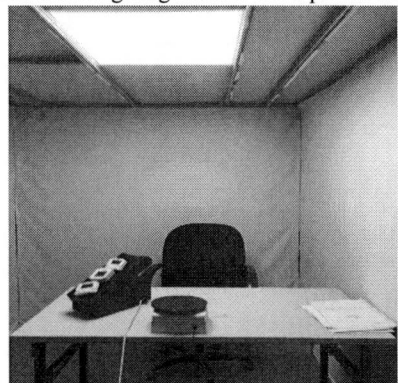

Figure 1 Experimental environment

Twelve subjects aged between 21 to 24 (Mean=22.58, std=±1.713) participated in this study, including 5 males and 7 females. All the participants reported corrected-to-normal vision and no photosensitivity epilepsy or eye diseases. Before the formal experiment, instructions were given to each participant to make them familiar with the experimental process and tasks.

The experimental conditions consisted of 9 series of selected lighting parameters in total with three illuminance levels (50lx/150lx/500lx) and three frequency levels (100Hz/300Hz/1000Hz). For the illuminance level, those parameters were chosen to represent typical indoor lighting environment including low, medium and high illumination. Similarly, the frequency parameters were chosen based on IEEE standard[12] in order to cover three different risk zones of stroboscopic. Note that other parameters were remained the same for each lighting condition through the whole experiment, which produced a consistent environment. Modulation depth being at 30% and correlated color temperature being at 4000K. Additionally, average color rendering index (Ra) was larger than 85. All parameters were calibrated before formal experiment begins.

Table 1 Experimental parameters

Parameter Type	Parameter	Value
Variables	Frequency (Hz)	100, 300, 1000
	Illuminance (lx)	50, 150, 500
	Modulation depths (%)	30
Constant	CCT (K)	4000
	Ra	>85
	UGR	<19

The whole experimental procedure took about 180min for each subject. Subjects need to traverse all 9 experimental conditions which were randomly performed. Both subjective scales and objective tests were used in the experiment. Subjects were asked to fill in two scales regarding to their assessment and feelings of lighting conditions. The D2 task, whose reproducibility and robustness have been well proven,

were chose to measure the attention of participants. Subjects were required to cross out all the "d"s with a total of two short dashes above and under the character. In order to avoid the learning effect, the letter arrangement of D2 tasks under each experimental condition were also different. After finishing D2 task, subjects were also required to fill out a scale for their feelings on this task. To evaluate the visual fatigue degree of participants, fatigue semantic scales were also used, including tired eyes, sore or aching eyes, irritated eyes, dry eyes, hot or burning eyes, double vision, blurred vision, dizzy and headache whose degree was rated by a five-point scale[13].

Table 2 Experimental procedure

No.	Process	Duration
1	Fill in the basic information	3min
2	Be familiar with the experimental procedure and tasks	5min
3	Dark adaption	3min
4	Begin Condition 1: environment adaption	3min
5	Fill in two scales regarding lighting condition assessment	4min
6	D2 test	5min
7	Fill in scale regarding D2 test assessment	3min
8	Fill in visual fatigue scale	3min
10	Repeat 3-8 until all 9 conditions finished	

Results

For all the subjective evaluation scales, a factor analysis was first conducted. Then about two or three factors were extracted from each scale. Meanwhile, all experimental data were analyzed by MANOVA. Note that all the data had passed the Bartlett's Test of Sphericity with significant level $p<0.01$ and all the scales showed adequate amount of inter-item consistency.

- **Subjective feelings**

To investigate participants' attitude and opinion about whole lighting condition, two subjective scales were adopted. The first one (recorded as S1) was composed of nine bipolar adjective word pairs regarding feelings about lighting environment. While the second one (recorded as S2) included six bipolar adjective word pairs focusing on feelings about lighting itself.

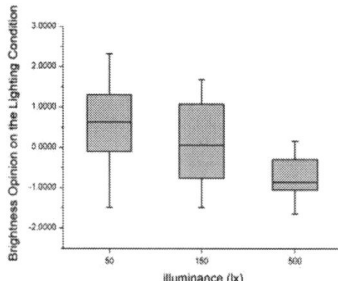

Figure 2 Effect of illuminance on "Brightness" factor

For S1 scale, two factors, "brightness" and "contentment" accounting for 76.5% of the total items, were extracted through factor analysis. It was found out that illuminance significantly affected brightness perception ($p<0.01$). As shown in Figure 2, the higher the illuminance, the brighter subjects consider the environment to be. Although this was a predictable result, it also proved the reliability and validity of the scales on the other hand.

Similarly, for S2 scale, three factors, "alertness", "relaxation" and "temporal light artefacts" accounting for 87.7% of the total items, were extracted. It was found out that illuminance significantly affected subjects' alertness and relaxation level (p<0.01). Figure 3 showed that the higher the illuminance, the more alert and less relaxed subjects tend to be. At 500lx, subject reported to be most alert and least relax. Note that this finding was also consistent with past studies.

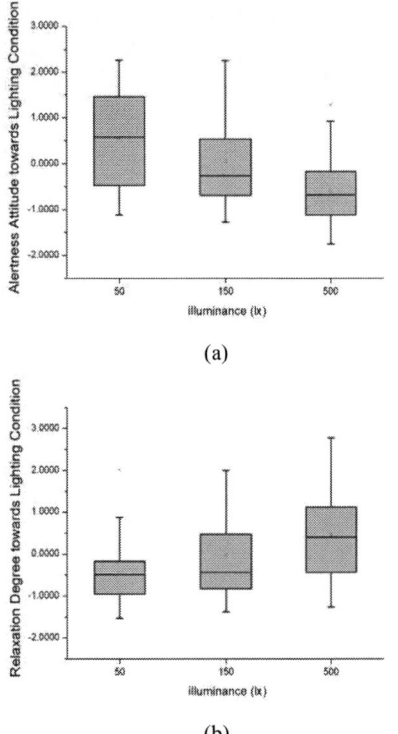

(a)

(b)

Figure 3　Effect of illuminance on "Alertness" factor (a) and "Relaxation" factor (b)

- **Attention**

Both D2 test and evaluation scale (recorded as S3) were used to measure subjects' attention degree. The scale included 10 items in response to the question "How do you feel upon finishing the D2 task" with a seven-point rating.

Base on Bates's[14] and Lee's[15] study, D2 task can be analyzed through some specific index such as total number of characters processed (TN), errors of omission (EO), errors of commission (EC), total correctly processed (TN-E), concentration performance (CP). One-way ANOVA was first used to analyzed the relationship between those variables and experimental conditions. However, the results showed no significance, which suggested that D2 test could not respond well to the attention level of subjects in present study.

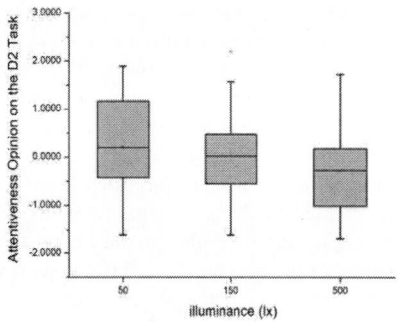

Figure 4　Effect of illuminance on "Attentiveness" factor

Then factor analysis was adopted to S3 scale. Two factors, "attentiveness" and "temporal light artefacts" accounting for 55.3% of the total items, were extracted. It was worth mentioning that illuminance significantly affected attentiveness (p=0.019<0.05). As shown in Figure 4, the higher the illuminance, the more focused subjects tend to be. This agrees with Wang et al. [10] and Perz et al.[11].

- **Visual fatigue**

Visual fatigue semantic scales (recorded as S4) was adopted to analyze how lighting condition affect subjects' visual perception. The KMO and Bartlett test results showed significance with p<0.01, which suggested that factor analysis could be conducted to S4 scale. Similarly, two factors "visual fatigue" and "temporal light artefacts" accounting for 78.7% of the total items, were extracted. The results indicated that illuminance significantly affected the perception of temporal light artefacts (p = 0.038<0.05), which was the main cause of visual fatigue.

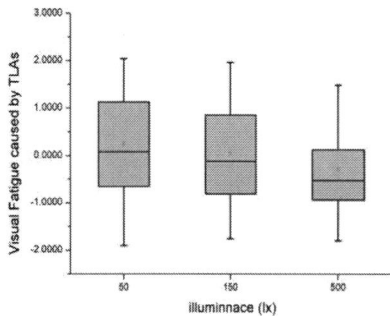

Figure 5　Effect of illuminance on "TLAs" factor

As shown in Figure 5, the higher the illuminance, the least aware of temporal light artefacts subjects tend to be. This result is somewhat different from previous researches. It was traditionally believed that even for non-visible flickering light, frequency only significantly affected subjects' perception of temporal light artefacts. However, our study found out that at least for frequency above 100Hz, illuminance is the dominant factor affecting subjects' perception of TLAs, while frequency only affected subjects' perception of motion. That means even at the same frequency, people in low illuminance environment are more likely to experience visual ghosting and blurring than high illuminance environment. In other words, this results proved that illuminance and frequency do have cross-effect, which verified our hypothesis from the beginning.

Conclusions

In this study, the combined effect of illuminance and light strobe (frequency) on human attention and visual fatigue was explored. By choosing both subjective evaluation and objective tasks, a series of experiments were performed in a simulated indoor lighting environment. All experimental data were analyzed through statistical methods.

For attention level, results indicated that illuminance significantly affected subjects' attention and alertness (both $p<0.05$). The higher the illuminance, the higher the alertness and attentiveness of the subject. At 500 lx, subjects were most alert and least relaxed. Meanwhile, subject also performed most concentrated under lighting condition of 500lx.

For visual fatigue degree, results showed that visual ghosting and blurring mainly caused by frequency were significant influenced by illuminance ($p=0.038<0.05$). The lower the illuminance, the more likely the subject was to experience visual fatigue. At 50lx, visual ghosting and blurring was most obvious for subjects regardless of 100Hz, 300Hz and 1000Hz.

Generally speaking, high illuminance level could make subjects feel more alert and concentrated no matter which frequency they are experienced, suggesting that illuminance is a more important factor when considering improve people's attention level. This could be particularly useful when designing indoor lighting environment. In addition, visual fatigue such as ghosting and blurring caused by frequency were more serious under lower illuminance, which reminds us that under low illumination indoor environment such as museum and restaurant, more attention should be paid to the light strobe, especially to the value of frequency.

Acknowledgments

This research was supported by National Key R&D Program of China (Project No. 2017YFB 0403700). We thank all the participants in the experiment.

References

1. CIE, "Visual Aspects of Time-Modulated Lighting Systems – Definitions and Measurement Models," 2016: *Cie Midterm Meetings & Conference on Smarter Light for Better Life.*
2. Wilkins, A., et al., "Fluorescent lighting, headaches and eyestrain*," Lighting Research Technology*, 1989. **21**(1): p. 11-18.
3. Fisher, R.S., et al., "Photic and pattern- induced seizures: a review for the Epilepsy Foundation of America Working Group." *Epilepsia*, 2005. **46**(9): p. 1426-1441.
4. de Bittencourt, P.R., "Photosensitivity: the magnitude of the problem." *Epilepsia*, 2004. **45**: p. 30-34.
5. Harding, G., et al., "Photic and Pattern - induced Seizures: Expert Consensus of the Epilepsy Foundation of America Working Group.," *Epilepsia*, 2005. **46**(9): p. 1423-1425.
6. Knez, I., "Affective and cognitive reactions to subliminal flicker from fluorescent lighting," *Consciousness and Cognition*, 2014. **26**: p. 97-104.
7. Bullough, J., et al., "Effects of flicker characteristics from solid-state lighting on detection, acceptability and comfort," *Lighting Research Technology*, 2011. **43**(3): p. 337-348.
8. Vogels, I., S. Sekulovski, and M. Perz. "Visible artefacts of LEDs," *in 27th Session of the CIE*, Sun City, South Africa, 9-16 July 2011.
9. Veitch, J.A. and S.L. McColl, "Modulation of fluorescent light: Flicker rate and light source effects on visual performance and visual comfort," *International Journal of Lighting Research Technology*, 1995. **27**(4): p. 243-256.
10. Wang, L., et al., *50.2: "*Invited Paper: Stroboscopic Effect of LED Lighting," 2015. p. 754-757.
11. Perz, M., et al., "Stroboscopic effect: contrast threshold function and dependence on illumination level," *Journal of the Optical Society of America A*, 2018. **35**(2): p. 309-319.
12. IEEE, "IEEE Recommended Practices for Modulating Current in High-Brightness LEDs for Mitigating Health Risks to Viewers," 2015.
13. Wang, Y., et al., "A model for evaluating visual fatigue under led light sources based on long-term visual display terminal work," *Lighting Research & Technology*, 2018. **50**(5): p. 729-738.
14. Bates, M.E. and E.P. Lemay, "The d2 Test of attention: construct validity and extensions in scoring techniques," *Journal of the International Neuropsychological Society,* 2004. **10**(3): p. 392-400.
15. Lee, P., et al., "Test–retest reliability and minimal detectable change of the D2 test of attention in patients with schizophrenia," *Archives of Clinical Neuropsychology*, 2017. **33**(8): p. 1060-1068.

Study of the Stroboscopic Effect Visibility Measure (SVM) based on Cognitive Performance

Xiaojie Zhao[1,2,4], Mengxin Li[1], Yandan Lin[2,3,4*], Wei Xu[1,2*]

1Institute for Electric Light Sources, Fudan University, Shanghai, China
2Engineering Research Center of Advanced Lighting Technology, Ministry of Education, Shanghai, China
3 Institute of Engineering and Applied Technology, Fudan University, Shanghai, China
4 Institute for Human Phenome, Fudan University, Shanghai, China
ydlin@fudan.edu.cn
weixu@fudan.edu.cn

Abstract

There have been many studies on stroboscopic effect from LEDs, which is considered unacceptable to affect psychological cognition. As a new measurement, Stroboscopic Effect Visibility Measure (SVM), can quantify the visibility of stroboscopic effect in general lighting applications. However, the relationship, if any, between SVM and human cognitive performance has not been specifically elucidated, such as spatial environment and work performance. Therefore, under six SVM values representing different lighting environments, through the D2 task, the rotating disk and the Landolt ring task, objects can perceive the stroboscopic effect in lighting environment, and use scales to make the self-cognitive judgement of spatial environment and work performance. The experimental results show that there is a significant relationship between the visibility of stroboscopic effect and cognitive performance. Subjects reported flicker, discrete spatial motion and bad environment perception in response to the higher visibility of stroboscopic effect. In addition, the visibility of stroboscopic effect also may have a significant impact on physiological indicators, subjects are more nervous during the D2 task, meanwhile the vision and attention are all decreased with a high SVM during the rotating disk.

1.Introduction

With the advantages of low power consumption, high efficiency, fast temporal response and so on, LED lighting has gradually been widely used for general lighting applications [1]. However, it also brings a new challenge to the fast time response of the driving current, that is, the change of the driving current changes into the change of the luminous output almost instantaneously [2]. Because the most driving currents are modulated, the output intensity is also modulated. And unintended modulation results from the supply of the lighting equipment with mains power, usually modulated by a at rectifier at 50Hz or 60Hz or twice these frequencies. Besides, drivers, dimmers or interference from other loads may also cause the lighting modulation.

The lighting modulation can change the perception of lighting environment which is undesirable and unacceptable for most general application. It has not been a new one and in 2016 the Technical Committee 1-83 of CIE has summarized all related studies in a Technical Note (TN) [1]. The TN describes the definitions for the perceptual effects produced by modulated light, including flicker, the stroboscopic effect and the phantom array effect. These effects are collectively referred to as Temporal Light Artefacts (TLAs) [1]. The TN present all the methods used to quantify the visibility of TLAs

and recommend using the Stroboscopic effect Visibility Measure (SVM) [3] to quantify visibility of the stroboscopic effect which is the key of this work. CIE defined it as a "change in motion perception induced by a light stimulus the luminance or spectral distribution of which fluctuates with time, for a static observer in a non-static environment [1]". And the newest measuring method, SVM, is defined as follows:

$$ \text{SVM} = \sqrt[3.7]{\sum_{m=1}^{\infty}\left(\frac{Cm}{Tm}\right)^{3.7}} \begin{cases} < 1 \ not \ visible \\ = 1 \ just \ visible \\ > 1 \quad visible \end{cases} $$

The measure consists of a Minkowski summation of the energy in the Fourier frequency components of the light waveform, normalized for human sensitivity. And Cm is the amplitude of the m-th Fourier component of the light waveform, and Tm is the visibility threshold, expressed in terms of modulation depth, for a sine waveform at the corresponding frequency [3]. The frequency range is 80-2000Hz, so at higher frequency more than 2000Hz the stroboscopic effect is not perceived. Perz *et al.* got the SVM under typical office condition with an illuminance of about 500lux on the rotating disk. And the speed of the moving target at the disk was fixed at 4m/s which comparable to an upper limit of hand movements' speed in the general lighting application [3]. The limit of SVM =1 defines the visibility threshold of the effect indicating that the observer will detect the stroboscopic effect with an average probability of 0.5 under critical conditions [4].

Therefore, it can be seen whether the stroboscopic effect can affect perception of the lighting application and activities. And there have been some researches mainly focusing on the healthy and the light modulation [5-6], for example, the light modulation can cause a decrease in visual seizures and migraine episodes [8]. it has an impact on affect and cognitive performance [9]. Rea and Ouellette found that the stroboscopic effect did not change the players' performance even under a very critical condition, but the spectators found the stroboscopic motion unacceptable [10]. And this study intends to discuss the cognitive performance in office environment with different SVM values. The goal is to explore the relation between the cognitive performance and SVM, as a new stroboscopic parameter. If any, it may support powerfully to accept the SVM quantifying the physiological cognitions for the stroboscopic effect. And our results can not only further deepen the significance of SVM, but also provide basic experiments for improving the model of SVM.

2. Methods

978-1-7281-5757-3/19 $31.00 © 2019 IEEE

2.1 Subjects

The subjects were randomly sampled from 19 to 23 years old (4 males,6 females). None of them knew the purpose of the experiment, but they were told what the stroboscopic effect was and presented with several examples of the effect before the experiment. In addition, they were informed that the test was harmless and it was in accordance with the ethics rules for research on human participants and they could leave it at any time they wished. They all read through and signed the consent form, confirming their eligibility for the study. All with normal orthoptists and no color blindness, color weakness, eye diseases and other head diseases.

2.2 Experimental setting

The experiment was carried out in the lighting laboratory (Fig 2.2.1) which was arranged as a general office. A desk of L1.04m×W0.73m and a chair were placed in the room. A beige sofa and a small table were placed opposite the desk, while the floor was covered with black carpets. In this experiment, a 600 mm×600 mm LED flat lamp was used to create six different lighting conditions with different SVM values. The power amplifier and waveform signal generator driving directly the LED were placed outside the room to achieve experimental conditions.

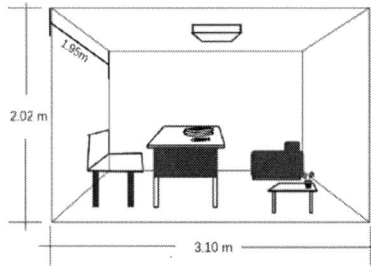

Fig 2.2.1. The simulated office environment

The LED lighting system was used to generate sine wave with different SVM values (Table 2.2.1). And strobe measuring instrument was used to measure the SVM values of the desktop to calibrate parameters. Other parameters are consistent after calibration through Lighting Passport Pro.

Table 2.2.1 The experimental parameters

Variable (SVM)	0.06	0.5	1	1.5	2.5	3.868
Invariable	Illuminance (lx):500 Color rending index (Ra):>85 CCT (K):4000					

2.3 Tasks

The first session is about subjective scale including the fatigue scale, Positive and Negative Affect Sales (PANAS) [11] and visual sensory semantic different scale. The fatigue scale (VAS-F) [12] was used to evaluate the fatigue degree through eighteen kinds of physical symptoms and seven kinds of eyes fatigue symptoms, such as painful eyes, dry eyes, blurred vision, drowsiness etc. Each symptom all had the five-level assessment. The emotional state was assessed in terms of the PANAS [11], with ten items measuring positive affect and ten items measuring negative affect. For each item, there are five grades from "almost no" to "extremely many". The visual

sensory semantic difference scale having sixteen bipolar, semantic, seven-level ratings was designed to investigate the perception for the general environment and lighting before and after each condition. After fully observing the environment and lighting, they were asked to evaluate the environment (comfortable, coordinating, bright and so on) and the lighting (flicker, satisfactory, relaxing and so on) according their subjective feelings.

The second session is the objective test including D2 task, the rotating disk task and the Landolt ring task. The D2 table consists of the letters d and P of 12 rows×47 columns, each with a vertical line above and below the letters [13]. The task was to find out all the letters d with two vertical lines added above and below the letter, and the completion time of each row is limited to 20 seconds. The observing rotating disk task used a disk with a black surface with a diameter of 20 cm and a white circle with a diameter of 1.5 cm positioned at 8 cm from the center of the disk [3]. The disk rotated at a speed of 4 mis, which is comparable to the maximum speed of hand movements in office. In addition, visual sensory semantic difference scale was also used to evaluate the self-perception of the two tasks. The Landolt ring was composed by unenclosed rings with 10 rows×10 columns, shown by a white card (50mm×50mm) printed on a white paper [14]. The measure was the number of rings per row for which the subjects correctly identified the orientation of the gap in the ring. The time is recorded to reflect the job performance.

2.4 Procedure

The light had been switched on for about 20min before the subjects entered the lab. Then they were instructed to be familiar the procedure. During the actual experiment, the subject was left alone, and detailed instructions were given over a loudspeaker. All the subjects were exposed to all experimental conditions in a random order of lighting conditions. Six experiment condition followed the same procedure for about three hours. The time were almost spent completing experimental tasks and filling different scales (Table 2.4.1).

Table 2.4.1. The experimental procedure

Time(min)	Operation
-20	Lights switched on
-20	Participant enters; Brief instructions
3	Adaptation to light environment
3	visual sensory semantic different scale
5	D2 test task
1	Observing rotating disk
3	visual sensory semantic different scale
1	The Landolt ring searching task
3	Fatigue Symptoms Questionnaire + VAS-F
3	Positive and negative affect sales (PANAS)
3	visual sensory semantic different scale
5	Rest

3. Results

3. 1 Cognition (environment & lighting)

For the scale of environment and lighting perception, there are two variables: SVM and time (before and after the experiment). The evaluation scores were set as dependent variables, the SVM and time as fixed factors by One-Way ANOVA of variance.

Time only had a significant difference on the item of "coordinating-uncoordinated"($p=0.004<0.05$), indicating that the perception has changed significantly before and after the experiment. After completing the task, the falling scores showed that people adapted to stroboscopic effect more quickly, meaning that the stroboscopic effect has the transient impact on cognition. And the results also showed that there was a significant difference between the conditions of SVM = 0.06 (stroboscopic invisibility) and SVM = 1.5/2.5 (stroboscopic visibility) ($p = 0.002/0.001 < 0.05$), meaning that the higher the SVM value, the bigger the scores, the more disharmonious the subjects reported (Fig 3.3.1).

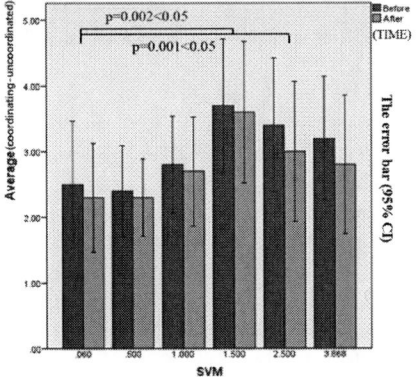

Fig 3.1.1. The error bar between the average score and SVM

For lighting perception, the item of "serious flicker-flat flicker" indicated a strong significance ($p < 0.001$) under six conditions. The ANOVA analysis also showed that there was a significant difference between SVM = 0.06/0.5 and SVM = 1.5/2.5/3.868 ($p < 0.001 < 0.05$), which told that the "flicker" cognition between invisible flicker (SVM < 1) and visible flicker (SVM > 1) accorded with the physical meaning of SVM. It can be seen from the figure (Fig 3.3.2) that with the increase of SVM, the severity of flicker felt by the subjects also increased on psychological cognition.

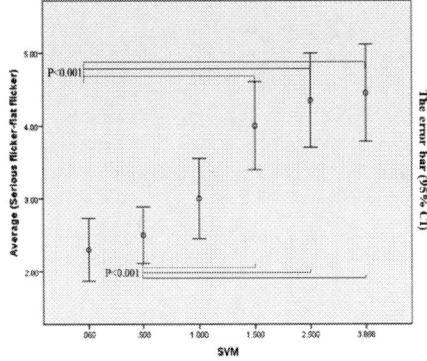

Fig 3.1.2 The error bar between the average score and SVM

3. 2 Cognition (D2 task)

Using One-Way ANOVA of variance, items in the scale were set as dependent variables and the SVM as fixed factor. The results showed that the scores of "relaxing-nervous" and "no flicker" were significant under different SVM ($p < 0.05$),

indicating that there were significant differences between the SVM and cognitive performance.

For the flicker cognition, the subjects reported less flicker under the invisible stroboscopic effect (SVM<1) than the visible stroboscopic effect (Fig 3.2.1). And it has the similar result with the cognition on the lighting perception.

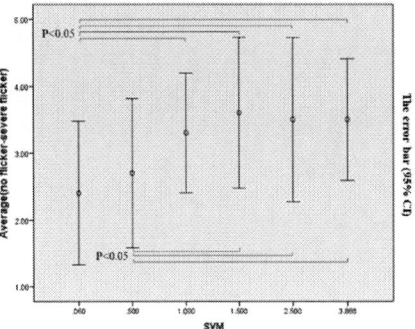

Fig 3.2.1 The error bar between the average score and SVM

with the increase of the SVM, the smaller average scores are, the more intense the subjects are, which is the result for cognitive performance of D2 test ($p<0.05$). However, there are apparent difference on the tension when the SVM is more than one, it can explain that the cumulative fatigue also has an error effect on the experimental results because of the long experimental time (Fig 3.2.2).

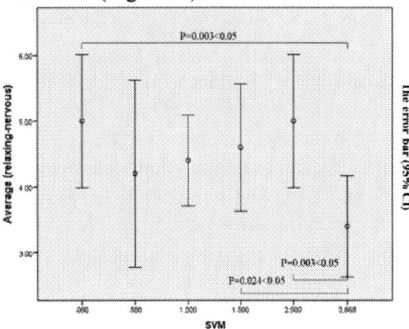

Fig 3.2.2 The error bar between the average score and SVM

3. 3 Cognition (observing the rotating disk)

Principal component analysis was used for all items of the visual sensory difference scale through SPSS and got three factors. The first factor is related to these items including "no flicker-severe flicker", "difficulty-simple", "relaxing-nervous", "eye fatigue-eye comfortable" and "dry eye-moist eye". The second factor includes "continuous motion-discrete motion", "concentrated-distracted", "blurred vision-clear vision" and "double shadows-no shadows (in front of eyes)". The third factor includes "dizzy-not dizzy" and "headache-not headache".

Therefore, using One-Way ANOVA of variance, the three factors were set as dependent variables and the SVM value as fixed factor. The results showed that the first two factors were significant under different SVM ($p<0.05$), indicating that there were significant differences in cognition of different stroboscopic environments through observing the rotating disk. That is, the higher SVM, the more serious the flicker, the more discrete the spatial motion. In addition, a significant difference in eye comfort, visual clarity and attention concentration under six conditions were found. The larger the SVM, the lower the eye comfort, visual clarity and attention concentration.

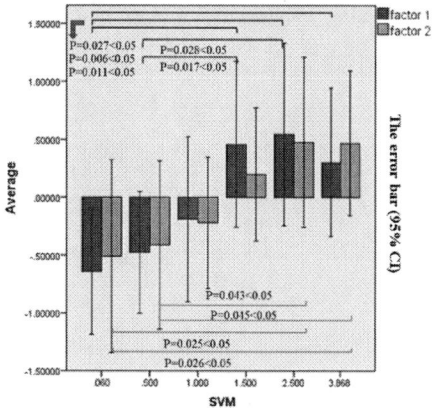

Fig 3.3.1 The error bar between the average score and SVM

In addition, no statistically significant effects were found for fatigue appraisal and the affect state measured by PANAS, nor the accuracy of the D2 test and the time of the Landolt ring searching task by One-way ANOVA analysis.

4. Conclusions and Discussion

In this study, there is a significant relationship between the visibility of the stroboscopic effect and cognitive performance. Subjects reported the stronger flicker, discrete spatial motion and worse environment perception in response to the higher visibility of stroboscopic effect. In addition, the visibility of stroboscopic effect also may have a significant impact on physiological indicators, subjects are more nervous during the d2 task, meanwhile the vision and attention are all decreased in the environment with a high SVM for the work performance of observing the rotating disk.

From the current experimental results, there is a superficial relationship between cognitive performance and SVM, which is worth further exploring and deepening the meaning of SVM combining the human cognition. This study has the affirmation for the practical meaning of SVM from the psychological cognition. Overall, as a new quantitative index, SVM is also a new parameter to characterize environmental stroboscopic effect.

This study only considers psychological cognition and job performance which are comprehensive indicators. In the future in-depth study, physiology will also be a very important factor to be excavated. Moreover, the conclusion of this study just showed the qualitative relationship between SVM and cognition performance, we always expect to build the quantitative model to instruct the lighting application.

Acknowledgments

This research is supported by National Key R&D Program of China (Project No. 2017YFB 0403700). We thank for all the participants in the experiment.

References

1. Commission Internationale de l'Eclairage. Visual Aspects of Time-Modulated Lighting Systems – Definitions and Measurement Models CIE TN 006:2016. Vienna: CIE, 2016.

2. Branas C, Azcondo F J, Alonso J M. Solid-State Lighting: A System Review[J]. IEEE Industrial Electronics Magazine, 2013, 7(4):6-14.

3. Perz M, Vogels I, Sekulovski D, Wang L, Tu Y, Heynderickx IEJ. Modeling the visibility of the stroboscopic effect occurring in temporally modulated light systems. Lighting Research and Technology 2014; 47(3): 281-300.

4. Perz M, Beeckman P, Sekulovski D. Acceptability criteria for the stroboscopic effect visibility measure. Proceedings of the CIE 2017 Midterm Meetings and Conference on Smarter Light for Better Life, Jeju, South Korea. CIE, Vienna: pp. 453-459

5. J. D. Bullough, K. S. Hickcox, T. R. Klein, A. Lok, and N. Narendran. Detection and acceptability of stroboscopic effects from flicker. Lighting Research and Technology 2010; 44: 477–483.

6. Shepherd AJ. Visual stimuli, light and lighting are common triggers of migraine and headache. Journal of Light and Visual Environment 2010; 34(2): 94-100.

7. Kozaki T, Hidaka Y, Takakura J Y, et al. Suppression of salivary melatonin secretion under 100-Hz flickering and non-flickering blue light[J]. Journal of Physiological Anthropology, 2018, 37(1).

8. Wilkins AJ, Veitch JA, Lehman B. LED lighting flicker and potential health concerns: IEEE standard PAR1789. In: Proceedings of Energy Conversion Congress and Exposition (ECCE), Atlanta, GA, 2010: 171–178

9. Bullough J, Sweater Hickcox K, Klein T, Narendran N. Effects of flicker characteristics from solid-state lighting on detection, acceptability and comfort. Lighting Research and Technology 2011; 43(3): 337-348.

10. Rea M S, Ouellette M J. Table-tennis under High Intensity Discharge (HID) Lighting[J]. Journal of the Illuminating Engineering Society, 1988, 17(1):29-35.

11. Watson D, Clark L A, Tellegen A. Development and validation of brief measures of positive and negative affect: The PANAS scales. Journal of Personality and Social Psychology 1988; 54(6): 1063-1070.

12. Shahid A, Wilkinson K, Marcu S, Shapiro CM. STOP, THAT and One Hundred Other Sleep Scales. New York: Springer, 2011.

13. Baron, I. S. Neuropsychological evaluation of the child. Oxford: University Press, 2004

14. Wesemann W. [Visual acuity measured via the Freiburg visual acuity test (FVT), Bailey Lovie chart and Landolt Ring chart][J]. Klinische Monatsblätter für Augenheilkunde, 2002, 219(9):660-667.

A summary to the personal consideration about the light's influence on myopia, procreation and even philosophy & society

Wenqing Fang, Chaopu Yang, Kaiqi Fang, Han Jin, Xu Zhang, Xiaojian Han, Chaolin Ma, Yuehui Zheng,
Fan Yang, Jiang Fu, Tuanqing Fang, Guoqing Fang, Taiyang Chen, Shanxiao Huang, Youming Zhang, Zhenquan Lai
Nanchang University
Nanchang, Jiangxi, P.R.China
fwq@ncu.edu.cn

Abstract

This paper summarizes the author's personal thoughts on some important effects of illumination, spectrum and 'the Non-visual effects of light' on axial myopia in recent years, as well as the important effects of 'Non-visual effects of light' on human reproduction and even philosophy and society. As far as possible, the author provides the conjecture basis, and also proposes the further verification scheme. Efforts are under way to prove the validity of three main conjectures in animal experiments. The most important speculations is that modern humans may be in a state of sexual excitement, if compared with the people 60 years ago. This wrong state has huge influence on Human reproduction, social customs and all other aspects. In addition, from the perspective of optics, based on the analysis of the concept of focal depth, it is concluded that in addition to dividing day and night based on the 'Non-visual effects of light' , the light should also be divided into near and far. Desktop reading and writing is using a near-range light, and 300Lx at this time may be too strong , which may lead to axial myopia. The author also believes that too much light in modern people may lead to a low-desire society and philosophically expounds the importance of "darkness". The author has verified that the light with a low emission of 0.2lx can control the phase of laying eggs in chickens. If it is extended to human ovulation, it will be very terrible, because the illumination of the moon on the ground is also 0.2lx. This partially proves that menstruation originated from the monthly changes in moon light, which may be a dose reference for light to affect human reproduction.

1. Introduction

Over the past 60 years, the rate of myopia in east Asia has soared[1], while the birth rate in China, Japan and South Korea has fallen sharply (the rate in these three countries are close to 1, far below the minimum of 2.05 for normal population replacement)[2]. Western societies such as Japan have entered low-desire societies, and China also has 200 million 'single dogs'. In China, compared with 20 years ago, the phenomenon of early sexual maturation of children has been widespread, and pregnancy are generally not smooth. Is the sudden change in the light environment an important factor in all this? Since 2015, the author has been studying these issues. Now, the results of these years' research are summarized as follows, with the new focus on myopia, hoping to attract the attention of the academic circle and the industry.

1.1 The dramatic evolution of the light environment for ordinary people in China in the past 60 years

1.1.1 Comparison and evolution of illumination, changes 10,000 times.

The unit of illumination, **Lx**, is derived from lighting a candle (cd) at 1m. Sixty years ago, there were still family spinning and weaving in rural areas of China. It is estimated that the illumination of women spinning at 5m away from the oil lamp (equivalent to 1cd) was only 0.04Lx. Compared with the 400Lx for reading and writing now, the illumination for close working and studying in the past 60 years has changed dramatically (by 10,000 times) ! **Fig. 2** also shows that in the past 60 years, as the light source changes from candle, incandescent, fluorescent, LED and mobile phone, the illumination of the close view has increased exponentially, but human beings do not realize that, they like brighter and brighter. But it's wrong to pursue brighter at night or at reading and writing distance.

1.1.2 Comparison and evolution of spectra, how to compare spectra?

To correctly compare the spectra of various light sources, it is necessary to use the same spectrometer for measurement, and to ensure that the spectral area is roughly the same. It is better to use incandescent lamp spectrum as a reference. As can be seen from Fig.1, the spectrum of fluorescent lamp is line type and discontinuous. It also lacks deep red and near-infrared light with wavelengths greater than 625nm. This lack of spectrum could be a big mistake for teen lighting, in other words, the spectrum of fluorescent lamp is quite unnatural. The best electric lamp at night is still DC incandescent lamp, this kind of lamp is rich in deep red and in near-infrared light, more and more evidence shows, this section of spectroscopic light has the health significance to protect the neuron and so on. Therefore, it cannot be carelessly eliminated for energy-saving. Full spectrum LED lighting should not only ensure the R9 index, but also try to supply the near-infrared light with IR LED or in other way, which can take into account both energy saving and health. This maybe the future development direction of LED lighting.

1.1.3 Comparison and evolution of the Non-Visual effects of light, changes 88,000 times.

We measured and estimated[3] the proportion of blue light of various light sources in table1:

978-1-7281-5757-3/19 $31.00 © 2019 IEEE

Table1: The proportion of the blue light means the ratio of the spectrum area between 446~477nm to the spectrum area between 400~700nm.Taking the candle light as reference and assuming its blue ratio to be 1, we get the proportions of blue light in various lights and displays.The spectra of displays were measured in white balance at 5400K

Lights	Incande- scent lamp	fluorescent lamp	Bright Moon	natural light at noon	Early LED table lamp
Blue ratio	1.4	8.8	8.8	8.8	20
Displays	LG CRT screen	HP CCFL screen	Dell laptop	Windows Pro Pad	iPhone 5s
Blue ratio	14	9.4	14	16	16

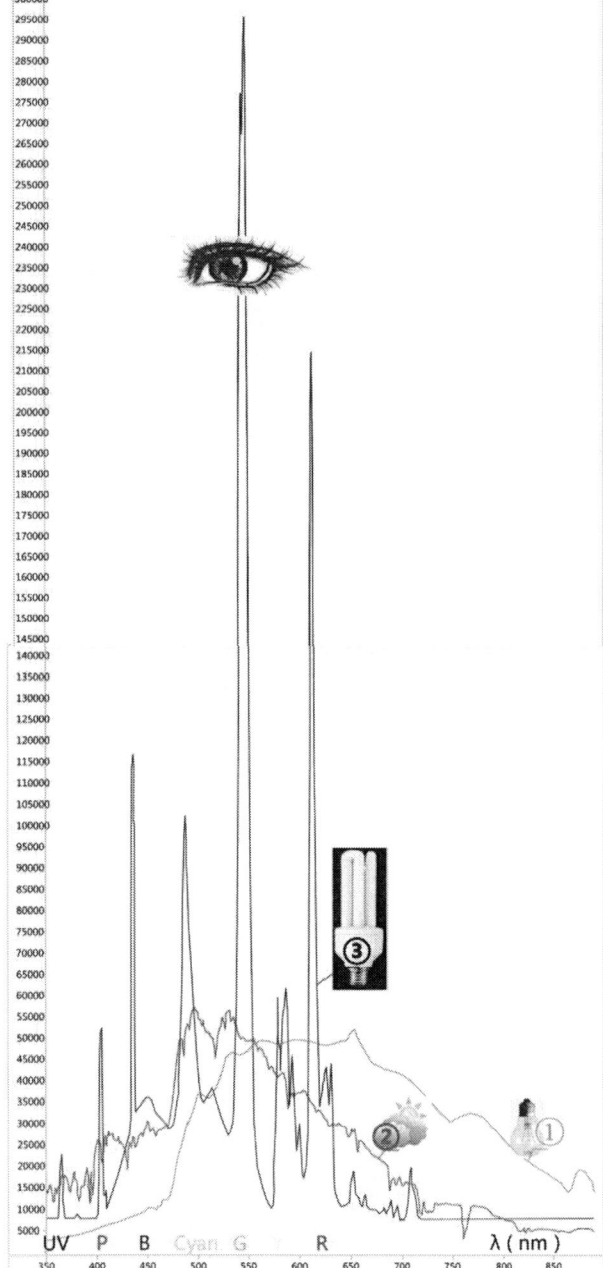

Fig.1 The spectrum of noon shade natural light, tungsten lamp light,fluorescent lamp must be measured with the same spectrometer （like Ocean USB2000+）before compared. When compared, the spectral area remains the same, which can highlight the "weird" spectrum of fluorescent lamps.

Usually, the Non-visual effect of light = proportion of blue light × light power × exposure time, if keeping the exposure time be equal, compared with the spinning female (under candle lamp) 60 years ago, the Non-visual effect of light received by modern people (under CCFL lamp) is 88,000 times that of predecessors! (10,000×8.8)

1.2 The two important properties of Non-visual effects of light. Is there anything wrong with the old lighting and display standard?

For the second time, in Oct.2019, CIE recommended "PROPER LIGHT AT THE PROPER TIME" to all the people in the world. So, what is the proper light at night? It should be "dark" first, "firelight" second, and "incandescent bulb light" the third. Obviously, the human need for light at night is contradictory. Psychologically, human beings will pursue brightness, but physically, they cannot have light. Another characteristic of the Non-visual effects of light which is easily overlooked by humans is that there is at least half a hour delay [4] after the blue ray in the light entering into our eyes. Therefore, we should not try to measure the Non-visual effects of light in-situ, and some old lighting and display standards should be re-evaluated. The so-called human-oriented lighting before may only meet the psychological needs of human beings for light.

2 Changes in the sexual state of people in last 60 years.
Is blue light a sexual stimulant? Does exposure to blue ray too much at night lead to early puberty? If blue light is a hormone drug, all lamps and screens and related projects should go through FDA?

Table 2: Lighting Index of Pig House recommended by American Society of Agricultural Bioengineering

Pig house type	Light intensity	light application time	remarks
Breeding house	≥100 lx	14-16 h/d	**Absolutely necessary for estrus**
Pregnancy house	≥50 lx	14-16 h/d	**Stimulate estrus**
Farrowing house	50- 100 lx	8	If no roast lamp, 24h lighting
Nursery house	50	8	24h lighting available
fattening house	50	8	

Notes: lower light intensity at night in farrowing and nursery houses

Table 2 is shock: The light intensity and exposure time for young people are larger than that listed for boar in the table 2. It is not through the skin but all through eyes for both human and pig. It is no use to wear cloths. This table also hints that a process "Light→Mel→hormones→Value (Sex) orientation→ general mood of society" may exist.

Table 2 shows that in humans, blue light is a sexual stimulant. Compared to humans 60 years ago, modern people are generally in a state of sexual excitement! This conclusion is very shocking, long time sexual excitement will leads to sexual fatigue, which affects fertility.

Wrong sexual state will not only affects health, but also affect people's values, namely there is close relation between the social atmosphere of modern society and (blue) light. This conclusion has no need to do animal experiments, it is objectively valid.

978-1-7281-5757-3/19 $31.00 © 2019 IEEE   158

Table 2 also shows that female animals are more likely to be sexually excited by light. It is not difficult to understand if we think of woman's clothing exposure in modern society.

The author published a paper "Will the blue light at night in the lamp and display destroy east Asia" in SSL China2016[5], in which the mechanism and evidence were elaborated in detail. This includes the evolution of the age of menarche over last 100 years. Too much blue light at night may lead to premature puberty.

This time, this paper indicates that humans are currently in a state of sexual excitement. Therefore, the author thinks that since blue light is a stimulant and a hormone drug, should all lamps and screens, including related project, pass FDA in the future?

3. The relationship between myopia and illumination, spectrum and Non-visual effects of light

The principle in the prevention and control of myopic is that the cone, rod cells and nerves must get accurate, comprehensive, balanced stimulation of the external image, and the use of the regulation should be the least; Based on this principle, it is proposed that the growth of the Axial Length may be directly related to the enlargement of the optical "Depth of Focus (DOF)" range, as shown in Fig.2. Then quantitative analysis of a variety of factors leading to a wider range of DOF are listed below, which is the first the desktop illumination. The absence of deep red light in the spectrum or the Non-visual effects of light can also cause the eye axis to lengthen.

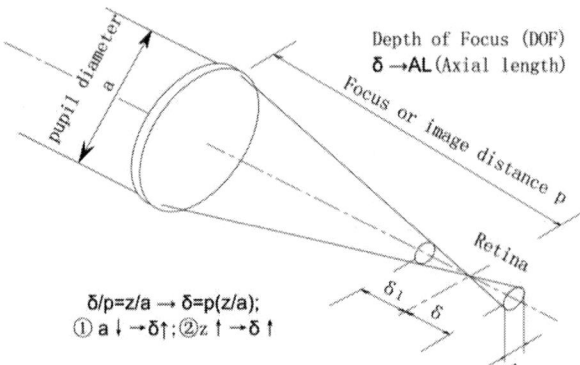

Fig.2 Depth of Focus (DOF) δ means the range that retina be allowed to move backwards while still keep seeing clearly, which may corresponding to the lengthen range of Axial Length ."Z" is the dispersion spot diameter . Z is inversely proportional to stadia

3.1 The relationship between illumination and axial myopia. Current desktop illumination requirements of more than 300Lx may be an important cause for the fast rise in the global rate of myopia

In **Fig.2,** the higher the illumination, the larger the pupil diameter **a**, and the wider the DOF, so, Axial Length will be allowed to lengthen. From Fig.2 it could also concluded that the nearer the reading distance, the wider the DOF. This means that the importance of illumination is equal to the

reading distance. In the prevention of myopic, people pay much attention to the reading distance, but completely ignored the effect of illumination.

In order to prevent and control myopic, based on the influence of illuminance on DOF, light must be distinguished between "near see light" and "far see light". We concluded that high illuminance greater than 38Lux may induce myopic, which means the current desktop lighting standards ($\geq$ 300Lx) may need to be reconsidered.

This view can be supported from five aspects: ①like the pupil response, as long as one can still see clearly, the retina may "dodge back" bright light, the distance allowed to dodge depends on the δ value in Fig.2, resulting in the eye axis longer ② In evolutionary terms, the human eye was not designed to see closely in bright light. Ancient people usually took a closer look in dark caves.③Physically, the natural light environment of human beings ranges from 3×10^{-4} Lx (the illumination generated by the sky on the ground in moonless night) to 10^5 Lx (direct sunlight outdoor), The powers are (-4, -3, -2, -1, 0, 1, 2, 3, 4, 5）, The intermediate value of the power is between 0 and 1, which corresponds to several Lx. ④As shown in Fig.3, with the rise of desktop illumination in east Asia, the rate of myopia also rises[6]. Is it a necessary relation? ⑤As shown in Fig.2,the negative effects of increased illumination are equal to the effects of decreased stadia.

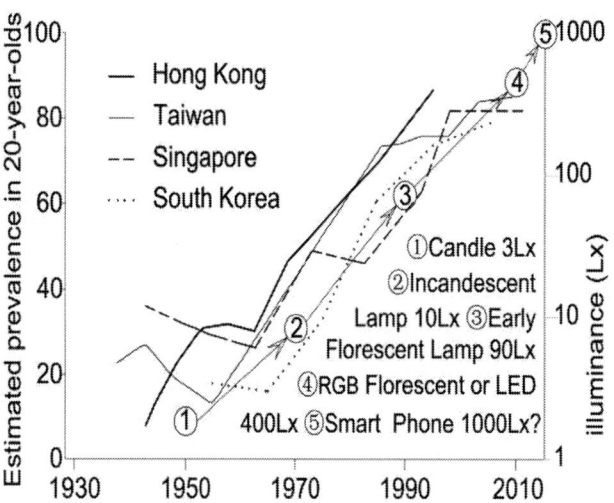

Fig.3: In the past 60 years, the evolution of myopia rate among 20-year-olds in Hong Kong, Taiwan, Singapore and South Korea, the relationship between myopia rate and illumination is added in this Fig., and the corresponding lamps, illumination value and popularity age are calculated or supplemented by the author according to his own experience.

We have numerically estimated the change in axial length induced by an increase in illumination. The related article may be published.

3.2 Relationship between spectra and axial myopia

Fig.3 shows that the rate of myopia has increased rapidly since the introduction of fluorescent lamps. Fig.1 shows that fluorescent lamps lack red light with a wavelength greater than 625nm. Since the refractive index of red light is less than that of blue light, the absence of red light will cause the retina to look behind the brain for the missing red light, thus inducing the axis of the eye to become longer.

Recently, a red light amblyopia therapeutic instrument has been widely spread in the optometry field. It can be used to shorten the eye axis and increase the thickness of choroid. This also supports the conjecture of this paper. Therefore, it would be better for semiconductor lighting to take the way of full spectrum technology, and may need to supplement near-infrared light (such as 850nm-880nm near-infrared).Or simply use LED lighting + incandescent lighting mode. We have numerically estimated that the lack of deep red would induce longer axial length. The related article may be published.

3.3 Relationship between non-visual effects of light and axial myopia.

To prevent and control myopia it is necessary to pay attention to the possible cause of light ripening shown in Fig.4, as fast-growing plants are more likely to break off. Sexual activity can also lead to calcification of the eyes[5]

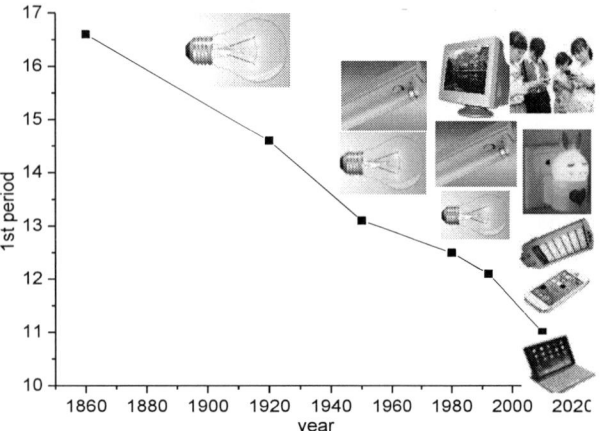

Fig.4: Menophania age and light source in Germany. This picture shows incandescent lamp may also have cause sexual precocity. At present, the occurrence of "smart phone addicts", the incorrect use of small LED night-lights and the popularization of LED lighting and displays may accelerate human precocity.

4 The philosophical relationship between light and dark, the origin of (competitive) consciousness, and the cause of low-desire society

Fig. 5 shows the importance of a balance of light and dark. The world is now entering a low-desire society, which may be linked to overexposure to light at night. Our lighting and display industry personnel should give priority to this problem. Light is the source of excitement, darkness is the source of desire, and earthlings competed for more light, but rarely compete for air. It is not enough to say that "all things grow by the sun". It would be more appropriate to say that this world originated from the change between brightness and darkness.

Fig.5 Human greed or enterprise, human competition consciousness, and even human consciousness may originate from the changes between brightness and darkness on the earth. Because there is darkness, we will pursue brightness, if the light is sustainable, human beings will have no desire, and society become low-desire!

5 The several ways in which light affects society

People's values and behaviors are related to their hormone levels, Fig.6 shows the most important and most basic hormone variation in lifetime. since blue light in light is

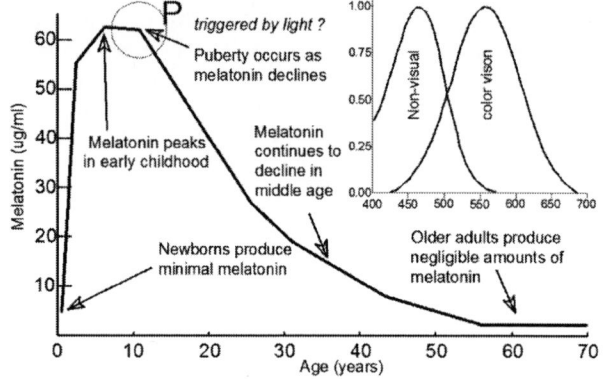

Fig.6: Changes of Mel in human's whole lifetime, Mel level of young people at night is about 10 times that in the daytime! At least, the phase of this huge change could be controlled by light, especially by blue ray in light. How powerful is the blue ray! But people normally select to ignore this light power, although they know that Mel is the basic hormone for all other (sex) hormones. So logically, we can speculate that the blue light is a powerful "hormone drug". Since there is a half hour delay for light to restrain Mel, and there is a even longer delay for Mel to affect all other hormones, and besides, due to humans inherently like "bright" like plants, Lighting scientists have difficult to supply the exact influence data of this "drug" to public, so humans still like this "drug" at night. For example, what is the "dose" reference that light start to affect humans? Can dim Moon light be the dose reference? Can the concentration of Mel be used to indicate the strength of desire?

Since blue light in light is a potent hormone drug, blue light is bound to affect society. Here are some of the main influences the author ponders:① Blue light is a kind of sexual stimulant, compared with 60 years ago, modern people are generally in a state of sexual excitement, which greatly affecting the people's physical and mental health and social ethos, destroying moral standards. ② Precocious puberty is a common phenomenon in modern society. Precocious puberty seriously affects physical and mental health, and ultimately affects the society. ③Too much and too disorder light affect biological clock, resulting in modern people upset, unable to concentrate on research, concentration is poor, everything is lack of wisdom. ④Because of sexual excitement will lead to sexual fatigue, so light will affect fertility, affect the birth of the population, affect the education of future generations.⑤ Because of lack of darkness, mankind will enter low-desire society.⑥ The change of dynasties in China shows that civilization is often destroyed by barbarians, which is because the emperor's son and grandson were exposed too much light at night to receive the best education, it is light that make them lack of testosterone and pioneering spirit. ⑦ Too much light may make humans neutral, which may be the source of homosexuality. It is also the source of low fertility. China semiconductor lighting people should attach great importance to the impact of light on human reproduction.

6. Experiments

As shown in Fig.7 We have designed a variety of experiments to verify our previous speculation

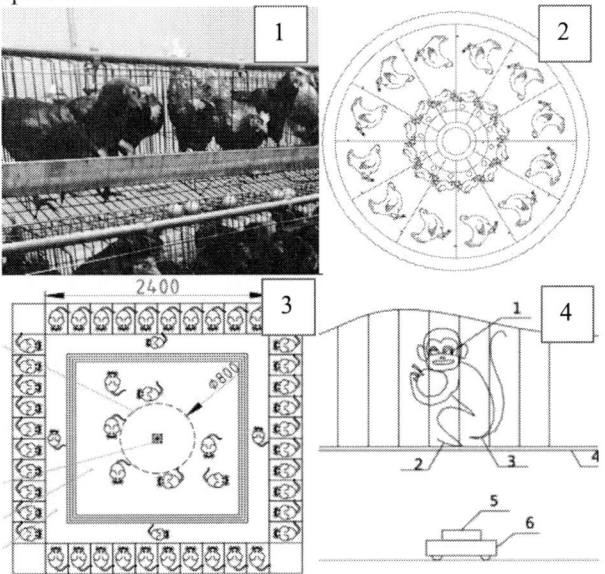

Fig.7: **Experiment 1**： proved that the phase of laying eggs could be regulated by very weak light. (Has been completed.) **Experiment 2:** use chicken to prove that 300Lux is a kind of bright light at close view, which can cause axial myopia(Under way.) **Experiment 3**: use mice to prove that overexposure to light would lead to a low-desire society. (Seek for global cooperation). **Experiment 4:** use monkey to proved that human menstruation originated from changes in the light of the moon, proving the power of modern lighting. (Seek global cooperation).

6.1 Completed experiments

6.1.1 High illumination (>300Lx) for reading and writing may be wrong

The conclusion is that under high illumination, with pupil narrowing, the center thickness of the lens will increase significantly, that is, under high illumination, more adjustment will be used, which is not conducive to the prevention and control of myopia. The instrument is an optical biological parameter measuring instrument (SW-9000) .The subjects were students from Nanchang university.

6.1.2 How much weak light can control the laying of eggs? Can it be extended to humans?

This experiment has proved that 0.2lx can control the phase of laying eggs in chickens. If it is extended to human ovulation, it will be very terrible, because the illumination of the moon on the ground is also 0.2lx. This partially proves that menstruation originated from the monthly changes in moon light, which may be a dose reference for light to affect human reproduction.

This experiment has proved that without light ,the egg will be laid with the same period and phase. That is, the egg laying cycle is not controlled by light, but the phase is controlled by light.

6.2 Ongoing experiment, ≥300Lx wrong?

Chickens in this experiment was divided into 5 groups, with 12 chickens in each group. Different illumination were used. Ophthalmic A ultrasound was used to measure the Axial Length. Finally, it is planed to prove that axial myopic may be related to strong reading and writing illumination, 300Lx is strong enough.

6.3 Experiments to be conducted. International and domestic cooperation sought

Experiment 3 in Fig. 7 is based on the famous experiment of Calhoun, John B. We will prove that it may be the light overdose that leads to a low-desire society, which is of great significance for today's low-desire society.

Experiment 4 will use simulated moonlight to change the phase of menstruation in monkeys. It will prove that menses originate from the monthly changes of moon light, proving that moon light is the dose reference of light affecting human reproduction.

These two experiments will have a huge impact on human society. We wish the whole world will work together.

Conclusions

① Compared with 60 years ago, modern people are generally in a state of sexual excitement, women are more sensitive to light, this state is not normal, which having a huge impact on human reproduction and society. ② while selecting illumination, it is necessary not only to distinguish day from night, but also to distinguish near from far. Reading and writing under bright light(≥ 300Lx) will use more adjustments. The rationality of the current illumination standard of desktop is questioned. ③ Non-visual effect is difficult to measure accurately, so the previous lighting display standards may not fully consider human physiological requirements for light, need to be reorganized.④Blue light may be a kind of hormone drug, all lighting and display products and engineering should go through FDA⑤It only takes 0.2Lux of light to change the phase of laying eggs. Can this be extended to humans?⑥Human consciousness may be originated from the change of light and dark, and the balance between light and dark should be highly valued. Too much light may lead to a low-desire society. The verification research in this aspect wants to seek global cooperation⑦The method of correct spectral comparison is provided. Full spectral illumination and supplement of deep red and near IR may be beneficial to prevention and control of myopia and other health problems.

Acknowledgments

This work was supported by National Key R&D Program of China （NO:2017YFB0403700）.Thanks to Dr. Zheng Fuhao from Wenzhou Medical University for his helpful discussion on myopia.

References

1. Dolgin E. "The myopia boom". *Nature*, 2015, Vol. 519, No.7543 (2015), pp. 276-278

2. Fuxian Yi, Jian Su, "2018: a historic inflection point -- China's population begins to decline " (*http://www.aisixiang.com*) (Chinese)

3. Kusiak,A., Brainard,G.,et al, Action spectrum for melatonin regulation in humans: Evidence for a novel circadian photoreceptor. *JNeurosci* 21, pp.6405-6412(2001)

4. Thapan, K.,*et al*, "An action spectrum for melatonin suppression: evidence for a novel non-rod, non-cone photoreceptor system in humans". *Journal of Physiology* Vol.535, No.1(2001), pp.261-267

5. Wenqing Fang, *et al*, "Will the blue light at night in the lamp and display destroy East Asia", *13ᵀᴴ,China International Forum on Solid State Lighting* (SSLChina 2016) , JEEE 2016

6. Ian G.Morgan, *et al.* "The epidemics of myopia: Aetiology and prevention". *Progress in Retinal and Eye Research*,Vol.62, Jan. 2018, pp. 134-149

7. Calhoun, John B, "Death Squared: The Explosive Growth and Demise of a Mouse Population". *Proc. R. Soc. Med.* 66: (1973), pp.80–88

Influences of Blue Component in White Light on Visual Discomfort

Yin Zhang*, Yan Tu*, Lili Wang*

*Joint International Research Laboratory of Information Display and Visualization, School of Electronic Science and Engineering, Southeast University, Nanjing, P.R. China

Contact Author Email: tuyan@seu.edu.cn

Abstract

An increasing population suffers from visual stress like eye strain, even myopia, which may be related with the high blue component of artificial light sources. The aim of this study is to investigate the influence of blue component on visual discomfort using EEG and EOG measurements. The results revealed that complex visual discomfort yielded under blue-enriched white light with EEG alpha activity enhanced near the central area and around parietal and occipital regions, as well as frequent blink. It suggests that the long-term exposure to commonly used light source with enriched blue component might relate with the produce of myopia.

1. Introduction

At present, people spend most of their time indoors, especially office workers who work indoors almost all day. Artificial light replaces natural light and becomes the main light source in our life. It is known to all that light affects humans in many aspects, such as human health and well-being, as well as behavior and visual discomfort [1-3]. The spectral variability between natural and artificial light changes our inherent biological system. An increasing number of studies have been performed to investigate the effects of light exposure on human functions based on neurobiological phenomena, such as melatonin suppression, pupil response and neural activity [1, 4, 5]. However, due to the complex physiological and psychological mechanisms, there is still no definitive conclusion.

Retina is the primary cell layer of human eyes receiving light exposure. The peak of retinal absorption is between 400 and 600 nm. Blue light has the most energy in the visible light. Compared with incandescent, the most prevalent light-emitting diodes (LEDs) and fluorescent lights emit more short-wavelength light. In addition, too much blue light is emitted from displays. It has been reported that short-wavelength light may be particularly hazardous to the retina [6]. Some standards organizations define blue-light hazard curve and specify the limitation of the exposure amount for blue light [7]. However, the potential for blue light hazard from light sources to cause retinal damage would require exposure at much higher intensities and for more prolonged exposures, it is typically not encountered in our daily life. The existing literature suggests that LEDs present no special concerns for the blue-light hazard over some other common sources (e.g., incandescent) in typical use cases [8]. While this is not to say that blue light is completely harmless to us, since light modulates the growth of human eyes and spending too much time indoors may impair children's vision and result in myopia. Thus, it is still worthy to investigate whether exposure to blue light could induce visual discomfort under common lighting conditions, which is important for the optimization of indoor light sources.

Visual discomfort is a complex phenomenon involving multiple aspects such as visual system and physiological activities. It has been proved that blink events can be used as an good indicator of visual discomfort [9]. Electroencephalogram (EEG) is often used to monitor brain activity which reveals corresponding changes of the CNS. According to the literature, there are four main brain rhythms extracted from EEG signals which closely relate to visual discomfort: delta, theta, alpha and beta brain waves [10]. Additionally, subjective evaluation is necessary for any assessment experiment since it directly quantify participants' perception.

Therefore, we investigated whether visual discomfort yield by blue component under commonly used light source. For the quantification of visual discomfort, the combination of subjective evaluation with objective measurement method (including electrooculogram (EOG) and EEG measurements) was used. We hypothesized that visual discomfort would be more severe with high amounts of blue component.

2. Method

2.1 Participants

Eleven participants (7 males, mean age: 23.90 ± 1.04 years) took part in the experiment. They underwent an optometric screening to exclude eye abnormality. All of them passed through the screening and their corrected visual acuity got 1.0 for monocular and binocular. Participants were not included if they had any ophthalmological and neurological abnormalities, as well as color blindness. Before the experiment, they gave their informed content to complete the experiment.

2.2 Experiment setup

There were two white lights performed in the experiment, one with low amounts of blue component (blue-less white light), one with high amounts of blue component (blue-enriched white light). Both light sources were created by THOUSLITE LED Cube which has 11 channels (2 white lights, 3 red, 3 green and 3 green monochromatic lights). The LED Cube was placed at a viewing booth to simulate a working environment, as shown in Figure 1. The spectrum, correlated color temperature (CCT) and photopic illuminance were measured at the eye level with SFIM-300 Spectral Flickering Irradiance Meter whose detector of pointed toward the participants' viewing direction, as shown in Figure 2.

978-1-7281-5757-3/19 $31.00 © 2019 IEEE

(a) Blue-less white light   (b) Blue-enriched white light

Figure 1. The simulated working environment

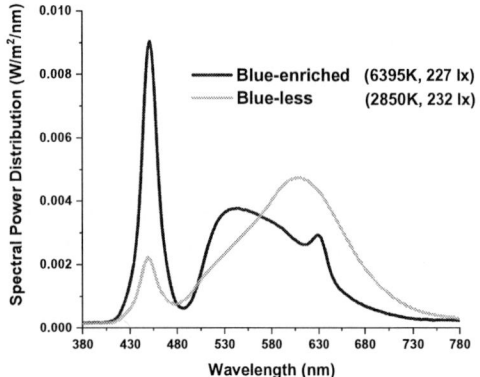

Figure 2. The measured spectrum, CCT and photopic illuminance for both experimental light sources

2.3 Experiment procedure

The experiment consisted of two sessions (i.e. two illuminants), and all participants involved the both sessions. The order of these two sessions was counterbalanced, such that half of the participants started with blue-less white light and vice versa. The interval period between the two sessions was at least two days to wash out the effects of previous light source. The day before the test, participants were prevented from drinking beverages containing alcohol or caffeine, and asked to maintain regular sleep.

After arriving at the laboratory, participants who reported any discomfort symptoms preceding the formal experiment were rescheduled to another day. Subsequently, participants filled out the questionnaire on the level of their current visual discomfort in terms of 5-point scales corresponding to none, slight, moderate, obvious and severe. Then, they were guided to sat in front of the viewing booth to perform visual tasks about one hour on a 7.9-inch iPad with the resolution 2048 × 1536. EEG and EOG signals were recorded during the first and last 5 minutes of the visual tasks. Once completing the visual tasks, participants were asked to fill out the questionnaire again.

2.4 Data acquisition and analysis

EEG signals were recorded at a sampling rate of 500Hz with the NeuroScan system. The reference electrode was set at a point halfway between electrodes Cz and CPz, when collecting data online. Twenty-seven active electrodes were positioned according to the International 10-20 system. Among these electrodes, two mastoids electrodes were used as the re-reference electrodes for the offline analysis, four electrodes recorded the EOG activity and twenty-one electrodes (Fp1, Fpz, Fp2, F3, Fz, F4, FC3,FCz, FC4, C3, Cz, C4, CP3, CPz, CP4, P3, Pz, P4, O1, Oz, O2) for EEG activity. Electrode resistance at each electrode was kept below 5 kΩ.

EEG signals were analyzed with MATLAB-based EEGLAB toolbox [11]. Raw EEG data were band filtered into 1-40Hz, then re-referenced to the average of left and right mastoids. Independent component analysis (ICA) was used to eliminate the irrelevant components like ocular artifacts and head movement. Power spectral density (PSD) was calculated by a fast Fourier transformation (FFT, Hanning 2-second window, 50% overlapped) and normalized divided by the total power. Four basic brainwave activities—δ (1-4Hz), θ (4-8 Hz), α (8-13Hz), and β (13-30Hz)—were extracted from EEG activities.

Blink events were extracted from EOG data with an automated pipeline (BLINKER) [12] based on MATLAB toolbox. In this study, blink rate, blink amplitude and blink duration were mainly analyzed.

3. Result and Discussion

In this section, paired-sampled T-test was performed to investigate whether there would be significant changes in the feeling of visual discomfort and EEG activities after completing visual tasks under each light source.

3.1 Changes under blue-less white light

Participants rated visual discomfort as significant higher after one-hour visual tasks under blue-less white light ($p < 0.001$), as shown in Figure 3. However, none of the four brainwave rhythms changed significantly for any electrodes at the end of the experiment when exposing to blue-less white light. In addition, blink events including blink amplitude, blink duration and blink rate had no significant difference between the start and end of visual tasks. It indicates that the visual discomfort produced under blue-less white light might be slight and unable to result in significant variation for EEG activity and blink events.

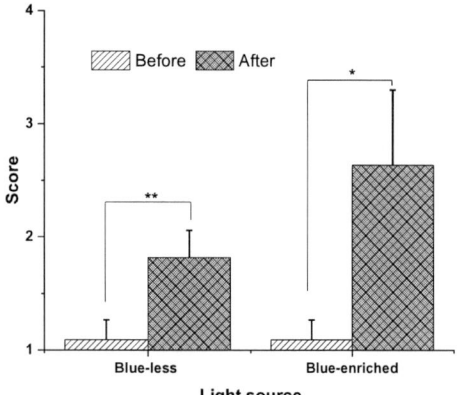

Figure 3. Subjective rating of visual discomfort under two light sources. $^*p < 0.05$, $^{**}p < 0.001$

3.2 Changes under blue-enriched white light

Visual discomfort yielded significant difference after the visual tasks under blue-enriched white light ($p = 0.001$). Participants felt moderate discomfort when they performed one-hour visual task (see Figure 3).

There were no significant changes for δ, θ and β brainwave activities at all twenty-one electrodes used in this study, only α rhythm have significant difference at some electrodes after the visual tasks under blue-enriched white light. The electrodes at which alpha rhythm varied significantly are listed in Table 1. It shows that the cortical regions near central area, and around parietal and occipital areas changed more significant. Figure 4 shows EEG alpha activity at the start and end of visual tasks. It illustrates that EEG alpha activity increased at the end of the experiment when exposing to blue-enriched light.

Table 1. Cortical regions and sensors that α rhythm varied significantly under blue-enriched white light ($^*p<0.05$)

	Frontal	Central	Parietal	Occipital
α rhythm	Fp1, Fpz, Fp2*, F3, Fz, F4*, FC3, FCz, FC4*	C3*, Cz*, C4*	CP3, CPz*, CP4*, P3, Pz*, P4*	O1, Oz*, O2*

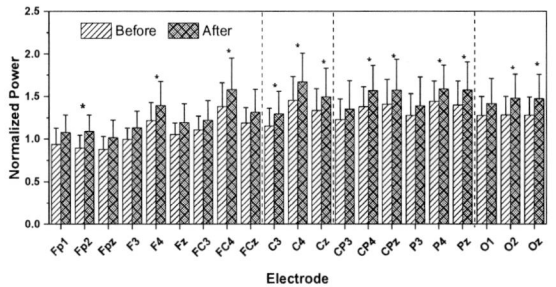

Figure 4. EEG alpha activity at the start and end of visual tasks under blue-enriched white light, $^*p<0.05$

Blink rate varied significantly (p=0.047) with frequent blinking at the end of the experiment under blue-enriched white light (see Figure 5).

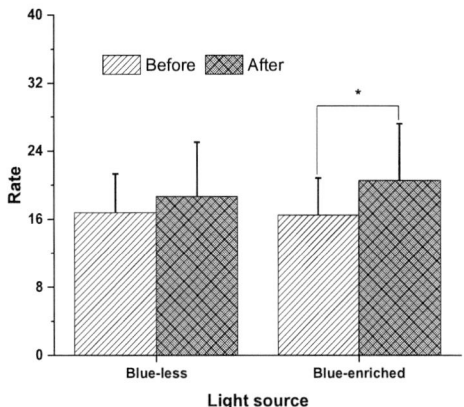

Figure 5. Blink rate at the start and end of visual tasks under two light sources. $^*p<0.05$

Compared with the blue-less white light, the perception of visual discomfort was relative more complex. Not only the subjective feeling of visual discomfort changed significantly,

but also the alpha rhythm of EEG activity and blink rate yielded obviously. It has been proved that blue light easily increased EEG power density in the alpha range compared to other monochromatic lights [13]. This study further evidenced that white light with higher blue component also enhanced EEG alpha activity, which is more easily detected at the cortical regions near central area, and around parietal and occipital areas. Some studies pointed that EEG alpha is a possible marker of fatigue-induced changes which is easily induced at parieto-occipital sites [14]. It is in accordance with our findings. In addition, visual discomfort is accompanied with frequent eye blink caused by glare [15]. Blink prevents the eye from dry and enables clear vision. White light with high amounts of blue component increased blink rate. It suggests that during short-term visual task, enriched blue component of white light is easily perceived by retina and result in increased blink rate.

4. Conclusions

The effects of blue component for white light on visual discomfort were investigated using EEG and EOG measurements. The results showed that only subjective perception of visual discomfort was produced under blue-less white light, no significant changes yielded for brain activity and blink events. However, the visual discomfort was complex under blue-enriched white light. Both physiological activity and visual system varied obviously, with EEG alpha activity enhanced near the central area and around parietal and occipital regions, as well as frequent blink. This might be a potential cause for the myopia.

Acknowledgments

This work was supported by National Key Research Program of China (2016YFB0401201), Natural Science Foundation of China (61505028).

References

1. C. Cajochen, M. Freyburger, T. Basishvili, C. Garbazza, F. Rudzik, C. Renz, K. Kobayashi, Y. Shirakawa, O. Stefani, and J. Weibel, "Effect of daylight LED on visual comfort, melatonin, mood, waking performance and sleep," Lighting Research & Technology, pp. 1477153519828419.
2. Z. Hamedani, E. Solgi, H. Skates, T. Hine, R. Fernando, J. Lyons, and K. Dupre, "Visual discomfort and glare assessment in office environments: A review of light-induced physiological and perceptual responses," Building and Environment, vol. 153, pp. 267-280, 2019.
3. T. Ru, Y. A. de Kort, K. C. Smolders, Q. Chen, and G. Zhou, "Non-image forming effects of illuminance and correlated color temperature of office light on alertness, mood, and performance across cognitive domains," Building and Environment, vol. 149, pp. 253-263, 2019.
4. J. de Zeeuw, A. Papakonstantinou, C. Nowozin, S. Stotz, M. Zaleska, S. Hädel, F. Bes, M. Münch, and D. Kunz, "Living in Biological Darkness: Objective Sleepiness and the Pupillary Light Responses Are Affected by Different Metameric Lighting Conditions during Daytime," Journal of biological rhythms, pp. 0748730419847845, 2019.
5. R. Nagare, M. S. Rea, B. Plitnick, and M. G. Figueiro, "Nocturnal Melatonin Suppression by Adolescents and

Adults for Different Levels, Spectra, and Durations of Light Exposure," Journal of biological rhythms, vol. 34, no. 2, pp. 178-194, 2019.

6. Y. Kuse, K. Ogawa, K. Tsuruma, M. Shimazawa, and H. Hara, "Damage of photoreceptor-derived cells in culture induced by light emitting diode-derived blue light," Scientific reports, vol. 4, pp. 5223, 2014.

7. I. E. S. o. N. America, "Recommended Practice for Photobiological Safety for Lamps and Lamp Systems–General Requirements," 2005.

8. J. D. Bullough, A. Bierman, and M. S. Rea, "Evaluating the blue-light hazard from solid state lighting," International Journal of Occupational Safety and Ergonomics, vol. 25, no. 2, pp. 311-320, 2019.

9. S. Benedetto, A. Carbone, V. Drai-Zerbib, M. Pedrotti, and T. Baccino, "Effects of luminance and illuminance on visual fatigue and arousal during digital reading," Computers in human behavior, vol. 41, pp. 112-119, 2014.

10. Y. Zheng, X. Zhao, and L. Yao, "The assessment of the visual discomfort caused by vergence -

accommodation conflicts based on EEG," Journal of the Society for Information Display, 2019.

11. A. Delorme, and S. Makeig, "EEGLAB: an open source toolbox for analysis of single-trial EEG dynamics including independent component analysis," Journal of neuroscience methods, vol. 134, no. 1, pp. 9-21, 2004.

12. K. Kleifges, N. Bigdely-Shamlo, S. E. Kerick, and K. A. Robbins, "BLINKER: automated extraction of ocular indices from EEG enabling large-scale analysis," Frontiers in neuroscience, vol. 11, pp. 12, 2017.

13. C. Cajochen, "Alerting effects of light," Sleep medicine reviews, vol. 11, no. 6, pp. 453-464, 2007.

14. C.-T. Lin, M. Nascimben, J.-T. King, and Y.-K. Wang, "Task-related EEG and HRV entropy factors under different real-world fatigue scenarios," Neurocomputing, vol. 311, pp. 24-31, 2018.

15. S. Berman, M. Bullimore, R. Jacobs, I. Bailey, and N. Gandhi, "An objective measure of discomfort glare," Journal of the Illuminating Engineering Society, vol. 23, no. 2, pp. 40-49, 1994.

Infer light diffuseness on light probes with different kinds of mesoreliefs

Yudi Wang, Ling Xia, Jinfeng Huang, Ruipeng Xu, Xiaofeng Liu*
College of IoT Engineering, Hohai University, Jiangsu Key Laboratory of Robotics and Intelligent Technology
Changzhou, China,
xfliu@hhu.edu.cn

Abstract

The appearance of objects varies enormously depending on light diffuseness. Conversely, the appearance of objects might be a cue to the light diffuseness. This would provide a solution for lighting designers to guess-estimate the light diffuseness levels based on the appearance of a light probe. However, the question remains whether rough probe provides additional cues about light diffuseness comparing with the smooth probes. If the answer to the last question is yes, whether different kinds of roughness performs the same. Furthermore, how the estimation of light diffuseness might be influenced by the illumination directions. To answer these questions, an appearance-matching experiment was performed, within which white spherical probes with four kinds of surface mesoreliefs were adopted and 5 illumination directions were tested. The results confirmed that observers' abilities to match the light diffuseness were improved by the use of probes with 3D texture, which means that observers can derive light diffuseness from the additional cues provided by the texture over the rough probe. However, the results showed that the roughness has no obvious effect on diffuseness perception at illumination direction of 90° or 120° owing to the much useful information provided by the terminator of the body shadow.

1. Introduction

Advances in technology promoted the development of solid-state lighting (such as LED), which advantaged over traditional light source in small size and flexible lighting arrangement. Alan Tulla (Tulla A, 2008) proposed that the development in solid-state lighting marked the arrival of the third stage of the lighting profession. In this stage, people's requirements for lighting are no longer just the visibility of objects, but also the various visual experiences. Therefore, the trend of lighting profession is to consider the interaction between the light and its surroundings, to meet people's comfort and the pursuit of visual perception, which requires a switch from thinking about light incident on planes to light arriving at eye (Cuttle, 2010). The intensity, the primary illumination and light diffuseness, which are three basic (low-order) properties of light field, can be sensed by the human visual system. Light diffuseness, describing the isotropy of a light distribution around a point in a space, can strongly influence the appearance of objects since shading, shadowing and vignetting vary with the diffuseness (Xia, Pont, & Heynderick, 2016; Xia, Pont, & Heynderickx, 2016).

Previous researches (Jan J Koenderink & Pont, 2007; Xia, Pont, & Heynderick, 2017) showed that observers estimated the light diffuseness much more coarsely comparing with other two light properties. However, studies (Xia, Pont, & Heynderick, 2014; Xia, Pont, & Heynderick, 2016; Xia, Pont, & Heynderickx, 2016) also showed that, generally, the observers are able to distinguish different diffuseness levels both on the

smooth and rough white spherical probes. Furthermore, Xia et al. (Xia et al., 2014) confirmed that observers could use illuminance flow over 3D objects as a cue to judge the light diffuseness. It means that the lighting designers could use the rough probe as a tool for a quick and coarse estimation of the light diffuseness levels in the space. Besides that, the appearance of the probe itself is also an indicator of how other objects would look like in such kind of diffuseness levels. However, the question remains whether different kinds of roughness make difference.

The estimation of light diffuseness might be influenced by illumination direction. For example, Pont et al. (Pont & Koenderink, 2007) found that the diffuseness and the illumination direction can influence the appearance of objects interactively. In their research, it was found that observers tend to overestimate diffuseness when the illumination direction was consistent with the viewing direction in a collimated light condition, because the images of the object appearance under diffuse lighting or frontal lighting seemed quite similar. Thus, such interactions might form a major problem for the "readability" of the illumination direction and diffuseness. In our research, the question follows how illumination direction affects the estimation of light diffuseness on the rough probes.

In this study, we attempt to clarify whether the use of rough probe could be a solution for the guess-estimate of light diffuseness and which kind of roughness would perform better. Furthermore, we are interested in how the illumination direction would influence the light diffuseness estimation. To answer the questions above, four kinds of surface mesoreliefs and five illumination directions were tested based on an appearance-matching experiment.

2. Experimental approach

Method

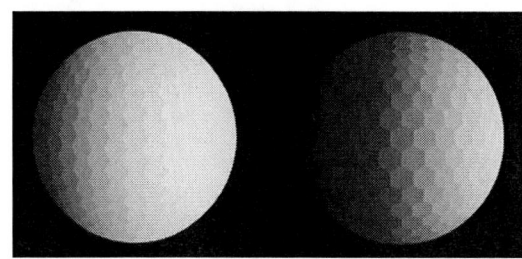

Figure 1 Experimental interface. A stimulus is displayed on the left side and a probe is displayed on the right side.

We conducted an appearance-matching experiment on a computer screen. The stimulus was displayed on the left side of the screen while an adjustable spherical probe was displayed on the right side. The probe on the right has the same mesorelief and illumination direction with the left stimulus but randomized diffuseness level (see Figure 1). Observers will be

978-1-7281-5757-3/19 $31.00 © 2019 IEEE

asked to adjust the light diffuseness on the probe to make it the same as the stimulus on the left. However, the placement of the stimulus and adjustable probe has a certain angle difference so that the surface structures of two spheres are not exactly the same, in case that participants simply match the image structures.

Stimuli

We chose four kinds of spherical probe types, namely smooth, flat, convex and concave (see Figure 2). "Smooth" means a sphere with the surface free from roughness. "Flat", represents a sphere with a faceted surface, which contains more edges comparing to smooth sphere. "Convex" stands for the sphere with small bumps. "Concave" has dimples on the surface, which is similar with golf ball. The bumps of the convex probe and dimples of the concave probe were in the same size and depth.

Five illumination directions were selected to investigate how the estimation of light diffuseness might be influenced by the illumination direction. The illumination angle varied between 0°, 30°, 60°, 90° and 120°, which cover three quarters of a hemisphere (see Figure 3).

We set 11 diffuseness levels between 0 to 1 with an interval of 0.1 for each type of probe surface and in each illumination angle. "0" corresponds to fully collimated light and "1" corresponds to fully diffuse light (Xia, Pont, & Heynderick, 2016; Xia, Pont, & Heynderickx, 2016). Fully collimated light comes from a single direction, such as direct sunlight, which tends to create a sharp contrast between light and dark on the objects. While fully diffuse light comes from all direction, and is characterized by a soft light with neither the highlight nor the shadow of collimated light, such as the snowfield in an overcast day.

Thus, totally, we rendered 220 (5 angles * 11 diffuseness * 4 probe types) stimuli on the left side of the screen to be estimated.

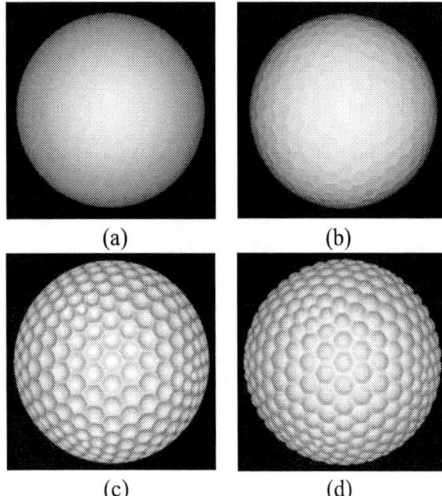

Figure 2 Examples of stimuli. Four kinds of surfaces are set up in the experiment: (a) smooth, (b) flat, (c) concave, (d) convex.

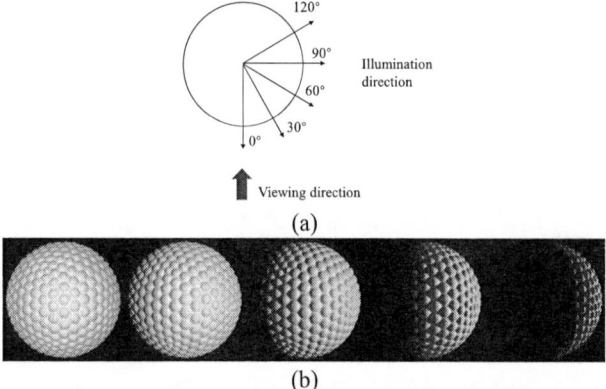

Figure 3 (a)Bird's eye view of the viewing direction of the observers and five illumination directions. (b) From left to right, the illumination direction varies from angles differing 0°, 30°, 60°, 90° and 120° from viewing direction.

Procedure

In the experiment, the order of 220 stimuli on the left side of the screen was given randomly for each observer. And for each trial, the initial diffuseness level on the probe was randomized. Participants could perform the adjustment with small steps (0.02 unit of the normalized diffuseness levels per press) on the probe by pressing "left" and "right" arrow keys, or take big steps (0.1 unit of the normalized diffuseness levels per press) by holding down the shift key at the same time. The diffuseness on the probe co-varied simultaneously during the adjusting by the observer.

To cover the whole range of possible diffuseness levels on the probes, we have prepared 1020 (5 angles * 51 diffuseness * 4 probe types) images. All the images were realistically rendered using Maxwell Software. We selected one stimuli from each mesorelief type as a demo to explain the task to naive observers before the experiment.

The experiment was performed on a Samsung display driven by display card NVIDIA Quadro P2000.

Participants

Twenty observers, ranged in age from 20 to 26, who were naive with regard to our research participated in the experiment. All observers had normal or corrected-to-normal vision.

3. Results

In order to have a more intuitive understanding of the estimated light diffuseness for each diffuseness levels on different surfaces, we have drawn boxplots of the data for four kinds of probe separately, which are shown in Figure 4. From the boxplots, it is clear that for all kinds of probe types, the estimated values and stimulus values correlate well. Nonetheless, we found that in comparison with other three types, overall variability of diffuseness settings on smooth probe increased. It indicates that, as we have expected, rough probes do provide more information for light diffuseness estimation than smooth ones.

Illumination direction:

- ▨ 0°
- ▨ 30°
- ▨ 60°
- ▨ 90°
- ▨ 120°

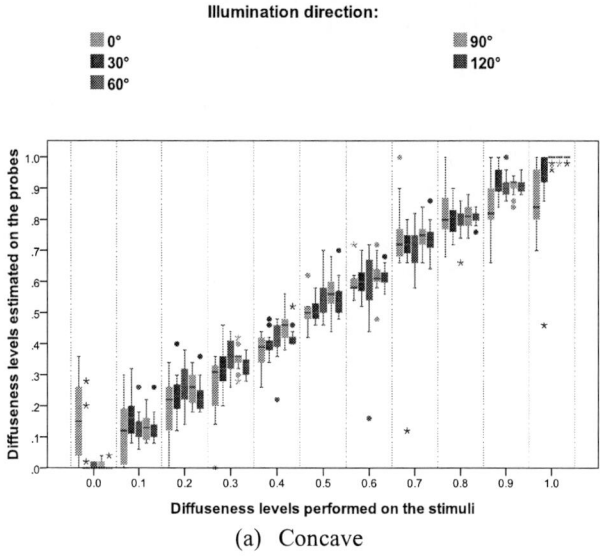

(a) Concave

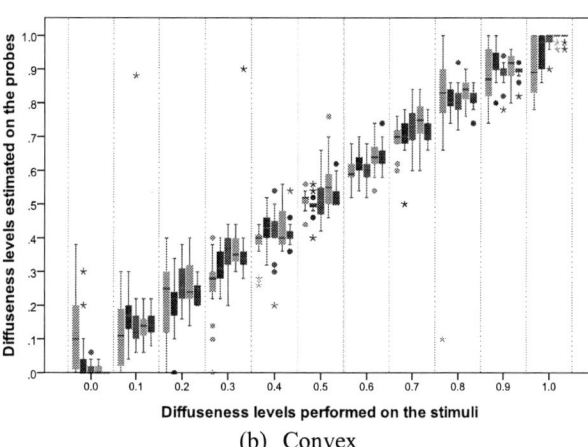

(b) Convex

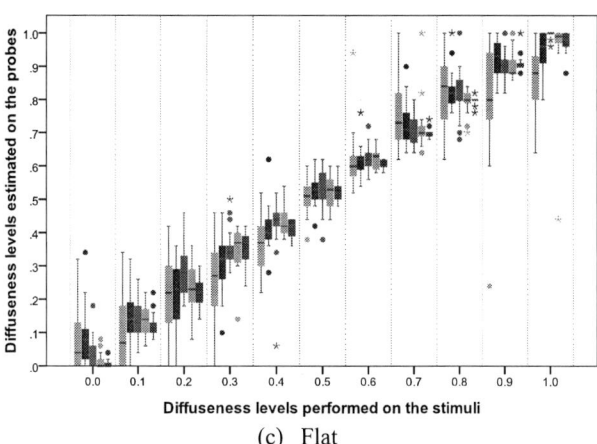

(c) Flat

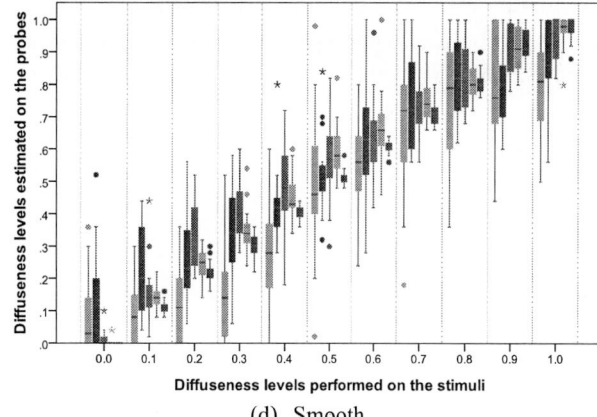

(d) Smooth

Figure 4 Boxplots of the estimation data for four kinds of probe types.

We analyzed the effect of probe type, illumination direction and the diffuseness levels on the estimated diffuseness on the right probe by performing an ANOVA analysis. The estimated diffuseness values were dependent variable while the probe type, illumination direction and the diffuseness levels were fixed factors. The results were shown in Table 1. Both illumination direction and diffuseness level had significant effects on light diffuseness estimation respectively and interactively. Furthermore, we found interaction between probe type and illumination direction. It means that, people were sensitive to the diffuseness changes on the stimuli. However, the performance of probe type might also depend on the angle difference between viewing direction and illumination directions.

Table 1. Results of ANOVA analysis for the effect of probe type, illumination direction and diffuseness on the estimated light diffuseness

Source	df	F	P
Probe type	3	1.842	0.137
Illumination direction	4	56.804	0.000
Diffuseness levels	10	6304.905	0.000
Probe type * Illumination direction	12	9.877	0.000
Probe type * Diffuseness	30	1.730	0.008
Illumination direction * Diffuseness	40	11.410	0.000

Additionally, we did linear regression analyses for the estimated diffuseness value for five illumination directions and on four types of probe separately. The regression efficiency was reflected by the coefficient of determination (R^2). The closer the R^2 value was to 1, the better effect of regression was. The results showed that R^2 value of smooth probe was lower on the whole than that of rough probes (see Figure 5), which confirmed the results in the boxplot and the result of ANOVA analysis. Furthermore, it reflected the interaction between probe type and the illumination direction. It was notable that as the angle between viewing direction and illumination direction increased, the R^2 value of smooth probe raised significantly (i.e. 0°, 30°, 60°, 90°). For illumination direction of 90°, and 120°, the R^2 value of smooth probe was almost as high as that of rough probes.

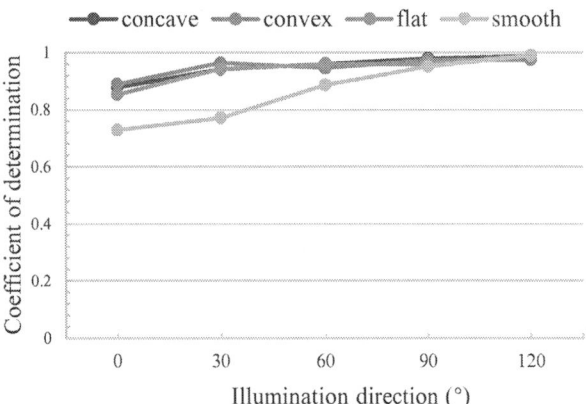

Figure 5 Coefficients of determination (R^2) as a function of illumination direction in the regression analysis results.

4. Discussion

The results confirmed that for both smooth and rough spherical probes, the observers had no difficulty to match the light diffuseness, which is consistent with previous studies. And as we have expected, the variance in the light diffuseness estimation was larger if the probe was smooth. It means that observers do use the light flow over rough probes as a cue to estimate diffuseness levels. However, we did not find significant difference between three kinds of mesorelief used in our experiments.

The results showed that, comparing three rough probes, the probe type had no significant effect on the estimated diffuseness values. However, the observers' performance on smooth probe was significantly lower than that of rough probe. It was consistent with previous research that the local shadow caused by 3D texture on rough probes could provide additional information to the light diffuseness. Nevertheless, further research still needed in order to find more effective mesoreliefs.

We found that the estimated diffuseness had the most variance for illumination angle of 0°. It might because that the contrast due to the mesorelief decreased as the angle difference between viewing angle and illumination angle decreases. We found that there was no difference between smooth surface and rough surface for the estimated diffuseness when the illumination angle was 90° or 120°. This result is consistent with Koenderink et al' finding that there would be a terminator of the body shadow on the surface of probe, which might be used as a cue to the light diffuseness (J. J. Koenderink & Pont, 2002).

However, in this experiment, we showed the stimuli in the form of 2D images. In reality, objects are three-dimensional, which provide more useful information concerning the light field, such as depth cues. Therefore, in further research, we are interested in investigate human being's interpretation of light diffuseness on 3D display instead of 2D. Futhermore, it would also be interesting to see how the estimated diffuseness could be influenced by illumination directions in 3D environment with information of depth cues provided.

5. Conclusions

In summary, we have conducted an appearance-matching experiment to investigate how sensitive human beings are to light diffuseness. Four kinds of mesorelief on the probes' surface were tested with one being smooth and three being rough. The illumination direction varies from 0° to 120° with 30° intervals. We confirm observers are sensitive to light diffuseness, but this sensitivity might be affected by surface mesoreliefs of the probe and illumination directions. The illumination flow on the roughness on the probe do provide additional cues to the perception of light diffuseness. However, further research still needs to be done in order to find more effective mesoreliefs. This research is an exploration of using rough probes to help people guess-estimate light diffuseness. It would provide a quick and intuitive method to have a rough measurement of the light diffuseness in the real scenes.

6. Acknowledgments

This work was partly supported by National Nature Science Foundation of China under grants No.61603123 and No.61703140; and by the China Fundamental Research Funds for the Central Universities under grand No.2017B02814; by the National Natural Science Foundation of Jiangsu under grant BK20170304; and by the Key Research and Development Program of Jiangsu under grants BE2017071, BE2017647 and BE2018004-04; and by the Open Research Fund of State Key Laboratory of Bioelectronics, Southeast University under grant 2019005;

References

1. Cuttle, C. (2010). Towards the third stage of the lighting profession. *Lighting Research & Technology, 42*(1), 73-93.
2. Koenderink, J. J., & Pont, S. C. (2002). *Texture at the terminator*. Paper presented at the Proceedings. First International Symposium on 3D Data Processing Visualization and Transmission.
3. Koenderink, J. J., & Pont, S. C. (2007). The visual light field. *Perception, 36*(11), 1595.
4. Pont, S. C., & Koenderink, J. J. (2007). Matching illumination of solid objects. *Perception & Psychophysics, 69*(3), 459-468.
5. Tulla A. (2008). Editorial Newsletter. *The Society of Light and Lighting, 1*(1)
6. Xia, L., Pont, S. C., & Heynderick, I. (2014). The visual light field in real scenes. *i-Perception, 5*(7), 613-629.
7. Xia, L., Pont, S. C., & Heynderick, I. (2016). Light diffuseness metric Part 1: Theory. *Lighting Research and Technology, 24*(5), 499-515.
8. Xia, L., Pont, S. C., & Heynderick, I. (2017). Separate and Simultaneous Adjustment of Light Qualities in a Real Scene. *i-Perception, 8*(1), 2041669516686089.
9. Xia, L., Pont, S. C., & Heynderickx, I. (2016). Light diffuseness metric, Part 2: Describing, measuring and visualising the light flow and diffuseness in three-dimensional spaces. *Lighting Research and Technology, 24*(5), 516-530.

The Effect of Lighting on Stereo Vision Test with Random-Dot Stereogram

Jinfeng Huang, Tingting Zhang, Yudi Wang, Jinwei Xie, Xiaofeng Liu*
Jiangsu Key Laboratory of Robotics and Intelligent Technology , College of IoT Engineering, Hohai University,
Changzhou, China,
xfliu@hhu.edu.cn

Abstract

Depth perception is the ability of people to evaluate the relative and absolute distance. Industry practitioners such as pilots, car drivers, athletes, tailors, and seafarers, always require better stereo vision and good depth perception ability to achieve high efficiency and good performance. Meanwhile, the lighting environment is another important factor that influence their working performance. This paper investigated the influence of ambient luminance and screen brightness on depth perception based on Random-Dot Stereograms. In the experiment, we set three ambient illumination and three screen brightness to test. We found depth perception of the participants was influenced by screen brightness (16 Lux ,8 Lux and 1.5 Lux), but not influenced by ambient illumination (124 Lux, 20 Lux, 2 Lux). Hence, we could conclude that the ambient illumination of 124 Lux, 20 Lux, 2 Lux is a safe lighting that does not influence depth perception and it is not suitable for operators to work with the screen brightness of 16 Lux ,8 Lux and 1.5 Lux.

1 Introduction

Depth perception is the ability of people to judge the relative distance between objects and the absolute distance from the object to the observer themselves. The depth perception ability can directly affects the performance and safety of operators (Sakata, Taira, Kusunoki, Murata, & Tanaka, 1997; Zhang, Cheng, Xia, & Liu, 2018) Depth perception can be affected by many factors, such as duration, viewing distance, stimulus size and luminance level(Amigo, 1963; Geib & Baumann, 1990).

Lighting is not only an essential element for human to perceive the world but also an important factor for depth perception(Hawes, Brunyé, & Mahoney, 2012). For example, the driver's judgment of the safe driving distance varies from morning to night. For surgeons, a proper lighting environment is extremely important to guarantee good visual ability. Furthermore, the effect of luminance on depth perception could change the athlete's attention, the accuracy of passing and the hit rate of shot to some extent(Hong, 2002; Hubona & Wheeler, 1999; Jiemin.Chen & hua.Zhen, 2000). Improper lighting conditions can blur people's sights and are likely to affect people's perception of depth. Therefore, it is of great significance to make clear how lighting affects depth perception.

In this paper, we investigated how screen brightness and ambient lighting influenced the depth perception performance during a Random-Dot Stereogram test. Random-Dot Stereogram test can exclude most of the monocular depth cues and it has high utilization. In these random-dot stereogram stimuli, 9 depth levels were designed, ranging from 51″ to 459″. Participants had to discriminate the direction of letter "E" under five different lighting environments.

2 Methods

a. Participants

20 college students (12 males and 8 females), with the average age of 22.3 years, participated in the experiment. All participants had normal or corrected-to-normal visual acuity. Participants signed the consent forms to make sure that they were voluntary.

b. Experiment design

The experiment was conducted in a lab, where day light could be completely shielded by curtains. An 27'' LCD display was used for presenting the stimuli, with a screen resolution of 2560×1440. The display was positioned on a table 75 cm in height. The distance from the front edge of the table to the center of screen was 1.5 m. The fluorescent lamps on the left is 1.8 meters from the center of display and the fluorescent lamps on the right is 0.2 meters from the center of display.

A within-subjects design was used, and two factors, Ambient luminance and Screen brightness were manipulated. The schematic diagram of the experiment is shown in Figure 1. It can be seen that there are six fluorescent lamps, positioned in two lines in the lab. The ambient illumination were designed in this experiment: 124 Lux (the right fluorescent lamps on), 20 Lux (the left lamps on), and 2 Lux (all the lamps off). The luminance was measured by a digital illuminometer in front of the screen. The screen brightness was set at the value of "50", whereas "0" represents the darkest and "100" means the brightest. When all the lamps were off, we set three different screen brightness: 100,50,0. The luminance in front of the screen was 16 Lux, 8 Lux, and 1.5 Lux, respectively.

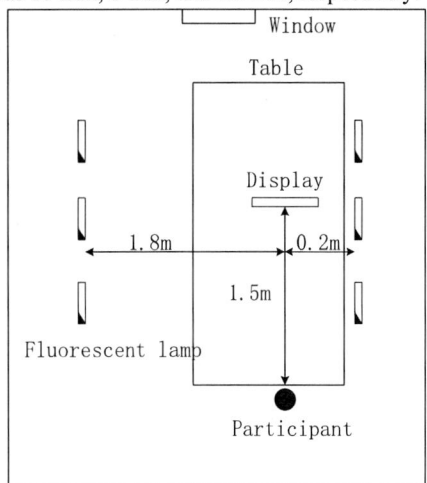

Fig.1 Experimental layout diagram

c. Stimulus

There were 9 levels of depth for the letter "E" in the Random-Dot Stereograms. For each depth, the letter "E" has 4 directions (up down, left, right). Hence, there were 36 stimuli in total. The image generation process is shown in Figure 2.

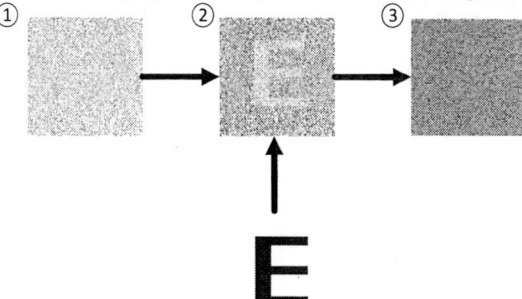

Fig.2 The process of stimulus generation

The pixel of picture is 400×400. Select a certain number of points and assign green randomly. The letter "E" covers the picture and attaches with as many dots as the previous green dots in red in the background area. For the graphics area, the number of pixel points n that need to be moved to right is calculated according to the depth of stereo acuity and Equation 1, and then these dots are assigned in red.

$$\delta y = a\, d \frac{p}{z^2} \qquad (1)$$

In equation 1, we need to transfer depth p into the parallax angle δy. d is the distance between the eyes of the subjects and z is the space of the fixed column from the eye, a is a constant whose value is 3437.75 (1 radian in arcminutes)(Gargantini, Facoetti, & Vitali, 2014). 9 levels of stereoscopic acuity were generated with the above method: 51″, 102″, 153″, 204″, 255″, 306″, 357″, 408″, 459″.

d. Procedure

The procedure of this experiment is shown in Figure 3. The participants first signed the consent form and filled in the form with personal information such as name, age, gender, nationality and so on. The experiment conductor explained the purpose and main content of the study. The subjects need to wear red-green filter glasses for Random-Dot Stereogram testing.

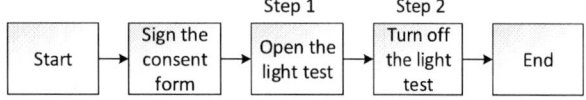

Fig.3 The procedure of the experiment

The experiment was separated into two sessions: dark session and bright session. In the bright session, either the left lamps or the right lamps were on. In the dark session, all the lamps were off, and the screen was the only light source. Figure 4 shows the procedure of the bright session and Figure 5 shows the dark session. Half of the participants first did the dark session and the others first did the bright session.

In Test 1, turn on the three fluorescent lamps on the left; and in Test 2, only turn on the three fluorescent lamps on the right. The screen brightness of the display in these two tests is at the value of "50". In Test 3,4, and 5, turn off all the fluorescent lamps and adjust the screen brightness of the display to "100", "50", and "0" respectively.

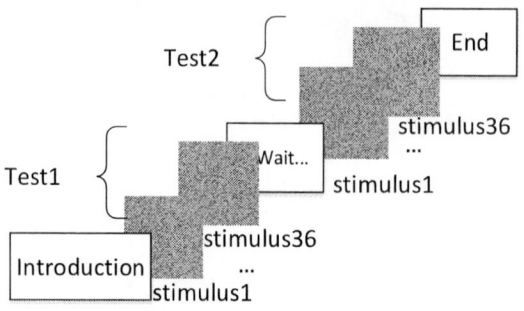

Fig.4 The procedure of the bright session

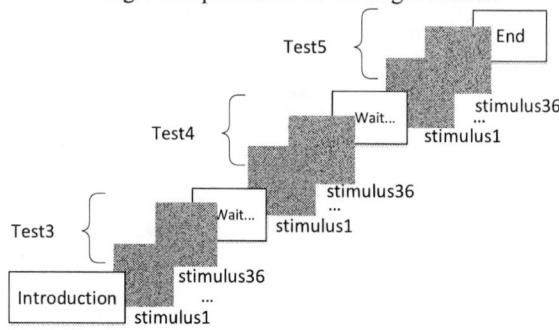

Fig.5 The procedure of the dark session

In each test, there were 36 different Random-Dot Stereograms. The stimuli were displayed randomly, and the participants had to report the direction of letter "E" by pressing "UPARROW", "DOWNARROW", "LEFTARROW", and "RIGHTARROW" keys on the keyboard. The participants could take a break between tests. The whole experiment lasted around 40 minutes.

3 Results

a. Data preparation

In the experiment, 4 participants' response time was very long, and they were not confident about their performance. Hence, these four people's data were excluded from the following analysis.

b. Accuracy

We calculated the mean reaction time and accuracy across depth acuity, ambient illumination and three screen brightness.

$$a_{\mathrm{pp}} = \frac{\sum_{i=1}^{4} ACC}{4} \qquad (2)$$

$$a = \frac{\sum_{i=1}^{16} a_{pp}}{16} \qquad (3)$$

The value of ACC is "1" or "0". "1" represent the answer is correct and "0" represents wrong answer. a_{pp} represents the accuracy of each depth acuity under different lighting conditions. a represents the average accuracy of 16 people in this visual acuity.

Figure 6(a) and (b) show that the accuracy decreases with decreasing ambient luminance and screen brightness. In Figure 6(c), we could not see obvious trend between the depth acuity and the accuracy.

978-1-7281-5757-3/19 $31.00 © 2019 IEEE

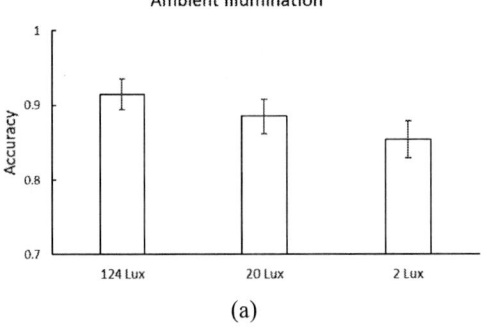

(a)

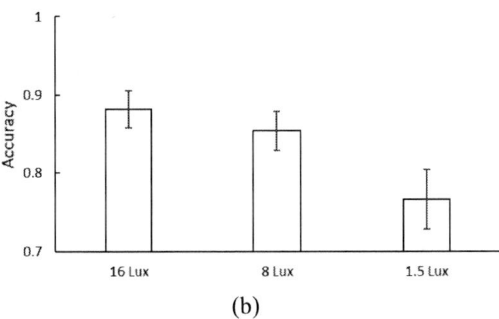

(b)

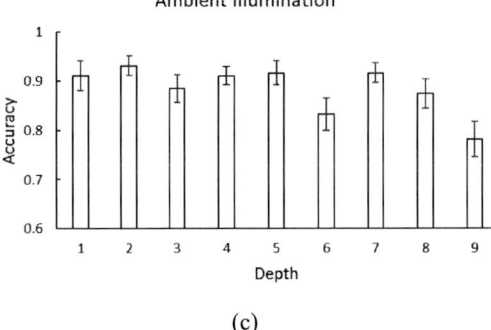

(c)

Fig.6 The accuracy of different illumination. (a) ambient illumination; (b) screen brightness; (c) the accuracy of stereo depth in ambient illumination

Repeated Measures ANOVA results indicated that ambient illumination and depth acuity had significant effect on accuracy ($F_{(4,60)}$ = 8.835, p < 0.01; $F_{(8,120)}$ = 7.052, p <0.01) but the interaction between ambient and depth acuity was not statistically significant ($F_{(32,480)}$ = 1.067; p = 0.066). When we changed the screen brightness, the three brightness (16 Lux, 8Lux, 1.5 Lux) and depth acuity were found to have significant effect on accuracy ($F_{(2,30)}$ = 8.037, p = 0.002; $F_{(8,120)}$ = 5.652, p < 0.01) but the interaction between screen brightness and depth acuity was not statistically significant ($F_{(16,240)}$ =1.164, p = 0.298). It was found that five mixed lights used in this study influence accuracy significantly ($F_{(4,60)}$ = 8.835, P < 0.01).

c. Reaction time

Reaction time refers to the duration between the stimulus appears on the screen and the participant presses the button.

$$rt = \frac{\sum_{i=1}^{4} RT}{4} \qquad (4)$$

$$Rt = \frac{\sum_{i=1}^{16} rt}{16} \qquad (5)$$

RT represents the reaction time of each stimulus in every level of stereoscopic depth. Rt means the average response time of 16 people in this visual acuity.

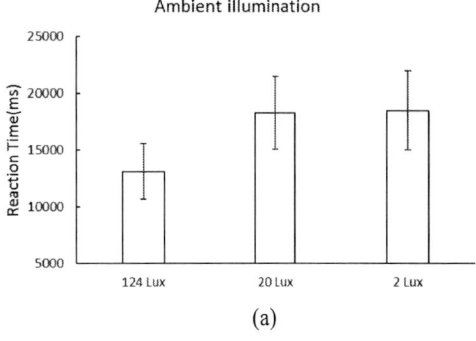

(a)

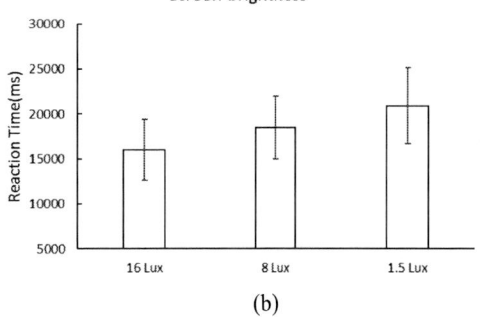

(b)

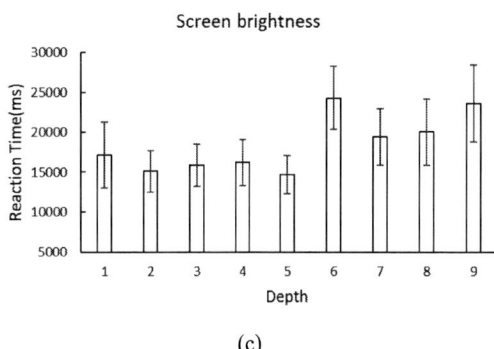

(c)

Fig.7 The reaction time of different illumination. (a) ambient illumination; (b) screen brightness; (c) the reaction time of stereo depth in different screen brightness

It is clear that the mean reaction time increases with decreasing ambient illumination or screen brightness, as shown in Figure 7(a) and (b). Figure 7(c) shows that the reaction time was higher when the depth acuity ranged between 306″ and 459″.

In addition, we found that depth acuity had significant effect on reaction time ($F_{(8,120)}$ = 5.699, p <0.01) but the ambient illumination and the interaction between ambient illumination and depth acuity were not statistically significant ($F_{(2,30)}$ = 2.654, p = 0.087; $F_{(16,240)}$ = 1.067; p = 0.091). When we changed the screen brightness, depth of stereoscopic acuity still had a significant effect on reaction time ($F_{(8,120)}$ = 3.483, p = 0.001) but different to ambient illumination, the screen brightness (16 Lux, 8Lux, 1.5 Lux) were found to have no effect on reaction time (F (2,30) = 0.88, p = 0.425), interaction between screen brightness and depth were not statistically significant (F(16,240) =1.519, p = 0.094). It was found that five mixed lights used in this study did not influence accuracy significantly (F (4,60) = 1.332, p = 0.268).

d. Correlation

Figure 8 shows the correlation between reaction times and accuracy.

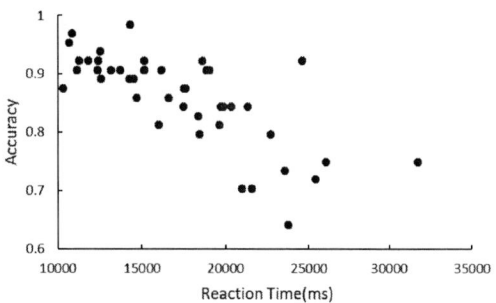

Fig.8 The correlation of accuracy and reaction time

The reaction time was found to be negatively correlated with the accuracy of random-dot stereogram (The Pearson's correlation coefficient r = -726, n = 45, p < 0.01).

4 Discussion

In the experiment, the effects of ambient luminance and screen brightness on depth perception in random-dot stereograms were investigated. First, the results showed that the accuracy decreases when the reaction time increases. This indicates that participants cannot see the current stimulus.

Second, the change of depth acuity will influence the reaction time and accuracy to some extent. Ambient illumination (124 Lux, 20 Lux, 2 Lux) and screen brightness (16 Lux, 8 Lux, 1.5 Lux) have an impact on accuracy of stereo depth, but ambient illumination has no significant effect on reaction time. Researchers have found that the depth threshold of human eye is 259″(Jiemin.Chen & hua.Zhen, 2000). From Fig.6(c) and Fig.7(c), we find that from the visual acuity of the 6th depth, the reaction time increases significantly, and the accuracy decreases significantly. This is because the 6th depth used in the experiment is 306″ and 7th, 8th,9th depth have greater sharpness than the visual depth threshold mentioned in literatures.

we expected that lighting would influence operators' reaction time. However, the results showed that reaction time did not change when the ambient illumination changed in 124 Lux, 20 Lux, 2 Lux. There could be several reasons. First, the difference between the ambient illumination was not large enough to influence participants' vision. The second reason could be that some participants have better visual acuity and the depth of stimuli used in this experiment was not small enough to influence their judgement.

5 Conclusions

In summary, reaction time and accuracy were influenced by depth of stereoscopic acuity, which was not changed by ambient illumination. Though the current study showed that reaction time and accuracy were not influenced by the ambient illumination, we cannot make the conclusion that ambient illumination does not influence reaction time. However, we can conclude that illuminance 124 Lux, 20 Lux, 2 Lux are safe lighting range. In a dark environment, manipulating the brightness of the display will have a certain impact on reaction time. Operators working under the safe lighting environment can maintain the accuracy of their depth perception. It is not suitable for operators to work with the screen brightness of 16 Lux ,8 Lux and 1.5 Lux. Further research is expected to investigate more details about the impact of luminance on depth perception.

Acknowledgments

This work was partly supported by the National Nature Science foundation of China : No.61603123, No.61703140; and by the China Fundamental Research Funds for the Central Universities under Grand No. 2017B02814, in part by the National Natural Science Foundation of Jiangsu under Grant BK20170304, and by the Key Research and Development Program of Jiangsu under grants BE2017071, BE2017647 and BE2018004-04; and by the Open Research Fund of State Key Laboratory of Bioelectronics, Southeast University under grant 2019005.

References

1. Amigo, G. (1963). Variation of stereoscopic acuity with observation distance. *Journal of the Optical Society of America, 53*(5), 630.
2. Gargantini, A., Facoetti, G., & Vitali, A. (2014). *A Random Dot Stereoacuity Test based on 3D Technology.* Paper presented at the International Conference on Pervasive Computing Technologies for Healthcare.
3. Geib, T., ., & Baumann, C., . (1990). Effect of luminance and contrast on stereoscopic acuity. *Graefes Archive for Clinical & Experimental Ophthalmology, 228*(4), 310-315.
4. Hawes, B. K., Brunyé, T. T., & Mahoney, C. R. (2012). Effects of four workplace lighting technologies on perception, cognition and affective state. *International Journal of Industrial Ergonomics, 42*(1), 122-128.
5. Hong, L. (2002). Depth perception of Football Athlete. *Journal of Yunyang teachers college, 22*(6), 75-77.
6. Hubona, G. S., & Wheeler, P. N. (1999). The Relative Contributions of Stereo, Lighting and Background Scenes in Promoting 3D Depth Visualization. *ACM Transactions on Computer-Human Interaction (TOCHI), 6*(3), 214-242.
7. Jiemin.Chen, & hua.Zhen, S. (2000). On the Effect of Illuminance on the Depth Vision of Men's Basketball Players and College Students. *Journal of Nanjing Sport Institute Social Science*(3), 108-111.
8. Sakata, H., Taira, M., Kusunoki, M., Murata, A., & Tanaka, Y. (1997). The TINS Lecture The parietal association cortex in depth perception and visual control of hand action. *Trends in Neurosciences, 20*(8), 350-357.
9. Zhang, T., Cheng, H., Xia, L., & Liu, X. (2018, 23-25 Oct. 2018). *The Influence of Indoor LED Lighting on Depth Perception in Real World.* Paper presented at the 2018 15th China International Forum on Solid State Lighting: International Forum on Wide Bandgap Semiconductors China (SSLChina: IFWS).

Design of a multi - wavelength high irradiance LED phototherapy system for LLLT

Weimin Li, Zhiliang Jin, Jialin Liu, Liquan Guo, Haiyang Wang, Daxi Xiong

Suzhou Institute of Biomedical Engineering and Technology, Chinese Academy of Sciences, New District, Suzhou, China

*Corresponding author: xiongdx@sibet.ac.cn

Abstract

Low level light therapy (LLLT) is a fast-growing technology used in noninvasive therapies which can stimulate healing and relieve pain and inflammation. The introduction of light-emitting diode (LED) devices has reduced many of the concerns associated with halogen lamp and lasers, such as expense and safety. However, many LED devices are bulky and not designed for home use. Besides, the effectiveness of the treatment by LLLT has significant variations in terms of dosimetry parameters for the used LED, such as wavelength, irradiance or power density, energy, etc. Therefore, we design a home-used multi-wavelength LED phototherapy system with the advantage of a small volume and high irradiance in this paper. By using AlN ceramic substrate and copper substrate as composite substrate, the vertical structure chip is tightly attached to the ceramic substrate circuit, then the package of the four-wavelength chip module is completed. In order to distribute the irradiance of all four different spectra over the desired target area, an optical mixing rod is used to collect the light, and the light is all reflected into the output face of the optical mixing rod through full internal reflection. Additionally, to form a light spot of 50mm in diameter at a distance of 5cm from the output surface, we design a pair of lens with a short focal length and large field of view based on Kohler illumination method. The parameters of the lens are designed by the software of ZEMAX. The hardware circuit uses the STM32F104 processor as controller and constant current drive mode to ensure the quality of LED light. An experimental prototype is established, and the irradiance and its uniformity, stability and safety of the phototherapy system are tested. The experimental results show that the maximum irradiance of the system is 115.4mW/cm², and the uniformity reaches 88%. After the phototherapy system works for 2 hours with maximum illumination, the irradiance deviation is 6.7% and the temperature rise of the system is 19℃.

1. Introduction

Low level light therapy (LLLT) is a fast-growing technology used in noninvasive therapies that requires simulations of healing, relief of pain and inflammation. Micheli *et al.* [1] found that 808 nm can relieve chronic joint pain or neuropathic pain. Pok *et al.* [2] investigated the application of 830 nm LED in peeling injury, dog bite, ulcer wound, secondary infection and lip edema infection. After treatment, all patients' wounds were completely healed. Kymplova *et al.*[3] found that phototherapy at 670 nm and 660 nm showed high healing effect and minimal secondary complications. In a study by Whelan *et al.*[4], they demonstrated that 670 nm phototherapy could increase the production of cytochrome C oxidase in primary neurons and reduce cyanide-induced apoptosis. Damien P Kuffler[5] summarized wound healing induced by specific wavelength and its mechanism of action, and found

that compared with cells irradiated at 830nm[6], cells irradiated at 633nm showed a higher levels of migration and ATP luminescence (an indicator of cell viability). The rate of ulcer healing induced by photobiomodulation of 637 and 956 nm was 49% higher than that of the controlled group, and the time of ulcer healing was 50-90% shorter[7]. Continuous irradiation for 2.5 minutes in a continuous mode at 810nm, 100mW and 15J/cm² can promote wound healing [8].

The traditional infrared physiotherapy apparatus uses infrared light bulb, halogen lamp as the light source, which has a wide spectrum and many useless wavelengths. Besides, they are absorbed by the skin epidermis and produce a lot of heat, posing a risk of skin burn. Compared with LED, halogen lamps have low efficiency and short service life. Meanwhile, the existing phototherapy apparatus is large in size, which is mostly used in hospitals and beauty salons. The high price is not suitable for household. Family phototherapy could be possible if the volume of the phototherapy apparatus could be reduced while maintaining the existing irradiance.

To solve this problem, a multi-wavelength LED phototherapy system with small size and high irradiance is proposed. The four wavelengths of 630nm, 660nm, 810nm and 940nm LED chips are highly integrated and packaged on the substrate, and the system is designed according to the LED module.

2. System design

2.1 Optical design

In this study, we consider integrating four kinds of chips into the light module and achieving uniformly distributed irradiance of 100mW/cm² on a target area. Within the target area, the uniformity of irradiance, defined as follows[9], shall be greater than 85%:

$$\eta = \frac{E_{\min}}{E_{\max}} \cdot 100\% \qquad (1)$$

Where $E_{\min}$ and $E_{\max}$ are represent the minimum and maximum irradiance respectively in this region.

Because the working distance is short and the system is required to be small, we have to consider a compact optical system design that can evenly distribute the irradiance of all four different spectra over the desired target area. First, Four LED chipsets with different spectra are encapsulated on a small light module. Second, the optical element have to be small enough.

2.2.1 LED light module design

LED chip is a high heat flux device, which generates a lot of heat and needs different heat channels to transfer to the atmosphere to achieve heat exchange with the outside world. The choice of substrate materials and thermosetting materials, the thickness control of the solidification layer, the control of baking temperature and time, and the design of the packaging substrate of the vertical structure chip with high thermal conductivity all play an important role in the packaging of the high power module. The packaging substrate is the most important part of the whole LED cooling system. In addition to carrying the chip, it is also the carrier that transfers the heat generated by the chip to the cooling device. AlN ceramic [10], a new material for LED packaging substrate, has the advantages of high thermal conductivity (its theoretical thermal conductivity can reach $180W\cdot m^{-1}\cdot K^{-1}$), high strength, low thermal expansion coefficient, low dielectric loss, high temperature resistance, chemical corrosion resistance and non-toxic environmental protection. Therefore, AlN ceramic and copper substrates are used as composite substrates. A four-series and four-parallel ceramic substrate printed circuit is completed by using screen printing technology and the circuit pattern made by positive and negative ink masks. Then, the vertical structure chip is super-tightly attached to the ceramic substrate circuit and the packaging of multi-wavelength chip module is completed.

As shown in Figure 1 (a), four different wavelength LED chipsets are encapsulated on the optical source module by COB package. Considering the dust-proof and safety of the chip, the surface of the chip is encapsulated with glass sheets. The light source is composed of 16 chips with a size of 1.0mm×1.0mm. Driven by a constant current source of 30W electric power, the power density of the chip is greater than 180 W/cm², and the whole chip size area is 4.4mm×4.4mm. The spectrum curve is shown in Figure 1 (b).

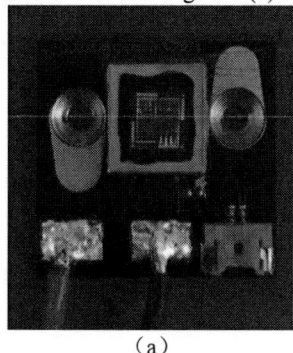

（a）

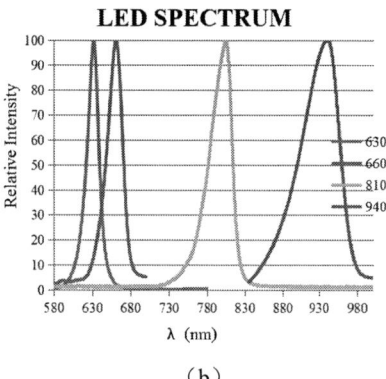

（b）

Fig. 1. (a) COB-packaged multi-wavelength LED module and (b) its four different spectral ranges.

2.1.2 Optical system design

In order to evenly distribute the irradiance of all four different spectra over the desired target area, an optical mixing rod is used to collect the output light from the LED module.The function of the optical mixing rod is to mix the light evenly[11]. The optical mixing rod is close to the glass surface of the light module, and the output light from light module is immediately collected. all the light is uniformly mixed at the output face of the rod via the total internal reflection. The distance between the LED bare die array and the input face of the mixing rod is 0.3mm, and there is a glass protective cover in the middle. Assume that the maximum aperture angle of the light from the LED module is 120°. Then, according to the size of the bare die array, that is, as mentioned above, the input surface size of the mixing rod can be calculated as 4.74mm×4.74mm. In addition, considering the feasibility of manufacturing rod, we take the cross section of bar as 5mm×5mm.

According to the definition of refractive index and led half angle ($\theta_{LED}= 60$ °), it is easy to get the maximum aperture angle of the light entering the mixing rod, that is, $\theta_{LED}= 35°$. Considering the compactness of the structure and the feasibility of making the rod, we choose the length as 20 mm according to experience, which also makes the rod reflect more and have higher uniformity. In conclusion, the size of the mixing rod is designed to be 5mm×5mm×20mm.

In order to form a spot with a diameter of 50mm at a distance of 5cm from the output surface, we use an aspheric lens group to project the light from the output surface of the mixing rod onto the target plane, forming a short-distance convergence, and then produce a uniform spot. Aspheric lens pair can reduce the complexity of optical system and improve the imaging quality. Based on Kohler illumination method, we design a short focal length and large field of view lens pair to meet the design requirements. The parameters of the lens pair are designed by ZEMAX. The settings are as follows: first, we regard the optical system in this work as an object side telecentric system. Secondly, in the design, the height of the object is considered in the field of view. Third, set the merit function (for example, using the default merit function of ZEMAX) to optimize the effective focal length of the system, minimize the root mean square radius, and run the optimization and modify the design parameters iteratively until their values reach the required range. The optical system design is shown in Figure 2.

Fig. 2. Schematic diagram of optical system

2.2 Hardware circuit design

The overall structure of multi-wavelength LED phototherapy system is shown in Figure 3. It mainly includes main controller, constant current driving mode, charge and discharge module.

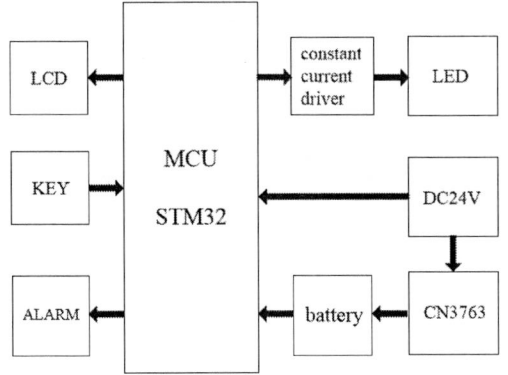

Fig. 3. System block diagram

The main controller and signal acquisition and processing module use STM32f104 processor as the controller, which has rich I/O interface, working frequency of 72MHz, and integrates high-precision A/D, PWM and other rich peripheral resources. It is very suitable for the design of multi-wavelength LED phototherapy system.

2.2.2 Constant current driver module

LED Driver

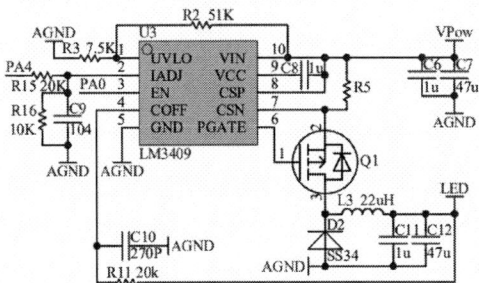

Fig. 4. Constant current source circuit

There are two driving modes of LED: constant voltage driving and constant current driving. The constant voltage source mode requires a current limiting resistor. According to the volt ampere characteristics of LED, a small voltage amplitude will cause a large fluctuation of LED current, coupled with the negative temperature effect of LED, the current fluctuation may lead to a vicious cycle of junction temperature and current, or even burn LED [11].Therefore, LED adopts constant current driving mode. Constant current drive can keep the amplitude of the circuit stable and produce good LED light quality. The constant current drive module is mainly composed of TILM3409, MOSFET Q1, inductor L3, rectifier diode D2, sampling resistor R3 and R5, as shown in Figure 4. LM3409 adopts the constant off time control method, which can realize the accurate constant current without external control loop compensation. It supports analog and PWM dimming, and provides programmable UVLO, low power shutdown and thermal shutdown functions. The MOSFET switch is controlled by a set of duty cycle signals. When Q1 is closed, input voltage C is charged through L3 and D2 is disconnected. When the switch is off, due to the

existence of inductance L3, it can be maintained for a short time, but gradually smaller. When the switch is closed, the load current and the capacitance current are provided by the inductive current; when the switch is open, the load current is the sum of the inductive current and the capacitance current. The output voltage is

$$V_o = V_{in} * D \tag{2}$$

Here, D is duty cycle, V_o is the output voltage, and V_{in} is the input voltage.

2.2.3 Charging and discharging module design

The design of the charging and discharging module is to realize the real portability, not to tie the user to the power socket, but also to avoid the sudden power failure which makes the phototherapy instrument unable to continue to work. The module uses CN3763 chip for charge and discharge management, adopts PWM step-down mode, and has constant voltage charging function. As shown in Figure 5 (a), when the external power supply is connected, the external voltage Vpp is high level, and the VPP is connected with the mode voltage VPD. At this time, the battery voltage bat and the mode voltage VPD are cut off, and the power supply charges the battery. As shown in Figure 5 (b), when the on key is pressed, the power supply stops supplying power to the battery, the mode voltage VPD is connected with the working voltage Vpow, and the phototherapy instrument starts to work. When there is no external power supply, VPP is at low level, BAT and VPD are on, when pressing the switch on light, VPD and Vpow are on, and phototherapy instrument starts to work.

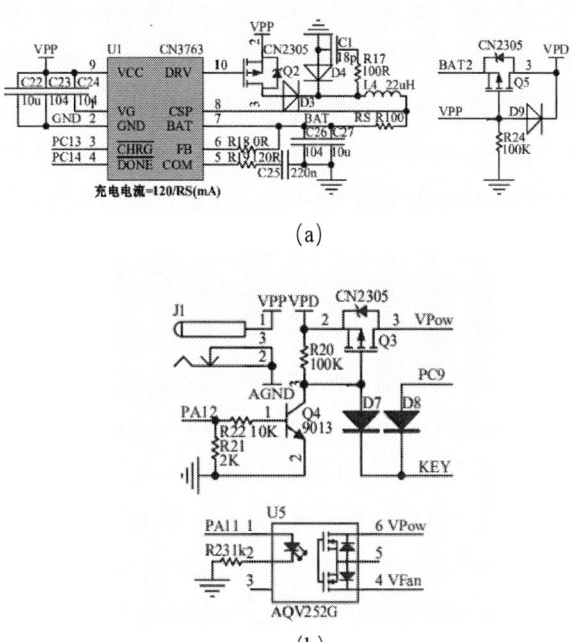

(a)

(b)

Fig. 5. (a) Charge and discharge management circuit; (b) one key switch circuit

3. Experimental test

In order to verify the proposed design method, based on the above optical scheme and control circuit, we designed a multi-wavelength LED phototherapy system prototype. In

978-1-7281-5757-3/19 $31.00 © 2019 IEEE   177

order to make the facula beautiful, the facula is shaped. The mixing rod and lens are made of quartz glass. The LED module is installed on a cooling module composed of an aluminum radiator and a fan. The total current of the four channels is 2.5A and the voltage is 10V. The size of the whole device is 95mm×84mm×144mm. The prototype picture is shown in Figure 6.

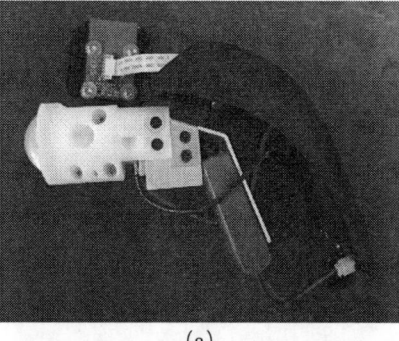

(a)

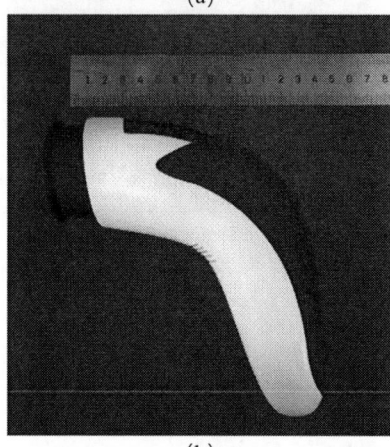

(b)

(c)

Fig. 6. Multi-wavelength LED phototherapy system prototype (a) internal structure of prototype (b) overall dimension of prototype (c) irradiation spot of prototype.

1) Optical irradiance and uniformity

In the experiment, we use thorlabs PM100D power meter and S120VC probe to measure the irradiance at 5 points on the target area 5cm away from the light source, as shown in Figure 7. The measured values are shown in Table 1.

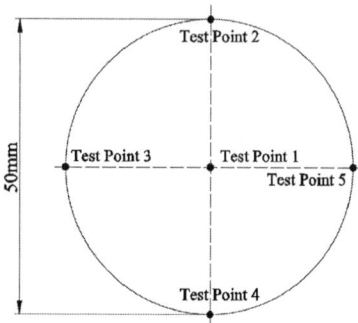

Fig. 7. Test points where the irradiance are measured.

Table 1. Measured Irradiance (in mW/cm²) at the Five Test Points Shown in Fig. 8, i.e., P1–P5

	P1	P2	P3	P4	P5	E	h
1	115.2	102.2	103.1	102.7	103.4	105.3	88.7%
2	115.4	102.9	102.3	102.7	101.8	105.0	88.2%
3	114.8	101.7	102.6	103.4	102.6	105.0	88.6%

It can be seen from table 1 that the irradiance of the whole facula range is above 100mW/cm², and the uniformity of irradiance is over 88%.

2) Stability test

In the stability test, we measure the change of the irradiance of the prototype every half an hour. In this test, we only measure the irradiance at P1. The measured values are shown in Table 2.

Table 2. Measured Irradiance (in mW/cm²) at the Test Point 1 Every Half Hour

	0	0.5h	1h	1.5h	2h
P1	115.2	109.7	107.5	107.5	107.5

From table 2, it can be seen that the irradiance of the phototherapy system is stable at 107mW/cm² after continuous operation of the maximum illumination for 2 hours. The attenuation of the irradiance is caused by the heating of the LED chip. It can be seen that the irradiance has been stable within 1 hour of continuous operation.

3) Safety test

In the safety test, measure the temperature rise of the phototherapy system for 2 hours. NTC thermistor NCP18XF-101E03RB is used to measure the resistance change of thermistor. The indoor temperature is 25℃, and the measured value is shown in Table 3.

It can be seen from table 3 that within one hour of continuous operation of the phototherapy system, the temperature rise of the system has been stable, with a temperature rise of 19 ℃. The cooling effect is good.

Table 3. Measured the Thermal Resistance of the LED Module Every Half Hour

	0.5h	1h	1.5h	2h

R (KΩ)	4.826	4.542	4.542	4.542
T(℃)	42.5	44	44	44

4. Conclusion

In this work, we design a multi-wavelength LED phototherapy system with small size and high irradiance, which highly integrate 630, 660, 810 and 940nm LED chips into one module. The optical system is composed of an optical mixing rod and a pair of aspheric lenses to achieve uniform distribution of light irradiance on the output surface. The circuit design of phototherapy system is proposed, and the power supply function of lithium battery is achieved. An experimental prototype is established. The irradiance and its uniformity, stability and safety of the phototherapy system are tested. The experimental results show that the maximum irradiance of the system was 115.4mW/cm^2, the uniformity reached 88%, After the phototherapy system works for 2 hours with maximum illumination, the irradiance deviation is 6.7% and the temperature rise of the system is 19℃. In conclusion, the performance of the system meets the application requirements, and the volume of the phototherapy system is greatly reduced, providing a promising research direction for the household phototherapy.

References

1. Micheli, Laura , et al. "Photobiomodulation therapy by NIR laser in persistent pain: an analytical study in the rat." *Lasers in Medical Science* 32.Supp1(2017):1-12.

2. Min, Pok Kee, and B. L. Goo. "830 nm light-emitting diode low level light therapy (LED-LLLT) enhances wound healing: a preliminary study." *Laser Therapy* 22.1(2013):43-9.

3. Kymplova, Jaroslava , L. Navratil , and J. Knizek . "Contribution of Phototherapy to the Treatment of Episiotomies." *Journal of Clinical Laser Medicine Surgery* 21.1(2003):35-39.

4. Whelan, Harry T., et al. DARPA soldier self care: rapid healing of laser eye injuries with light emitting diode technology. MEDICAL COLL OF WISCONSIN MILWAUKEE, 2004.

5. Kuffler, and P. Damien . "Photobiomodulation in promoting wound healing: a review." *Regenerative Medicine* (2015) : rme.15.82.

6. Houreld, Nicolette , and H. Abrahamse . "Low-Intensity Laser Irradiation Stimulates Wound Healing in Diabetic Wounded Fibroblast Cells (WS1)." *Diabetes Technology & Therapeutics* 12.12(2010):971-978 .

7. Schubert, Vivianne . "Effects of phototherapy on pressure ulcer healing in elderly patients after a falling trauma." Photodermatology, *Photoimmunology & Photomedicine* 17.1(2001).

8. Krynicka, I , et al. "The role of laser biostimulation in early post-surgery rehabilitation and its effect on wound healing. " Ortopedia Traumatologia Rehabilitacja 12.12(2010):67-79.

9. Moreno, Ivan , Maximino Avendaño-Alejo, and R. I. Tzonchev . "Designing light-emitting diode arrays for uniform near-field irradiance." *Applied Optics* 45.10(2006):2265-2272.

10. Werdecker, W. , and F. Aldinger . "Aluminum Nitride-An Alternative Ceramic Substrate for High Power Applications in Microcircuits." *IEEE Transactions on Components, Hybrids, and Manufacturing, Technology* 7.4(1984):399-404.

11. W.J.Cassarly, Nonimaging optics: concentration and illumination, in OSA Handbook of Optics, 3rd edition, Vol. 2, McGraw-Hill, New York (2010), 39.1-39.51.

Multi-chip dynamic white light emitting diode with high level photobiological safety and good color fidelity

Jingxin Nie[1], Zhizhong Chen[1*], Fei Jiao[1,2], Chengcheng Li[1], Yifan Chen[1], Jinglin Zhan[1], Yiyong Chen[1], Tongjun Yu[1], Xiangning Kang[1], Yongzhi Wang[3], Shunfeng Li[3], Guoyi Zhang[1,3] and Bo Shen[1]

1 State Key Laboratory for Artificial Microstructure and Mesoscopic Physics, School of Physics, Peking University, Beijing 100871, China. E-mail: zzchen@pku.edu.cn

2 State Key Laboratory of Nuclear Physics and Technology, School of Physics, Peking University, Beijing 100871, China

3 Dongguan Institute of Optoelectronics, Peking University, Dongguan, 523808, Guangdong, China

Abstract

Blue light hazard has been the main concern of photobiological safety of light emitting diode (LED) and blue light hazard efficiency of radiation (BLHER) is used to evaluate the extent of blue light hazard. In this work, we used four monochromatic LEDs with the color of purple, azure, green, and red to mix light without blue LED, in order to reduce the BLHER of hybrid white light. Also, the mixed white light kept high level of color fidelity when the correlated color temperature (CCT) ranges from 2700 to 7000 K. Color rendering index (CRI), color fidelity score (Rf), and color gamut score (Rg) are used to evaluate the color fidelity. We simulated the spectrum and parameters of hybrid white light based on different peak wavelengths, full width at half maxima (FWHMs), and relative ratios of four LEDs. Finally, we fabricated a four-channel LED to obtain tunable white light with high level of photobiological safety and good color fidelity, which BLHER is 0.185 at 7110 K, CRI, Rf and Rg are larger than 85.0, 78.7 and 87.4, respectively.

1. Introduction

White LEDs are usually consisted of blue LEDs and yttrium aluminum garnet (YAG) phosphors, or red-green-blue multi-chip LEDs [1,2]. As a result, blue light is one of the main components in white LEDs. However, blue light has been confirmed to damage the retina, disrupt circadian rhythm, and even cause breast cancer [3–5], which is called blue light hazard. The extent of blue light hazard of a lighting source is evaluated by blue light hazard efficiency of radiation (BLHER), proposed by International Electrotechnical Commission (IEC) [6]:

$$\eta_B = \frac{\sum_\lambda S(\lambda)B(\lambda)\Delta\lambda}{\sum_\lambda S(\lambda)\Delta\lambda} \qquad (1)$$

where η_B is the value of BLHER, $S(\lambda)$ is the spectral irradiance of a lighting source, and $B(\lambda)$ is the blue light hazard spectral weighting function, which is shown in Fig.1.

There were many researches focusing on the blue light hazard and trying to decrease η_B. Some groups utilized blue light blocked filter to decrease the components of blue light, but this method caused the deviation of chromaticity coordinates from Planckian locus and the poor performance of color fidelity [7]. Using LEDs or OLEDs to fabricate warm white light with low correlated color temperature (CCT) was another way [8,9]. However, tunable CCT is significant in modern smart lighting, considering the non-visual effect on emotion, work efficiency, and depression [10,11]. Also, some groups proposed multi-chip LEDs to obtain hybrid white light. The peak wavelengths and FWHMs of multi-chip LEDs could

be optimized, considering the tunable CCT, color fidelity, and blue light hazard of hybrid white light [11–13]. However, because blue LED is indispensable in mixing white light, the η_B of white LED is approximately 20 percent higher than that of daylight at the same CCT [14].

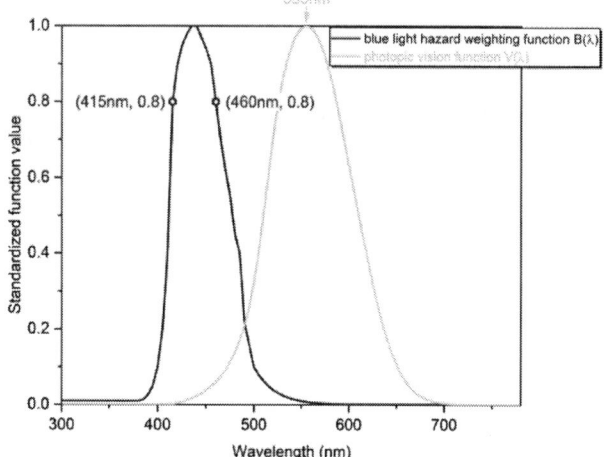

Fig.1 Function value of $B(\lambda)$ and $V(\lambda)$. The blue line represents $B(\lambda)$. The peak value of $B(\lambda)$ appears at 435-440 nm, and $B(\lambda)$ is above 0.8 at the range of 415-460 nm. The green line represents $V(\lambda)$. The peak value appears at 555 nm.

CRI has been used widely to evaluate color fidelity of lighting source [15,16], however, recent studies have found some shortcomings of CRI [17,18]. So Rf and Rg, which were proposed by Illuminating Engineering Society (IES) and adopted in the IES TM-30 [17,19], were also used to evaluate color fidelity in this study.

Luminous efficiency of radiation (LER) can evaluate the efficiency of a lighting source, depending on the SPD of lighting source and photopic vision function [12]:

$$\text{LER} = \frac{K_m \sum_\lambda S(\lambda)V(\lambda)\Delta\lambda}{\sum_\lambda S(\lambda)\Delta\lambda} \qquad (2)$$

where $S(\lambda)$ is the spectral irradiance of a lighting source. $V(\lambda)$ is the photopic vision function, shown in Fig. 1. The peak value of $V(\lambda)$ is at the wavelength of 555 nm. K_m is a constant, equals 683 lm/W, which is the maximum LER of a lighting source.

In this study, in order to avoid the wavelength where blue light hazard is serious ($B(\lambda)$>0.8), we used a purple LED with the peak wavelength less than 415 nm and an azure LED with the peak wavelength more than 460 nm to replace the blue LED. A yellow-green LED and a red LED were also necessary,

978-1-7281-5757-3/19 $31.00 © 2019 IEEE

taking account of the tunability of CCT and color fidelity of hybrid white light. The peak wavelengths and FWHMs of four monochromatic LEDs were optimized and figured out theoretical simulated results with the restricted conditions (CRI>85, Rf>80, Rg>90, LER>150 lm/W, η_B <0.20) when CCT ranged from 2700 to 7000 K. Then we selected four LEDs to mix light in the experiment and figured out the parameters of practical hybrid white light. In the experiment, combined with the tunability of CCT from 2704K to 7110K, CRI was above 85.0, Rf was above 78.7, Rg was above 87.4, and η_B was less than 0.185. It meant that a four-chip white LED with tunable CCT, good color fidelity, and high level of photobiological safety was achieved.

Method and Experiment

According to Grassmann Color Law [20], in order to occupy more color space in CIE 1931 in mixing light, conventional multi-chip LEDs select LED whose chromaticity coordinates locate on the vertex of color space, as shown in Fig.2. Typical RGB LEDs are consisted of blue (450 nm), green (520 nm), and red (630 nm), as shown in Fig.2 [1]. According to $B(\lambda)$ in Fig.1, we avoid the 415-460 nm, where $B(\lambda)$ is above 0.8. Considering the color space in Fig.2 and $B(\lambda)$ in Fig.1, we figured out the range of peak wavelengths of four monochromatic LEDs in simulation. The peak wavelengths are 390-415 nm, 460-480 nm, 510-555 nm, and 600-650 nm, respectively.

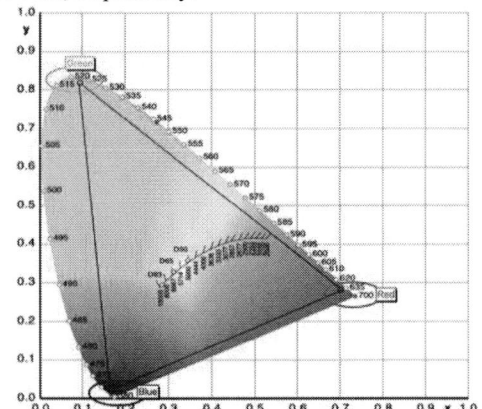

Fig.2 CIE 1931 color space and Grassmann Color Law

The spectral power distribution (SPD) of monochromatic LED can be calculated by a mathematical model, using peak wavelength and FWHM [21]:

$$\begin{cases} S(\lambda, \lambda_P, \Delta\lambda_{0.5}) = \dfrac{g(\lambda, \lambda_P, \Delta\lambda_{0.5}) + 2g^5(\lambda, \lambda_P, \Delta\lambda_{0.5})}{3} \\ g(\lambda, \lambda_P, \Delta\lambda_{0.5}) = e^{-\left(\frac{\lambda-\lambda_P}{\Delta\lambda_{0.5}}\right)^2} \end{cases} \quad (3)$$

where $S(\lambda, \lambda_P, \Delta\lambda_{0.5})$ is the SPD of monochromatic LED, λ_P is the peak wavelength, and $\Delta\lambda_{0.5}$ is the FWHM. The unit of wavelength is the nanometer. The relationship between peak wavelengths and FWHMs of commercial monochromatic LEDs is shown in Fig. 3.

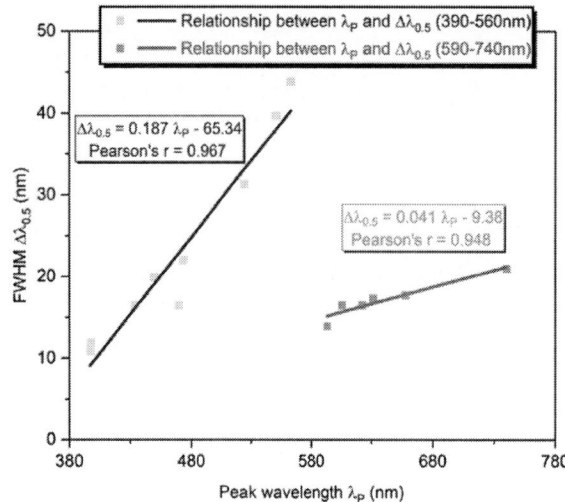

Fig.3 The relationship between λ_P and $\Delta\lambda_{0.5}$ of monochromatic LEDs. The cyan and orange points represent the peak wavelengths and FWHMs of some monochromatic LEDs in commercial markets, and cyan points represent the LEDs whose peak wavelengths are in the range of 390-560 nm, while orange points represent 590-740 nm. Blue and red lines are the linear fitting curves of cyan and orange points. The equations of linear fitting curves and their correlation coefficients are showed in the figure.

According to the linear fitting equations in Fig.3, $\Delta\lambda_{0.5}$ depends on λ_P in the simulation:

$$\Delta\lambda_{0.5} = \begin{cases} 0.187\lambda_P - 65.34, & 390 \, nm \leq \lambda_P \leq 560 \, nm \\ 0.041\lambda_P - 9.38, & 590 \, nm \leq \lambda_P \leq 740 \, nm \end{cases} \quad (4)$$

The method of calculating relative ratios of four LEDs in mixing light is as follows:

I. The SPDs of four monochromatic LEDs are supposed as $S_1(\lambda), S_2(\lambda), S_3(\lambda),$ and $S_4(\lambda)$. The relative ratios of four LEDs are $r_1, r_2, r_3,$ and r_4. Then the SPD of the mixed white light is:

$$S_W(\lambda) = r_1 \cdot S_1(\lambda) + r_2 \cdot S_2(\lambda) + r_3 \cdot S_3(\lambda) + r_4 \cdot S_4(\lambda) \quad (5)$$

II. The chromaticity coordinates (x_W, y_W) of the mixed white light are calculated as:

$$\begin{cases} x_W = \dfrac{\sum_{380}^{780} S_W(\lambda)\bar{x}(\lambda)\Delta\lambda}{\sum_{380}^{780} S_W(\lambda)\bar{x}(\lambda)\Delta\lambda + \sum_{380}^{780} S_W(\lambda)\bar{y}(\lambda)\Delta\lambda + \sum_{380}^{780} S_W(\lambda)\bar{z}(\lambda)\Delta\lambda} \\ y_W = \dfrac{\sum_{380}^{780} S_W(\lambda)\bar{y}(\lambda)\Delta\lambda}{\sum_{380}^{780} S_W(\lambda)\bar{x}(\lambda)\Delta\lambda + \sum_{380}^{780} S_W(\lambda)\bar{y}(\lambda)\Delta\lambda + \sum_{380}^{780} S_W(\lambda)\bar{z}(\lambda)\Delta\lambda} \end{cases} \quad (6)$$

III. We set T as the target CCT of the mixed white light and we can calculate the SPD of black-body radiation $S_B(\lambda, T)$ [22]. Then the chromaticity coordinates $(x_{B,T}, y_{B,T})$ of black-body radiation are calculated as [23]:

$$\begin{cases} x_{B,T} = \dfrac{X_{B,T}}{X_{B,T} + Y_{B,T} + Z_{B,T}} \\ \quad = \dfrac{\sum_{380}^{780} S_B(\lambda, T)\bar{x}(\lambda)\Delta\lambda}{\sum_{380}^{780} S_B(\lambda, T)\bar{x}(\lambda)\Delta\lambda + \sum_{380}^{780} S_B(\lambda, T)\bar{y}(\lambda)\Delta\lambda + \sum_{380}^{780} S_B(\lambda, T)\bar{z}(\lambda)\Delta\lambda} \\ y_{B,T} = \dfrac{Y_{B,T}}{X_{B,T} + Y_{B,T} + Z_{B,T}} \\ \quad = \dfrac{\sum_{380}^{780} S_B(\lambda, T)\bar{y}(\lambda)\Delta\lambda}{\sum_{380}^{780} S_B(\lambda, T)\bar{x}(\lambda)\Delta\lambda + \sum_{380}^{780} S_B(\lambda, T)\bar{y}(\lambda)\Delta\lambda + \sum_{380}^{780} S_B(\lambda, T)\bar{z}(\lambda)\Delta\lambda} \end{cases} \quad (7)$$

IV. We set the chromaticity coordinates of the mixed white light to be the same as those of black-body radiation:

$$\frac{\sum_{380}^{780} S_W(\lambda)\bar{x}(\lambda)\Delta\lambda}{\sum_{380}^{780} S_W(\lambda)\bar{x}(\lambda)\Delta\lambda + \sum_{380}^{780} S_W(\lambda)\bar{y}(\lambda)\Delta\lambda + \sum_{380}^{780} S_W(\lambda)\bar{z}(\lambda)\Delta\lambda} = x_{B,T}$$
$$\left(\frac{\sum_{380}^{780} S_W(\lambda)\bar{y}(\lambda)\Delta\lambda}{\sum_{380}^{780} S_W(\lambda)\bar{x}(\lambda)\Delta\lambda + \sum_{380}^{780} S_W(\lambda)\bar{y}(\lambda)\Delta\lambda + \sum_{380}^{780} S_W(\lambda)\bar{z}(\lambda)\Delta\lambda} = y_{B,T} \right) \quad (8)$$

That is to say:

$$\begin{cases} \sum_{380}^{780} S_W(\lambda)\bar{x}(\lambda)\Delta\lambda \cdot (x_{B,T}-1) + \sum_{380}^{780} S_W(\lambda)\bar{y}(\lambda)\Delta\lambda \cdot x_{B,T} + \sum_{380}^{780} S_W(\lambda)\bar{z}(\lambda)\Delta\lambda \cdot x_{B,T} = 0 \\ \sum_{380}^{780} S_W(\lambda)\bar{x}(\lambda)\Delta\lambda \cdot y_{B,T} + \sum_{380}^{780} S_W(\lambda)\bar{y}(\lambda)\Delta\lambda \cdot (y_{B,T}-1) + \sum_{380}^{780} S_W(\lambda)\bar{z}(\lambda)\Delta\lambda \cdot z_{B,T} = 0 \end{cases} \quad (9)$$

There are four unknown variables r_1, r_2, r_3, and r_4 in calculating relative ratios of four LEDs. Set one of the four unknown variables as the unit. For example, set $r_4 = 1$ and there are only three independent unknown variables. According to Eq. (5) and Eq. (9), we can figure out two linear equations about r_1, r_2, r_3, and r_4, which can decrease two independent variables. At last, there are only one independent unknown variable. Optimize the only independent variable with the restricted conditions that CRI is above 85, Rf is above 80, Rg is above 90, LER is above 150 lm/W, and minimize the η_B. In this way, we can decrease the complexity of computing when calculate the relative ratios of four LEDs with restricted conditions about color fidelity and photobiological safety.

Results and Discussion

We simulated the parameters of the mixed white light, consisted of four monochromatic LEDs, whose peak wavelengths were in the range of 390-415 nm, 460-480 nm, 510-555 nm, 600-650 nm, respectively. The peak wavelengths of each LED were with 1 nm increments in simulation. We required that when the CCT of the mixed white light ranged from 2700 to 7000 K, CRI was always above 85, Rf was above 80, and Rg was above 90. Also, in order to decrease the blue light hazard, we required η_B of the mixed white light to be less than 0.20 at 7000 K. And we also required LER above 150 lm/W, considering the efficiency of lighting source. Fig.4 shows the combinations of peak wavelengths of four monochromatic LEDs and the parameters of corresponding mixed white light.

The ideal combinations of peak wavelengths of four monochromatic LEDs are at the ranges of 396-401 nm, 464-474 nm, 545-553 nm, and 611-623 nm, respectively. According to Fig.4, the simulated LER and η_B are quite related to the peak wavelength of purple LED. Generally, with the same LER, the shorter the peak wavelength is, the smaller the η_B is. Also, with the same η_B, the shorter the peak wavelength is, the larger the LER is. Because the lighting source is supposed to decrease η_B and increase LER, it is important to select the peak wavelength of purple LED and the ideal peak wavelength is supposed to be small at the range of 396-401 nm.

According to the combinations of peak wavelengths that met the requirements, we selected four monochromatic LEDs in the experiments. Fig.5 shows the SPD of this four LEDs. In the experiment, the peak wavelengths were 396.8, 469.5, 550.4, and 622.6 nm, respectively. The FWHMs were 11.9, 19.5, 39.7, and 16.5 nm, respectively. Then we calculated the ratios of four LEDs, using Eq. (5)-(9), and measured the SPD of the mixed white light from 2700 and 7000 K and figured out the parameters. The experimental results are shown in Fig.6.

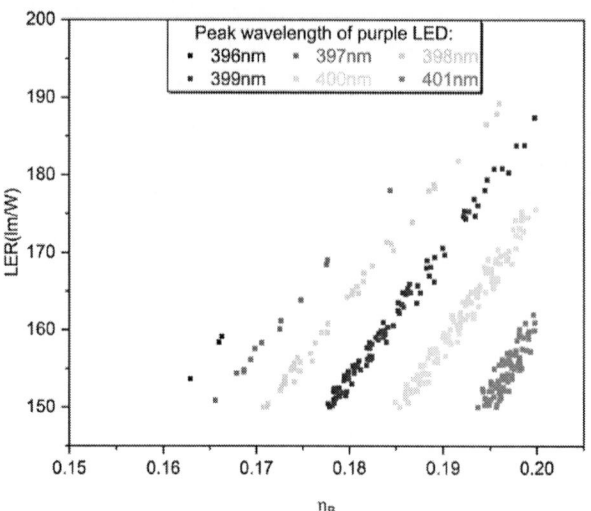

Fig.4 The simulated LER and η_B of mixed white light at 7000 K, consisted of four monochromatic LEDs with different peak wavelengths, and meet the restricted conditions that CRI>85, Rf>80, Rg>90, LER>150 lm/W, η_B <0.20. Each point represents a combination of four peak wavelengths, and the position of each point in the figure depends on the optimal LER and η_B of mixing these four LEDs. Different colors represent different peak wavelengths of purple LEDs in mixed white light.

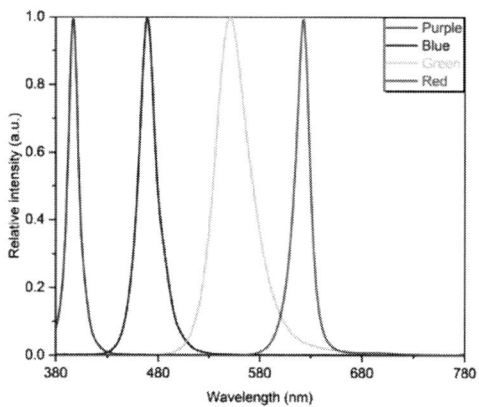

Fig.5 The standardized SPDs of purple, blue, green, and red monochromatic LEDs in the experiment.

Fig. 6a shows the SPD of the mixed white light in the experiment. When the CCT changes from 2704 to 7110 K, there is more radiation power of short wavelength in the SPD. The intensity is at a very low level at about 430 nm, which is in the range where blue light hazard is serious. Fig. 6b shows the color fidelity and blue light hazard of the mixed white light. With the restricted conditions in simulation that CRI was above 85, Rf was above 80, and Rg was above 90, while in the experiment, CRI was above 85.0, Rf was above 78.7, and Rg was above 87.4, indicating good color fidelity of the mixed white light. The experimental results were a little worse than the simulated results, may because the peak wavelengths and FWHMs of monochromatic LEDs in the experiment were different from those of ideal simulated LEDs. The comparison

978-1-7281-5757-3/19 $31.00 © 2019 IEEE

of η_B between the mixed white light, black-body radiation, and typical daylight is shown in Fig. 6b [22]. Generally, high CCT means more radiation power at short wavelength, which may lead to more serious blue light hazard [12,13]. So, we focused on η_B at 7000 K. At this CCT, the η_B of typical daylight is 0.215, while η_B of the mixed white light in this work is 0.185, which is about 14% less than typical daylight. In contrast, η_B of conventional white LEDs is about 20 percent larger than the typical daylight at the same CCT [14]. The experimental results show high level of photobiological safety of the mixed white light.

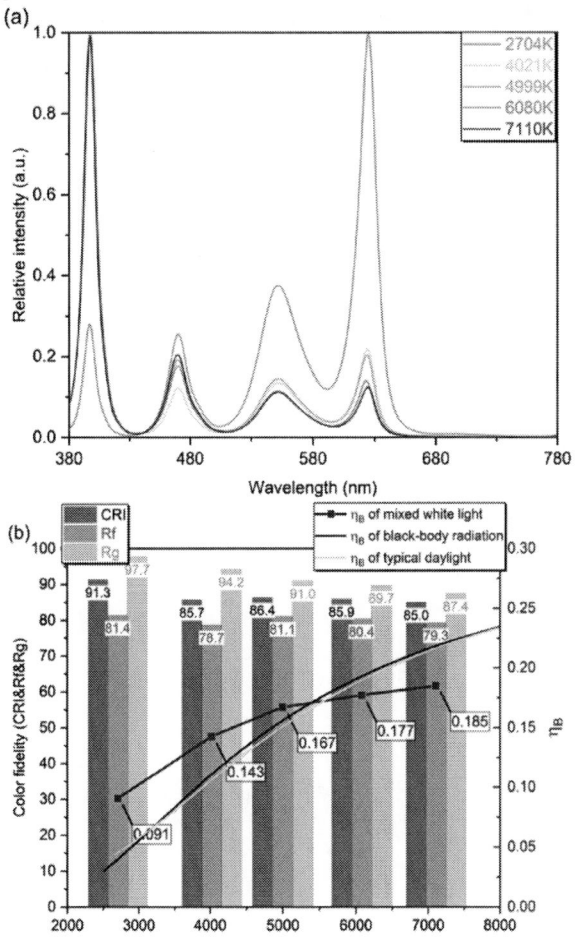

Fig.6 (a) The SPD of the mixed white light at 2704, 4021, 4999, 6080, and 7110 K. (b) CRI, Rf, Rg, η_B of the mixed white light from 2704 to 7110 K in the experiment. The gray, red, and orange bars represent CRI, Rf, and Rg of the mixed white light. The color fidelity results correspond to the left axis. The blue, black, and cyan lines represent η_B of mixed white light, black-body radiation, and daylight at different CCT. The η_B results correspond to the right axis.

In this work, we simulated the parameters of the mixed white light, consisted of four monochromatic LEDs with different wavelengths and FWHMs, using mathematical model of monochromatic LED in Eq. (3) and Eq. (4). Based on the combinations of wavelengths in simulated results, we selected four LEDs and fabricated the mixed white light with tunable

CCT from 2700 to 7000 K, considering the significance of color fidelity and blue light hazard. In the further study, we will optimize both peak wavelengths and FWHMs of monochromatic LEDs. Some specific FWHMs for a certain wavelength LED would be developed. On the other hand, the biological experiments are required to verify the effect of this lighting source on decreasing blue light hazard.

Conclusions

In summary, we proposed a new method to reduce the blue light hazard sharply, combined with good color fidelity and tunable CCT. In the simulation, we select four monochromatic LEDs whose peak wavelengths are located in 390-415 nm, 460-480 nm, 510-555 nm, 600-650 nm, respectively. The mixed white light in simulation is restricted with the visual and non-visual conditions of CRI>85, Rf>80, Rg>90, LER>150 lm/W, and η_B<0.20 when the CCT changes from 2700 to 7000 K. In the experiment, we fabricated four-chip LEDs to obtain hybrid white light with good color fidelity with CRI>85.0, Rf>78.7, Rg>87.4 and high level of photobiological safety with η_B of 0.185 at 7110 K.

Acknowledgments

This work was supported by the National Key Research and Development Program under Grant No. 2017YFB0403104, National Natural Science Foundation of China under Grant No. 61674005, Beijing Municipal Science & Technology Commission under Grant No. Z161100001616010, Science and Technology Major Project of Guangdong Province under Grant No. 2016B010111001, and Science and Technology Planning Project of Henan Province under Grant No. 161100210200.

References

1. S. K. Ng, K. H. Loo, Y. M. Lai, and C. K. Tse, "Color control system for RGB LED with application to light sources suffering from prolonged aging," IEEE Trans. Ind. Electron. **61**(4), 1788–1798 (2014).

2. P. Pust, V. Weiler, C. Hecht, A. Tücks, A. S. Wochnik, A. K. Henß, D. Wiechert, C. Scheu, P. J. Schmidt, and W. Schnick, "Narrow-band red-emitting Sr[LiAl3 N4]:Eu2+ as a next-generation LED-phosphor material," Nat. Mater. **13**(9), 891–896 (2014).

3. M. M. Benedetto and M. A. Contin, "Oxidative Stress in Retinal Degeneration Promoted by Constant LED Light," Front. Cell. Neurosci. **13**(April), 1–11 (2019).

4. R. G. Stevens, G. C. Brainard, D. E. Blask, S. W. Lockley, and M. E. Motta, "Breast cancer and circadian disruption from electric lighting in the modern world," CA. Cancer J. Clin. **64**(3), 207–218 (2014).

5. A. J. K. Phillips, P. Vidafar, A. C. Burns, E. M. McGlashan, C. Anderson, S. M. W. Rajaratnam, S. W. Lockley, and S. W. Cain, "High sensitivity and interindividual variability in the response of the human circadian system to evening light," Proc. Natl. Acad. Sci. **116**(24), 201901824 (2019).

6. International Electrotechnical Commission, *IEC 62471: 2006 Photobiological Safety of Lamps and Lamp Systems* (2006).

7. H. Chen, R. Zhu, Y.-H. Lee, and S.-T. Wu, "Correlated color temperature tunable white LED with a dynamic color filter," Opt. Express **24**(6), A731 (2016).

8. J. H. Jou, H. H. Yu, F. C. Tung, C. H. Chiang, Z. K. He, and M. K. Wei, "A replacement for incandescent bulbs: high-efficiency blue-hazard free organic light-emitting diodes," J. Mater. Chem. C **5**(1), 176–182 (2017).

9. H. Wang, C. Zang, G. Shan, Z. Yu, S. Liu, L. Zhang, W. Xie, and H. Zhao, "Bluish-Green Thermally Activated Delayed Fluorescence Material for Blue-Hazard Free Hybrid White Organic Light-Emitting Device with High Color Quality and Low Efficiency Roll-Off," Adv. Opt. Mater. **7**(9), 1–9 (2019).

10. E. F. Schubert and J. K. Kim, "Solid-State Light Sources Getting Smart," Science (80-.). **308**(May), 1274–1279 (2005).

11. Y. Gao, H. Wu, J. Dong, and G. Q. Zhang, "Constrained optimization of multi-color LED light sources for color temperature control," in *2015 12th China International Forum on Solid State Lighting (SSLCHINA)* (IEEE, 2015), pp. 102–105.

12. J. Zhang, W. Guo, B. Xie, X. Yu, X. Luo, T. Zhang, Z. Yu, H. Wang, and X. Jin, "Blue light hazard optimization for white light-emitting diode sources with high luminous efficacy of radiation and high color rendering index," Opt. Laser Technol. **94**, 193–198 (2017).

13. J. Zhang, B. Xie, X. Yu, X. Luo, T. Zhang, S. Liu, Z. Yu, L. Liu, and X. Jin, "Blue light hazard performance comparison of phosphor-converted LED sources with red quantum dots and red phosphor," J. Appl. Phys. **122**(4), 043103 (2017).

14. F. Behar-Cohen, C. Martinsons, F. Viénot, G. Zissis, A. Barlier-Salsi, J. P. Cesarini, O. Enouf, M. Garcia, S. Picaud, and D. Attia, "Light-emitting diodes (LED) for domestic lighting: Any risks for the eye?," Prog. Retin. Eye Res. **30**(4), 239–257 (2011).

15. J. H. Oh, S. J. Yang, and Y. R. Do, "Healthy, natural, efficient and tunable lighting: Four-package white LEDs for optimizing the circadian effect, color quality and vision performance," Light Sci. Appl. **3**(2), e141 (2014).

16. Z. He, C. Zhang, H. Chen, Y. Dong, and S.-T. Wu, "Perovskite Downconverters for Efficient, Excellent Color-Rendering, and Circadian Solid-State Lighting," Nanomaterials **9**(2), 176 (2019).

17. A. David, P. T. Fini, K. W. Houser, Y. Ohno, M. P. Royer, K. A. G. Smet, M. Wei, and L. Whitehead, "Development of the IES method for evaluating the color rendition of light sources," Opt. Express **23**(12), 15888 (2015).

18. K. W. Houser, M. Wei, A. David, M. R. Krames, and X. S. Shen, "Review of measures for light-source color rendition and considerations for a two-measure system for characterizing color rendition," Opt. Express **21**(8), 10393 (2013).

19. S. Jost, C. Cauwerts, and P. Avouac, "CIE 2017 color fidelity index Rf: a better index to predict perceived color difference?," J. Opt. Soc. Am. A **35**(4), B202 (2018).

20. D. H. Krantz, "Color measurement and color theory: I. Representation theorem for Grassmann structures," J. Math. Psychol. **12**(3), 283–303 (1975).

21. Y. Ohno, "Spectral design considerations for white LED color rendering," Opt. Eng. **44**(11), 111302 (2005).

22. T. Matsumoto and M. Tomita, "Modified blackbody radiation spectrum of a selective emitter with application to incandescent light source design," Opt. Express **18**(S2), A192 (2010).

23. W. Davis, "Color quality scale," Opt. Eng. **49**(3), 033602 (2010).

The Influence of Lighting on Human Circadian Rhythms

Hung-Wei Chen[1], Chien-Yu Chen[1], Pei-Jung Wu[2]

1. Graduate Institute of Color & Illumination Technology,
National Taiwan University of Science & Technology,
No. 43, Section 4, Keelung Road, Taipei 10607, Taiwan

2. College of Information and Distribution Science,
National Taichung University of Science & Technology,
No.129, Sec. 3, Sanmin Rd,North Dist.,Taichung City 404, Taiwan

Author e-mail address: chencyue@mail.ntust.edu.tw

Abstract

The mainly of the physiological measurement mode is measuring EEG and HRV, as an indicator to explore the impact on subject's non-subjective of the lighting, and using psychological questionnaires as subjective evaluation of the subject's physiological effects.

The main purpose of this research is to look for a tunable intelligent light suitable for sleep. Subjects were a total of 30 (14 male and 16 female), mean age 23.83±0.86 years old. All subjects were required to subject to no light sleep, Good-night light sleep, and low-color-temperature light sleep. Their EEG and ECG are detected during the whole processing. Then through the analysis of HRV and EEG, observed the subject's quality of sleep. Experimental results of Good-night light is the subject's sleep for the δ wave impact of the significant rise. HRV part, nHFP strength portion are higher than no light sleep and low-color-temperature lighting sleep. This proved Good-night light really help to improve the quality of sleep.

1.Introduction

Following the fierce competition in the society and the accelerating pace of life, the economy and industry have approached the state of long working hours. Nevertheless, long working hours could result in inadequate break time for employees. Besides, working time of most people is divided into day and night, where people who work at night show worse efficiency than those working during the day [1]. In this case, a short break during the long working hours could effectively enhance night-shift employees' concentration and work efficiency.

Moreover, chaotic circadian rhythm caused by jet lag and shifts could result in insomnia or health problems. Most people would treat such problems with drugs; however, there are various side effects in drug treatment so that many alternatives have emerged. Illumination is one of such alternatives. Light is essential in human life, and lighting influences human health and life all the time. Lighting gets into human bodies through eyes that it does not simply affect eyes, but also physiology and psychology. Influencing the circadian rhythm of a human body through illumination could improve the sleep quality and health [2].

2.The effect of light on the human body

Light therapy is a non-pharmacological treatment that uses bright light for physiological adjustment of the human body. Actinotherapy has a wide range of effect, the most common is the use of actinotherapy to adjust the time of melatonin

secretion in the human body, and then adjust the circadian clock. A typical actinotherapy experiment is to ask subject to sit in front of the bright light source in the morning for about an hour. Use bright light to suppress the secretion of melatonin and adjust the secretion time of melatonin to achieve the purpose of adjusting circadian clock. Daylight is a healthy light source, but people who stay indoors for a long time cannot receive enough sunlight. A healthy light source must be changed according to the needs of the human body, to form a dynamic lighting source for a full day. Use indoor lighting to regulate the non-visual biological response of the human body and regulate the melatonin secretion. [3]

2.1 Light therapy research

K. Richter et al. [4] used a 10,000 lux light unit for actinotherapy experiments, and joint activities followed after actinotherapy. According to the results of the experiment, the subjects who received actinotherapy had a significant improvement in sleep quality. Brainard et al. [5] measured the amount of melatonin secreted in the blood of the subject by measuring the light of different wavelengths. Blue light with a wavelength of 446-477 nm was found to be the primary source of suppression of melatonin secretion. Figueiro [6] also uses actinotherapy with a wavelength of 470 nm to treat the old adults to reduce sleep disorders and depression. However, in the process of actinotherapy using only blue light, the light source receiving through the retina can also cause visual side effects such as glare, visual fatigue, and eye discomfort.

2.1 The effect of light on sleep

Canazei et al.[7] proposed a dynamic lighting scheme for physiological experiments on female employees who need to work shifts. Because the sleep quality, mood, and stress of shift workers are worse than those of normal life. In order to adjust the staff's circadian rhythm, experiments were carried out using dynamic changes in the working environment light source. The results of the experiment showed that the staff who used the dynamic light source at work had a significant improvement in mood and sleep quality compared to the staff using general lighting. Currently, dynamic lighting sources are most commonly used in classrooms and offices but less frequently used for night sleep. Since dynamic light has an impact on physiology, this study explores whether it is equally effective in improving sleep.

978-1-7281-5757-3/19 $31.00 © 2019 IEEE

3.Methods

An adjustable LED system, with changing colour temperature and illuminance with time, is used as the sleep light source in this experiment. The participants are requested to fill in the questionnaire before and after the experiment, and the physiological signals of brainwave and electrocardiogram are fully recorded during the sleep experiment.

3.1 Experimental lighting

This dynamic lighting system is used for relaxing and falling asleep. The experiment is divided into three stages, namely, three stages of no-light sleep, sleep-sleeping sleep and low-color temperature illumination, and experiments are carried out on different days. The first experimental light source is the intelligent illumination source produced by Liverage biomedical inc. In order to fix the brightness and the illumination frequency, there are three kinds of light sources, the conversion frequency is 30 seconds for one color, and the transformation is performed by using a fade-in and fade-out manner, the conversion frequency is as shown, and the first light source is yellowish. The illuminance is 22.5 Lux and the brightness is 1530 Lum. The second source is reddish, the illumination is 5.3 Lux and the brightness is 394.5 Lum. The third source is purple, and the illumination is 6.8 Lux and the brightness is 361.0 Lum. The second experimental light source is an LED OA panel lamp produced by Donglin Co., Ltd., which is an LED panel lamp with color temperature and illumination.

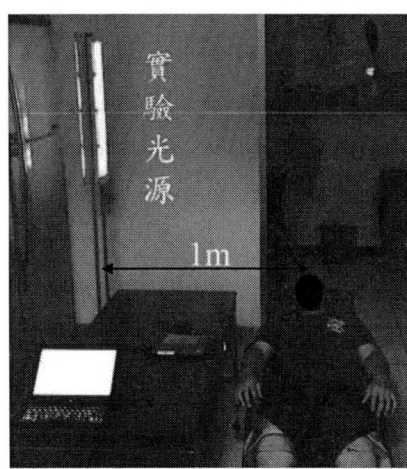

Figure 3.1 Experimental environment

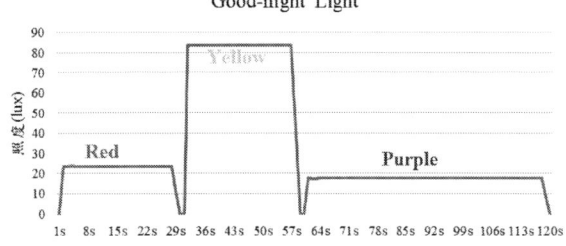

Figure 3.2 Light conversion frequency

Table 3.1 Good-night parameters

	Light source 1 (yellow)	Light source 1 (red)	Light source 1 (purple)
Illuminance (1m)	22.5 lux	5.3 lux	6.8 lux
CCT	1881K	0K	0K
Spectrum			

Table 3.2 LED panel light information

Illuminance	9.1 lux
CCT	2767K
Spectrum	

3.2 Experimental procedure

Total 30people, including 14 males and 16 females, aged 23.83±0.86 years old, participate in this study. Each participant is informed, before the experiment, to avoid drinks with caffeine and tea and have 8 full-hours of sleep; ones with heart and heart rhythm diseases are excluded. The experiment is preceded from 7-11pm with constant indoor temperature 26±1°C and the illumination 1m away from a participant's head. Wire wearing and explanation are proceeded and The Karolinska Sleepiness Scale [8] is filled in before the experiment. The participants are fully measured the EEG and ECG during the lighting sleep for 1hr and then fill in The Karolinska Sleepiness Scale and the psychological evaluation questionnaire again after the experiment. The physiological and psychological analyses are eventually compared and discussed.

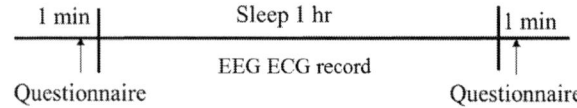

Figure 3.1 Experimental process

3.2.1 Physiological detection

Elecgtroencephalography (EEG) and heart rate variability (HRV) are measured as the physiological measurement in this experiment. the electrode points O1, O2, T7, T8, C3, and C4 of the EEG are the major measuring points, where O1 and O2 of occipital lobe are the visual zone, the major observation points in the light and sleep experiment, T7 and T8 of temporal cranial zone are the emotional zone, where the slow-wave intensity is more obvious [9], and C3 and C4 of parietal lobe are the central reference points [10]. The bipolar induction is utilized for the measurement.

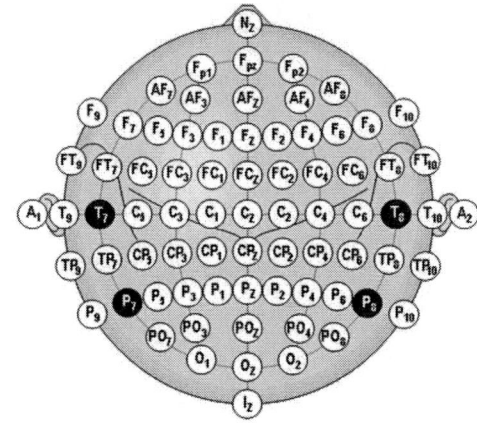

Figure3.3 Electrode locations of International 10-20 system for EEG recording [11]

3.3 Statistical Method

In this study, statistical analysis of brain wave data was performed using SigmaStat 3.5 (Systat Software, Inc, San Jose, CA). Statistical significance was evaluated using a paired-t test ($p < 0.05$). Data from the two groups were compared to determine if there were statistical differences after the subjects slept in the two lighting conditions for 1 hour.

4. Result

This study used two different physiological indicators of brain wave and heart rate variability to study, and found that the two physiological indicators have the same result, the brain wave part, taking O1 and O2 points as an example, because O1 and O2 are visual zones. The influence of light changes is most obvious. As shown in Figure 5.1 and Figure 2, the vertical axis is the intensity of the δ wave, and the horizontal axis is the time axis. At 25 to 30 minutes, the δ wave brain of the sleep light sleep experiment can be seen. The wave intensity increased significantly, while in the HRV part, nHFP did increase at 20-30 minutes, and L/H decreased. As shown in Fig. 3, both physiological indicators showed an increase in sleep quality.

In the no-light sleep experiment and the low-color temperature sleep experiment, the delta wave intensity of the brain wave part increased slightly, but in the HRV part, the

nHFP and L/H of the lampless sleep experiment and the low color temperature sleep experiment part were not large. Changes, the trend is close to steady, which means that the light source during sleep does not change much for HRV, and if the quality of sleep is greatly improved, nHFP will also rise sharply, making the subject's parasympathetic nerve active and entering a relaxed state.

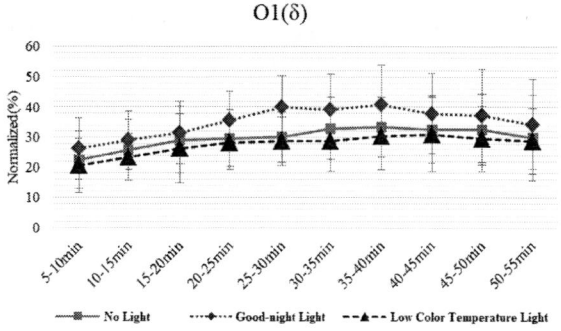

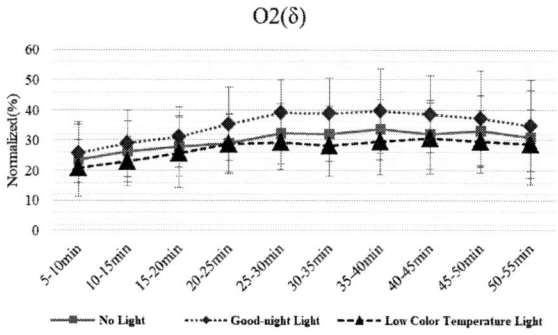

Figure4.1、 O1 wave δ wave brain wave intensity percentage trend graph

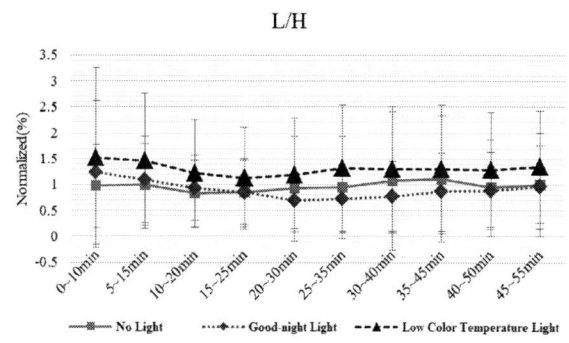

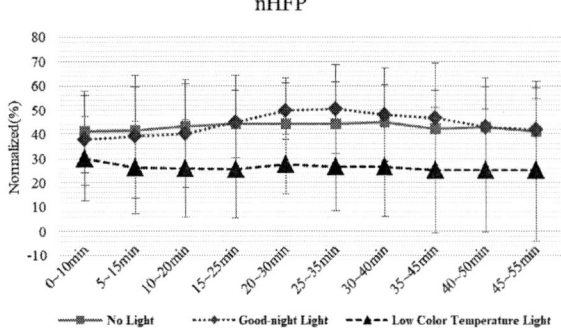

Figure 4.2、HFP intensity and L/H change trend chart

5. DISCUSSION

The brainwave delta analysed in this study stands for the physiological indicator of sound sleep, as delta presents the most direct relationship with sleep [12-13]. The results show that the delta energy intensity 30min before the use of the dynamic light is larger than it without lighting, revealing the higher sleep quality with the dynamic light in the first 30min of sleep. Furthermore, regarding the analysis of sleep with heart rate variability, the sleep stage appears when the sympathetic activity is reduced and the parasympathetic activity is increased [14]. The sympathetic activity under the dynamic light is higher than it without lighting; since most people do not have lights on during the sleep that they might be nervous when sleeping with lighting to result in higher sympathetic activity.

6. CONCLUSION

This study intends to discuss whether the dynamic lighting with changing illuminance could optimize sleep for a short break so that people could fall asleep fast. The research results reveal that the dynamic lighting indeed could enhance falling asleep within 30min, but it presents a better effect without lighting after 30min. In this case, it attempts to receive the best sleep quality by using the dynamic light for the beginning 30min of sleep and then turning the light off. As the number of participants in this experiment is few, there has not been a result with statistical differences. It is expected to increase the number of people to prove the significant effect of lighting on sleep so as to help people with long working hours but inadequate breaks have short breaks to enhance the work efficiency and spirit.

References

[1] TELEKY · L. 1943."Problems of Night Work-Influences on Health and Efficiency." Industrial Medicine and Surgery Vol.12 No.11 pp.758-779; 9 ref.18

[2] Falchi, Fabio, et al. "Limiting the impact of light pollution on human health, environment and stellar visibility." Journal of environmental management 92.10 (2011): 2714-2722.

[3]Duffy, J. F., & Wright Jr, K. P. 2005. Entrainment of the human circadian system by light. Journal of biological rhythms, 20(4), 326-338.

[4] Foster, R. G., & Roenneberg, T. 2008. Human responses to the geophysical daily, annual and lunar cycles. Current biology, 18(17), R784-R794.

[5] Vandewalle, G., Maquet, P., & Dijk, D. J. 2009. Light as a modulator of cognitive brain function. Trends in cognitive sciences, 13(10), 429-438.

[6] Provencio, I., Rodriguez, I. R., Jiang, G., Hayes, W. P., Moreira, E. F., & Rollag, M. D. 2000. A novel human opsin in the inner retina. Journal of Neuroscience, 20(2), 600-605.

[7] Hattar, S., Liao, H. W., Takao, M., Berson, D. M., & Yau, K. W. 2002. Melanopsin-containing retinal ganglion cells: architecture, projections, and intrinsic photosensitivity. Science, 295(5557), 1065-1070.

[8] SHAHID, A. et al. 2012 Karolinska sleepiness scale (KSS). STOP, THAT and One Hundred Other Sleep Scales. Springer New York, 209-210

[9] Van Hese, Peter, et al. "Automatic detection of sleep stages using the EEG."Engineering in Medicine and Biology Society, 2001. Proceedings of the 23rd Annual International Conference of the IEEE. Vol. 2. IEEE, 2001.

[10] Carskadon, Mary A., et al. "Guidelines for the multiple sleep latency test (MSLT): a standard measure of sleepiness." Sleep 9.4 (1986): 519-524.

[11] MALMIVUO, J. PLONSEY, R. 1995 Bioelectromagnetism: principles and applications of bioelectric and biomagnetic fields. Oxford University Press.

[12] MERICA, H. BLOIS, R. GAILLARD, J.-M. 1998 Spectral characteristics of sleep EEG in chronic insomnia. European Journal of Neuroscience, 10.5, 1826-1834.

[13] TAROKH, L.; CARSKADON, M. 2010. Developmental changes in the human sleep EEG during early adolescence. Sleep, 33.6,801.

[14] John Trinder; Jan Kleiman; Melinda Carrington. (2001). "Autonomic activity during human sleep as a function of time and sleep stage." J. Sleep Res. 10, 253-264

The impact of E_{mel} range of dynamic lighting on alertness, fatigue and sleeping quality

Yu Liu[1], Ming Ronnier Luo[1*], Peijung Wu[2], Binyu Teng[1]

[1]State Key Laboratory of Modern Optical Instrumentation, Zhejiang University, Hangzhou, CHINA

[2]College of Information and Distribution Science, National Taichung University of Science and Technology, Taichung Chinese Taipei

*m.r.luo@zju.edu.cn

Abstract

Our earlier works found dynamic lighting to have some positive effect on alertness and circadian response comparing with static light. This study further investigates different CCT ranges and melanopic illuminance (E_{mel}) or circadian stimulus (CS) of dynamic lightings. Sixteen subjects participated the experiment including 4 lightings, one each day from 20:30 to 23:00, and 4 days apart between each lighting. Four different lightings include two CCT ranges (from 6000K to 12000K and from 3000K to 9000K), each having a low and a high E_{mel} level. Six testing methods were included task performance (the D2 test, go/nogo), Critical Flicker Frequency (CFF), ocular fatigue questionnaire, sleepiness (Karolinska sleepiness scale), biochemical responses (melatonin and cortisol). The results showed that subjects were statistically significant more alert under high E_{mel} lightings from the results of task performance, CFF, and both biochemical responses. As for the KSS and ocular-fatigue results, subjects felt sleepier and less fatigue under both low E_{mel} and low CCT range lightings. The strong duration effect was revealed from the results of KSS, ocular fatigue and biochemical responses.

1. Introduction

The investigation of the alertness, sleeping quality relating to well-being caused by different lighting parameters such as correlated colour temperature (CCT), illuminance levels, spectra of particular wavelength is a hot topic in the lighting research. Dynamic lighting research with fast light changes is

Dynamic lighting research applies the continuous lightings changed in a constant rate in a range of CCT or illuminance. Hu et al.[3] carried out the first research to change CCT from 6500 to 12,000K against a static light at 12,000K. They found dynamic light induced more alertness than static light. In our team, Zheng et al.[4] did 4 phases of dynamic lightings, a) CCT range from 6,000 to 12,000K with 2 hours per cycle, b) CCT range from 6,000 to 12,000K with 4 hours per cycle, c) static light at 9000K and d) static light at 6000K. The results showed that dynamic light is more effective than static light, and the faster light is more effective than the slower light. Wang et al.[5] investigated the static light for the young subjects aged from 8 to 10 and found the higher CS lighting will induce high work performance. Ye et al.[6] extended the earlier Zheng's work to look at CCT range and faster and slower lighting. The results showed a wider range of CCT [6000K, 12000K] than that of [4000K, 10000K] to induce more alert and gave better performance but different CCT cycle time had little effect.

The hypothesis in this study is that a higher E_{mel} or CS and higher CCT range of dynamic lighting will lead to a higher work performance (higher alertness). The goal of the present study is to investigate the dynamic lighting effect of CCT range and E_{mel} or CS levels on alertness, fatigue, and circadian responses. These two lighting parameters were not studied together previously. The results can also be used to understand the capability of multi-channels LED system to control the dynamic lighting.

2. Experimental Setup

now also possible due to the advance of spectrum tunable LED lighting system. The lighting parameters in these systems such as illuminance, correlated colour temperature (CCT), Duv, colour rendering metric, metamerism index. Furthermore, the new ipRGC-influenced responses including E_{mel}, E_{rod}, E_s, E_m, E_l proposed by the CIE[1], or Circadian Stimulus (CS) by Rea et al.[2].

2.1 Lighting conditions

The experiment was carried out in an office. Its window was sealed by a heavy curtain. Ten LEDcube® units from Thouslite company were installed in the ceiling. Eleven LED channels in each unit were included to achieve the desired spectral power distribution by using the supplied software. Figure 1 shows the lighting environment for the experiment.

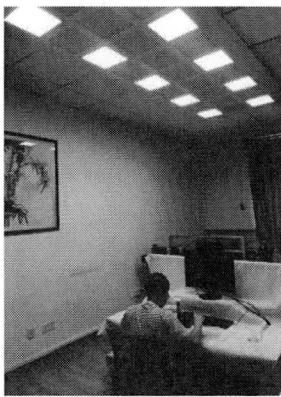

Figure 1. The experimental environment

The lighting conditions in the experiment were carefully controlled. Figure 2 shows the relationship between E_{mel} and CCT for the high and low E_{mel} lightings in the experiment. There are 60 lightings that represent each lighting. Figure 2 showed that E_{mel} had about similar range between the high and low E_{mel}. The corresponding Circadian Stimulus (CS) values were also calculated. In general, the E_{mel} and CS have a strong positive correlation. To avoid visual flickering when lighting changes, 60 lightings were changed in the range of 6000K in a period of 2.5 hours and the Duvvalues between successive lightingswere kept as small as possible and their lightings change smoothly.

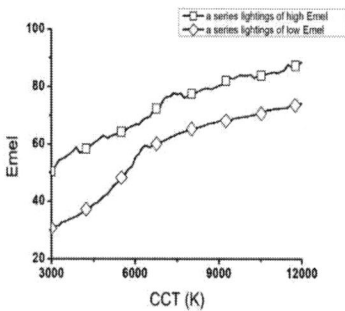

Figure 2. The dynamic light settings between the E_{mel} and CCT

Table 1 provides the lighting conditions for each phase.There were two CCT ranges (from 3000K to 9000K and from 6000K to 12000K) and two CS levels(High and Low). The illuminance at desk were kept 500lux, and a cornea illuminance of 160 lux. AJETI-Specbos 1211 spectroradiometerwas used an it was fixed on the desk for measuring all the lights.

Table 1. The specification of the Lighting conditions used

Lighting Code	CCT range	CCT Mean±δ	E_{mel} Mean±δ	CS Mean±δ
3-9 L	3000-9000K	6024± 1734	51.0± 12.7	0.34± 0.09
3-9 H	3000-9000K	6185± 1733	68.0± 8.7	0.52± 0.04
6-12L	6000-12000K	8964± 1732	70.6± 4.9	0.45± 0.04
6-12H	6000-12000K	9017± 1730	79.2± 5.6	0.55± 0.01

2.2 Experimental methods

Sixteen subjects(8 males and 8 females; 23.1±1.4 years) took part in the study. All subjects were recruited to closely follow to have no habits to intake caffeine or alcohol, and substance abuse or tobacco use.

The study was carried out in a 3-weeks period. Each subject attended the experiments in all 4 lightings. Each lighting started from 20:30 to 23:00. There are 4 days between lightings to avoid any undesired effects caused by the interaction from the previous lighting. All the subjects were instructed to keep a regular sleep-wake pattern during the whole experimental period. They all sleptabout 8 hours. Table 2 shows the experiment procedure. Subjects were first adapted

in the dark for 30 minutes. And in the following one and half hours, the data collection was repeated every half hourand read books between the sampling. Only go/no-go and D2 taskswere performed from 22:30 to 23:00. In addition, the Actiwatch supplied by Spectrum PRO device; Philips Respironics Inc., Murrysville, PA, USA was also used to study the sleeping quality.

A comprehensive set of tests was used to evaluate the 4 lightings in terms of task performance, ocular fatigue, sleeping quality. Six types of tests including biochemical, physiobiological, questionnaire, tasks were used to evaluate the performance of 4 lightings. These tests were 1) task performance (d2 and Go/no go), 2) Karolinska Sleepiness Scale questionnaire(KSS), 3) Critical Flicker Frequency (CFF), 4) Virtual reality symptom questionnaire (VRSQ) (all the test symptoms are listed in Table 3), 5) Biochemical responses (Melatonin and Cortisol), and 6) activity monitor(Actiwatch). In total, 23 scores were collected.

Table 2. The experiment procedure

Time	Activity
20:30-21:00	Relaxation and adaptation
21:00-21:10	CFF measurement, questionnaires, saliva sample
21:10-21:30	Reading
21:30-21:40	CFF measurement, questionnaires, saliva sample
21:40-22:00	Reading
22:00-22:10	CFF measurement, questionnaires, saliva sample
22:10-22:30	Reading
22:30-22:40	CFF measurement, questionnaires, saliva sample
22:40-23:00	Reading

Table 3. The test symptoms included in the VRSQ test

Ocular Symptoms
Tired eyes
Sore/aching eyes
Blurred vision
Difficult to focus

3. Results

IBM SPSS Statistics 22 software® was used to analyse the data. All the data were normalized by subtracting the results from the first one. For example, For the questionnaires and CFF scores, their baseline scores were subtracted which filled in after 30 minutes' dark adaptation.Saliva data were normalized by dividing that of dark baseline data. Main repeated factors "light condition" and "time" of mixed-model analysis were applied in questionnaire and CFF.There were too many results to be reported here. The main findings are summarized below.

Task performance:For both D2 and go/no go performances, high E_{mel} lightings showed more alert effect than low

E_{mel}conditions, i.e. subjects performed faster and less errorunderhighE_{mel} lightings. Comparing the low and high range CCT lightings, the former lightings performed faster and less error under latter lightings.

Sleepiness from KSS score:There was a significant strong effect onduration of time, i.e. a subject' sleepiness increase for the longer duration (see Figure 3(a)). Aclear trendwas also foundthat lowE_{mel}lightings would be sleepier than high E_{mel} lightings, and the 3000K-9000K CCT range showed sleepierthan the 6000-12000K range.

Critical Flicker Frequency (CFF):High E_{mel} lightings showed less fatigue than low E_{mel} lightings. In addition, High CCT range showed less fatigue than low CCT range.

VRSQ questionnaires:For all ocular symptoms tested, there was a clear duration effect, i.e. the longer the time goes on, subjects' eyes felt more intense,e.g more difficult to focus (Figure 3(b)). The results showed that high E_{mel} lightings felt easier to focus (less fatigue) than low E_{mel} lightings. And high CCT range lightings felt lessfatigue than low CCT lightings.The other symptoms showedlittle
significance.

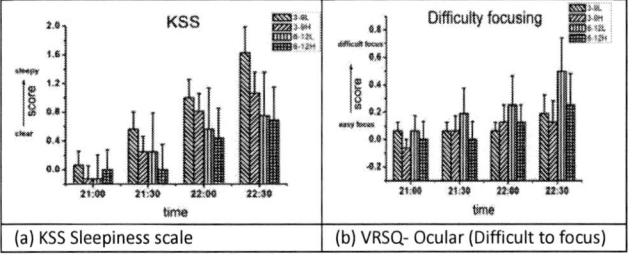

Figure 3. Typical results to illustrate the testing outcome (all of them showed a clear time duration effect.)

Biochemical test: The saliva samples collected were used to obtain Melatonin and Cortisol concentrations. There was a significant duration effect on melatonin secretion, i.e. its concentration at 22:30 was significantly higher than that at 21:00 and that at 21:30.Participants have significantly lower melatonin concentration under 3-9H and 6-12H than 6-12L and higher cortisol consentrationunder 3-9H and 6-12H.It was showed that for both sets of biochemical results agreed well for high E_{mel} lightings to be more alert than low E_{mel}lightings.

Sleeping quality: The results from the Actiwatch were found to bestatistically insignificant between the 4 lightings tested due to insufficient data, i.e. only one subject can be tested a day. Further study is required to use Actiwatch to study sleeping quality.

4.Conclusions

This study explored the impact of dynamic light with different CCT range and E_{mel} (or CS)on subjects'responses in terms of cognitive abilities, sleepiness and ocular fatigue. The results showed significance effect of duration of time on KSS, ocular fatigue and biochemical responses, indicating subjects' sleepiness increased and alertness decreased during experiment. As hypothesized, the results showed subjects reported more alert and ocular fatigue under both highE_{mel} and high CCT range lightings from the earlier findings.The present results showed thata general support for the former but opposite results from the latter. Furthermore, it was found in this study is the lightings with low E_{mel} have less influence on human circadian responses.

Acknowledgement

The authors like to thank the support from the National Key R&D programme of China (Project No. 2017YFB0403700).

References

1. CIE S 026/E:2018 CIE System for Metrology of Optical Radiation for ipRGC-Influenced Responses to Light, Vienna, CIE BHu.
2. Rea and Fiueiro, Bierman (2005). A model of phototransduction by the human circadian system, Brian Research Reviewer, 50(2), 213-228.
3. Hu NC, Feng YC, Lin YH. Effects of alertnessfor static and dynamical lighting at post-noon:CIE Conference, Hangzhou, China, 2012: pp.124–129.
4. S. Q. Zheng, M. R. Luo and M. Ye, (O) The effect of CCT-changing dynamic light on human alertness, Melbourne, CIE 2016 Proceedings, pp300-304.
5. Wang ML, Luo MR. Effect of lighting parameters on work performance using comprehensive methods: International Commission on Illumination Meeting, Melbourne, Australia,3–5 March, 2016: pp. 304–309.
6. Ye M, Zheng SQ, Luo MR. The impact of dynamic light with different CCT ranges and frequencies on human alertness, China International Forum on Solid State Lighting. 2017.
7. Pesudovs, K. (2005). The development of a symptom questionnaire for assessing virtual reality viewing using a head-mounted display, Optometry and vision science 82(7): 571.

Applying LEDs as Therapeutic Light Sources for Anti-microbial Treatment: An Experimental Study

Tianfeng Wang[1,2], Jianfei Dong[1,]*and Guoqi Zhang[2]

[1]Suzhou Institute of Biomedical Engineering and Technology, Chinese Academy of Sciences
No.88,Keling Road, Suzhou, China
[2]Department of Microelectronics, Delft University of Technology, Netherlands
*jfeidong@hotmail.com

Abstract

Microbial infection is one of the most common diseases in the world, which gives rise to morbidity and mortality. At present, the main treatment is using antibiotic drugs. However, the side effects range from fever and nausea to major allergic reactions cannot be ignored. Moreover, the increasing drug resistance also necessitates a new safe alternative therapeutic approach against the microbial infections. As non-antibiotic methods, photodynamic therapies (PDT) and anti-microbial blue light (ABL) therapies have been investigated in this field. However, one challenge of the PDT is the introduction of photosensitizers to the specific pathogens rather than the host cells. In contrast, ABL therapies inactivate the microbes without the involvement of exogenous photosensitizers. The general mechanism of ABL therapies is that ABL can excite the endogenous photosensitizers, and trigger the accumulation of cytotoxic reactive oxygen species (ROS), which in turn leads to cell damage. In this study, we investigated the inhibitive capability of the 405nm LED light source on Candida *albicans* (*C. albicans*). *C. albicans* is the most common pathogen of fungal infections. The intracellular ROS level of *C. albicans* was detected after the ABL irradiation. The results obtained from this study demonstrated that irradiation of 405nm LED significantly inactivated the *C. albicans*. The viability of *C. albicans* was reduced to 1% after 25 minutes of ABL exposure of 50mW/cm^2. On the other hand, the intracellular ROS was increased by the ABL irradiation. This study demonstrates the effectiveness of applying 405nm LED to threat the infection caused by *C. albicans*, as the result of the stimulated accumulation of the intracellular cytotoxic ROS by this light.

1. Introduction

Microbial infection is one of the most common disease in the world, which gives rise to morbidity and mortality. It can be divided in to bacterial infections and fungal infections. *C. albicans* is the most common pathogen of the fungal disease of human, including fungal infections. The *C. albicans* is widely found in nature and commonly occurs as a superficial infection on mucous membranes.

At present, the main treatments of fungal infections including the candidal infection, are the antibiotics drugs, anti-fungal drugs and oral azole agents [1]. However, the side effects such as fever, nausea and major allergic reactions cannot be ignored. Moreover, *C. albicans* is showing an increasing resistance for all these drugs, which leads to the necessity of finding alternative ways to treat fungal infections.

As non-antibiotic methods, photodynamic therapies (PDT) and anti-microbial blue light (ABL) therapies have been investigated in this field. A major advantage of these light-based is the avoidance of the drug resistance. However, one challenge of the PDT is the introduction of photosensitizers to the specific pathogens rather than the host cells. In contrast, ABL therapies inactivate the microbes without the involvement of exogenous photosensitizers. Another advantage of ABL is the safety in the therapeutic process. Compare to germicidal ultraviolet irradiations, it is well accepted that ABL is much less harmful to the host cells.

The commonly accepted mechanism of ABL therapies is that ABL can excite the endogenous photosensitizers, and trigger the accumulation of reactive oxygen species (ROS). ROS is a phrase used to describe a number of reactive molecules and free radicals derived from molecular oxygen including superoxide; hydrogen peroxide; hydroxyl radical; hydroxyl ion; and nitric oxide. These molecules, produced as byproducts during the mitochondrial electron transport of aerobic respiration or by oxidoreductase enzymes and metal catalyzed oxidation, have the potential to cause a number of deleterious events [2].

Both lasers and LEDs have been applied on this topic. Compared with lasers, LEDs, especially in visible light range, are much more affordable and safer than lasers, and can be easily integrated into an array to treat much larger surfaces [3]. They are hence appealing alternatives for lasers in various light therapies. The red and infrared LED therapy can be useful in pain release [4,5]. In [6], the green LED showed beneficial effects in accelerating the healing process of third-degree burns in rats. The LED based blue light has also been widely investigated and proofed to be effective against the microbial infections. For instance, in [7], the antimicrobial effect of 415nm LED on *C. albicans* was studied in vitro and in vivo. The *C. albicans* was compared with human cells which indicates that the inactivation rate of *C. albicans* by 415nm light is 42-fold faster than that of human keratinocytes. The applications of 405nm LED blue light were also widely investigated. An existing study proved that 405nm has an obvious inactivating effect on of *C. albicans* and other microbes [8]. In another study [9], it was reported that the 405nm LED light was able to entirely eliminate two species of candida strains with 30 minutes LED blue light irradiation with an irradiance of 280mW/cm^2. The anti-microbial effect between different blue light wavelengths, 405nm and 470nm, were compared in different studies, and all of them demonstrated that 405 nm was a much more effective wavelength than 470nm [10,11].

Although the inhibiting effect of 405nm ABL has been studied, the cellular ROS level after the ABL irradiation has not yet been explicitly measured. Since the ROS is the mainly accepted cytotoxic trigger of the cells damage, the measurement of the level of ROS is necessary in studying the mechanism of ABL on treating the microbial infections.

Fig.1 The 405nm LED module.

Fig.2 The ABL source and measurement of its irradiance

By summarizing all together, the ABL therapy is a new, safe and effective treatment on microbial infections. The present work is aim to study the effectiveness and mechanism of 405nm LED blue light on the inhibition of *C. albicans*.

2. Design of the LED Anti-microbial Light

The design of the LED light used in the anti-microbial experiments follows the procedures proposed in [3]. In brief, the 405nm light source is made of a 4 by 4 LED array (Kingbright ATDS3534UV405B, with a peak wavelength of 405 nm and a FWHM of 15 nm), arranged in a 1.5 by 1.5 cm square. Fig. 1 shows the design of the LED chip. Fig. 4 depicts the simulated irradiance distribution from the 4-by-4 405nm LED array on a plane 13cm away from it. The spacing

between each two adjacent LEDs is 0.5cm. The half angle of this type of LED is 60°. The average irradiance in the 6cm-diameter central circle, e.g. within a 60mm petri dish, is 49.39mW/cm2. The relative variation of the irradiance in this circle is 6.56%.

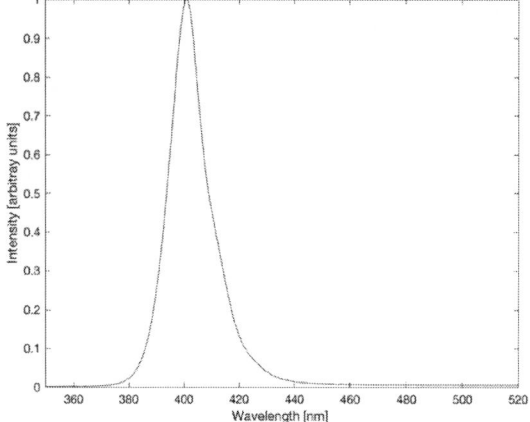

Fig.3 The actual spectra of the 405nm LED light source.

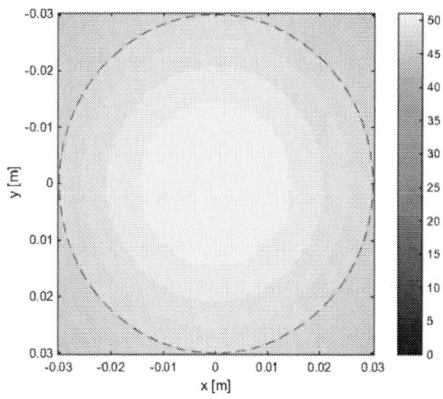

Fig.4 Simulated irradiance (mW/cm^2) distribution from the 405nm LED module onto a target plane 13cm away from it.

The actual spectrum of LED light source was measured using a Maya2000Pro spectrometer (Ocean Optics, US), which is shown in Fig. 3. The actual light power density of the LED was measured using a PM100D power/energy meter with S120VC probe (Thorlabs, US). The 50mW/cm^2 light power density was calibrated by adjusting the height of the LED source to 13cm as shown in Fig. 2. The uniformity of the light was also checked by the power meter. Within a spot with a 35mm diameter, the difference of the power was less than 10%.

3. Anti-microbial experimental results

The viability of fungi was estimated by colony counting in terms of colony-forming units (CFU). After the concentration of fungi reached 10^7 cells/ml, the fungal suspension was diluted by 10^4 fold with sterilized water. Then it was spread on the tryptic soy agar (TSA) and divided it into a control group and a treatment group. The treatment group was irradiated by 405nm blue light for various time, while the control group was kept in the dark. Both the treatment group and the control group were cultured overnight at 26^{0}C before the colony counting. Triplicate experiments were performed at each sampling time. The raw data were processed to produce the mean and standard deviation of the viability rate at each sampling time.

The changing viability of the *C. albicans* over time is shown in Fig. 5. In the first 10 minutes irradiation, no obvious decreasing was observed. The viability of *C. albicans* only decreased by 12.37%. In contrast, after 10 minutes irradiation, the viability of *C. albicans* decreased in a much higher rate. After 20 minutes irradiation, only 6.24% *C. albicans* survived, and after another five minutes irradiation the survival rate decreased to 0.56%. This can be considered as the entirely eliminated, which demonstrated that the 405nm blue light is very effective in the inhabitation of the *C. albicans*. In Fig. 6, the photos of the colonies directly showed the inhibition of the cells. There are only two colonies after 25 minutes of irradiation, comparing to more than 150 when not irradiated. All results indicated that the 405nm blue light is very effective in inhibiting the *C. albicans*.

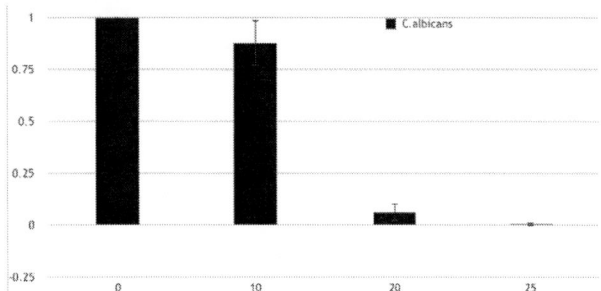

Fig.5 The viability of the *C. albicans* irradiated by various time.

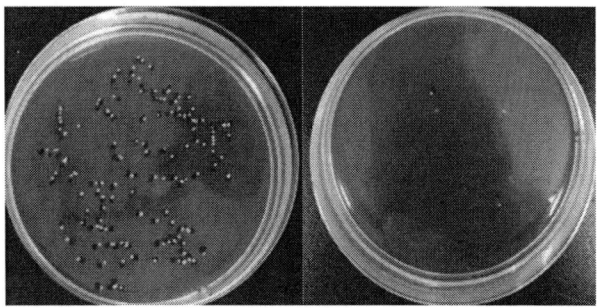

Fig.6 The fungal colony of before and after the ABL irradiation.

One phenomenon that observed in the experiment is the trend of the viability has an obvious change after 10 minutes irradiation, which can be attributed to the mechanism of ABL therapies, the ROS accumulating. In fact, there is a certain threshold of the ROS level. Low concentrations of ROS do not cause damage to cells and play an important role in cell signaling and homeostasis [12]. However, once the concentration of ROS exceeds the threshold, it becomes harmful and causes a series of oxidative damage to the cells. In the experiment, the ROS started to increase when the cells started to be exposed. However, the ROS level still did not reach the threshold after 10 minutes, which explains why there is no obvious decreasing of the viability of *C. albicans*. Then, the ROS increased and exceed the threshold and immediately starts to kill the cells, which leads to a faster killing rate.

4. Mechanisms of the anti-microbial Light and measurement of the generated ROS by light

Changes in intracellular ROS levels were determined by measuring the oxidative conversion of cell permeable non-fluorescent probe 2',7'- dichlorodi-hydrofluorescein diacetate (DCFH-DA) to fluorescent dichloro-fluorescein (DCF) in a VL0L0TD0 Varioskan LUX multimode microplate reader (Thermo Fisher, US). The concentration of the fungi was estimated by its absorption at 450 nm by U-3900H Ultraviolet spectrophotomete (Hitachi, Japan) before the measurement of the ROS level. The fungal absorption was controlled between 6.5-7. The fungi were centrifugated and separated with the medium, then dissolved with the DCFH-DA diluted by 500-fold with phosphate buffer saline (PBS). After incubated at 37^{0}C in a shaker for 30 minutes, centrifugate the suspension three times to remove the redundant probes. Then the suspension was seeded into a 96-well plate, and divided into a treatment group and a control group. The treatment group was irradiated by blue light where the control group was covered in the dark. Finally, the fluorescence was measured by the microplate reader. Triplicate experiments were performed at each sampling time. The raw data were processed to produce the mean and standard deviation of the viability rate at each sampling time.

The results of the ROS assay demonstrate a linear monotonically increasing of the ROS level. After 15 minutes irradiation, the ROS level was more than two times of the control group. In the end of the experiment, after 30 minutes irradiation, the ROS level was increased over 3 times of control group.

The results of the ROS level assays demonstrated that the cells inhabitation happened because of the high intracellular ROS level. Combing the previous results, the obvious decreasing of the viability started after 10 minutes irradiation. Thus, we can deduce that the lethal ROS level that can significantly inhibit the *C. albicans* is about two times of that of the control group.

5. Conclusions

In this work, the inactivating effects of the *C. albicans* by the irradiation of 405nm were studied, and the ROS level of the irradiated cells were tested. The results indicate that, in vitro, the 405nm LED was very effective in inhibiting the *C. albicans,* and the cellular ROS level was significantly increased after the irradiation. Furthermore, the harmful ROS level that can significantly inhibited the fungi was estimated to be two times of the control group. In summary, ABL is an effective, potential treatment for fungal infections. A potential extension of our current work is to further investigate the optimal dosage of treating the microbial infections in clinical

978-1-7281-5757-3/19 $31.00 © 2019 IEEE

experiments, and to model the dynamic changes of the viability of the fungi.

Acknowledgments

This work is supported by the National Key R&D Program of China (No. 2017YFC0108500, Subproject No. 2017YFC0108502), and is also supported by the research funding of Suzhou Institute of Biomedical Engineering and Technology (No. Y757031705).

References

[1] Sobel, Jack D., "Vaginitis." *New England Journal of Medicine,* Vol. 337, No. 26 (1997), pp.1896-1903.

[2] Held, Paul. "An introduction to reactive oxygen species measurement of ROS in cells." *BioTek Instruments Inc.,* Application Guide (2012), pp. 1-2.

[3] Dong J. and Zhang Z., "Design of LED Light for Stimulating Cells in the Study of Light Therapies," *2018 15th China International Forum on Solid State Lighting: International Forum on Wide Bandgap Semiconductors China (SSLChina: IFWS)*, Shenzhen, 2018, pp. 74-77.

[4] Figueira, I. Z., AP C. Sousa, A. W. Machado, F. A. L. Habib, L. G. P. Soares, and A. L. B. Pinheiro, "Clinical study on the efficacy of LED phototherapy for pain control in an orthodontic procedure." *Lasers in medical science*, Vol. 34, No. 3 (2019), pp. 479-485.

[5] Panhoca, V.H., Lizarelli, R.D.F.Z., Nunez, S.C., de Andrade Pizzo, R.C., Grecco, C., Paolillo, F.R. and Bagnato, V.S., "Comparative clinical study of light analgesic effect on temporomandibular disorder (TMD) using red and infrared led therapy." *Lasers in medical science*, Vol. 30, No. 2 (2015), pp. 815-822.

[6] de Vasconcelos Catão, M.H.C., Nonaka, C.F.W., de Albuquerque, R.L.C., Bento, P.M. and de Oliveira Costa, R., "Effects of red laser, infrared, photodynamic therapy, and green LED on the healing process of third-degree burns: clinical and histological study in rats." Lasers in medical science, Vol. 30, No. 1 (2015), pp. 421-428.

[7] Zhang, Yunsong, Yingbo Zhu, Jia Chen, Yucheng Wang, Margaret E. Sherwood, Clinton K. Murray, Mark S. Vrahas, David C. Hooper, Michael R. Hamblin, and Tianhong Dai. "Antimicrobial blue light inactivation of Candida albicans: In vitro and in vivo studies." *Virulence* Vol. 7, No. 5 (2016), pp. 536-545.

[8] Trzaska, Wioleta J., Helen E. Wrigley, Joanne E. Thwaite, and Robin C. May. "Species-specific antifungal activity of blue light." *Scientific reports,* Vol. 7, No. 1 (2017), pp. 4605.

[9] Tsutsumi-Arai, Chiaki, Yuki Arai, Chika Terada-Ito, Yusuke Takebe, Shinji Ide, Hirochika Umeki, Seiko Tatehara, Reiko Tokuyama-Toda, Noriyuki Wakabayashi, and Kazuhito Satomura. "Effectiveness of 405-nm blue LED light for degradation of Candida biofilms formed on PMMA denture base resin." *Lasers in medical science,*(2019), pp. 1-8.

[10] Guffey, J. Stephen, and Jay Wilborn. "In vitro bactericidal effects of 405-nm and 470-nm blue light."

Photomedicine and laser therapy, Vol.24, No. 6 (2006), pp. 684-688.

[11] Enwemeka, C.S., Williams, D., Enwemeka, S.K., Hollosi, S. and Yens, D., "Blue 470-nm light kills methicillin-resistant Staphylococcus aureus (MRSA) in vitro." *Photomedicine and laser surgery*, Vol. 27, No. 2 (2009), pp. 221-226.

[12] Li, Z., Xu, X., Leng, X., He, M., Wang, J., Cheng, S. and Wu, H., "Roles of reactive oxygen species in cell signaling pathways and immune responses to viral infections." *Archives of virology,* Vol. 162, No. 3 (2017), pp. 603-610.

Influence of Dynamic-Brightness Environment on Ocular Physiological Parameters

Shanshan Zeng[1], Wentao Hao[2], Ya Guo[1], Xiangyu Qu[3], Ke Wei[4], Shanshan Tang[5], Rongrong Wen[2], Jianqi Cai[1, 5*]

1 Institute for standardization of public safety, China National Institute of Standardization, Beijing, 100191, China
2 Kunshan Company of Human Factor Engineering Research and Development Center, Suzhou, 215333, China
3 College of Optical and Electronic Technology, China Jiliang University, Hangzhou, 310018, China
4 Mianyang Institute of Product Quality Supervision and Inspection, Mianyang, 621051, China
5 School of life science, Beijing Institute of Technology, Beijing, 100081, China
*Corresponding author: caijq@cnis.ac.cn.

Abstract

In this study, we were sought to compare the influence of lighting environment with constant brightness and dynamic brightness on human ocular physiological parameters. We chose 20 participants for the human factor experiments. To measure the influence of lighting environment on ocular physiological parameters, three ocular physiological parameters were used in the current study: far point accommodation, the 4th and the 12th terms of aberrations. It was found that far point accommodation presented more significant response to the lighting environment with constant brightness than that with dynamic brightness, and sphere aberrations presented insensitive to the brightness variation. Both aberrations and far point accommodation could reflect the influence of lighting environment on human eye, while far point accommodation presented higher sensitivity to the lighting environment compared to sphere aberrations.

Keywords: ocular physiological parameters, dynamic brightness environment, constant brightness environment, sphere aberrations, far point accommodation

1.Introduction

Indoor lighting is dependent on both natural sunlight and artificial light. Natural sunlight is much more complicated than traditional luminaires or screens, since its optical performance varies with time. Characterized by time-varying optical performance, natural sunlight is a kind of dynamic light [1-3]. Previous researches have performed numerous studies on photo-biological effect of dynamic light [4-6]. Optical performance of dynamic lighting includes brightness, color and variation. Participants in dynamic lighting environment are likely to present corresponding physiological responses such as sleep quality, psychological stress, visual performance, alertness, and mental load. These responses originate from the interaction between photon and intrinsic photosensitive retina ganglion cell (ipRGC) [7], and are related to circadian rhythm [8-10].

Studies on the influence of dynamic lighting were mainly focused on the field of non-visual effect [11-13]. Visual effects of brightness on ocular physiological parameters have scarcely been investigated. It is known that lighting environment brightness affects pupil size directly. Previous researches have shown that pupil size is related to visual acuity, refractive error, contrast sensitivity, and brightness is likely to influence ocular physiological parameters [14-16]. Consequently, ocular physiological responses are likely to be different in dynamic lighting with time-varying brightness and in static lighting with constant brightness. Pupil size is controlled by a section of ciliary muscle. Another section of ciliary muscle controls crystalline lens deformation by accommodation of zonular fibers. Accommodation could reflect the ability of this section of ciliary muscle. Stretching or relaxation of ciliary muscle is likely to cause deformation of the crystalline lens, and lens deformation tends to induce variations in aberrations, especially high order aberrations [17-20].

In this study, accommodation (ACC) and high order aberrations (HOAs) were used to describe ocular physiological responses to dynamic-brightness environment. Subjects participanted in the human factor experiments. Luminaires with constant brightness and time-varying brightness were used as the indoor lighting source respectively. We were sought to investigate ocular response of human eye to dynamic-brightness environment.

2.Materials and methods

In this study, human factor experiments were performed on 20 participants. Detailed information of participants were shown in Table 1. Informed consent forms were collected from all participants.

Two teams of lamps were used in this study: lamps with constant brightness and lamps with dynamic brightness. Lamps with constant brightness were made in ordinary process, while lamps with dynamic brightness were equipped with implanted chips which control the output illuminance. Optical performances of the two kinds of lamps were shown in Table 2 and Fig. 1.

TABLE I
DETAILED INFORMATION OF PARTICIPANTS.

Item	Information
Participant total number	• 36
Age distribution	• 23~35 (average 30)
Diopter distribution	• 40%: 1.00D ~ -1.00D • 40%: -1.00D ~ -3.00D • 20%: -3.00D ~ -5.00D
Visual acuity distribution	• 40%: 0.8 • 60%: 1.0
Anisometropia above 2.5D	• <5%
Intraocular pressure range	• 14~20
Sex	• 50%: Male • 50%: Female

TABLE II
DETAILED INFORMATION OF THE LAMPS.

LED classroom lamp	Optical performance
Light quantity (lm)	2146
Luminous Efficiency (lm/W)	60.57
CRI	94.7
CCT (K)	4667
Beam angle (degree)	67.9
Power factor	0.972

Human factor experiments were performed on each

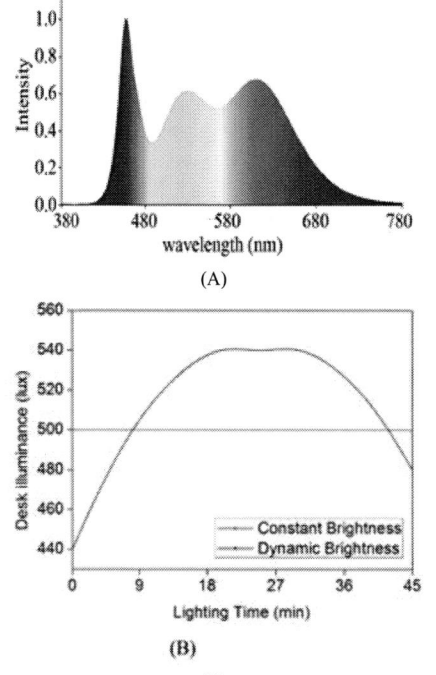

Fig. 1. (A) Spectra power distribution (SPD) and (B) time-illuminance curve of the lamp.

participant twice: the first time in the lamp with constant brightness, and the second time in the lamp with dynamic brightness. Human factor experiments contained measurements on ACC and HOA prior to and following the visual task. Instruments for the measurements are in Table 3. Experiment steps, visual task photo and task content were shown in Fig. 2.

Ciliary accommodation was collected by NIDEK AR-1S

TABLE III
DETAILED INFORMATION OF MEASUREMENT INSTRUMENTS..

Measurement	Instrument
ACC	NIDEK AR-1S (Calibrated by The Measurement Test Research Institute of Beijing, calibration certification No. DC18J-QQ000362)
HOA	NIDEK OPD Scan III (Calibrated by The Measurement Test Research Institute of Beijing, calibration certification No.JC17C-AB010010)

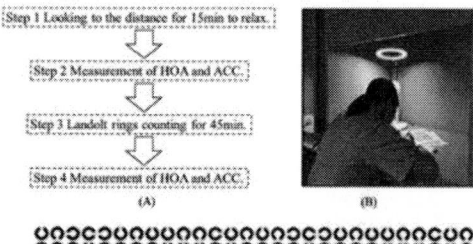

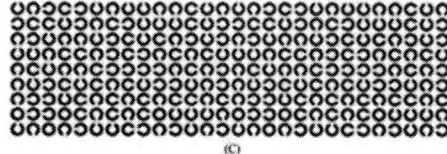

Fig. 2. (A) Measurement process, (B) visual task photo and (C) Landolt rings.

(Calibrated by The Measurement Test Research Institute of Beijing, calibration certification No. DC18J-QQ000362). Measurements were performed binocularly. Each participant was asked to attach the forehead and chin to the specified location on the instrument, and glared at the target in the screen binocularly under natural condition. Accommodation data were collected automatically by the instrument. Training trial was ran for each participant before data collection until the task was well understood. Participants were not allowed to blink during the measurement to avoid the breaking of ciliary muscle status.

Ocular aberrations were recorded using NIDEK OPD Scan III (Calibrated by The Measurement Test Research Institute of Beijing, calibration certification No.JC17C-AB010010). Measurements were performed on right and left eyes respectively. Each participant was asked to attach the forehead and chin to the specified location on the instrument and glared at the target in the screen monocularly under natural condition. Ocular aberration data of the 1st~35th Zernike terms were collected automatically by the instrument. Training trial was ran for each participant before data collection until the task was well understood. The procedure was repeated after the training trial, and the collected data were in the form of Zernike expansions. Participants were allowed to blink during the measurement to avoid aggravating of ocular fatigue caused by an extended inter-blink interval. Finally we used the 3rd~14th terms aberrations.

3. Results

A. Accommodation

ACC describes accommodation of ciliary muscle. Ciliary accommodation contains two important parts: near point accommodation and far point accommodation. Regarding near point accommodation, ciliary muscle fibers adjust the crystalline lens from the current status to the tensest status. Regarding far point accommodation, ciliary muscle fibers adjust from the current status to the completely loose status. Significant difference was found prior to and following the visual task in far point accommodation (Table 4). No significance was found in near point accommodation. Results of significance tests were similar for visual tasks in constant brightness lighting and dynamic brightness lighting.

Variation of far point accommodation (far point ΔACC) during the visual task reflects ocular physiological response to fatigue in the corresponding lighting environment. Participants present significant difference in constant brightness and in dynamic brightness. Significance exists in participants (Table 5,

TABLE IV

SIGNIFICANCE TEST OF DIFFERENCE BETWEEN ACCOMMODATION VALUES PRIOR TO AND FOLLOWING THE VISUAL TASK. * REPRESENTS P<0.05, AND ** REPRESENTS P<0.01..

Difference between 0min and 120min	Significance
Near point ACC in constant brightness	t=-0.12, p=0.906
Far point ACC in constant brightness	t=-3.017, p=0.007**
Near point ACC in dynamic brightness	t=-0.084, p=0.934
Far point ACC in dynamic brightness	t=2.966, p=0.008**

TABLE V

SIGNIFICANCE TEST OF DIFFERENCE BETWEEN ACCOMMODATION VALUES PRIOR TO AND FOLLOWING THE VISUAL TASK. * REPRESENTS P<0.05, AND ** REPRESENTS P<0.01..

Far point ΔACC difference	Significance
Constant & Dynamic brightness	t=2.338, p=0.03*

Fig. 3). It was implied that the lighting environment with dynamic brightness could induce less influence on ocular physiological parameters compared to the lighting environment with constant brightness.

For near point accommodation, difference was not significant prior to and following the visual task. Variation of near point accommodation was not obvious to be observed. It was indicated that ciliary adjustment toward tension status was unlikely to be influenced by the accumulation of ocular fatigue.

Variation of far point accommodation was significant prior to and following the visual task. Far point accommodation describes ciliary adjustment toward relaxed status, and this section of accommodation was proved to be obviously affected by ocular fatigue. Myopia is characterized by weak adjustment ability of ciliary muscle fibers. Without intervening of ocular fatigue, decrease of far point accommodation is likely to induce myopia. Participants in both constant brightness and dynamic brightness presented significant difference in far point accommodation prior to and following the visual task.

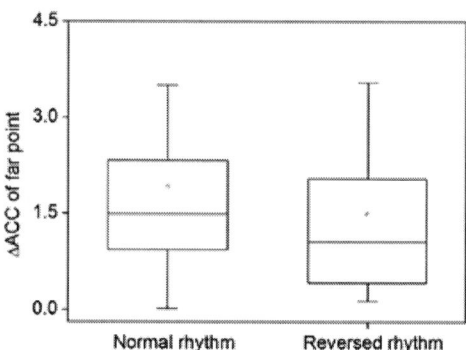

Fig. 3. Significance test of difference between far point ΔACC values in lighting environments with constant brightness and dynamic brightness. There was no significance (p=0.41).

B. Aberrations

Aberration reflects distortion of the image in retina. Zernike polynomial is widely used to describe aberration of human eye. In this study, we collected ocular aberrations of the 1st~35th terms aberrations. High order aberrations (the 3rd~14th terms) were often used to describe ocular function in ophthalmology. Aberration values of the 3rd~14th terms presented variations shown in Fig. 4. The 3rd~5th terms were listed separately form the 6th~14th terms due to different magnitude orders.

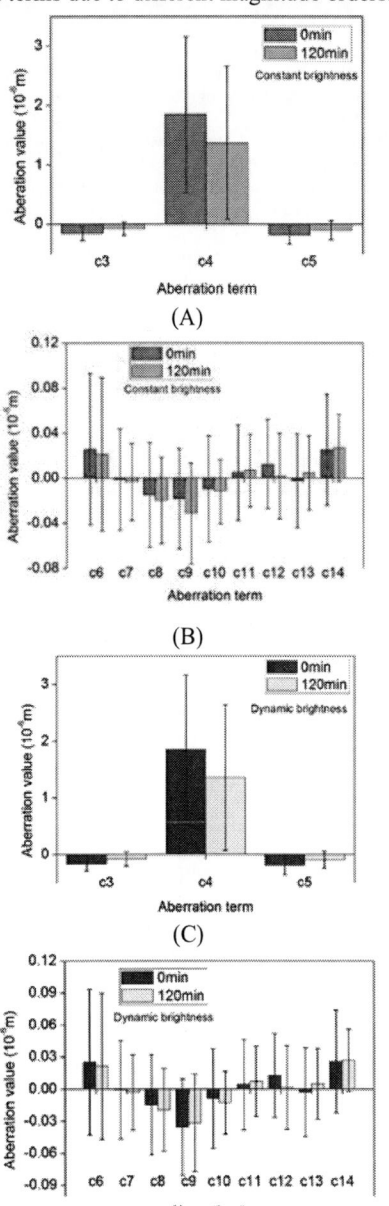

Fig. 4. Aberrations of participants in different lighting environments prior to and following visual tasks. (A) The 3rd~5th terms aberrations of participants in the lighting environment with constant brightness. (B) The 6th~14th terms aberrations of participants in the lighting environment with constant brightness. (C) The 3rd~5th terms aberrations of participants in the lighting environment with dynamic brightness. (D) The 6th~14th terms aberrations of participants in the lighting environment with dynamic brightness.

Aberrations of the 3rd~14th terms presented complicated variations. Among all terms, the 4th and the 12th terms were sphere aberrations and independent from image directions. In this study, we focused on sphere aberrations to analyze.

TABLE VI
SIGNIFICANCE TEST OF ABERRATION VARIATION IN LIGHTING ENVIRONMENTS WITH DIFFERENT BRIGHTNESS. ** REPRESENTS P<0.01.

Significance of aberration difference	4th term	12th term
Normal bio-clock in dynamic brightness	t=7.304, p=0**	t=6.773, p=0**
Reversed bio-clock in dynamic brightness	t=3.275, p=0.005**	t=0.223, p=0.826

Significance tests were performed on the 4th and the 12th terms aberrations (Table 6).

Significance tests were also performed on aberration variations for the 4th and the 12th terms aberrations. Aberration variations in constant brightness environment and in dynamic brightness environment presented no significant difference (Table 6). It was implied that lighting environment with dynamic brightness was likely to induce physiological response variations more in ΔACC than in sphere aberrations. Aberration variations could reflect the influence of ocular fatigue, but the accuracy was not enough to reflect lighting environment influence on ocular fatigue accumulation.

Sphere aberrations reflect image distortion in retina. Perception of visual fatigue presents blurred vision and perception of eye soreness as well as dryness. These two factors jointly decide visual fatigue degree. Perception of eye soreness as well as dryness originates from nerve signal caused by ciliary muscle triggering, so this factor of visual fatigue relates to muscle stretching force. Blurred vision reflects that image does not fall in retina. This is because that focus distance of lens could not be adjusted by accommodation of ciliary muscle. Consequently, visual fatigue is dominated by ciliary muscle accommodation ability. Crystalline lens is likely to deform during the process of accommodation, and lens shape is related to accommodation regularity. However, lens shape could hardly be measured by instruments. It is necessary to figure out physiology parameters that can reflect lens shape. In this study, both the 4th and the 12th terms aberrations present significant difference prior to and following the visual tasks. However, the differences seem independent from the lighting environments and the circadian rhythm of participant. Compared to aberrations, the far point accommodation presents higher sensitivity in reflecting ocular reaction to lighting environment.

4.Conclusion

Artificial lighting source plays an important role for indoor lighting. Compared to sunlight with time-varying brightness, most artificial luminaires and screens present constant brightness independent from time. In this study, we were sought to investigate ocular response of human eye to dynamic-brightness environment. Two kinds of brightness environments were used in the experiment: dynamic brightness and constant brightness. Participants were instructed to execute the visual task. Variations of physiological parameters were collected and analyzed prior to and following the visual task. It was indicated that lighting environment with dynamic brightness was likely to help decrease lighting influence on human eye compared to constant brightness environment. Lighting environment with dynamic brightness was likely to cause lower ocular response to the lighting environment by decreasing ocular fatigue accumulation.

Acknowledgment

This work was supported by National Key Research and Development Plan (grant no. 2017YFB0403700) and the Fundamental Research Fund of China National Institute of Standardization (grant no.512019Y-6685).

References

[1] J.C.Lam, D.H.W. Li, D.H.W. "An analysis of daylighting and solar heat for cooling-dominated office buildings." Solar Energy, vol. 65, pp 251-262. 1999.

[2] P.J. Littlefair. "A comparison of sky luminance models with measured data from Garston, United Kingdom". Solar Energy, vol. 53, pp. 315-322. 1994.

[3] D.W. Li, J. Lam, C.S. Lau. "A Study of Solar Radiation Daylight Illuminance and Sky Luminance Data Measurements for Hong Kong". Architectural Science Review, vol. 45. no. 1, pp. 21-30. 2002.

[4] K. Choi, H.J. Suk. "Dynamic lighting system for the learning environment: performance of elementary students." Optics Express, vol. 24, no. 10, pp. A907. 2016.

[5] M. Konstantoglou, A. Tsangrassoulis. "Dynamic operation of daylighting and shading systems: A literature review." Renewable & Sustainable Energy Reviews, vol. 60, pp. 268-283. 2016.

[6] H. Chen, R. Zhu., Y.H. Lee. "Correlated color temperature tunable white LED with a dynamic color filter." Optics Express, vol. 24, no. 6, pp. A731. 2016.

[7] D.M. Berson, F.A. Dunn, M. Takao. "Photo-transduction by retinal ganglion cells that set the circadian clock." Science, vol. 295, no. 5557, pp. 1070-1073. 2002.

[8] S. Lei, H.C. Goltz, X. Chen. "The relation between light-induced lacrimation and the melan-opsin-driven post-illumination pupil response." Investigative Ophthalmology & Visual Science, vol. 58, no. 3, pp. 1449. 2017.

[9] J. Aboshiha, N. Kumaran, A. Kalitzeos. "A quantitative and qualitative exploration of photoaversion in Achromatopsia." Investigative Ophthalmology & Visual Science, vol. 58, no. 9, pp. 3537-3546. 2017.

[10] R. Lasauskaite, C. Cajochen. "Influence of lighting color temperature on effort-related cardiac response." Biological Psychology, vol. 132, pp. 64-70. 2018.

[11] W.J. Van Bommel. "Non-visual biological effect of lighting and the practical meaning for lighting for work." Applied Ergonomics, vol. 37, no. 4, pp. 461-466. 2006.

[12] T. Sexton, E. Buhr, R.N. Van Gelder. "Melanopsin and mechanisms of non-visual ocular photoreception." The Journal of Biochemistry, vol. 287, no. 3, pp. 1649-1656. 2012.

[13] L.E. Gimenez, S.A. Vishnivetskiy, F. Baameur. "Manipulation of very few receptor discriminator residues greatly enhances receptor specificity of non-visual arrestins." The Journal of Biochemistry, vol. 287, no. 35, pp. 29495-29505. 2012.

[14] V. Petternel, C.M. Köppl, I. Dejaco-Ruhswurm. "Effect of accommodation and pupil size on the movement of a posterior chamber lens in the phakic eye." Ophthalmology, vol. 111, no. 2, pp. 325-331. 2004.

[15] E. Fernández, W. Drexler. "Influence of ocular chromatic aberration and pupil size on transverse resolution in ophthalmic adaptive optics optical coherence tomography." Optics Express, vol. 13, no. 20, pp. 8184-8197. 2005.

[16] J. Mckelvie, B. Mcardle, C. Mcghee. "The influence of tilt, decentration, and pupil size on the higher-order aberration profile of aspheric intraocular lenses." Ophthalmology, vol. 118, no. 9, pp. 1724-1731. 2011.

[17] J. Castejón-Mochón, N. López-Gil. "Ocular wave-front aberration statistics in a normal young population." Vision Research, vol. 42, no. 13, pp. 1611-1617. 2002.

[18] N. López-Gil, V. Fernández-Sánchez. "Accommodation-related changes in monochromatic aberrations of the human eye as a function of age." Investigative Ophthalmology & Visual Science, vol. 49, no. 4, pp. 1736-1743. 2008.

[19] N. López-Gil. 2007. "Effect of third-order aberrations on dynamic accommodation." Vision Research, vol. 47, no. 6, pp. 755-765. 2007.

[20] N. Lopez-Gil. "The change of spherical aberration during accommodation and its effect on the accommodation response." Journal of Vision, vol. 10, no. 13, pp. 12. 2010.

Study on 3D thermal transport in micro-LEDs on GaN substrate at the level of kW/cm²

Zhizhong Chen[a*], Chengcheng Li[a], Fei Jiao[a,b], Jinglin Zhan[a], Yifan Chen[a], Yiyong Chen[a], Jingxin Nie[a], Tongyang Zhao[a], Xiangning Kang[a], Shiwei Feng[d], Guoyi Zhang[a,c], Bo Shen[a]

[a]State Key Laboratory for Artificial Microstructure and Mesoscopic Physics, School of Physics, Peking University, Beijing 100871, China. Email: zzchen@pku.edu.cn; Tel: +86 131 6738 8462
[b]State Key Laboratory of Nuclear Physics and Technology, School of Physics, Peking University, Beijing 100871, China
[c]Dongguan Institute of Optoelectronics, Peking University, Dongguan, 523808, Guangdong, China
[d]School of Electronic Information and Control Engineering, Beijing University of Technology, Beijing 100124, China

Abstract

In this work, different sizes of micro-light-emitting diodes (μLEDs) were fabricated on the sapphire and GaN substrates. The thermal characteristics of μLEDs were studied by the forward voltage method, thermal transient measurement, and infrared (IR) thermal imaging. The μLEDs on the GaN substrate showed an approximately 10°C lower junction temperature and smaller amplitude of the K factors than those on the sapphire substrate under the current injection level of 4 kA/cm². IR thermal imaging results showed the uniform temperature distributed on the GaN substrate. The thermal transient measurement showed that the thermal resistances of the mesa, epilayer, and the interface of GaN/substrate were reduced significantly for μLEDs on the GaN substrate. This means that a high-quality GaN crystal and homogeneous interface corresponded to little scattering for phonons. The APSYS simulation indicated that the high thermal and electrical conductivity of the GaN substrate played a key role in the low junction temperature and uniform temperature distribution. A small-sized μLED combined with a GaN substrate can become a perfect candidate for high-power applications and visible light communication.

Introduction

Many researches show excellent performances of micro-light emitting diodes (μLEDs) at high injection level of kA/cm²[1-4]. This injection level is usually required in the applications for visible light communication (VLC), pumping sources of organic lasers, high brightness micro-display, cell manipulation, and so on [5-8]. Some studies reported that the low junction temperature (Tj) was attributed to the high injection performances for small size μLEDs [1, 3, 9]. In large chip LED or nanorod LED, high thermal resistance is supposed since the small thermal cross-section prohibit the thermal dissipation for small size LEDs [10, 11]. Careful calculations show that the ratio of sidewall surface area to active volume does not increase significantly when the diameter is larger than 20 μm [12]. The thermal dissipation for the huge power density of tens of kW/cm² is an exact problems for μLEDs. On the other hand, LEDs on a GaN freestanding substrate can also sustain the current densities of kA/cm²[13-15]. Rashidi et al. fabricated micro-LEDs on a freestanding GaN substrate [15] and obtained an excellent electro-optical modulation performance for nonpolar μLEDs. However, they did not concern the thermal dissipation in the μLEDs.

In the previous work, spectral shift method was used to measure the Tj [1]. The Tj is the averaged value of the device.

The Tj distribution cannot detect by the technique. Infrared (IR) thermal imaging can provide the lateral temperature distribution at the micron precision and the temperature resolution reach to 20 mK [16]. However, the temperatures are still the averages in vertical direction by IR technique. The vertical thermal dissipation route and the interface effects are still unknown. The forward voltage (Vf) is sensitive to the Tj [17]. After the temperature coefficient K is measured, the junction temperature can be obtained in real time according to the voltage. A transient thermal measurement based on the forward voltage method was developed to analyze multilayer LED package structures [10,18-20]. By analyzing the cumulative structure function and/or the differential structure function, different components of the LED packages can be recognized. However, there are few reports on the structure functions on the multilayer LED chips.

Here the μLEDs were fabricated on sapphire and GaN substrates with the diameters from 10 to 160 μm. the forward voltage method, IR thermal imaging, and transient thermal measurements were carried out on these μLEDs. The temperature distributions in the micro-LEDs were also simulated with the APSYS Packages of Crosslight software. Three dimensional (3D) thermal transport in the chips and packages of the μLEDs was discussed under the injection level of kA/cm².

Experiment

The 2 inch GaN substrate were provided by Sino Nitride Co., which the thickness is about 400 μm and the threading dislocation is about 5×10^6 cm⁻². The GaN substrate and sapphire substrate are loaded into the reactor in same run. The emission wavelength were set as 385 nm. Conventional photolitho-graphy and inductively coupled plasma (ICP) etching technology were used to obtain micron-sized pillars with diameters of 10 – 160 μm. The samples were named after their substrates and pillar sizes. For example, "fs-20μm" and "ref-20μm" refer to 20 μm LEDs fabricated on the GaN substrate and sapphire substrate, respectively. KOH solution was applied to repair etching damage on the sidewalls of the micro-LEDs. The chip was directly bonded onto the Al based printed circuit board (PCB) using Ag epoxy. For thermal transient measurement, the chips were bonded to the heat sink using thermal grease. Figure 1 shows optical microscopy images and electroluminescence (EL) spectrum of the μLEDs on the GaN substrate.

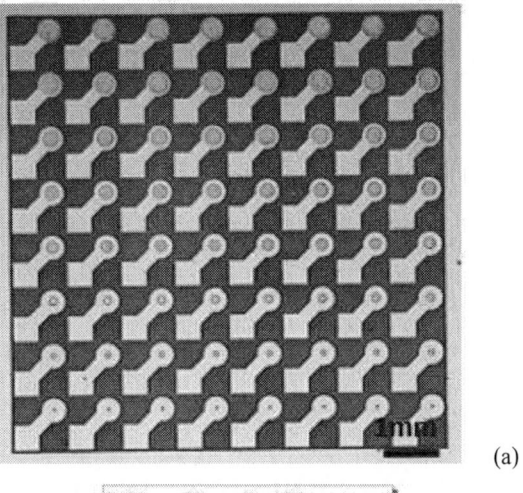

(a)

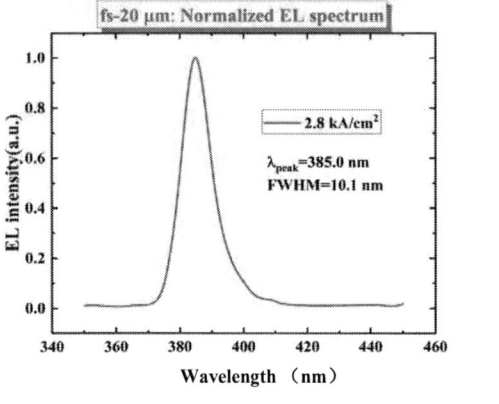

(b)

Fig.1 (a) Optical microscope image of the μLEDs on GaN substrate. (b) EL spectrum measured under 2.8 kA/cm2 for fs-20 μm LED. The peak wavelength is 385.0 nm.

The EL spectra were collected with a SSP 6612 LED Multiple Parameters Tester with a coupled spectrometer and charge coupled device (CCD) detection system in an integrated sphere. The forward voltage method were performed on the μLEDs by an oven and a Keithley 2601A system SourceMeter. The oven temperature was changed from room temperature to 120 °C to obtained the K factor. The transient thermal measurements were performed with TTE-400 Light Emitting Diodes Thermal Test Equipment by Beijing University of Technology.The switch time between the heating and test current was 1 μs. The temperature distribution images of the μLEDs were captured by an FLIR System SC5700 camera with a working wavelength in the 2.1 − 5.5 μm range and a temperature resolution of 20 mK. The thermal infrared camera recorded the images with a frequency of 115 Hz (1 ms integration time, 7.6 ms dead time) and a spatial resolution of about 3.0 μm per pixel. The temperature distributions in the micro-LEDs were simulated with an APSYS package of Crosslight software [21].

Results and Discussion

Fig.2 shows the K factors for different sizes of μLEDs on GaN and sapphire substrates. The sensor current keeps on 10

μA, by which the heat generated can be neglected. The 10 μA of the current corresponds to the 0.09-0.80 A/cm2 of the current density for μLEDs with the diameter from 120 to 40 μm. In the small injection level, forward bias is below the turn-on voltage. The Shockley-Read-Hall (SRH) recombination is dominant in the active layer. Because the forward bias decreases with the Tj increasing, the K factor is usually negative. For convenience, K factor mentioned follow represent the absolute values of those negative values. The K factors of μLEDs on freestanding GaN is higher than those on sapphire substrate. The K factors of μLEDs are from 2.33 to 2.55 mV/℃ on sapphire, while the values of μLEDs on freestanding GaN are from 1.92 to 2.05 mV/℃. The K factor decreases when the crystal quality are improved by patterned sapphire substrate [22]. The threading dislocation density (TDD) is about 5×10^{6} cm^{-2} in the LED epilayer on GaN substrate, which is two order lower than that on sapphire substrate. It is reasonable that smaller amplitude of the K factors on GaN substrate are attributed to its high crystal quality. As to the K factors variation with the diameter, an additional voltage may be appears when the polarization field increases due to more strain [23]. In large size μLEDs, the strain is less relaxed than that in small size ones. With the temperature increases, the strain in MQW increases due to the thermal mismatch between GaN and InGaN. This trending indicates the smaller K factor on GaN substrate can partially due to the strain relaxation in MQWs [24].

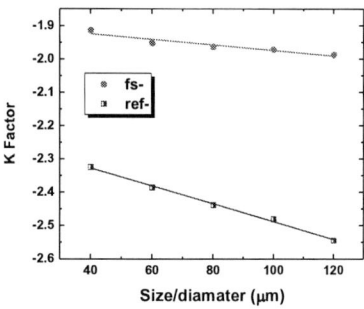

Fig.2 Temperature coefficient, K factor of μLEDs with different diameters on GaN and sapphire substrates.

After the temperature coefficient K is measured, the junction temperature can be obtained in real time according to the voltage. The vertical components and interfaces in μLED chips and packages can be obtained by thermal transient measurements. The heat flow is from the pn junction to the bottom heat-sink. Fig. 3 shows the typical cumulative functions for 40 and 120 μm LEDs, where the C_{th} and R_{th} terms represent the thermal capacitance and the spreading thermal resistance, respectively. The step-shaped curves indicate the thermal components and interfaces. The steep segments correspond to the the relative high heat capacitance and the low spreading thermal resistance, which are contributed by the components of the LED devices. On the contrary, the flat segments can be attributed to the thermal interfaces between the components.

978-1-7281-5757-3/19 $31.00 © 2019 IEEE   202

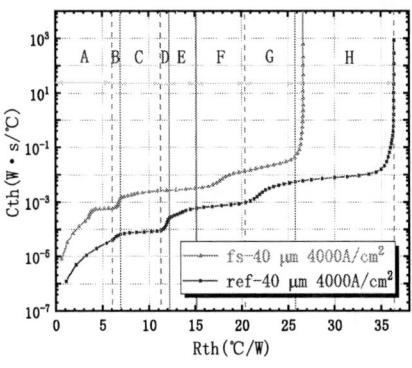

(a)

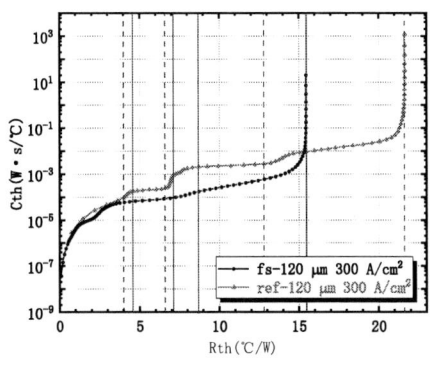

(b)

Fig.3 Cumulative structure functions (a) for the samples of ref-40 μm and fs-40 μm at 4 kA/cm² and (b) for the samples of ref-120 μm and fs-120 μm at 300 A/cm².

The curves in Fig. 3 are divided into eight parts, as marked by the capital letters A–H for the samples of ref-40 μm and ref-120 μm. The curves for the fs-40 μm and fs-120 μm samples are also marked by "A–H", but the corresponding spreading thermal resistance values that these letters represent are different than those for the reference samples. In our previous work, "A-H" segments were assigned to p-mesa and multiple quantum wells (MQWs) interface, interface of epilayers and substrates, substrate, Ag epoxy, interface of Ag epoxy and PCB(Al), PCB(Al), thermal grease interfaces of PCB/heat-sink and heat-sink [25]. Table 1 lists the spreading thermal resistances of the components and their interfaces, drawn from the structure functions. It is found that the total thermal resistances are 26.6 and 36.4℃/W for fs-40 μm and ref-40 μm samples at 4 kA/cm². The values are 15.4℃/W and 21.6℃/W for fs-120 μm and ref-120 μm samples at 300 A/cm². It is obvious that the total thermal resistances on GaN substrate are always lower than those on sapphire substrate at same current density. As the bold-marked in Table 1, segments A, C, G and H are attributed to the differences between different substrates. Segment A means the p-mesa and MQWs components. The lower TDD for GaN substrate corresponds to less phonon scattering by TD or interfaces, especially for large size μLEDs. Because the less phonons scatter at the homogeneous interface, the thermal resistance

of segment C reduces to half of the reference one. As to segments G and H, the thermal resistance decreases on GaN substrate may be due to the smaller spreading thermal resistance and the capacitance for high-quality GaN, which leads to a higher temperature and a more uniform temperature distribution in Segments G and H than those for sapphire substrate.

Table 1 Spreading thermal resistances of the components and their interfaces drawn from structure functions

segments	Type	thermal resistance of the segments (°C/W)			
		fs-40 μm	ref-40 μm	fs-120 μm	ref-120 μm
A	Mesa and MQWs interface	**3.43**	**6.08**	**0.64**	**4.00**
B	GaN epilayer	0.54	0.87	0.50	0.57
C	Interface of epilayers and substrates	**2.40**	**4.32**	**1.05**	**2.12**
D	Substrate	0.69	0.94	0.51	0.53
E	Ag epoxy	2.93	2.93	4.75	1.58
F	Interface of Ag epoxy and PCB(Al)	6.40	5.24		4.09
G	PCB(Al)	**2.72**	**5.40**	**1.48**	**2.67**
H	thermal grease interfaces of PCB/heat-sink and heat-sink	**7.52**	**10.59**	**6.39**	**6.14**
A-H	Total	26.64	36.38	15.45	21.62

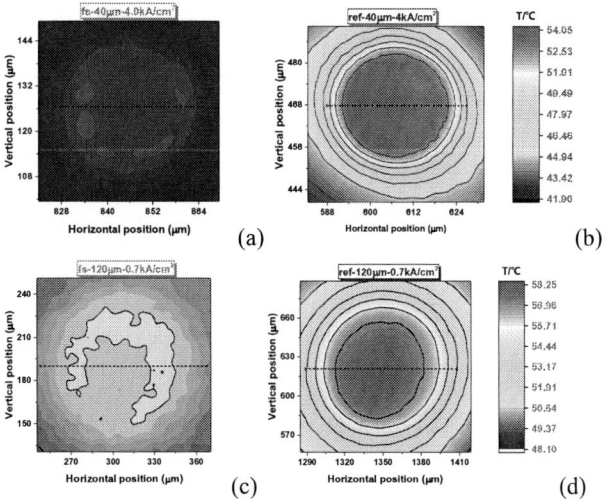

Fig.4 Thermal images of the (a) fs-40 μm and (b) ref-40 μm samples operated under 4 kA/cm², and the (c) fs-120 μm and (d) ref-120 μm samples operated under 0.7 kA/cm².

Beside the vertical distribution, the lateral Tj distribution are measured on the μLED by IR thermal imaging, as shown in Fig.4. Fig. 4a and b shows the IR thermal images of fs-40 μm and ref-40 μm samples operated at 4 kA/cm². The temperature scale bars are put on the right side of the thermal images in the same range from 41.9 to 54.1 °C for comparison. Fig.4c and 4d shows the IR thermal images of fs-120 μm and ref-120 μm samples operated at 0.7 kA/cm². It can be clearly seen that the temperature is higher on the mesa than the surrounding parts. The temperature distribution on the mesa is relatively uniform. The average temperatures are 43.0 and 51.8 °C on the mesa for the fs-40 μm and ref-40 μm

978-1-7281-5757-3/19 $31.00 © 2019 IEEE

samples, respectively. As to 120 μm LEDs operated at 700 A/cm^2, these values are 55.9 and 51.8 °C for the fs-120 μm and ref-120 μm samples, respectively. Moreover, the temperature differences between the mesa and surroundings are 1 °C for 40 and 120 μm LEDs on GaN substrate, and 12 and 7 °C for 40 and 120 μm LEDs on sapphire substrate. This means that the generated heat in the junction could be rapidly dissipated to the GaN epilayers, GaN substrate, and heatsink for the GaN freestanding samples. It is also observed that the maxima of the temperature distribute at the edge of p-mesa for freestanding GaN samples, while they located at the center of the mesa for one on sapphire substrate.This may be due to the different current and thermal spreading, which will be discussed in the simulation part.

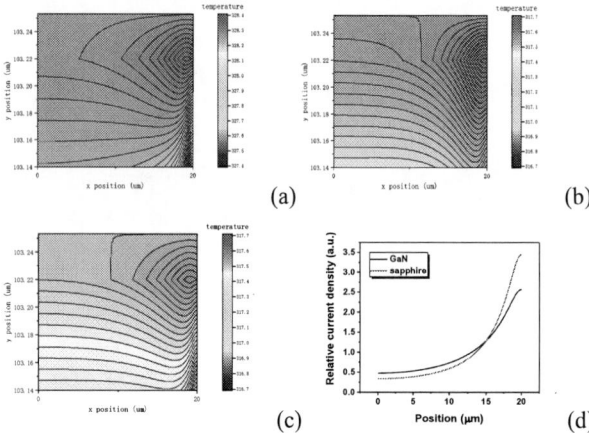

(a)

(b)

(c)

(d)

Fig. 5 Simulated temperature distribution of the 40 μm LEDs on sapphire substrate at 4 kA/cm^2 with the thermal conductivities of substrates: (a) 41.9 W/mK as sapphire and (b) 170W/mK as GaN, and (c) the 40 μm LEDs on GaN substrate at 4 kA/cm^2.

The temperature distributions are simulated for μLEDs on different substrates using APSYS software. The μLEDs structure is built up according to the experimental structure. The parameters of each material are set to the defaults in the software. When the thermal conductivity (TC) of the sapphire substrate is set as that of freestanding GaN. The total average values of Tj are reduced from 55.9 to 45.3 °C, which are similar to the real freestanding values, as shown in Fig.5(a-c). The shape of the temperature platform also changes to "M" for the fs-40 μm sample, which the maximum locates at the edge of the mesa. After compared the isotherm densities in Fig.5b and 5c, it was lower for freestanding GaN than that for the high TC sapphire substrate. This indicates that electrical conductive for GaN substrate makes the current spreading more uniform than insulating sapphire, which causes a more uniform temperature distribution. The current density distributions of the ref-40μm and fs-40μm samples are simulated under a current density of 4 kA/cm^2, as shown in Fig. 5d. It is found that the current density keeps increasing from the center to the edge of the mesa for both samples. The ratios of the current density at the edge of the mesa to the average one (4 kA/cm^2) are 3.45 and 2.57 for the ref-40 μm

and fs-40 μm samples, respectively. A higher ratio means that the current crowding is more serious.

Conclusions

In summary, different sizes of μLEDs were fabricated on the sapphire and GaN substrates. The 3D thermal characteristics of μLEDs were studied by the forward voltage method, thermal transient measurement, and IR thermal imaging. The μLEDs on the GaN substrate showed an approximately 10°C lower junction temperature and smaller amplitude of the K factors than those on the sapphire substrate under the current injection level of 4 kA/cm^2. IR thermal imaging results showed the uniform temperature distributed on the GaN substrate. The thermal transient measurement showed that the thermal resistances of the mesa, epilayer, and the interface of GaN/substrate were reduced significantly for μLEDs on the GaN substrate. This means that a high-quality GaN crystal and homogeneous interface correspond to little scattering for phonons. A small-sized μLED combined with a GaN substrate can become a perfect candidate for high-power applications and visible light communication.

Acknowledgments

This work was supported by the National Key Research and Development Program under Grant No. 2017YFB0403601, National Natural Science Foundation of China under Grant No. 61674005, Beijing Municipal Science & Technology Commission under Grant No. Z161100001616010, Science and Technology Major Project of Guangdong Province under Grant No. 2016B010111001, and Science and Technology Planning Project of Henan Province under Grant No. 161100210200.

References

1. Z. Gong, S. M. D. Dawson et al., "Size-dependent light output, spectral shift, and self-heating of 400 nm InGaN light-emitting diodes," J. Appl. Phys. 107(1), (2010), pp. 013103.

2. Q. Q. Jiao, Z. Z. Chen, et al., "Capability of GaN based micro-light emitting diodes operated at an injection level of kA/cm ^{2} ," Opt. Express 23(13), (2015) pp.16565–16574.

3. S. C. Huang, H. Li, et al., "Superior characteristics of microscale light emitting diodes through tightly lateral oxide-confined scheme," Appl. Phys. Lett. 110(2), (2017) pp. 021108.

4. C.C.Li, Z.Z.Chen, et al., "Operating behavior of micro-LEDs on a GaN substrate at ultrahigh injection current densities", Opt. Express 27(16), (2019) pp.A1146-1155.

5. R. X. G. Ferreira, E. Xie, M. D. Dawson, et al., "High bandwidth GaN-based micro-LEDs for multi-Gb/s visible light communications," IEEE Photonics Technol. Lett. 28(19), (2016) pp.2023－2026 .

6. J. Herrnsdorf, Y. Wang, N. Laurand, et al., "Micro-LED pumped polymer laser: A discussion of future pump

sources for organic lasers," Laser Photonics Rev. 7(6), (2013) pp.1065–1078 .

7. J. Herrnsdorf, J. J. D. McKendry, et al., "Active-matrix GaN micro light-emitting diodedisplay with unprecedented brightness," IEEE Trans. Electron Devices, 62,(2015) pp.1918–1925.

8. A. H. Jeorrett, S. L. Neale, M. D. Dawson, et al., " Optoelectronic tweezers system for single cell manipulation and fluorescence imaging of live immune cells," Opt. Express 22(2), (2014) pp.1372–1380.

9. P. F. Tian, J. J. D. McKendry, M. D.Dawson, et al., "Size-dependent efficiency and efficiency droop of blue InGaN micro-light emitting diodes," Appl. Phys.Lett. 101(23), (2012) pp.231110.

10. L. Q. Yang, J. Z. Hu, et al., "Thermal Analysis of GaN-Based Light EmittingDiodes With Different Chip Sizes", IEEE Trans.Device Mater. Reliab., 8(2008) pp. 571–575.

11. Q. M. Li, K. R. Westlake, et.al., "Optical performance of top-down fabricated InGaN/GaN nanorod light emitting diode arrays", Opt. Express, 19(2011) pp.25528–25534.

12. H.-S. Chen, D.-M. Yeh, et al.,"Mesa-size-dependent color contrast in flip-chip blue/green two-color InGaN/GaNmulti-quantum-well micro-light-emitting diodes", Appl. Phys. Lett., 89(2006) pp.093501.

13. A. David, N. G. Young, et al., "All-optical measurements of carrier dynamics in bulk-GaN LEDs: Beyond the ABC approximation," Appl. Phys. Lett. 110(25),(2017) pp.253504 .

14. A. David, C. A. Hurni, et al., "Electrical properties of III-Nitride LEDs: Recombination based injection model and theoretical limits to electrical efficiency and electroluminescent cooling," Appl. Phys. Lett. 109(8), (2016) pp.083501 .

15. A. Rashidi, M. Nami, et al., "Differential carrier lifetime and transport effects in electrically injected III-nitride light-emitting diodes," J. Appl. Phys. 122(3), (2017) pp.035706 .

16. E. Tamdogan, G. Pavlidis, S. Graham and M. Arik, "A Comparative Study on the Junction Temperature Measurements of LEDs With Raman Spectroscopy, Microinfrared (IR) Imaging, and Forward Voltage Methods", IEEE TRANSACTIONS ON COMPONENTS PACKAGING AND MANUFACTURING TECHNOLOGY 8(11), (2018) pp.1914-1922.

17. Y. Xi and E. F. Schubert, "Junction-temperature measurement in GaN ultraviolet light-emitting diodes using diode forward voltage method," Appl. Phys. Lett., 85 (2004)2163–2165.

18. A.V.Aladov, K.A.Bulashevich, et al., " Thermal resistanse and nonuniform distribution of electroluminescence and temperature in high-power AlGaInN light-emitting diodes ", St. Petersburg Polytechnical University Journal: Physics and Mathematics, 1(2015) pp.151–158.

19. Bo-Hung Liou, Chih-Ming Chen, et al., "Improvement of thermal management of high-power GaN-basedlight-emitting diodes", Microelectronics Reliability 52 (2012) pp.861–865.

20. Shih-Yi Wen, Hung-Lieh Hu, et al., "A novel integrated structure of thin film GaNLED with ultra-low thermal resistance ", OPTICS EXPRESS, 22(S3)(2014) pp.A601.

21. APSYS (2018 version), Crosslight Software, Inc., Burnaby, Canada, http://www.crosslight.com.

22. P. C. Tsai, Ricky W. Chuang et al., "Lifetime Tests and Junction-TemperatureMeasurement of InGaN Light-EmittingDiodes Using Patterned Sapphire Substrates ", JOURNAL OF LIGHTWAVE TECHNOLOGY, 25(2), (2007) pp.591.

23. D. A. Browne, B. Mazumder, Y. Wu, and J. S. Speck, " Electron transport in unipolar InGaN/GaN multiple quantum well structures grown by NH 3 molecular beam epitaxy," J. Appl. Phys. 117,(2015) pp.185703 .

24. J. Z. Li, Y. B. Tao, Z. Z. Chen, X. Z. Jiang, X. X. Fu, S. Jiang, Q. Q. Jiao, T. J. Yu, and G. Y. Zhang, "Quasi-homoepitaxial GaN-based blue light emitting diode on thick GaN template," Chin. Phys. B 23(1), (2014) pp.016101 .

25. Chengcheng Li, Zhizhong Chen, et al., "Effects of interfaces and current spreading on the thermal transport of micro-LEDs for kA-persquare-cm current injection levels", RSC Adv., 9(2019) pp.24203

Influence of the Charge Transfer on the Lifetime of Quantum-Dot Light-Emitting Diodes

Yue Liang[1,2], Chongyu Shen[1], Junfei Chen[2], Weiye Zheng[2], Zheng Xu[2], Jay Guoxu Liu*[1]

[*1]Shineon (Beijing) Technology Co.
Ltd,3/F, Building#3, Digital Planet, No.58, 5[th]Jinghai Road, BDA, Beijing, China 100167
yueliang@shineon.cn
[2]Key Laboratory of Luminescence and Optical Information, Ministry of Education, Beijing Jiaotong
University, Beijing 100044, China
* jayliu@shineon.cn

Abstract

The operating lifetime of quantum-dot light-emitting diodes (QLEDs) is a critical parameter for quantum-dot display which is becoming an emerging display technology. To pinpoint the causes of device degradation, we demonstrated an enhanced reliability of all-solution processed QLEDs by introducing an insulating interfacial layer between the ETL and the QDs. It is confirmed that a PMMA interfacial layer can delay the electron transfer and reduce the nonradiative recombination, which in turn slow down the degradation of QDs-ZnO layer. In comparison with the standard QLEDs, device fabricated with PMMA of 1.0 mg/ml shows longest lifetime of 28.7h, which improves the lifetime by 50%. We studied the effect of electrons transfer on the degradation mechanisms, which lays the foundation for further improvement of the device life.

Keywords: Quantum-dot, QLED, Electroluminescence, Degradation mechanisms, Charge transfer, Interfacial layer.

Introduction

Quantum dots (QDs) are semiconductor quasi nano-crystals which have quantum confinement effect. QDs become one of the ideal light source materials due to their advantages of high photoluminescence quantum yield (PLQY)[1, 2], narrow emission peak[3, 4] and the solution-based processability[5, 6]. An electroluminiscent quantum dot display is a display device that uses QD light-emitting diodes (QLED) which can produce pure monochromatic red, green, and blue light. These displays are similar to active-matrix organic light-emitting diode (AMOLED) and MicroLED displays, in that light would be produced directly in each pixel by applying electric current to the inorganic nano-particles. Due to the superior properties of QDs, QLEDs have a lot of outstanding advantages such as high purity, good stability and outstanding luminous efficiency, which have been explored for two applications of far-reaching technological importance, displays and lighting[7]. Emissive quantum dot displays can achieve the same contrast as OLED displays with "perfect" black levels in the off-state and are capable of displaying wider color gamut than OLEDs with some devices approaching full coverage of the BT.2020 color gamut[8].

QDs are known to be degraded via various molecular interactions with oxygen and water, which can enhance or degrade their optical performance[8-13]. The QDs can be degraded by various mechanisms, such as thermal heating, moisture, oxygen, and UV light[14, 15]. The known degradation mechanisms of QLEDs include electrode corrosion, material degradation, charge accumulation/leakage, and Joule heating[16]. Charge balance is a key factor. Holes and electrons are injected from both electrodes, and then transferred to the QD emissive layer (EML) through the charge transport layer (CTL) to form excitons and emit light, which affects device lifetime via various processes, as well as device efficiency. However, many QLEDs suffer from charge imbalance due to the different injection and transport rates of the holes and electrons[17]. To meet the requirements for display, QLEDs need to have high external quantum efficiency (EQE) by optimizing the multilayer structure and balancing the carrier injection[18, 19]. The recently reported maximum external quantum efficiency (EQE) for red, green and blue QLEDs are 21.6%, 22.9% and 19.8%, respectively[20]. For lighting applications, threshold values for both brightness and EQE have been enhanced[21]. However, the lifetime of QLEDs still cannot meet the commercial requirement and needs to be further improved. To our best knowledge, experimental study on the QLED degradation mechanisms is very limited. To productize this technology, researchers have to solve the life degradation issue of QLEDs by understanding its degradation mechanisms [22].

Our experimental study is on the QLED lifetime values related to the charge transfer. It has been reported that the charge transfer efficiency of the interface between the emission layer (EML) and the electron transporting layer (ETL) influences the operational instability of QLEDs significantly[19]. We constructed a device that is inserted an insulating layer (polymethyl methacrylate, PMMA), which holds back electron transfer. By changing the thickness and concentration of PMMA, we can adjust the charge transfer efficiency of the interface between QD and ZnO, and therefore to identify the relationship between charge transfer and device lifetime.

Experimental Details

The QLED devices were fabricated on commercially available patterned ITO glass substrates with the sheet resistance less than 15 Ω/sq. Prior to deposition of functional layers, the ITO substrates were successively cleaned in ultrasonic bath of detergent, deionized water, ethyl alcohol, deionized water for 30 min, and dried by nitrogen. After UV ozone treatment for 10 min, PEDOT:PSS solution (Heraeus-Clevios P VP Al 4083, filtered through a 0.22 μm MCE filter) was spin coated onto the ITO-coated glass substrates at 4000 rpm and baked at 150°C for 15 min in the air. Then, the PEDOT:PSS-coated substrates were transferred into nitrogen-filled glove box to sequentially deposit other functional layers. The hole transport layer TFB was spin coated at 4000 rpm for 40 seconds and baked at 120°C for 20 min. After that, the 15 mg/ml red QDs (CdZnSe/ZnS/OT, core and shell ~12.2 nm, obtained from Mesolight Inc., QY ~85%) dissolved in octane solution were spin coated at 3000 rpm for 40 seconds, followed by baking at 100°C. After cooling the samples for 10 min, 60 µl PMMA was spin coated on QD EML at 3000 rpm and baked at 80°C for 10 min. The 30 mg/ml ZnMgO was spin coated at 3000 rpm and baking at 70°C for 20 min. Finally, the samples were transferred to a high-vacuum evaporation chamber to sequentially deposit Al anode at a base pressure of 5×10^{-4} Pa.

Results and Discussion

We studied the effect of electrons transfer on the degradation mechanism by choosing a popular device structure composed of a transparent anode, a polymeric hole injection layer (HIL), a polymeric hole transporting layer (HTL), an emission layer (EML) assembled of red QDs, an electron transporting layer (ETL), and a top cathode[23], as shown in Fig. 1a. The energy level diagram of devices is shown in Fig. 1b. PMMA with different thickness is inserted into the interface of QD and ZnO to change the charge transfer efficiency at the interface. Fig. 1c shows the J-V-L characteristics of the QLEDs. Device fabricated with PMMA of 1.0 mg/ml shows the highest luminance of 90068 cd/m2 (at 7.8 V). Under the same voltage, the current density of the devices decreases with the increase of PMMA concentration, while their luminance values exhibit a clear difference. It suggests that the electron transfer of device is reduced with the increasing concentration of PMMA, and electrons and holes can combine in the QD layers more efficiently. As a result, the current efficiency increases from 6.93 to 15.30 cd/A with PMMA concentration increasing from 0 to 1.0mg/ml, which indicate that the PMMA layer with an appropriate concentration benefits to improve the performance of QLED. The representative EL spectra of devices based on different PMMA concentrations at 8V are presented in Fig. 1f. The peaks of the spectra are at 628nm. The full width at half maximum (FWHM) of the EL spectrum is 20nm. No parasitic peaks from other layers are observed for all devices throughout the entire voltage range. With the change of

Table 1 Comparison of the performance of different solution concentrations of PMMA

Solution Concentration of PMMA (mg/ml)	EL peak[a] (nm)	FWHM (nm)	V_T[b] (V)	Lmax[c] (cd/m2)	CEmax[d] (cd/A)	EQEmax[e] (%)
0	628	20	2.6	27540	6.93	5.13
0.5	628	20	2.2	88593	10.80	8.40
1.0	628	20	2.1	90068	15.30	11.50
1.5	628	20	2.6	29756	7.59	5.64

a Electroluminescence peak at maximum luminance.
b Measured voltage when luminance was 1 cd/m2.
c Maximum luminance.
d Maximum current efficiency.
e Maximum external quantum efficiency.

PMMA concentration, there is no significant change in EL spectrum. These phenomena indicate that electrons and holes can be effectively confined in the QD active layers during device operation. The detailed device performance data are summarized in Table 1.

The introduction of an insulating interfacial layer between the ETL and the QDs is known to be effective in controlling charge balance. As Fig.2a shows, the introduction of a nanometer-thin PMMA layer between the ZnO and QD layers to delay electron transfer greatly improved the lifetime. Device fabricated with PMMA of 1.0 mg/ml shows the longest lifetime of 28.7h. In order to verify the effect of PMMA addition on interfacial charge transfer, PL decay at different PMMA concentrations was tested, which is shown in Fig. 2b. The PL emission of the QDs decays slower by increasing the concentration from 0 mg/ml to 1.5 mg/ml, signifying that the nonradiative recombination is alleviated with PMMA layer. Therefore, it is deduced that the PMMA layer can decrease electron transfer, which reduces the nonradiative recombination, corresponding to the enhanced PL intensity and the enlarged PL lifetime. The enhanced PL intensity favors the high brightness and efficiency in QLEDs, and therefore, the reduced electron transfer by introducing PMMA is also one of the derivations for the improvement of device performance and lifetime, which explains the reason why insulating interfacial layer can further improve the device lifetime.

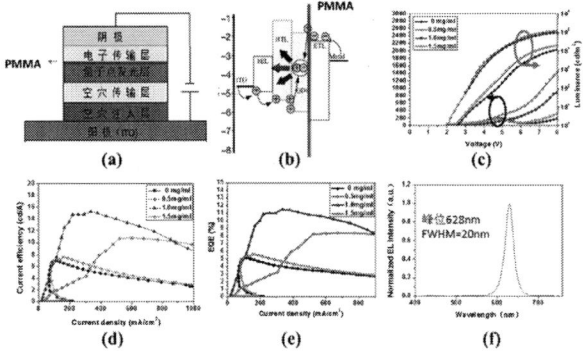

Fig.1 (a) Schematic of the QLED device structure, (b) Energy level diagram of QLEDs, (c) J-V-L curves, (d) CE-J characteristics, (e) EQE-J characteristics and (f) E-L spectrum.

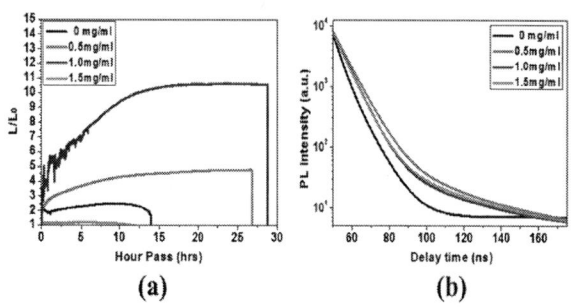

Fig.2 The performance of QLEDs with different solution concentrations of PMMA, (a)L/L$_0$-t, (b) TRPL.

Table 2. Lifetime of different solution concentrations of PMMA

Solution Concentration of PMMA (mg/ml)	0	0.5	1.0	1.5
Lifetime (h)	13.9	26.7	28.7	14.9

Conclusions

Due to the high free carrier density in the ETL, the holes in the QD are trapped on the interface, which leads to the non-radiative recombination of the interface. This exacerbates the interface aging and affects the normal injection of carriers.

In summary, we have demonstrated a life time performance enhancement of all-solution processed QLEDs by introducing a PMMA insulating interfacial layer between the ETL and the QDs. Device fabricated with PMMA of 1.0 mg/ml shows the highest luminance of 90068 cd/m2 (at 7.8 V). The maximum CE and EQE achieved are 15.3 cd/A and 11.5%, respectively. Meanwhile, the lifetime has been almost doubled from 13.9h to 26.7h. It is confirmed that charge transfer of the interface between the EML and ETL influences the operational instability of QLEDs significantly, and the lifetime is positively correlated with charge transfer.

Acknowledgments

This research was supported by the National Key Research and Development Program of China under Grant No. 2016YFB0401702, the National Natural Science Foundation of China under Grant No. 61775013, No. 11474018 and No. 61704007, the Fundamental Research Funds for the Central Universities under the Grant No. 2017JBZ105 and No. 2017RC034.

References

[1] Y. Yang *et al.*, "High-efficiency light-emitting devices based on quantum dots with tailored nanostructures," *Nature Photonics,* vol. 9, no. 4, pp. 259-266, 2015.

[2] X. Jin *et al.*, "Bright alloy type-II quantum dots and their application to light-emitting diodes," *Journal of Colloid and Interface Science,* vol. 510, pp. 376-383, Jan 15 2018.

[3] M. M. Yan *et al.*, "Enhancing the Performance of Blue Quantum-Dot Light-Emitting Diodes Based on Mg-Doped ZnO as an Electron Transport Layer," *IEEE Photonics Journal,* vol. 9, no. 2, pp. 1-8, 2017.

[4] Y. Fu, W. Jiang, D. Kim, W. Lee, and H. Chae, "Highly Efficient and Fully Solution-Processed Inverted Light-Emitting Diodes with Charge Control Interlayers," *Acs Applied Materials & Interfaces,* vol. 10, no. 20, p. 17295, 2018.

[5] X. Dai, Y. Deng, X. Peng, and Y. Jin, "Quantum-Dot Light-Emitting Diodes for Large-Area Displays: Towards the Dawn of Commercialization," *Adv Mater,* vol. 29, no. 14, Apr 2017.

[6] Y. Zou *et al.*, "A General Solvent Selection Strategy for Solution Processed Quantum Dots Targeting High

Performance Light-Emitting Diode," *Advanced Functional Materials,* vol. 27, no. 1, 2017.

[7] H. Shen, W. Cao, N. T. Shewmon, C. Yang, L. S. Li, and J. Xue, "High-efficiency, low turn-on voltage blue-violet quantum-dot-based light-emitting diodes," *Nano Lett,* vol. 15, no. 2, pp. 1211-6, Feb 11 2015.

[8] S. R. Cordero, P. J. Carson, R. A. Estabrook, G. F. Strouse, and S. K. J. J. o. P. C. B. Buratto, "Photo-Activated Luminescence of CdSe Quantum Dot Monolayers," vol. 104, no. 51, pp. 12137-12142.

[9] W. G. J. H. M. v. S. Dr, P. L. T. M. F. Dr, A. A. B. Dr, H. C. G. P. Dr, and A. M. P. D. J. ChemPhysChem, "Blueing, Bleaching, and Blinking of Single CdSe/ZnS Quantum Dots," vol. 3, no. 10, pp. 871-879, 2002.

[10] M. Jones, J. Nedeljkovic, R. J. Ellingson, A. J. Nozik, and G. J. J. o. P. C. B. Rumbles, "Photoenhancement of Luminescence in Colloidal CdSe Quantum Dot Solutions," vol. 107, no. 41, pp. 11346-11352.

[11] J. Müller, J. M. Lupton, A. L. Rogach, J. Feldmann, D. V. Talapin, and H. J. P. R. L. Weller, "Monitoring Surface Charge Movement in Single Elongated Semiconductor Nanocrystals," vol. 93, no. 16, p. 167402.

[12] C. Carrillo-Carrion, S. Ca?rdenas, B. M. Simonet, and M. J. A. C. Valcarcel, "Selective Quantification of Carnitine Enantiomers Using Chiral Cysteine-Capped CdSe(ZnS) Quantum Dots," vol. 81, no. 12, pp. 4730-4733.

[13] K. Pechstedt, T. Whittle, J. Baumberg, and T. J. T. J. o. P. C. C. Melvin, "Photoluminescence of Colloidal CdSe/ZnS Quantum Dots: The Critical Effect of Water Molecules," vol. 114, no. 28, pp. 12069-12077.

[14] K. Z. Xing, N. Johansson, G. Beamson, D. T. Clark, J. L. Brédas, and W. R. Salaneck, "Photo‐oxidation of poly(p‐phenylenevinylene)," vol. 9, no. 13, pp. 1027-1031, 2010.

[15] G. J. Herrera and J. E. W. J. S. Metals, "Photoemission study of the thermal and photochemical decomposition of a urethane-substituted polythiophene," vol. 128, no. 3, pp. 0-324.

[16] C. Garditz, A. Winnacker, F. Schindler, and R. J. A. P. L. Paetzold, "Impact of Joule heating on the brightness homogeneity of organic light emitting devices," vol. 90, no. 10, p. 103506.

[17] H. Moon, C. Lee, W. Lee, J. Kim, and H. J. A. M. Chae, "Stability of Quantum Dots, Quantum Dot Films, and Quantum Dot Light-Emitting Diodes for Display Applications."

[18] Y. Shirasaki, G. J. Supran, M. G. Bawendi, and V. Bulović, "Emergence of colloidal quantum-dot light-emitting technologies," *Nature Photonics,* vol. 7, no. 1, pp. 13-23, 2012.

[19] A. Perumal, H. Faber, N. Yaacobi ≒ Ross, P. Pattanasattayavong, and D. D. C. J. A. M. Bradley, "High-Efficiency, Solution-Processed, Multilayer Phosphorescent Organic Light-Emitting Diodes with a Copper Thiocyanate Hole-Injection/Hole-Transport Layer," vol. 27, no. 1, pp. 93-100, 2015.

[20] L. Wang et al., "Blue Quantum Dot Light-Emitting Diodes with High Electroluminescent Efficiency," ACS Appl Mater Interfaces, vol. 9, no. 44, pp. 38755-38760, Nov 8 2017.

[21] J. H. Chang, D. Hahm, K. Char, and W. K. Bae, "Interfacial engineering of core/shell heterostructured nanocrystal quantum dots for light-emitting applications," *Journal of Information Display,* vol. 18, no. 2, pp. 57-65, 2017.

[22] S. Chen *et al.*, "On the degradation mechanisms of quantum-dot light-emitting diodes," *Nat Commun,* vol. 10, no. 1, p. 765, Feb 15 2019.

[23] L. Qian, Y. Zheng, J. Xue, and P. H. Holloway, "Stable and efficient quantum-dot light-emitting diodes based on solution-processed multilayer structures," *Nature Photonics,* vol. 5, no. 9, pp. 543-548, 2011.

Effect of Mechanical Stress on the Electrical Characteristics of Different Type IGBT Chips

Yihui Zhang[1,2], Jinyuan Li[1,2], Yinghan Liu [1,2], Guanbin Wu[3]

1. State Key Laboratory of Advanced Power Transmission Technology, Beijing, 102209, China
2. Global Energy Interconnection Research Institute Co. Ltd., Changping District, Beijing, 102209, China
3. State Grid Shandong Electric Power Company, Jinan, Shandong, China

Abstract

IGBT power semiconductor devices are widely used in modern power systems, and their electrical characteristics directly affect the performance of converters. Firstly, the influence of mechanical stress on the electrical characteristics of MOSFET and IGBT chips is summarized. The way of mechanical stress generation in IGBT module is summarized from two aspects of chip technology and packaging technology, and the electrical characteristics of different types of IGBT chips are also discussed. The effects of mechanical stresses in different directions on the electrical characteristics of different types of IGBT chips are analyzed.

1. Introduction

At present, power devices play a more and more important role in power system, electric vehicle and other applications. With the increasing demand for the capacity of power devices, new chips and packaging forms emerge as the times require. The current level of the chip is improved by designing a new chip structure, and the current level and heat dissipation performance of the module are improved by designing a new packaging structure. In order to improve the current level and heat dissipation ability of power module, IGBT packaging mode without bonding wire is produced. In this way, the sub-module is directly connected between two copper plates, and the two-sided heat dissipation structure can reduce the junction temperature of the chip. Because the thermal expansion coefficients of silicon and copper are different, when the chip is subjected to thermal stress, the chip will be subjected to horizontal mechanical stress due to the expansion of the copper plate and the shrinkage of the chip. When the temperature difference is 200 C, the thermal stress of the device is about 500 MPa, of which the thermal expansion coefficient of silicon is 3 x10-6/K and that of copper is 17 x10-6/K. The Young's modulus of silicon is 170 Gpa. Such a large stress will affect the electrical characteristics and interconnection reliability of power equipment.

1In reference [1], the mode of mechanical stress generation in IGBT is studied, and the variation of static and dynamic parameters of trench IGBT under stress which is perpendicular to channel current direction is analyzed. Reference [2] compares the static and dynamic parameters of planar gate IGBT and trench IGBT under mechanical stress, and points out that the influence of mechanical stress on the electrical characteristics of IGBT is mainly concentrated in the channel and drift region of IGBT. Reference [3] has studied the electrical characteristics of

LDMOS under uniaxial mechanical stress. It is pointed out that the long channel nLDMOS is greatly affected by mechanical stress, and the transverse and vertical electric fields have important influence on the strain effect. Reference [4] analyses the variation of on-state electrical and off-state characteristics of LIGBT under stress parallel to the channel direction. It is pointed out that gate voltage and collector voltage have great influence on strain effect of IGBT.

Before the 1980s, the research on improving the chip's current-passing capability was based on the "silicon limit", and the curve between on-state resistance and breakdown voltage was studied. Recently, it has been proposed to increase carrier mobility by applying mechanical stress, and then reduce the on-state resistance of the chip to reduce heat generation. Before that, the drift resistance was reduced by increasing the doping concentration in the drift region, and the breakdown voltage of the device was reduced by increasing the doping concentration in the drift region. By applying stress, the breakdown voltage of the chip will not be changed because the carrier doping concentration in the drift region is not changed. Therefore, the decoupling between on-state resistance and breakdown voltage is realized.

In 2002, Intel introduced process-induced strain into silicon MOSFET for the first time, and began to play an important role in the mainstream VLSI semiconductor industry [5,6]. Literature [7] proposes that mechanical stress can be applied to the chip by embedded SiGe substrate and deposited polysilicon buried layer to reduce the on-state resistance. Until recently, experimental evidence for drift-loss-based integrated power transistors (through the region of MOS capacitors and breaking the silicon limit) was presented. The device is called "Xtremos" and its implementation is compatible with standard CMOS process flow [8] [9]. Reference [10] analyses the increase of carrier mobility of integrated power transistors under the action of strained silicon and the variation of electrical characteristics. The influence of strained silicon on the electrical characteristics of n-type and p-type MOSFETs is analyzed in reference [11]. In reference [7], a strained silicon process is described, which allows N-type and P-type MOSFETs to independently orient strain by adjusting the stress of N-type and P-type sealing films. Compared with biaxial strain, p-type MOSFET of uniaxial strained silicon has significant advantages. It has greater mobility enhancement for given strain, and mobility enhancement occurs in large vertical field. Reference [12] comprehensively studied the effects of these four mechanical stresses on the NBTI reliability behavior of silicon PMOSFET. For the first time, uniaxial and biaxial compressive strains can improve the NBTI reliability of p-type silicon MOSFET. In contrast, the reliability of the biased temperature instability test (NBTI) of uniaxial and biaxial tension strain Si-PMOSFET is

Project Supported by Science and Technology Project of State Grid of China (SGGR0000GLJS1800176).

reduced. These observations can be explained by considering the interaction between carriers and Si-H bonds on the interface and the change of gate current through the interface.

In this paper, the influence of mechanical stress on the electrical characteristics of MOSFET and IGBT chips is summarized at first. In section 2, the mode of mechanical stress generation is analyzed from two aspects: chip technology and packaging technology. In section 3, the piezoresistive effect of IGBT is analyzed and the influence of mechanical stress on the electrical characteristics of IGBT is analyzed (current force). The influence of piezoresistive effect on electrical characteristics is analyzed.

2. The formation of chip mechanical tress

2.1 Chip Techinology

At present, the technology of silicon integrated circuits driven by CMOS devices has entered the nano-scale. After reducing to the nano-scale, the thickness of gate dielectrics decreases gradually. This will lead to the problems of increasing leakage current, increasing power density and decreasing mobility of collector, which deteriorate the performance of devices due to physical limitations. This requires new technologies to improve transistor performance. Silicon strain technology can improve carrier mobility and break the "silicon limit".

One of the important aspects is to take measures to improve the mobility of carriers in the channel to compensate for the degradation of mobility caused by the Coulomb interaction caused by high channel doping and the enhancement of effective electric field intensity and interface scattering caused by the thinning of gate dielectrics. At present, strained silicon technology is widely used. Compared with the same size bulk Si MOSFET, the strained silicon MOSFET manufactured by the existing silicon production line can reduce power consumption by one third, increase speed by 30%, and increase device package density by 50%. The cost of production increased by only 2%. Strained silicon refers to an ultra-thin strain layer with only a few nanometers thickness. Strained silicon is used to replace the original high-purity silicon to make the internal channels of transistors. By stretching the distance between atoms in transistors, the number of atoms per unit length can be reduced. When electrons pass through these regions, the resistance they encounter decreases, thus improving the transistor performance.

Strain formation in MOSFET channels can be achieved by process steps, differences in natural lattice constants on materials, and packaging. From the action area of strain, it can be divided into global strain (biaxial strain) and local strain (uniaxial strain). According to the type of applied stress, it can be divided into tension strain and compression strain. Tensile strain is formed by growing Si layer on SiGe substrate, and compressive strain is formed by growing SiGe layer on SiGe substrate. Stress-enhanced carrier mobility can be achieved mainly through two ways: reducing the effective mass and reducing the scattering probability. The mobility can not increase indefinitely. When the stress increases to a certain value, the mobility enhancement will reach saturation.

2.2 Packaging Process

Because the power module must achieve high output power density, it is very important to reduce the thermal resistance of the power module. Recently, a new module structure is proposed, that is, power devices are directly welded between two copper plates. The junction temperature of the double-sided cooling structure is reduced both in transient and steady state. Because the thermal expansion coefficients (TCE) of copper and silicon are very different, the thermal stress of this module is larger than that of the traditional module. In this case, due to the expansion and contraction of the copper plate, the stress on the chip surface occurs along the horizontal direction (X axis). The discontinuity of surface morphology and mismatch of thermal expansion coefficient caused by chip technology and packaging process; the thermal stress caused by the frequent conversion between hot and cold environments during the temperature cycle test; and the high temperature of each component during welding in the packaging process. In the process, the thermal expansion coefficient and thermal conductivity of different materials are different, which results in a certain stress in the module.

For pressed IGBT devices, pressure perpendicular to the direction of the device will be applied in use, which will lead to stress in the vertical direction of the chip.

3. Effect of Pressure on Electrical Characteristics of IGBT

The variation of voltage in MOS channel and drift layer is larger than that of built-in potential and P substrate. The voltage variation of P substrate is small because the piezoresistive coefficient of the hole is 2% of that of the electron. Therefore, the effect of mechanical stress on the on-off voltage is mainly due to the change of partial electron mobility of MOSFET. Compressive stress will enhance the electron injection into the drift layer from the MOS channel. Therefore, in the case of bipolar devices, the mechanical stress changes the ratio of current component (electron/hole) and carrier mobility. From the above analysis, it can be seen that mechanical stress can improve carrier mobility, improve conductivity by increasing carrier mobility, and then reduce the on-state resistance of IGBT, affecting the electrical characteristics of IGBT. The effect of mechanical stress on carrier mobility is shown in (1):

$$\mu_{n-st} = \mu_n \cdot [1 - (\pi_{11} \cdot \sigma_{xx} + \pi_{12} \cdot \sigma_{yy})] \qquad (1)$$

The dynamic parameters of IGBT are greatly influenced by pressure, because the switching off current of IGBT can be divided into two stages: the first stage is switching off the inverted channel, resulting in a rapid drop of current; the second stage is a longer duration, which can lead to the trailing current of IGBT. The first stage is called MOSFET turn-off, and the second stage is called transistor turn-off. Because the first stage mainly depends on the electronic current, it is greatly affected by the stress. According to reference [1] - [2], the electrical parameters of IGBT are affected by mechanical stress as shown in Table 2.

Table 3.1 Mechanism of Pressure Influencing Electrical Parameters

electrical parameters	influencing mechanism

breakdown voltage	Voltage withstanding is mainly related to the depletion region formed at the low doping concentration N-in the drift region, so the withstand voltage level is related to the band gap width, while the mechanical stress has very little effect on the band gap width, so the breakdown voltage is little affected by the mechanical stress and almost unchanged.
Collector leakage current	Depending on the electron mobility, the electron mobility is greatly affected by mechanical stress.
saturation voltage	Depending on the electron mobility, the electron mobility is greatly affected by mechanical stress.
Collector current	Depending on the electron mobility, the electron mobility is greatly affected by mechanical stress.
Turn-off time	Depending on the electron mobility, the electron mobility is greatly affected by mechanical stress.
trailing time	Depending on the hole mobility, the effect of mechanical stress on the hole mobility is less significant.
Saturated current	Depending on the electron mobility, the electron mobility is greatly affected by mechanical stress.

Next, the effects of mechanical stresses on the electrical characteristics of IGBT chips with planar and grooved grids are analyzed respectively.

When IGBT is blocked, the change of band gap caused by mechanical stress is very small (only 30 meV under 500 MPa stress). Therefore, for planar IGBT and trench IGBT, the breakdown voltage variation caused by the change of bandgap is not significant.

4. Variation of Electrical Characteristics of Trench IGBT

For the stress of [110] and [$\bar{1}$10] directions, they are perpendicular to the current direction, including channel current and drift current. Because the piezoresistive coefficient of electrons is much larger than the piezoresistive coefficient of holes, the effect of mechanical stress on the electronic current will be higher than that of the hole current. For IGBT with high voltage level, on-state resistance mainly occurs in n-drift region. At this time, if the mechanical stress is applied, the electron mobility will be reduced, the resistivity of channel and drift region will be increased, so the on-state voltage drop will be increased, the I_C will be reduced, the saturated current will be reduced, and the turn-off time will be increased, and the tail current has little change. And the stress will also give the proportion of electrons/holes in the alternating current, which will increase the injection of electrons from the channel to the drift layer.

For the pressure in [001] direction, it is parallel to the current direction. Therefore, the pull force will increase the electron mobility in the channel and drift region, and decrease the resistivity in the channel and drift region. As a result, the on-state voltage drop decreases, increases, the saturated current increases, the turn-off time decreases, and the tail current has little effect.

Table 4.1 Effect of Stresses in Different Directions on Electrical Characteristics of Trench IGBT

the direction of the stress	relation with Current.	trend of electrical parameters change
[001]	parallel to channel current	$V_{ce(sat)}$ decreases; I_C increases saturation current increases
[001]	parallel to drift current	turn-off time decreases; tailing time unchanged I_{ces} increases
[110] [$\bar{1}$10]	vertical to channel current	$V_{ce(sat)}$ increases; I_C decreases saturation current decreases
[110] [$\bar{1}$10]	vertical to drift current	turn-off time increases; tailing time unchanged I_{ces} decreases

5. Variation of Electrical Characteristics of Planar IGBT

For the stress in [110] direction, the current perpendicular to the drift region and parallel to the channel current will increase the channel carrier mobility and decrease the drift region carrier mobility. However, because the piezoresistive coefficient in parallel is larger than that in vertical, the increase of channel carrier mobility is dominant, so the on-state voltage drop decreases, the I_C increases, the saturated current increases, the turn-off time decreases, and the tail current has little effect.

For the stress in [$\bar{1}$10] direction, the current perpendicular to the drift region and perpendicular to the channel current will decrease the channel carrier mobility and the drift region carrier mobility, so the on-state voltage drop will increase, I_C decrease, the saturation current will decrease, the turn-off time will increase, and the tail current will not change much.

For the force in [001] direction, it is perpendicular to the channel current and parallel to the drift region current, which will reduce the carrier mobility in the channel and increase the carrier mobility in the drift region. However, because the piezoresistive coefficient in parallel is larger than the vertical piezoresistive coefficient, the drift region mobility increases mainly, so the tensile force will make the carrier mobility in the drift region increase. The on-state voltage drop decreases, I_C increases, saturation current increases, turn-off time decreases, and tail current has little effect.

Table 5.1 Effect of stress in different directions on electrical characteristics of planar IGBT

the direction of the stress	relation with Current.	trend of electrical parameters change
[001]	vertical to	$V_{ce(sat)}$ decreases

Direction	Condition	Effect
	channel current	I_C increases saturation current increases
	parallel to drift current	turn-off time decreases tailing time unchanged I_{ces} increases
[110]	parallel to channel current	$V_{ce(sat)}$ decreases I_C increases saturation current increases
[110]	vertical to drift current	turn-off time decreases tailing time unchanged I_{ces} increases
[$\bar{1}$10]	vertical to channel current	$V_{ce(sat)}$ increases I_C decreases saturation current decreases
[$\bar{1}$10]	vertical to drift current	turn-off time increases tailing time unchanged I_{ces} decreases

6. conclusion

In this paper, the research status of the influence of mechanical stress on the electrical characteristics of power devices is reviewed. The source of stress is introduced from the aspects of chip design and packaging technology. The influence of mechanical stress on the electrical characteristics of chip is analyzed from the influence of pressure on carrier mobility and energy band. The influence of Mechanical stress on the electrical characteristics of chip and on the IGBT of planar gate and groove gate in different directions is also discussed. The change rule of electrical characteristics under mechanical stress is summarized.

Acknowledgment

This work is supported by Science and Technology Project of State Grid of China "the Effect of Pressure and Transient Overcurrent on Performance and Reliability of Voltage-bonded IGBT Chip Devices" (SGGR0000GLJS1800176). Meanwhile, thanks for the experimental support of State Key Laboratory of Advanced Power Transmission Technology.

References

[1] Usui, M. , Ishiko, M. , Hotta, K. , Kuwano, S. , & Hashimoto, M. . (2005). Effects of uni-axial mechanical stress on igbt characteristics. Microelectronics Reliability, 45(9-11), 1682-1687.

[2] Wu, W. , Liu, S. , Zhu, J. , & Sun, W. . (2018). [IEEE 2018 IEEE 30th International Symposium on Power Semiconductor Devices and ICs (ISPSD) - Chicago, IL (2018.5.13-2018.5.17)] 2018 IEEE 30th International Symposium on Power Semiconductor Devices and ICs (ISPSD) - Comprehensive investigation on mechanical strain induced performance boosts in LDMOS. (pp.60-63).

[3] Thompson, S. E. , Armstrong, M. , Auth, C. , & Cea, S. . (2004). A logic nanotechnology featuring strained-silicon. Electron Device Letters IEEE, 25(4), 191-193.

[4] Iwai, H. (2009). Roadmap for 22nm and beyond (Invited Paper). Roadmap for 22 nm and beyond (Invited Paper).

[5] Thompson, S. E. , Armstrong, M. , Auth, C. , & Cea, S. . (2004). A logic nanotechnology featuring strained-silicon. Electron Device Letters IEEE, 25(4), 191-193.

[6] Moens, P. , Bauwens, F. , Baele, J. , & Vershinin, K. (2006). XtreMOS : The First Integrated Power Transistor Breaking the Silicon Limit. International Electron Devices Meeting.

[7] Moens, P. , Bauwens, F. , Desoete, B. , Baele, J. , & Tack, M. . (2007). Record-low on-Resistance for 0.35 μm based integrated XtreMOSTM Transistors. Power Semiconductor Devices and IC's, 2007. ISPSD '07. 19th International Symposium on. IEEE.

[8] Moens, Roig, Clemente, Wolf, D. , Desoete, & Bauwens. (2007). Stress-Induced Mobility Enhancement for Integrated Power Transistors. IEEE International Electron Devices Meeting. IEEE.

[9] Hoyt, J. L., Nayfeh, H. M., Eguchi, S., Aberg, I., Xia, G., & Drake, T., et al. (2002). Strained silicon MOSFET technology. International Electron Devices Meeting..

[10] Wu W, Liu C, Sun J, et al. Experimental Study on NBTI Degradation Behaviors in Si pMOSFETs Under Compressive and Tensile Strains[J]. IEEE Electron Device Letters, 2014, 35(7):714-716.

[11] F.Baccar, S.Azzopardi, L.Theolier, et al. Electrical characterization under mechanical stress at various temperatures of PiN power diodes in a health monitoring approach[J]. Microelectronics Reliability,2013.

[12] Belmehdi Y, Azzopardi S, Deletage J Y, E. . (2010). Electromechanical characterization of "flying" Planar Gate Punch Through IGBT bare die. Energy Conversion Congress and Exposition. 2010. ECCE 2010. IEEE. IEEE.

[13] Belmehdi, Y. , Azzopardi, S. , Deletage, J. Y. , & Woirgard, E. . (2009). Assessment of uni-axial mechanical stress on Trench IGBT under severe operating conditions: a 2D physically-based simulation approach. Energy Conversion Congress and Exposition, 2009. ECCE 2009. IEEE. IEEE.

[14] Belmehdi, Y. , Azzopardi, S. , Deletage, J. Y. , & Woirgard, E. . (2009). Assessment of uni-axial mechanical stress on Trench IGBT under severe operating conditions: a 2D physically-based simulation approach. Energy Conversion Congress and Exposition, 2009. ECCE 2009. IEEE. IEEE.

[15] Gallon, C., Reimbold, G., Ghibaudo, G., & Bianchi, R. A. (2004). Electrical analysis of mechanical stress induced by sti in short mosfets using externally applied stress. IEEE Transactions on Electron Devices, 51(8), 1254-1261.

Optimal thermal design of LED automotive headlamp with the response surface method

Zhibin Tang[1,4], Jiajie Fan[1,2,3*], Wei Chen[2], Yutong Li[1,2], Moumouni Guero Mohamed[1], Ru Li[4]

[1] College of Mechanical and Electrical Engineering, Hohai University, Changzhou 213022, China;
[2] Changzhou Institute of Technology Research for Solid State Lighting, Changzhou 213161, China
[3] EEMCS Faculty, Delft University of Technology, Delft 2628, the Netherlands
[4] Changzhou Xingyu Automotive Lighting System Co, Ltd. China.
*Corresponding: E-mail: jay.fan@connect.polyu.hk, jiajie.fan@hhu.edu.cn. Tel: +86-519-85191933.

Abstract

Light Emitting Diode (LED) is gradually being applied in automotive lighting systems as the new generation of light source. Recently, increasing demands are adapt on automotive headlamps, including not only being functionally perfect, cost-effective and durable, but also with fashion, energy saving and environmental protection. LED light source has the advantages of long lifetime, high efficiency and energy saving, compact size etc., therefore, it has been widely used in headlamps, turn signals, brake lights, position lights, etc., and will soon become the mainstream of the automotive light source market. Although it has higher energy conversion efficiency, LED headlamp is still suffering difficulties on the thermal management. In this paper, we optimize the designs of heatsink for a commercialized LED automotive headlamp module by using the response surface method. Firstly, the temperature distribution of the test sample is simulated by the finite element (FE) method. Then, we use the response surface method to optimize the design parameters of heatsink on its thermal dissipation capacity. The results indicate that: (1) Through optimizing the structure of the heat sink with the thickness, spacing and height of fins as 3mm, 9mm and 60mm respectively, the LED junction temperature drops by 2.9%; (2) By changing the materials of heat sink and PCB substrate as 6063 aluminum alloy and AlN respectively, the LED junction temperature lowers down by 11.9%.

Keywords: LED; Automotive headlamp; Thermal management; Finite element; Response surface method

1. Introduction

As a new type of light source, Light-emitting Diode (LED) has the following advantages: (1) small size, easy to design and meet the requirements of automotive lightweight; (2) good photoelectric performance and quick response to effectively increase the active safety of the car [1]; (3) it has excellent characteristics of energy saving, environmental resistance and vibration[2-3]. At present, LED light sources are mainly used in automotive brake lights, turn signals, reversing lights, tail lights, and instrument lights [4]. However, the application of LED in automobiles still has technical limitations, for instance, the thermal effect caused by poor heat dissipation is one of the main bottlenecks [5]. When it works in a high temperature environment for a long time, the life of LED will be greatly shortened [6], and the optical performance will be greatly affected [7], which has a great negative impact on the safe driving of the car. Therefore, it is important to solve the thermal management problem for LED automotive lighting.

In terms of thermal management for LEDs, Luo et al. [8] proposed a new heat dissipation method with the steam chamber coupling finned heat sink. Ma et al. [9] developed a heat dissipation method for the front surface of cooling used in high power LED lamp. Ye et al. [10] proposed a new way to realize the thermal management by using phase change technology. Yang et al. [11] studied the heat transfer characteristics of different layer crack shapes and positions on the LED package interface by constructing different interface spallation models. Shaeri et al. [12] enhanced the heat dissipation performance of the heat sink through the fin opening. S. W. Jang et al. [13] used the micro jet to dissipate the vehicle lights' heat, and discussed the effect of heat sink shape on heat dissipation. S. J. Park et al. [14] studied the effects of fin shapes on the heat dissipation of the lamp, and obtained the best heat dissipation effect with the radioactive radial fins.

Generally, the main objective of thermal management for LED automotive headlamp is to maintain the LED junction temperature lowest for the stability, reliability and longevity of LED. In this study, the FEA simulation with CFD software and the response surface method based on the Design-Expert program were used to optimize the thermal design of heatsink for a commercialized LED automotive headlamp module.

2. Modeling and simulation

2.1 3D modeling

The high power LED automotive headlamp used in this study is LUW HWQP, which delivers high brilliant white light for automotive forward lighting .This type of LED can work in the temperature range from -40 to 150°C, the maximum permissible junction temperature is 130°C, the light-emission area is less than 1mm^2, and the typical correlated color temperature (CCT) is 5600K. Its colorimetric properties conform to the ECE regulation. Figure 1 presents the 3D model of this headlamp.

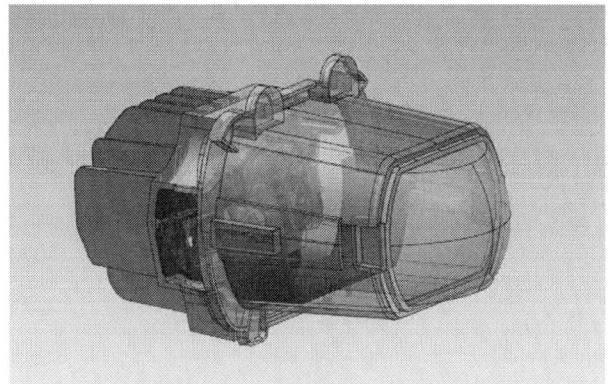

Figure1. 3D modeling of the LED automotive headlamp used in this study.

2.2 Simulation result of original model

The temperature of each component in the initial model is determined by the FEA simulation. First, the air module is built according to the size of the heat sink module. Table 1 and 2 show the simulation parameter and the boundary condition settings, respectively. Finally, the mesh numbers of air domain, heat sink, PCB substrate, and LED are 8, 3, 2, 1 respectively and we used 150 steps iterative to solve. Figure 2 shows the simulation result under the boundary condition set in Table 2.

Table 1. Simulation parameter setting

Heat sink material	Die-cast Al
PCB substrate material	Aluminum plate
LED model	LUW HWQP

Table 2. Boundary condition setting

Pressure on the upper and lower sides	0Pa
Six-sided temperature	25℃
LED thermal power	3w*70%

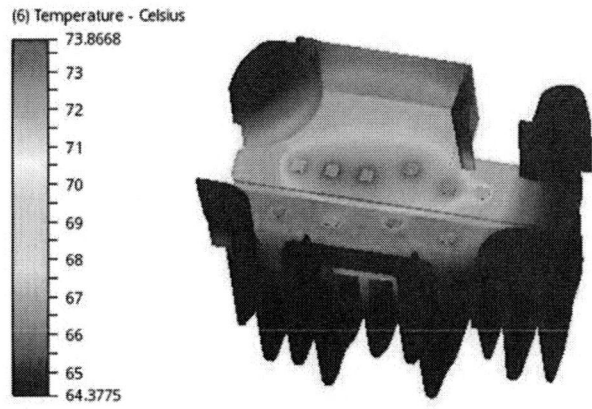

Figure 2. The simulated temperature distribution of original LED headlamp model.

2.3 Effect of heat sink material on the heat dissipation

Heat sinks are designed using materials with high thermal conductivity, such as aluminum alloys and copper. In the auxiliary heat sink design for LED headlamps, the heat dissipation effect can be achieved through increasing the air convection between the interior lamp and the base of heat sink. Thus, the thermal characteristics of heat sink have a direct influence on the heat transfer of headlamp. To understand the influence rule of heat sink on the LED junction temperature distribution, we used the steady-state FEA method to simulate the temperature distributions of heat sinks with different materials. Table 3 shows the parameter settings for different heat sink materials. Figure 3 lists the simulated characteristic temperatures.

Table 3. Parameter settings for different heat sink materials

Materials	Cu	Die-cast Al	6061 Al alloy	6063 Al alloy
Thermal conductivity (W/m·K)	398	110	154	200

Considering the aspects of weight and material cost, the copper with high thermal conductivity is used as the material of lamppost, and the aluminum alloy with high thermal conductivity is usually used as the heat sink base. As copper has highest thermal conductivity up to 398W/m·K at room temperature, it can reduce the LED's junction temperature to 72.020℃. Because the density of copper is three times higher than that of aluminum, the weight of copper-based heat sink with the same structure is three times higher than that of aluminum-based heat sink. At the same time, the price of copper is higher than aluminum, so aluminum is generally applied as a raw material for LED heat sink. However, the hardness of pure aluminum is too low to prepare force-resistant products, it is also not suitable to be as a heat sink material. Compared to pure aluminum, aluminum alloys possess better physical properties and a simpler preparation process with a lightweight. Figure 4 shows the FEA simulation result of using the 6063 Al alloy material in heat sink.

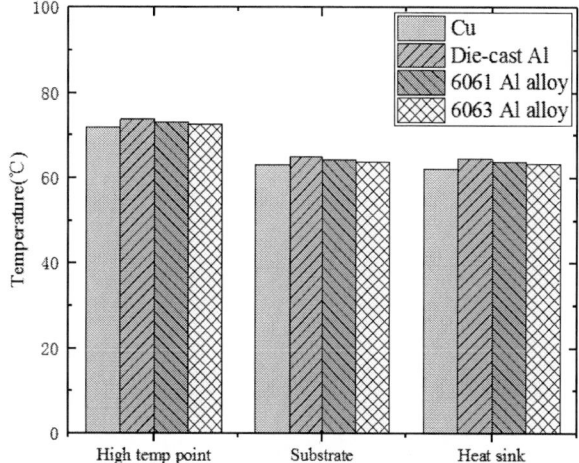

Figure 3. The simulated characteristic temperatures of four conditions with different heat sink materials.

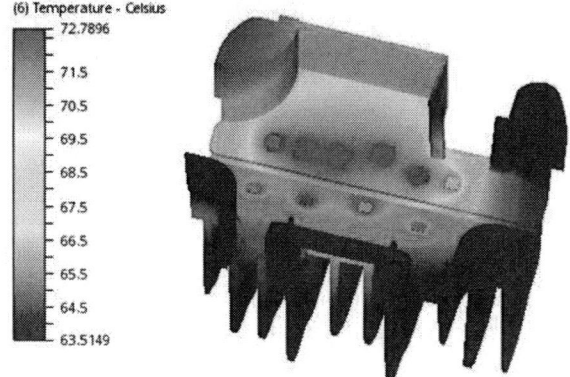

Figure 4. The simulated temperature distribution of model with 6063 Al alloy heatsink.

2.4 Effect of substrate material on the heat dissipation

During the heat dissipation process of headlamp, the substrate not only mechanically supports LEDs, but also acts as the heat transfer from the LEDs to the external radiator, which is a very important heat dissipation channel. Therefore, the substrate material should not only provide structural strength, but also achieve the high thermal conductivity requirement. Traditional FR-4 PCB can meet the demand of cooling at low power applications, however, with the increase of LED power, it has difficulty to meet the demand. Thus, the aluminum-based metal circuit board (MCPCB) and ceramic substrate (Al3O2, AlN) appear. To make clear the impact of different substrate materials on the heat dissipation, we compared four substrate materials, including FR-4, MCPCB, Al3O2 and AlN by simulation calculation on thermal dissipation characteristic of LED headlamps. Table 4 shows the thermal conductivity of each substrate material used in simulation.

Table 4. The thermal conductivity of each substrate material used in simulation

Materials	FR-4	MCPCB	Al_3O_2	AlN
Thermal conductivity (W/m·K)	16.9	50	29	170

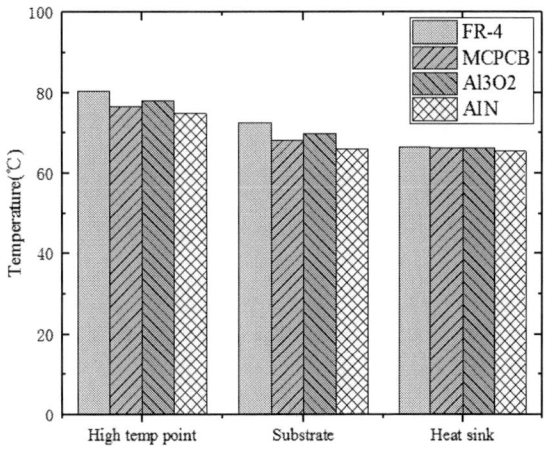

Figure 5 .The simulated characteristic temperatures of four conditions with different substrates.

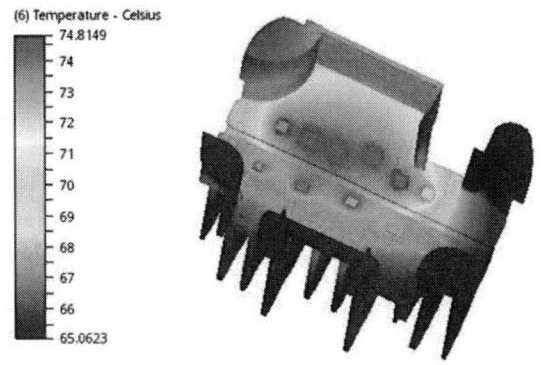

Figure 6 .The simulated temperature distribution of model with AlN substrate.

The simulation results shown in Figure 5 indicate that the AlN substrate with highest thermal conductivity can lower the LED junction temperature around 5.573 ℃, which builds essential basis for selecting AlN as the better substrate material in high power LED headlamp. Figure 6 shows the temperature distribution simulation result with the AlN substrate.

3. Optimization with Response Surface Method

Response surface method (RSM) is a useful method for optimization modeling processed by the mathematical analysis and statistical techniques. It combines the experimental design with regression analysis to implement systematically the experiments within the defined area. After collecting the required response values, it conducts the regression analyses to collate the relations between response values and parameters for optimal solution found within the planned experiments. Thus, we applied the RSM design optimization in this paper to estimate the parameters and confirm the influences with respect to target value.

3.1 Design of Experiments

There are many variables that can be optimized, such as fin spacing, fin thickness, fin height, dimension of heat sink, airflow velocity, etc. In this paper, the fin thickness, spacing and height of heat sink were considered. When process factors (independent variables) satisfy an important assumption that they are measurable, continuous, and controllable by experiments, we carried out the RSM procedure as follows:

(1) Perform a series of experiments for adequate and reliable measurement of the response of interest.

(2) Develop a mathematical model of the second-order response surface with the best fit.

(3) Determine the optimal set of experimental parameters producing the optimum response value.

(4) Represent the direct and interactive effects of the process parameters (factors) were through two and three-dimensional plots.

In order to determine if a relationship existed between the factors and the responses investigated, the collected data was statistically analyzed by using the regression analyses [15]. We employed a regression design to model the response as a function of a few continuous factors.

3.2 Box-Behnken test results and analysis

On the basis of changing the heat sink material and the PCB substrate material, three significant factors affecting the thickness, spacing and height of the heat sink fins were considered as factors. The Box-behnken test factors and levels are shown in Table 5. Where A, B, and C represent the thickness, spacing, and height of the heat sink fins, respectively. The test results are shown in Table6.

Table 5. Factors and levels of Box-Behnken of radiator structure design.

Factors	-1	0	1
A Fin thickness/mm	1	2	3
B Fin spacing/mm	5	7	9
C Fin height/mm	20	40	60

Table 6. Results and analysis of Box-Behnken experiments of radiator structure design.

Serial number	A	B	C	Y/Temperature
1	3	5	40	76.126
2	2	7	40	71.271
3	2	7	40	71.161
4	1	7	20	79.609
5	1	9	40	70.407
6	2	7	40	71.245
7	2	9	60	65.242
8	2	7	40	71.154
9	2	9	20	81.094
10	1	5	40	73.525
11	2	7	40	71.283
12	1	7	60	69.806
13	3	7	60	65.623
14	2	5	20	84.273
15	2	5	60	70.049
16	3	7	20	81.185
17	3	9	40	71.503

Using the Design Expert software to fit the results in Table 6, the regression equation is obtained:

$$Y = 71.22 + 0.14A - 1.97B - 6.93C$$
$$- 0.38AB - 1.44AC - 0.41BC$$
$$+ 0.28A^2 + 1.39B^2 + 2.55C^2$$

Performing the variance analysis on the above regression equation. The results are shown in Table 7. It can be seen that the established model with P<0.0001 is significant, which indicates that the model has high reliability. After variance analysis, the primary and secondary relationships of three factors on the temperature of the radiator can be obtained: C>A>B, that means the heat sink fin height > the fin thickness > the fin pitch. The one-time coefficient C has a most significant effect on the results. The cross-terms AB and BC have no significant effect on the results. The quadratic coefficients A^2 and C^2 have extremely significant effects on the results.

Table 7. Variance analysis of response surface test results.

Source	Squares	Df	Square	Value	Prob
Model	463.10	9	51.46	46.41	<0.0001
A	0.15	1	0.15	0.13	0.7257
B	30.93	1	30.93	27.89	0.0011
C	384.21	1	384.21	346.54	<0.0001
AB	0.56	1	0.56	0.51	0.4985
AC	8.29	1	8.29	7.48	0.0291
BC	0.66	1	0.66	0.60	0.4648
A^2	0.33	1	0.33	0.30	0.6028
B^2	8.12	1	8.12	7.32	0.0304
C^2	27.45	1	27.45	24.76	0.0016
Residual	7.76	7	1.11		
Lack of Fit	7.75	3	2.58	688.86	<0.0001
Pure Error	0.015	4	3.784E-003		
Cor Total	470.86	16			

Figure 7 shows the diagnostics result in Design Expert, in which the closer of data to the line means the higher reliability. According to this result, we can confirm that this model has high significance.

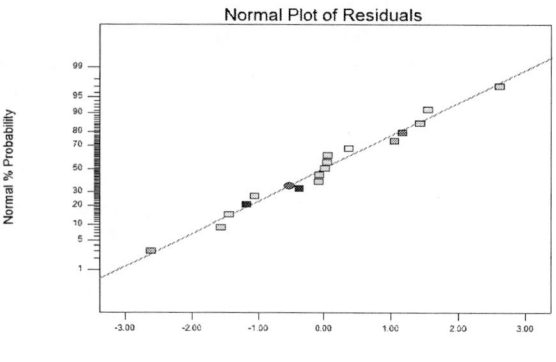

Figure 7. Reliability diagnosis chart

3.3 Response surface results and analysis

The influence of the fin thickness, spacing and height on the LED junction temperature are obtained and shown in Figure 8-10.

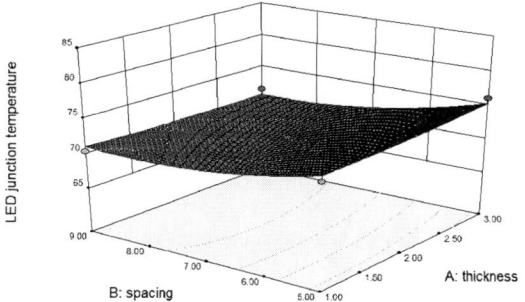

Figure 8. The effect of heat sink fin thickness and spacing of on the heat dissipation.

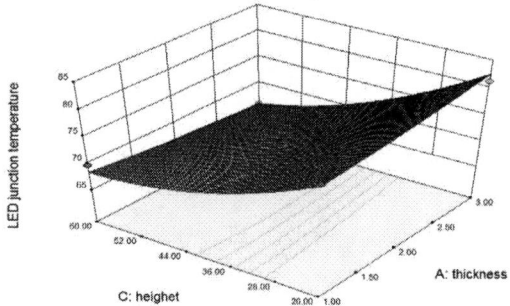

Figure 9. The effect of heat sink fin thickness and height of on the heat dissipation.

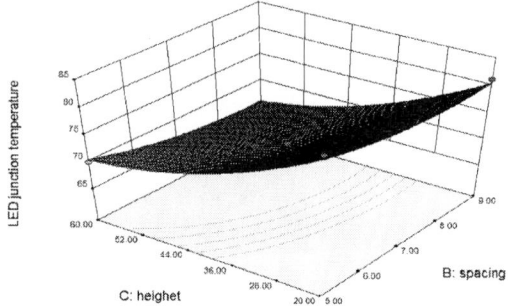

Figure 10. The effect of heat sink fin spacing and height of on the heat dissipation.

The above results indicate that: (1) When the height of the fins is fixed, with the thickness and spacing of the fins increasing, the junction temperature of the LED's firstly decreases and then increases. The fin thickness between 2mm and 3mm and the fin spacing between 7mm and 9mm are more suitable; (2) When the distance of the heat sink is fixed, with the height of the fins increasing, the junction temperature of the heat sink LED is getting higher and higher, indicating that the higher heights of the heat sink fin, the better effects of the heat dissipation; (3) When the thickness of the fins is fixed, with the pitch and height of the fins increasing, the junction temperature of the LED's firstly decreases and then increases, indicating that the spacing and height of the heat sink fins have strong interaction, which has a great influence on the junction temperature of the LED. According to the optimization conditions of the Box-Behnken software prediction model, the results show that when the thickness of heat sink fin is 3 mm, the spacing is 8.98 mm, and the height is 60 mm, the LED junction temperature can be controlled at the lowest.

3.4 Simulation and verification

According to the above analysis, we used the optimized heat sink fin design. In the optimization model 1, the heat sink material is selected as die-cast aluminum and the PCB substrate material is aluminum substrate. After setting the boundary conditions and solving the mesh, the simulation result of optimized is shown in Figure 11. Moreover, setting the heatsink material as 6063 aluminum alloy and the PCB substrate material as AlN in the optimization model 2, the simulation result of optimized is shown in Figure 12.

Figure 11. The simulated temperature distribution of the optimized model 1.

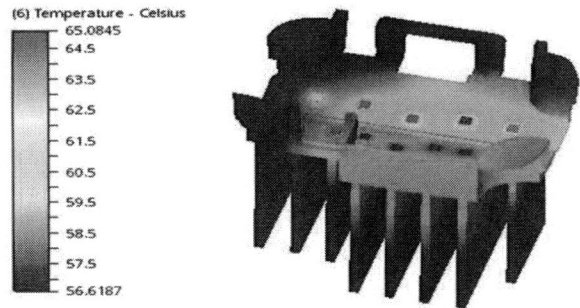

Figure 12. The simulated temperature distribution of the optimized model 2.

Table 8 shows the simulated maximum temperatures of the heat sink, substrate, and LED. The results indicate that compared with the initial model, the LED junction temperature caused by the optimization of the fin structure drops by 2.9%. Due to the optimization of the heat sink material, PCB substrate material and fin structure, the LED junction temperature drops by 11.9%. It indicates that the response surface method is an effective tool to optimize the design of heatsink for LED automotive headlamp.

Table 8. Model temperature distribution after simulation.

	The heat sink Temperature/℃	PCB substrate Temperature / ℃	LED Junction Temperature /℃
Initial model	64.378	68.197	73.867
Optimization model 1	61.995	63.219	71.684
Optimization model 2	56.619	56.487	65.085

Note: The initial model represents the original LED headlamp model with the heat sink material as cast aluminum, the PCB substrate material as aluminum substrate. The optimized model 1 is the model only with the structure optimization. The optimization model 2 represents the model by replacing the heat sink material as 6063 aluminum alloy and the PCB substrate material as AlN.

4. Conclusions

In this paper, we optimize the heat dissipation design of a commercialized LED automotive headlamp module by using the response surface method and the FEA simulation. The structure of heatsink is optimized with the thickness, spacing and height of fins as 3mm, 9mm and 60mm respectively. Meanwhile, the 6063 aluminum alloy and AlN are selected as new heat sink and PCB materials. The results indicate that: compared to the initial model, the LED junction temperature contributed by the structure optimization drops by 2.9%; the LED junction temperature due to the material optimization drops by 11.9%. This improvement can provide a new design for this type LED automotive headlamp module with more effective thermal management.

Acknowledgements

The work described in this paper was partially supported by the National Natural Science Foundation of China (51805147), the Six Talent Peaks Project in Jiangsu Province (GDZB-017) ,the Fundamental Research Funds for the Central Universities (2017B15014) and the Jiangsu Planned Projects for Postdoctoral Research Funds (Grant No.2018K200C).

References

1. HONG E, NARENDRAN N. A method for projecting useful life of LED lighting systems [C] // Proceedings of the SPIE.2004,5187:93-99.DOI:1117/12.509682.
2. FU Jianhua, LI Wenqiang, ZHOU Hua,et al. Applications of LED technology in auto lighting [J].China Illuminating Engineering Journal, 2010,21(3):64-69(in Chinese).
3. KIM L, CHOI J H, JANG S H,et al. Thermal analysis of LED array system with heat pipe[J]. Thermochimica Acta, 2007,455(1):21-25.

4. Li Wei, Yang Zhenfeng. LED technology in automotive lighting applications [J]. South Agricultural Machinery, 2017(8): 142-143.

5. Liu Yanchao, Fu Guicui, Gao Cheng, et al. Research on high-power LED heat dissipation for lighting [J]. Electronic Devices, 2009, 31 (6): 1716-1719.

6. Li Qin, Zhu Minbo, Liu Haidong, et al. Electronic equipment thermal analysis technology and software applications [J]. Computer Aided Engineering, 2005, 14(2): 50-52.

7. Ding Tianping, Guo Weiling, Cui Bifeng, et al. The effect of temperature on the spectral characteristics of power LEDs [J]. Spectroscopy and Spectral Analysis, 2011, 31(6): 1450-1453.

8. Luo X,Hu R,Guo T,et al. Low thermal resistance LED light source with vapor chamber coupled fin heat sink [C] // 2010 Proceedings 60th Electronic Components and Technology Conference (ECTC).NY,USA:IEEE, 2010: 1347－1352.

9. Ma L,Yang Y,Liu J. Cooling of high power LED through ventilating ambient air to front surface of chip [J].Heat Mass Transfer,2013,49(1): 85－94.

10. Ye H,Mihailovic M,Wong C K Y,et al. Two－phase cooling of light emitting diode for higher light output and increased efficiency[J]. Applied Thermal Engineering. 2013.52(2): 353－359.

11. Yang Daoguo, Mo Yuezhu, Nie Yao, and so on. High-power LED package interface Analysis of the effect of layer cracking on interface heat transfer performance [J]. Electronic Components and Materials, 2016, 35(8): 76-80.

12. SHAERI M R, YAGHOUBI M. Thermal enhancement from heat sinks by using perforated fins [J]. Energy conversion and management, 2009, 50(5) : 1264.

13. JANGSW, LEEYL. A study on the thermal performance of Automotive LED head lamps with synthetic jet [J].International Journal of Applied Engineering Research, 2016, 11(5): 3294-3298.

14. SANGJP, LEE Y L. Study on the development of high-Efficiency,long-life LED fog lamps for the used car market[J]. Transactions on Electrical and Electronic Materials,2014, 15(4): 201-206.

15. Thuy Khanh Trinh, Lim Seok Kang, Application of Response Surface Method as an Experimental Design to Optimize Coagulation Tests, 559-1, Daeyeon-3-dong, Nam-gu, Busan 608-739, Korea.

Optical and Thermal Designs of LED Matrix Module used in Automotive Headlamps

Wei Chen[1], Jiajie Fan[1,2,3*], Gaojin Qi[1], Chengzhong Sun[1], Weiqiao Yang[1], Suming Cao[1]

1. Changzhou Institute of Technology Research for Solid State Lighting, Changzhou, 213161, China
2. College of Mechanical and Electrical Engineering, Hohai University, Changzhou, 213022, China
3. EEMCS Faculty, Delft University of Technology, Delft, the Netherlands
*Corresponding: jay.fan@connect.polyu.hk, Tel: +86-189-5155-4115.

Abstract

With higher efficiency, brightness and reliability than traditional light sources, LED has been widely applied in automotive headlamps. The smaller volume and higher power density of a LED package drive the design of automotive headlamps with LED matrix model, such as adaptive driving beam (ADB) system. This paper proposed a matrix automotive headlamp module with rectangle lens array to balance the contradiction between improving the visual condition of driver in the nighttime driving and reducing the impact of glare on others from the opposite direction. Firstly, according to the requirements on light distribution provided by the GB 25991 regulation, the optimized optical design, including configuration of LEDs and structure of optical system, was developed based on the optical simulations. Secondly, the thermal management of the module was conducted by using the fluid finite element simulation through considering both the natural and forced convections. The results show that: 1) a LED matrix with specially designed rectangle lens array that contains different size of lenses can produce the independent and nonoverlapping rectangular light spots as required; 2) the heatsink with optimal designed pin fins can provide more effective thermal dissipation in convection.

Keywords: LED matrix; Automotive headlamp; Multiple light beam; Thermal management

1. Introduction

In recent years, lighting-emitting diodes (LEDs), as an advanced solid-state lighting (SSL), draw worldwide attention and have progressed rapidly in automotive lighting because of its superiority over other conventional light sources such as halogen lamps, Xenon HID lamps [1-4]. With high wall-plug efficiency (WPE) and long service life which will continue to be advanced, LEDs have increasing advantages in energy conservation and environment-friendly [5]. Furthermore, since LED doesn't equip electric ballast, and it can achieve the changing operating state immediately without time delay, LED headlamp is safer for driving at night. High-performance LED headlamp can provide drivers with wonderful visual conditions, which will be hopeful to reduce traffic accidents. Besides, the small size of LED makes the design of automotive headlamp more flexible, digital, compact and distinctive.

The LED matrix always contains an LED array that can produce diversified beam mode by controlling the operating state of each LED package. The more LEDs the matrix has, the more diversified the beam can be. Treating each LED in a matrix as a pixel, high resolution matrix automotive headlamp has the function of generating appropriate beam mode to avoid uncomfortable glare for other road users, and to improve safety when driving on the poor road conductions such as curve, rainy day, etc. This LED matrix automotive headlamp will become the major trend in the ADB system.

When the LED matrix automotive headlamp is applied, two essential challenges need to be considered in the design: (1) the design of optics, such as collimating lens and free-from lens, that can produce rectangular spot with a small beam angle. (2) thermal management with a powerful but small heatsink is needed to lower the junction temperature of LED matrix. From the specifications of automotive LEDs, it is known that the maximum junction temperature of LED is around 150°C. However, taking luminous efficacy and design margin into consideration, the requirement on junction temperature is always no more than 120°C. In addition, the automotive headlamp always is close to the engine, causing the ambient temperature very high. The GB/T 10485 standard regulates the ambient temperature as 60°C for the experimental validation, but some automobile manufacturers restrict the ambient temperature to 70°C. In short, the optical and thermal optimizations are two essential considerations in the design of LED matrix automotive headlamp.

Although there are many studies on the optical design and thermal management for LED automotive headlamp [1, 6-9], however, few researches on matrix automotive headlamp with simultaneously considering both the optical design and thermal design. In this paper, optical system and heatsink of matrix module are designed and verified by simulation. The remainder of the paper is organized as follows: Section 2 and Section 3 introduce the optical and thermal design respectively. Section 4 proposes the concluding remarks.

2. Optical design

In this part, the optical system of LED matrix automotive headlamp is designed, and the light distribution is simulated to meet the requirements in the GB 25991 standard.

2.1 LED light source

A commercially available phosphor converted white LED package (Model: LUXEON Neo) is selected in automotive headlamp, as shown in Figure 1. With small volume (luminous surface is 0.5mm^2), high luminous flux density (over 100 lm), low resistance (6.2°C/W), this LED is suitable to be used as the light source of matrix automotive headlamp. The maximum current, maximum voltage and correlated color temperature of this LED are 750mA, 3.51V and ~5800K, respectively. Two emission peaks include the emitted light from the LED chip, another is the converted light from the phosphors, as shown in Figure 1 [10]. Figure 2 plots the relative luminous intensity distribution of selected LED [11].

2.2 Design of optical system

As shown in Figure 3 (a), the design of optical system

mainly contains the following parts: 45 LUXEON Neo LEDs, a rectangle lens array containing 45 rectangle lenses and an aspherical lens. The rectangle lens can convert the light from the LED as a rectangular spot. As shown in Figure 3 (b), the length, width and thickness of the ambient lenses in the rectangle lens array are 4mm, 4mm and 2mm, respectively. The central lenses in the rectangle lens array are smaller to avoid the overlap of rectangle spot. The focal lengths of all rectangle lenses are 2mm, and the focus is designed to be located in the center of corresponding LED's luminous surface. The diameter, thickness and focal length of the aspherical lens are 120mm, 0.5mm and 80mm and it operates to converge light from the rectangle lens array.

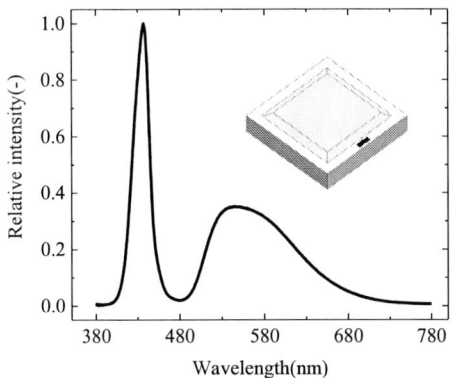

Figure 1. The spectral power distribution of the LUXEON Neo (Inset: Schematic diagram of the LED)

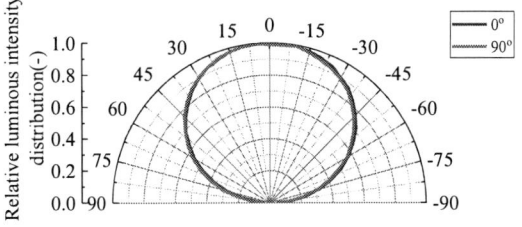

Figure 2. The relative luminous intensity distribution of the LUXEON Neo

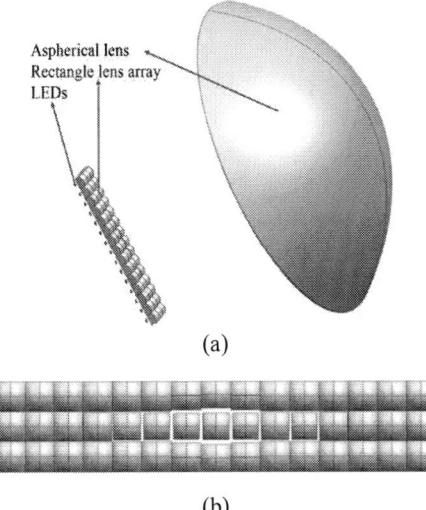

Figure 3. (a) schematic diagram of optical system and (b) front view of the rectangle lens array

2.3 Optical simulation and validation

The optical simulation based on the ray trace method used in this study was implemented in the LucidShape 2.0 to verify the designed light distribution performance of optical system mentioned above. The output luminous flux of each LED is 120lm with the current of 500mA and the voltage of 3.2V. The material of rectangle lens array and aspherical lens are all polycarbonate (PC) with refractive index of 1.586. The distance between the optical system and the measured screen is set as 25m. The simulated illuminance distribution on the measured screen is showed in Figure 4(a). The beam angle in horizontal direction is higher than 30°, and the beam angle in vertical direction is higher than 8°, which are all above the value regulated by GB25991. Table 1 compares other requirements on the high beam by GB25991, which also indicates that the optical performances of the designed optical system meets all requirements of GB25991. Figure 4(b) presents the driver's view on the road, when this designed headlamp is mounted in the automobile.

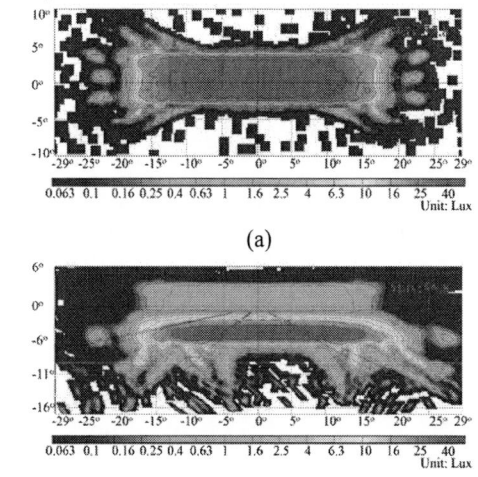

Figure 4. The simulated light distribution of LED matrix automotive headlamp (a) illuminance on measured screen; (b) driver's view

Table 1. The verification of simulated light distribution performance

Items	Minimum values (Lux)	Maximum values (Lux)	Simulated values (Lux)
E_{max}	48	240	57.02
HV	$0.8E_{max}$	-	49.71
HV-1125L HV-1125R	24	-	47.59
HV-2250L HV-2250R	6	-	47.55

To avoid dazzling of the pedestrians and the drivers in opposite side, the designed optical system can provide the rectangle "dark areas" by powering-off the corresponding LED. Figure 5 and Figure 6 show the light distributions when a column of LEDs is powered-off and a row of LEDs is powered-off respectively. Comparing to the other areas, distinct differences are appeared and the illuminance of "dark areas" is dozens of times smaller than other areas.

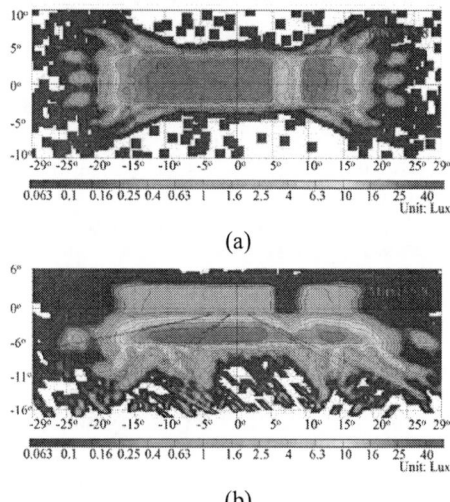

(a)

(b)

Figure 5. The simulated light distribution of LED matrix automotive headlamp when a column of LEDs is powered-off: (a) illuminance on measured screen; (b) driver's view

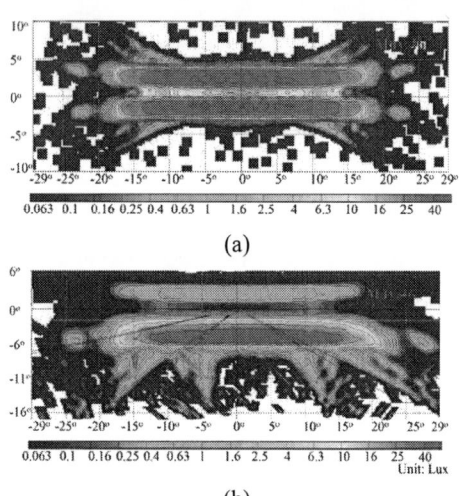

(a)

(b)

Figure 6. The simulated light distribution of LED matrix automotive headlamp when a row of LEDs is powered-off: (a) illuminance on measured screen; (b) driver's view

The efficiency of the optical system can be calculated by Equation (1):

$$\eta = \frac{\phi_{ls}}{\phi_{scn}} \times 100\% \tag{1}$$

in which η, Φ_{ls} and Φ_{scn} are the efficiency, the luminous flux emitted from the light sources and the luminous flux accepted by the measurement screen. When all LEDs are powered-on, the luminous flux accepted by the measurement screen is 1670lm, and the calculated luminous efficiency is 30.93%.

3 Thermal design

In this part, the optimization design of heatsink with pin fins for LED matrix automotive headlamp is provided, and the effect of wind direction and wind velocity on the thermal management is considered.

3.1 Design of heatsink

The LED matrix module, containing the LEDs, printed circuit board (PCB) and heatsink with pin fin, is showed in Figure 7. The material of heatsink is aluminum alloys (6061). To explore the optimal structure of heatsink, 25 heatsink schemes were designed by considering two factors: 1) the height (H) of the pin fin (from 36-60mm with an interval of 6mm); 2) the distance (X) between two adjacent pin fins (from 5-9mm with an interval of 1mm). The surface areas of all schemes are the same, so, the diameters (D) of pin fins in each scheme are different as list in Table 2.

Table 2. The D of pin-fin under different H and X

D \ X H	5	6	7	8	9
36	2.45	3.11	4.08	5.57	7.26
42	2.11	2.68	3.52	4.82	6.30
48	1.85	2.36	3.10	4.24	5.56
54	1.65	2.10	2.76	3.79	4.97
60	1.49	1.90	2.49	3.42	4.49

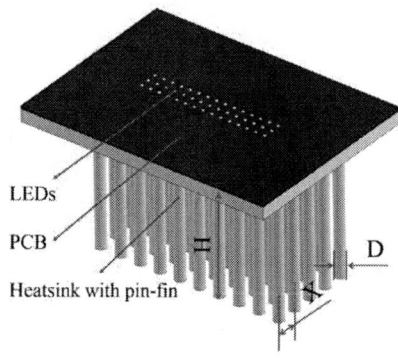

Figure 7. Schematic diagram of LED matrix module

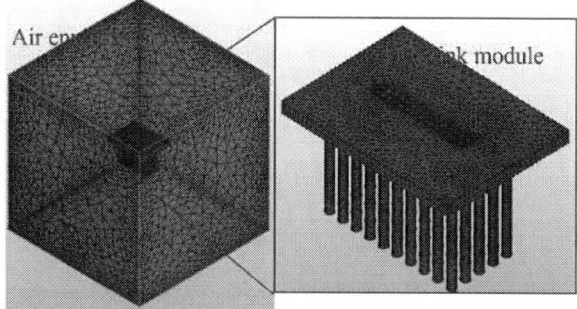

Figure 8. The meshing model used in FEA simulation

3.2 Thermal simulation and validation

The finite element analysis (FEA) based on the Autodesk 2018 software is used to simulate the temperature distribution of the LED matrix module. The material parameters used in this simulation are showed in Table 3. The density of air is changed with the environmental temperatures calculated by software. The meshing model containing the air environment is showed in Figure 8.

978-1-7281-5757-3/19 $31.00 © 2019 IEEE   222

Table 3. Material parameters used in this study

Materials	Density (Kg/m³)	Specific heat (J/Kg•°C)	Thermal conductivity (W/m•°C)
Air	/	1004	0.0256
LED	6150	490	30
PCB	6905	397	197
6061 aluminum alloys	2823	963	142

3.2.1 Simulation results under natural convection

Figure 9 plots the junction temperature (T_j) distribution under different X and H of heatsink design schemes. The T_j decreases with the increase of H. T_j decreases firstly and then increases as the X increases. Among all schemes, the T_j can be controlled as lowest (~185°C) when the heatsink design with the X of 9mm and the H of 60mm. As shown in Figure 10, The air flow surrounding the LED matrix automotive headlamp is caused by the temperature difference of air.

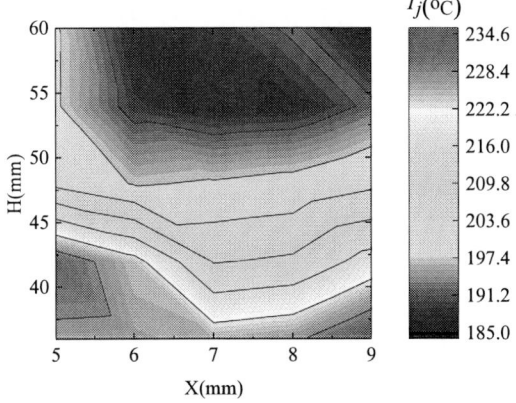

Figure 9. The simulated junction temperature distribution under different X and H

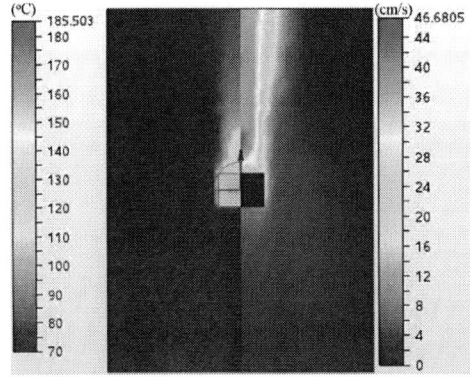

Figure 10. The simulated temperature (left) and wind velocity (right) distributions under the X of 9mm and the H of 60mm

3.2.2 Simulation results under the forced convection

Above results demonstrate that the heatsink under natural convection is difficult to meet the junction temperature requirement of LED, thus, the forced convection is considered in this part. A commercial fan is selected and the inner diameter (R), outer diameter (r) and the blast volume (q) are 29mm, 18mm and 2.68 cubic feet per minute (CFM) respectively. The wind velocity (V) of the fan is calculated by the Equation (2):

$$V = \frac{q}{S} = \frac{4q}{\pi \left(R^2 - r^2 \right)} \qquad (2),$$

where S is the area of the wind outlet. According to the calculation by Equation (2), the wind velocity is ~3 m/s.

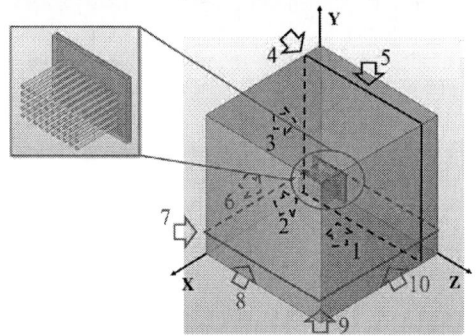

Figure 11. Schematic diagram of wind direction setting

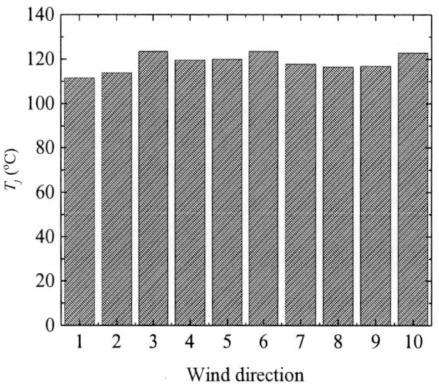

Figure 12. The simulated junction temperatures under different wind directions

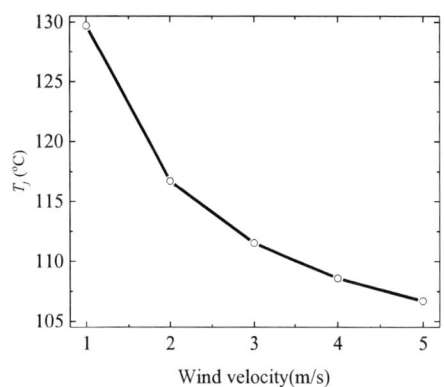

Figure 13. The simulated junction temperatures under different wind velocities

Firstly, the wind direction is considered to guide the installation of fan in the best design scheme with the natural convection. The setting of wind direction is shown in Figure 11, in which the angle between direction 1, direction 2, direction 3, direction 4, direction 5 and the Y-axe are 0°, 45°, 90°, 135°, 180° respectively; the angle between direction 6, direction 7, direction 8, direction 9, direction 10 and the Z-axe are 0°, 45°, 90°, 135°, 180° respectively. -Y direction is the direction of gravity. The simulated T_j results under different wind directions are showed in Figure 12. In all direction, the direction 1 with the lower T_j (111.52°C) coincides with the direction of air flow caused by the temperature change (shown in Figure 10 right). Figure 13 shows the simulated T_j changes with the wind velocity. With the wind velocity increasing, the T_j continues to drop, but the drop trend flattens out. Comprehensively considering the cost and practicality, 3m/s is a suitable wind velocity for this matrix automotive headlamp.

4. Conclusion

In this paper, an LED matrix automotive lamp is proposed to realize ADB system efficiently. The optical and thermal simulations of LED matrix automotive lamp are conducted by the LucidShape 2.0 and Autodesk CFD 2018 respectively. The results indicate that: 1) the proposed design of LED matrix automotive lamp in this study meets all the optical and thermal requirements; 2) rectangle lens that possesses the merit of easy to manufacture for its facile structure can convert the round light spot to rectangle light spot, however the efficiency is needed to improve; 3) with increasing the height of pin fin, the heat dissipation effect of heatsink improves; 4) the opposite direction of gravity is the best installation direction for a fan, since the lower junction temperature occurs in this direction.

Acknowledgments

The work described in this paper was partially supported by the National Natural Science Foundation of China (51805147), the Six Talent Peaks Project in Jiangsu Province (GDZB-017) and the Fundamental Research Funds for the Central Universities (2017B15014).

Reference

1. E. D. Jung and Y. L. Lee, "Development of a heat dissipating LED headlamp with silicone lens to replace halogen bulbs in used cars," Applied Thermal Engineering, vol. 86, pp. 143-150, 2015.
2. C. Yu, J. Fan, C. Qian, X. Fan, G. Zhang, C. Yu, et al., "Luminous flux modeling for high power LED automotive headlamp module," in International Conference on Electronic Packaging Technology, 2017, pp. 1389-1395.
3. M. H. Crawford, "LEDs for Solid-State Lighting: Performance Challenges and Recent Advances," IEEE Journal of Selected Topics in Quantum Electronics, vol. 15, pp. 1028-1040, 2009.
4. J. Zhou, X. M. Long, J. G. He, L. Fang, and X. Li, "System-Level Thermal Design for LED Automotive Lamp-Based Multiobjective Simulation," IEEE Transactions on Components Packaging & Manufacturing Technology, vol. 7, pp. 591-601, 2017.
5. X. Long, J. He, J. Zhou, L. Fang, X. Zhou, F. Ren, et al., "A review on light-emitting diode based automotive headlamps," Renewable & Sustainable Energy Reviews, vol. 41, pp. 29-41, 2015.
6. J. Jiao, "Etendue concerns for automotive headlamps using white LEDs," Proceedings of SPIE - The International Society for Optical Engineering, vol. 5187, pp. 234-242, 2004.
7. X. Zhu, Z. Qian, W. Han, and C. Chen, "Optical design of LED-based automotive headlamps," Optics & Laser Technology, vol. 45, pp. 262-266, 2013.
8. M. S. Huang, C. C. Hung, Y. C. Fang, W. C. Lai, and Y. L. Chen, "Optical design and optimization of light emitting diode automotive head light with digital micromirror device light emitting diode," Optik - International Journal for Light and Electron Optics, vol. 121, pp. 944-952, 2010.
9. Y. Lai, N. Cordero, F. Barthel, F. Tebbe, J. Kuhn, R. Apfelbeck, et al., "Liquid cooling of bright LEDs for automotive applications," Applied Thermal Engineering, vol. 29, pp. 1239-1244, 2007.
10. https://www.sendspace.com/pro/dl/6fdquf.
11. https://www.sendspace.com/pro/dl/a70ggd.

Comparison of Life Testing Standards for LED Lighting Products

Zhu Zhike, Cao Suming, Cai Shasha, Shi Tingting, Lian Yuanhui
Changzhou Institute of Technology Research for Solid State Lighting, Changzhou 213161, China
Place your company address here No.9,TIANAN CYBER PARK, Changzhou,Jiangsu,China
zkzhu @sklssl.org, smcao@sklssl.org

Abstract

Due to its high luminous efficacy, small size, low energy consumption, high reliability and long lifetime, LED has become the 3rd generation light source, after incandescent lamp and fluorescent lamp. In 2017, the LED market reaches a new high point of 653.8 billion RMB, and penetrates to new applications such as plant and animal cultivation, visible light communication, medical and health care, etc.Conventional test methods are no longer suitable for LED luminaires.LED modules and luminaires lifetime test follows US DoE Energy Star, IES LM-79, IES LM-80, IES LM-82, IES LM-84, IES TM21, IES TM28 etc. which require a 6000hrs test duration. This has affected the rapid development of the LED application industry, so it is particularly urgent to develop and select appropriate accelerated life test methods for LED products.

This paper compares three kinds of fast/accelerated life testing methods for LED products, summarizes the advantages and disadvantages of each method, and provides some choices for life testing of LED products.

1. General test requirements

Comparing the three standards, they are consistent in environment, power voltage and stability determination, and different in the minimum number of samples required for the test and the installation method.

1.1 Test environment requirements

Unless otherwise specified, LED samples shall be tested for performance parameters in stable operation at ambient temperature of 25°C ± 1°C and relative humidity not exceeding 65% in an airless convection environment.

1.2 Power voltage requirements

Unless otherwise specified, all tests shall be carried out at rated voltage and rated frequency.If the manufacturer gives a voltage range, the test is carried out at the maximum voltage.During the stability of thermal balance, the power supply voltage is stable within ±0.5% of the rated value; During the measurement of the optical color electrical parameters and the acceleration test, the power supply voltage is stable within ±0.2% of the rated value; the total harmonic content of the power supply voltage shall not exceed 3%.

1.3 Stability judgment condition

During stabilization period the luminaire should work normally during stabilization. The light output (light flux or illuminance) is measured every 1 min. When the difference between the maximum value and the minimum value of the light output and the average value of the 15 min is less than 1% for 15 minutes, it is considered that the lamp is stable and can be measured. Measurements can also be made if stability is not achieved within 150 min, but the fluctuations should be described.

1.4 Minimum sample size

In the CSA 020-2013 standard, the minimum sample size of LED spotlights, LED downlights and LED bulbs is 12, and the minimum sample size of LED streetlights and LED tunnel lights is 3.

In the GB/T 33720-2017 standard, the minimum sample size of LED spotlights, LED straight tube lights, LED downlights and LED bulbs is 12, and the minimum sample size of LED street lights and LED tunnel lights is 5.

In the GB/T 33721-2017 standard, the minimum size of samples for LED lamps is 3.

1.5 Installation method

The GB/T 33721-2017 standard does not specify the sample installation method. The CSA 020-2013 standard and the GB/T 33720-2017 standard stipulate that the sample is installed with the light-emitting surface facing down.

2. Test procedure

Comparing the three standards, there are their respective focuses on the test procedures..

2.1 Test procedure for CSA 020-2013

The test procedures proposed in this standard are (1) sampling, (2) determining the accelerated test temperature, (3) accelerated attenuation test method, and (4) determining whether the accelerated test is up to standard.

2.1.1 Sampling

Inspection batch composition: In this standard, the same materials, components and light sources are used. At the same time, the same driving power source, heat sink, and products with the same structure form which are continuously produced on the same production line. The test sample is randomly selected from the parent of the same inspection batch and at least two or more inspection batch.

2.1.2 Determine the accelerated test temperature

Select a batch of samples according to the sampling requirements, select one test sample from it; measure the luminous flux of the sample at room temperature, and measure the optical properties (light flux or illuminance) of the sample in the test chamber. The test chambers were sequentially set to the four test temperatures listed in Table 1,and the samples were kept at least 60 minutes at each test temperature to achieve a stable state of thermal balance, and the optical properties of the samples were measured at that temperature. At the end of each measurement, the sample should be taken out of the temperature chamber and left standing at room temperature to measure the luminous flux when the sample reaches a stable state of thermal equilibrium.

The difference between the measured luminous fluxes of the samples outside the temperature test chamber shall be within 5%, and the difference between the measured optical properties of the samples at the four set test temperatures in the temperature test chamber shall be within 10%. The test temperature (except 25°C) was reduced by 5°C as the accelerated test temperature for the accelerated decay test. The highest accelerated test temperature that satisfies the above conditions shall be used in the accelerated test, but shall not be higher than 55°C and shall not be higher than the nominal maximum operating temperature of the product.

The measurement of the optical properties of the sample shall be carried out in a test chamber and each measurement shall be completed within 15 minutes.

The serial number	Test temperature (°C)	Allowable accelerated test temperature (°C)
0	25	N/A
1	40	35±3
2	50	45±3
3	60	55±3

Table 1 Determine the accelerated test temperature

To simplify the selection of accelerated test temperature, the accelerated decay test can be performed by specifying the allowable accelerated test temperature of a sequence in table 2.However, the test shall be carried out under the test temperature of No. 0 and the selected serial number in Table 2, and the difference between measured the luminous fluxes of the samples outside the temperature test chamber shall be within 5%, and 2 in the temperature test chamber.The difference between the optical properties measured at the two set test temperatures in the temperature chamber should be within 10%.

2.1.3 Accelerated attenuation test method

The initial luminous flux is measured first, and then the sample is placed in a test chamber for accelerated test. The relative humidity of the test environment does not exceed 65%, and the sample is continuously operated at the rated voltage.

After the accelerated test begins,when the sample reaches the stable state of thermal balance, the difference between the temperature of the lighting product casing and the ambient temperature shall be the same as the difference between the casing temperature and the ambient temperature at room temperature, and the error shall not exceed ±3°C.

After the acceleration test, the sample was taken out and placed at room temperature to measure the luminous flux after reaching the thermal balance.The percentage of the measured luminous flux and the initial luminous flux of the sample is the accelerated optical flux maintenance.

The same measuring instrument shall be selected for parameter measurement during the test, and the same test conditions shall be selected to ensure the consistency of measurement results. The number of interruption of the accelerated test shall not exceed 2 times, and each time shall not exceed 8 hours. The load time shall be accumulated only after the heat balance of LED lamps is stable.

2.1.4 Determine whether the accelerated test is up to standard

The minimum expected life of a sample with a nominal life of 25,000 hours, 30,000 hours and 35,000 hours is determined to reach its nominal life after the cumulative load time shown in table 2, table 3 and table 4, respectively, and

the maintenance rate of accelerated optical flux is not less than 95%.

Accelerated test is required for all samples in the sample. The criteria for judging the minimum expected life of LED lighting products to reach its nominal value are: the number of samples that have not reached the nominal life is not greater than the number shown in Table 2, Table 3 and Table 4. If the LED lighting product undergoing the accelerated decay test has a sudden quenching of the sample, the sample is considered to have a catastrophic failure, not counting the number of unacceptable samples in the accelerated decay test.In the number of samples specified in this standard, If a catastrophic failure occurs to any sample in the quantity specified in this standard, the accelerated decay test shall be deemed invalid.

Accelerated test temperature(°C)	Cumulative load time(h)	Judging condition（Number of samples that did not reach the nominal life）
35	2550	0
45	1800	0
55	1300	0

Table 2 25000 hours minimum expected life accelerated test conditions for LED lighting products

Accelerated test temperature(°C)	Cumulative load time(h)	Judging condition（Number of samples that did not reach the nominal life）
35	3050	0
45	2150	0
55	1550	0

Table 3 30000 hours minimum expected life accelerated test conditions for LED lighting products

Accelerated test temperature(°C)	Cumulative load time(h)	Judging condition（Number of samples that did not reach the nominal life）
35	3500	0
45	2450	0
55	1750	0

Table 4 40000 hours minimum expected life accelerated test conditions for LED lighting products

2.2 Test procedure for GB/T 33720-2017

The test procedures proposed in this standard are (1) pre-treatment test, (2) accelerated test, (3) data processing, (4) qualification determination, and (5) life expectancy. The total time of pretreatment test and acceleration test is 2000h.

2.2.1 Pre-treatment test

1)Determine the number of samples and test the luminous flux before sample pretreatment;

2)Install according to the regulations, adjust the temperature of the test chamber to the accelerated test environment condition at a rate of change not exceeding 1°C/min;

3)After the sample reaches thermal equilibrium in an accelerated test environment, the sample is continuously ignited for 500 h. If a sample has a fatal failure, record the time to failure of the sample in the report;

4)After the end of 500h, the remaining samples are allowed to stand under the test conditions for more than 2 hours, and the initial luminous flux and color coordinates are tested after heat balance;

5)A sample with an initial luminous flux lower than 70% of the pre-treatment luminous flux is regarded as a failed sample.

After the pretreatment test, the sample should meet the following conditions at the same time, otherwise the test is suspended:

1)The number of non-failed samples meets the minimum number of samples specified in the standard, and is not less than 70% of the total number of samples;

2)The chromaticity coordinates of the non-failed sample shall be in the tolerance quadrilateral at the nominal color temperature of the sample in Table 5-1 and Table 5-2;

3)Any interruption time of the pretreatment test should be less than 8h, and the total interruption time is less than 24h.

Nominal color temperature	2700K		3000K		3500K		4000K	
Chromatogram coordinates	x	y	x	y	x	y	x	y
Center point	0.4577	0.4098	0.4339	0.4032	0.4078	0.3929	0.3818	0.3796
(x,y) tolerance quadrilateral vertex coordinates	0.4811	0.4315	0.4561	0.4259	0.4302	0.4171	0.4003	0.4034
	0.4561	0.4259	0.4302	0.4171	0.4003	0.4034	0.3737	0.3879
	0.4373	0.3892	0.4149	0.3820	0.3895	0.3708	0.3671	0.3583
	0.4591	0.3941	0.4373	0.3892	0.4149	0.3820	0.3895	0.3708
Center point	0.2614	0.5267	0.2490	0.5206	0.2364	0.5125	0.2249	0.5030
Chromatogram coordinates	u'	v'	u'	v'	u'	v'	u'	v'
(x,y) tolerance quadrilateral vertex coordinates	0.2667	0.5382	0.2535	0.5325	0.2408	0.5254	0.2274	0.5157
	0.2535	0.5325	0.2408	0.5254	0.2274	0.5157	0.2164	0.5054
	0.2575	0.5154	0.2458	0.5090	0.2336	0.5003	0.2237	0.4912
	0.2696	0.5207	0.2574	0.5154	0.2458	0.5090	0.2336	0.5003

Table 5-1 Chromaticity coordinate requirements

Nominal color temperature	4500K		5000K		5700K		6500K	
Chromatogram coordinates	x	y	x	y	x	y	x	y
Center point	0.3613	0.3669	0.3446	0.3551	0.3287	0.3425	0.3123	0.3283
(x,y) tolerance quadrilateral vertex coordinates	0.3737	0.3879	0.3550	0.3752	0.3375	0.3619	0.3205	0.3475
	0.3550	0.3752	0.3375	0.3619	0.3205	0.3475	0.3027	0.3310
	0.3514	0.3480	0.3366	0.3372	0.3221	0.3255	0.3067	0.3118
	0.3671	0.3583	0.3514	0.3480	0.3366	0.3372	0.3221	0.3283
Center point	0.2163	0.4943	0.2098	0.4863	0.2038	0.4777	0.1978	0.4679
Chromatogram coordinates	u'	v'	u'	v'	u'	v'	u'	v'
(x,y) tolerance quadrilateral vertex coordinates	0.2164	0.5054	0.2091	0.4971	0.2025	0.4885	0.1964	0.4790
	0.2091	0.4971	0.2025	0.4885	0.1964	0.4790	0.1902	0.4679
	0.2171	0.4838	0.2113	0.4762	0.2058	0.4678	0.2002	0.4579
	0.2237	0.4912	0.2171	0.4838	0.2113	0.4762	0.2058	0.4678

Table 5-2 Chromatographic coordinate requirements

2.2.2 Accelerated test

The non-failed samples were subjected to an accelerated test, and the time was calculated from zero after heat balance for a total of 1500 h, intermediate tests were performed at 900 h and 1200 h, and final tests were performed at 1500 h.

1) Install according to the regulations, adjust the temperature of the test chamber to the accelerated test environment condition at a rate of change not exceeding 1 °C/min;

2) After the sample reaches thermal equilibrium in an accelerated test environment, the sample is continuously ignited for 900 h. If a sample has a fatal failure, record the time to failure of the sample in the report, the test of the sample is interrupted, and the sample without fatal failure is allowed to stand under the test condition for more than 2 hours, and the initial luminous flux and color coordinates are tested after the heat balance;

3) Repeat 1);

4) Continue the test to 1200h, the same step as 2);

5) Repeat 1);

6) Continue the test to 1500h, the same step as 2);

Any interruption time of the pretreatment test should be less than 24h, and the total interruption time is less than 72h.

2.2.3 Data processing

The luminous flux maintenance rate of the sample at a specific time is calculated by the formula (1) and recorded in the test report.

$$\eta_\Phi = \frac{\Phi_v}{\Phi_{v0}} \tag{1}$$

Where：

$\eta\Phi$ -------- Luminous flux maintenance rate at specific times；

Φv -------- Luminous flux at a specific time;

$\Phi v0$ ------- Initial luminous flux.

During the accelerated test, if the sample is fatally invalid, the luminous flux maintenance rate of the sample is regarded as zero.

The average value of the luminous flux maintenance rate of the sample was calculated by the formula(2)and recorded in the test report.

$$\overline{\eta_\Phi} = \frac{\sum\limits_{i=1}^{n} \eta_{\Phi i}}{n} \tag{2}$$

Where：

$\overline{\eta\Phi}$ --------Average value of luminous flux maintenance rate at a specific time；

$\eta\Phi i$--------The luminous flux maintenance rate of the i-th sample at a specific time;

n --------The number of samples during the accelerated test.

2.2.4 Qualification determination

When the test results satisfy the following conditions at the same time, the test is passed.

a) During the accelerated test, the average luminous flux maintenance rate of all samples is not less than 95% at 900h, 1 200h and 1500h.

b) During the accelerated test, the chromaticity coordinates of not less than 90% of the samples meet the requirements of ANSI_ANSLG C78.377-2011 at the 900h, 1 200h and 1500h time points. That is, in the tolerance quadrilateral at the nominal color temperature of the sample shown in Table 5-1 and Table 5-2.

2.2.5 Life expectancy

The estimated luminous flux maintenance life of the LED lighting products in this test at ambient temperature of 25°C± 5°C(that is, the time required for the luminous flux maintenance of the LED lighting products to reduce to 70% of the initial value after lighting) can be claimed as 25000h.

2.3 Test procedure for GB/T 33721-2017

According to the simulation process, the test time of this standard is 6000h direct method, but as long as certain conditions are met, the 1000h accelerated test method can be used.

2.3.1 Test method selection

Depending on whether the luminaire uses an LM-80 test report and determines the ts' position of the LED module, the measured parameters are consistent with the LM-80 report (including the temperature of the module in the luminaire ts' and current If' and the LM-80 report Ts, If), and whether the luminaire uses the secondary optical material given in Table 6, the applicable test method is determined according to the procedure specified in Figure 1.

Note: ts and If are the LED module solder joint temperature and input current value in the LM-80 report, respectively, ts' and If' are the LED module solder joint temperature and input current value in the LED fixture.

Material	Claimed life ≤25000h Light attenuation estimation/%	25000h< Claimed life ≤35000h Light attenuation estimation/%	35000h< Claimed life ≤50000h Light attenuation estimation/%
Glass	0	0	0
Silica gel	1	2	3
Polystyrene（PS）	6	8	11
Polycarbonate（PC）	5	7	10
Polymethylmethacrylate（PMMA）	4	6	9

Note 1: The light decay value is based on an estimate of the material under normal operating temperature conditions (typically not exceeding 80 °C).
Note 2: Only the materials commonly used on luminaires that attenuate under normal operating temperature conditions are listed in this table.
Note 3: 35000h and 50000h life claims are not applicable to direct-illuminated luminaires for outdoor use without transmissive cover. A light decay value of 50000 h was used for 25000 h.
Note 4: PS is not suitable for outdoor lighting.
Note 5: The 50000h life expects to provide at least 9000h of LM-80 data.

Table 6 Chromaticity coordinates require typical estimates of secondary optical material light decay ΔLo

The luminaire operates at the manufacturer's stated operating conditions and performance operating temperature tq ± 2°C. If the manufacturer has multiple claims for tq, the test should be conducted at the highest claimed tq.

The input current of the LED module in the luminaire is measured, and the If' of the module unit is calculated according to different circuits to determine the correspondence between the If' and the module current If given by the LM-80 report. The result of the determination may be:

1）If' ≤ If

2）If' ＞If

The temperature of the ts' point of the LED module in lamp is measured. The result of the measurement may be:

a) The temperature of the ts' point of the lamp does not exceed the minimum temperature of ts in the LM-80 test report, ie ts' ≤ tsmin, or the temperature of the ts' point of the lamp is equal to any of the other two temperatures in the LM-80 report;

b) The temperature of the ts' point of the lamp exceeds the minimum temperature of the ts point in the LM-80 test report, but does not exceed the other maximum temperature of the ts point in the LM-80 test report, ie tsmin< ts' < tsmin/max ;

c) The temperature of the ts' point of the lamp exceeds the maximum temperature of the ts point in the LM-80 test report, ie ts' > tsmax.

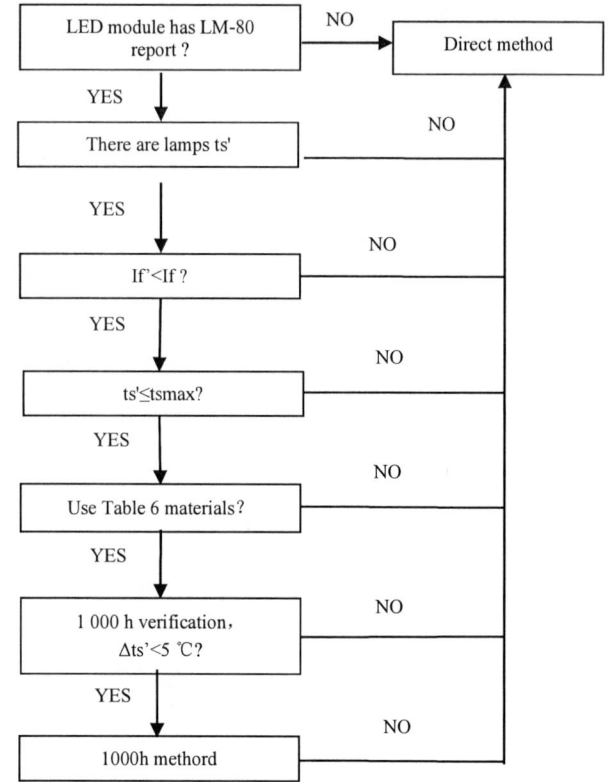

Figure 1 Test method selection flow chart

2.3.2 1000h Method

2.3.2.1 1000h Verification test

The luminaire shall be tested for 1000 h according to the conditions specified in Table 7. After the test, the luminaire shall meet the various indicators specified in Table 7.

Considering the influence of the drift of the ambient temperature within the tolerance range during the actual experiment, the calculation of Δts' needs to introduce the compensation of the drift of the ambient temperature ts before and after the 1000 h test, and the temperature rise of the LED solder joint Δts' = Δts' (1000h)- ts'(0h)+ Δta, where the ambient temperature drift Δta = ta(0h) - ta(1000h).

2.3.2.2 According to the LM-80 data, determine the 6000h lumen maintenance ratio L1 of the LED module in the fixture

a）When the temperature of the LED module ts' point in the luminaire belongs to a) above, the LM-80 data is directly used as the luminous flux data of the LED module in the luminaire, that is, the 6000h lumen maintenance rate at the relevant temperature in the LM-80 report is taken as L1.

b）When the temperature of the ts' point of the LED module in the lamp belongs to b) above, the luminous flux maintenance ratio L1 of 6000h at the ts' temperature is calculated by the interpolation method by the formula (1).

$$L_1 = L_{LED}(t_s', 6khs) = L_{LED}(t_{s1}) + \frac{(t_s' - t_{s1}) \times [L_{LED}(t_{s2}) - L_{LED}(t_{s1})]}{t_{s2} - t_{s1}} \tag{1}$$

Where LLED(ts1) and LLED(ts2) are the two solder joint temperatures ts1 and ts2 (eg 55 °C and 85 °C) given in the LM-80 report with the ts' point temperature at the test time of 6000 h Through maintenance rate.

c）When the temperature of the ts' point of the LED module in the luminaire belongs to c) above, the data of the LM-80 cannot be used to determine the luminous flux maintenance ratio of the LED module in the luminaire as L1.

Verification parameter	Claimed life≤ 25000h	25000h< Claimed life≤ 35000h	35000h< Claimed life≤ 50000h
Ambient temperature and relative humidity	40°C±2°C, 65%±5%	50°C±2°C, 65%±5%	60°C±2°C, 65%±5%
Test time/h	1000	1000	1000
Luminous maintenance rate	>93%	>94%	>95%
LED solder joint temperature change △ts'	<5°C	<5°C	<5°C
Power input power change	<3%	<3%	<3%
Chromaticity drift（△u' v'）	0.004 （thinking）	0.004 （thinking）	0.004 （thinking）

Table 7 Test conditions and assessment indicators

2.3.2.3 Determine the 6000h lumen maintenance rate L1' of the LED module in the fixture according to the claimed life of the luminaire

According to the influence of the light output attenuation of the luminaire during the life of the luminaire, the optical attenuation of the secondary optical material, and the aging of the heat dissipation structure on the attenuation of the light output, the minimum luminous flux maintenance ratio L1 of the LED module related to the claimed lifetime T of the luminaire is obtained by the formula (2).

$$L_1' = e^{\frac{6000}{T}\ln\left[L_{lum}(T)+\Delta L_O+\Delta L_S\right]} \tag{2}$$

where，

Llum(T) ------- -The luminaire claims a lifetime maintenance rate of T, such as 0.7;

△Lo -------- Secondary optical material light output decay estimates over time; wherein Table 6 gives a typical light output degradation estimate for a portion of the secondary optical material;

△Ls -------- The LED light output is attenuated due to aging of the heat dissipation structure. If △ts'< 5°C, △Ls is negligible. If △ts'≥5°C, it is carried out according to the method specified by the direct method. The △ts' value can be obtained from the 1000h verification test, and the verification test conditions are as shown in one of the above 2.3.2.

Note: When the product claims to have a life of no more than 35000h, use the existing formula to calculate. When the product claims a lifetime of more than 35000h and does not exceed 50,000h, change the "6000" in the existing formula to "9000".

2.3.2.4 Number of samples

The minimum number of test samples is 3.

2.3.2.5 Qualification judgment

When L1'>L1, it is considered that the manufacturer claims that the life is acceptable, and vice versa.

2.3.3 Direct method

2.3.3.1 Test conditions

The luminaire operates at the manufacturer's stated operating conditions and performance operating temperature tq ±2°C. If the manufacturer has multiple claims for tq, the test should be conducted at the highest claimed tq.

2.3.3.2 Test time t

The luminaire operates for at least 6000 hours at the manufacturer's stated working conditions and performance operating temperatures. For luminaires claiming a life of 50000h, the test time can be increased to 10000h or longer, given the time and conditions allowed by the test equipment. The luminous flux should be measured every 1000h, and the initial 1000h measurement interval can be shortened.

The actual working time of the LED luminaire should be accurately recorded. For time accuracy, video surveillance, current monitoring, and other methods can be used to determine actual working hours.

2.3.3.3 Number of samples n

The number of test samples is related to the maximum estimated life, as shown in Table 8. The number of test samples is at least three.

2.3.3.4 Calculate the luminous flux maintenance life Lp

A) Normalization:

The luminous flux data for each measurement point is normalized to 0h.

B) Average:

Perform arithmetic mean calculation on multiple sets of data obtained by normalization.

C) Select the data for the fit:

----- When the measurement period is 6000h, the data after 1000h is used for fitting.

----- When the measurement period is 6000h~10000h, the last 5000h data is used for fitting.

----- When the measurement period is greater than 10000h, the data of the second half of the total measurement period is taken as the fitting data. For example, if the test period is 13000h, then the data used is between 6500h and 13000h. If there is no data at 50%, the data from the previous point in time is used in the data fit. For example, the measurement period is 13000h, and every 1000h read, data points between 6000 and 13000h should be used.

D) Data fitting and calculation of lumen maintenance life:

Assuming that the lumen lumen attenuation follows the natural exponential law, the initial constant B and the decay rate constant α are obtained by fitting with equation (3).

$$\Phi(t) = B \times e^{-\alpha t} \tag{3}$$

Where：

$\Phi(t)$-----lumen output after normalized ;

t -----test time, the unit is hour (h);

The lumen maintenance life is calculated using equation (4).

$$L_p = \frac{1}{\alpha} \ln\left(\frac{B}{p}\right) \qquad (4)$$

where：

Lp----- lumen output after normalized ;

P-----lumen maintenance maintains the lumen maintenance rate; this value is claimed by the manufacturer, if not specified, the default is 70%.

Note 1: When $\alpha>0$, the exponential fitting curve shows a downward trend, and Lp>0 is obtained at this time; when $\alpha<0$, the exponential fitting curve shows an increasing trend, and Lp<0 is obtained at this time.

E) The maximum estimable value of lumen maintenance life Lp'

According to the amplification factor x obtained in Table 8 according to the number of test samples n, calculate the maximum estimable value of the lumen maintenance life according to equation (5):

$$L_p' = t \times x \qquad (5)$$

where：

t-----Test time in hours (h).

For example, the number of test samples is six, and the corresponding amplification factor x is 5. When the test time t is 6000h, the maximum estimated lifetime of the lumen maintenance life is 30000h.

Number of test samples n	Magnification factor x
3	3
4	4
5~6	5
7~9	5.5
10+	6

Table 8 Relationship between the maximum estimable value of lumen maintenance life and the number of test samples

F) Determine the predicted lumen maintenance life

The predicted luminous flux maintenance life is determined according to the calculation of the luminous flux maintenance life Lp and the maximum estimable value Lp' of the luminous flux maintenance life. The specific steps are shown in Fig. 2.

Note: When the calculated luminous flux maintenance life Lp>0, and Lp $\leq$ Lp', the predicted luminous flux maintenance life does not exceed Lp; the calculated luminous flux maintenance life >0, and Lp >Lp', then predict The lumen maintenance lifetime does not exceed Lp'; the calculated lumen maintenance lifetime does not exceed Lp'; the calculated lumen maintenance lifetime Lp < 0, the predicted lumen maintenance lifetime does not exceed Lp'.

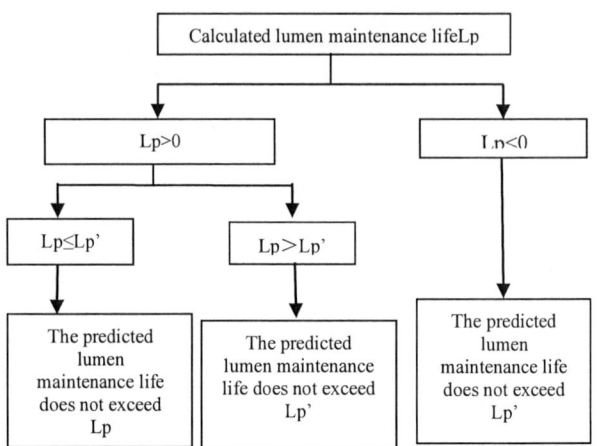

Figure2 shows the process of predicting the lumen maintenance rate

G）Qualification judgment

During the test, accidental failure samples were recorded in the report but were not included in the calculation. The company's claimed lumen maintenance life should not exceed the predicted lumen maintenance life.

Conclusions

By comparing the three life standards, this paper summarizes their advantages and disadvantages. Among them, the advantages of GB/T 33720-2017 standard are shorter test time, simple test method and theoretically strong life evaluation. The disadvantage is that the life expectancy is only 25,000 hours. The advantages of the CSA 020-2013 standard are that the test time is short, there are many life expectancy estimates, and the life evaluation is theoretically strong. The disadvantage is that the accelerated test temperature selection process is cumbersome. The advantage of the GB/T 33721-2017 standard is the test time. Short, the number of samples is small, the disadvantage is that there are many restrictions to meet the test, once it is not satisfied, you need to use the direct method of 6000 hours.

References

1. GB/T 33720-2017 LED lighting products luminous flux attenuation accelerated test method.
2. GB/T 33721-2017 LED lamp reliability test method.
3. CSA 020-2013 LED lighting products accelerated attenuation test method.

A Comparative Study of the Lifetimes of High-End and Low-Cost Off-Line LED Drivers Under Accelerated Test Conditions

F. Keil, K. Hofmann
Technische Universität Darmstadt
Integrated Electronic Systems Lab, Merckstr. 25, Darmstadt, Germany
Ferdinand.Keil@ies.tu-darmstadt.de

Abstract

This work presents a comparative study between a set of high-end, name-brand LED off-line drivers and a second set of low-cost drivers. To determine the lifetime of the devices in each sample an accelerated lifetime test was used. The degradation of important parameters of the devices during the test, as well as their lifetime, were recorded. A failure analysis reveals two important failure causes: galvanic corrosion and failed metallized film capacitors. Galvanic corrosion is found to be more frequent in the sample of low-cost devices. It is proposed that this is due to lack of proper cleaning of the assembled circuit board. Finally, the question whether the high-end sample has a significantly longer lifetime compared to the low-cost sample is assessed through a proper statistical test. A higher mean lifetime for the high-end sample cannot be confirmed.

Introduction

LED lighting's share of the general lighting market keeps increasing and is expected to reach 70 % by 2020 [1]. Besides higher energy efficiency – compared to traditional light sources – another major appeal of LED lighting is its supposedly superior lifetime. Recent generations of solid-state industrial luminaires and street lights are supposed to last up to 100.000 hours. As it is impossible to test such long lifetimes this high in real-time, accelerated tests have to be used. The LEDs themselves have been subjected to in-depth research regarding their lifetime under various environmental conditions. However, it is proposed in recent work that luminaire lifetime is limited by other components, in particular the driver [2].

In this work a sample of 13 drivers for industrial use from well-known manufacturers – hereinafter labeled the high-end sample – is compared to one of 10 drivers from no-name or contract manufacturers – hereafter considered the low-cost sample. Both samples were subjected to an accelerated test and their lifetimes were recorded. A failure analysis is performed and the failure modes are compared to the recorded data. The most frequent failure causes that have been identified are failed metal-oxide-varistors, transistors, integrated circuits and passive components shorted due to galvanic corrosion. Galvanic corrosion and the failure of metallized film capacitors are discussed in detail. Finally, a statistical approach is devised to show whether the high-end sample has a significantly longer mean lifetime than the low-cost one.

Test Setup

A complex test setup for up to 30 devices under test (DUT) was implemented for this study. It consists of two climate chambers of 408 l volume and wiring to support 15 DUTs in each chamber. The wiring is connected to test equipment to facilitate online monitoring of all important parameters of the

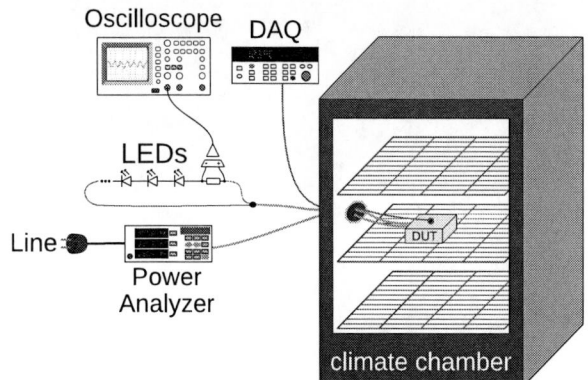

Fig. 1 Illustration of the used test setup.

DUTs while the accelerated life test is running. The setup is illustrated in Figure 1.

On the primary side a power analyzer records voltage, current, power, power factor, and current distortion. On the secondary side the DC current and AC current ripple are monitored by using an oscilloscope and a galvanically isolated differential amplifier connected across a shunt resistor. The temperature of the DUT at the t_C-point is measured using a data-acquisition unit. The setup is fully automated using a lab computer running custom Python software and has been operating without major issues for more than two consecutive years.

Temperature-Humidity Bias Test (THB Test)

Early studies on LED driver reliability focused on the lifetime of the integrated aluminum electrolytic capacitors [3, 4]. However, more recent studies have indicated that different components might be lifetime limiting [5, 6]. These components, like metallized film capacitors or semiconductor devices, are susceptible to humidity and thus their degradation can be effectively accelerated in a temperature-humidity bias (THB) test. A test at 85 °C and 85 % relative humidity as specified in the JESD22-A101D [7] standard is very common in the industry and was used in this study. It therefore makes it possible to compare the results presented here with those published by other authors.

Another reason for employing a THB test is that LED drivers can survive for a very long time in mild environmental conditions. In a previous study it was found that lifetimes in excess of 15.000 hours can be achieved even in a high-temperature operating life test at 85 °C dry heat (not yet published).

978-1-7281-5757-3/19 $31.00 © 2019 IEEE

Galvanic Corrosion

The majority of solder processes today are of the no-clean type, meaning that assembled PCBs will not undergo a cleaning process after they have been assembled [8]. The active components (e.g. resin, weak organic acids) contained in the no-clean fluxes used in these processes need to be heated to a certain temperature to deactivate them. This might not always

Fig. 2 Galvanic corrosion on a PCBA from the low-cost sample.

happen due to uneven heat distribution during soldering [9]. It has been demonstrated for weak organic acids (WOA), that they are hygroscopic in humid environments and dissolve readily [10]. The presence of these contaminants can trigger two major failure mechanisms: galvanic corrosion between different metals due to the water layer the contaminants attract; electrolytic corrosion along electric fields as the aforementioned solution works as an electrolyte [11, 12]. Cleaning of the PCB assembly (PCBA) after soldering can prevent this, however this process adds cost to the production. Usually the cleaning process is omitted in the manufacturing of low-cost LED drivers for this very reason, explaining the increase in corrosion damage in the THB test. A drastic example of corrosion on the PCBA of a low-cost driver is presented in Figure 2. Green, porous copper oxide can be seen around the highlighted PCB trace. The solder joint and the end-cap of the ceramic capacitor also show signs of corrosion. On the ceramic body a white residue has formed, which might lead to a short across the capacitor subsequently.

Metallized Film Capacitors

Metallized film capacitors are widely used in LED drivers for filtering and bypassing. Due to the self-healing effect inherent in their construction, they are even irreplaceable in some applications, e.g. primary side EMI filtering. Compared to aluminum electrolytic capacitors they are less susceptible to high temperatures. Therefore, they can increase reliability in applications where heat is suspected to be a major cause of failure. However, moisture has been identified as one of the most critical stressors for metallized film capacitors [13]. It can be reasoned, that the increased use of metallized film capacitors in LED drivers – for which circumstantial evidence was found when drivers were disassembled during this work – might lead to higher failure rates in humid conditions.

Humidity ingress into metallized film capacitors leads to three major failure mechanisms [14]: electrode corrosion,

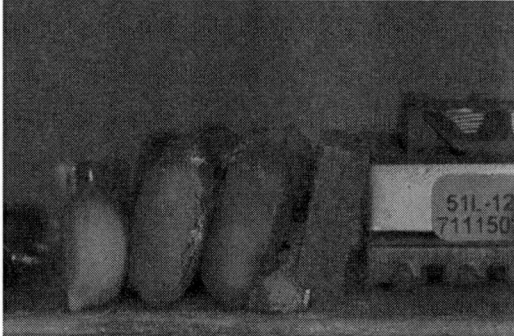

Fig. 3 Cracked and melted metallized film capacitors right next to main bridge rectifier.

causing an increase in series resistance; reduction of the insulation resistance of the dielectric, causing increased leakage currents; corona discharge at local electric field maxima which vaporizes the metallization, causing a decrease in capacitance. The first two mechanisms result in increased power dissipation in the component and further increases its temperature. They can even cause thermal run-away, which leads to catastrophic destruction of the component by either melting or even explosion. Both effects could be identified on the PCBAs of the DUTs after the test. An exemplary case is shown in Figure 3: the windings of two capacitors melted and emerged from the cracked casings, flowing under and shorting adjacent components (the main rectifier bridge in this case).

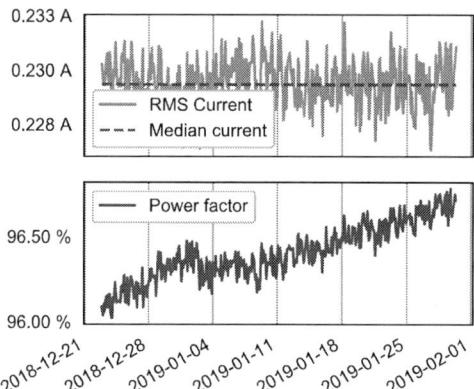

Fig. 4 Main side current and power factor for the DUT pictured above.

17 of the 28 tested LED drivers show at least cracks in the plastic case of a metallized film capacitor. The damage is worse for components on the line side of the circuit. It was also noticed that degradation of capacitors on the line side causes characteristic parametric changes to the power drawn by the DUT. Most notably is an increase of the measured power factor as shown in Figure 4 for the DUT mentioned before. This is readily comprehensible since a load that is capacitive in nature has a lower power factor and if a capacitor on the line side of the circuit degrades and its capacitance decreases, the power factor of the DUT increases.

Destruction of the film capacitor does not always lead to a lights-out failure of the LED driver, as film capacitors are often used in non-critical parts of the circuit (e.g. EMI filters). However, this kind of failure still leads to parametric changes that might violate regulatory requirements or datasheet specifications of the device.

Statistical Approach

As aforementioned, this work tries to answer whether high-end LED drivers have a significantly higher lifetime compared to low-cost drivers when subjected to THB conditions. This question can be restated as follows: is the mean lifetime of the sample of high-end drivers higher than that of the sample of low-end drivers. Given the data from the test in THB conditions this can be asserted using the common t Test for Two Independent Samples. Before the test is applied to the data, its assumptions have to be checked in order to achieve a reliable test result.

Test Assumptions

The t Test for Two Independent Samples is based on three assumptions [15]:
1. Each sample has been chosen randomly from the population it represents.
2. The data of the populations underlying the samples is distributed normally.
3. The variances of the populations from which the samples were chosen are equal (homogeneity of variance).

The drivers used in the experiments were supplied by several companies manufacturing LED lighting and selected from their current portfolio of LED luminaires. Thereby,

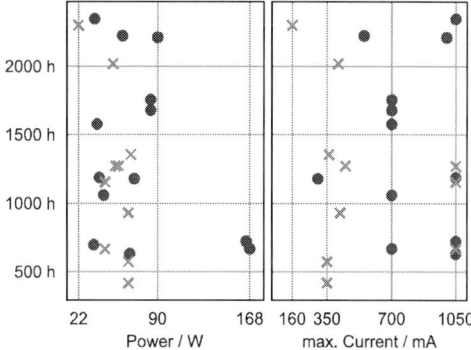

Fig. 5 Scatter plot of important parameters of the high-end (circle) and low-cost (cross) samples.

randomness is ensured by experimental design. Additionally, after the experiment was concluded two scatter plots were generated, comparing the lifetime with the maximum output power and the maximum output current of the LED drivers. The plots are shown in Figure 5. It can be clearly seen that a broad range of devices are part of the experiment in terms of maximum power as well as maximum output current. In addition, there is no obvious correlation between these parameters and the lifetime of the respective drivers, further demonstrating the randomness of the sample.

Although statistical tests for normality do exist and were applied to the data as well, a graphical test is presented here. The reasoning is that due to the small sample size the power of

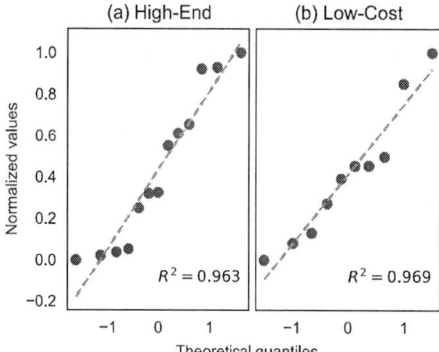

Fig. 6 Probability plot for both samples.

these statistical tests is questionable. Additionally, the graphical method fosters a more intuitive understanding of the data and generally conveys more information than a test with binary outcome [16]. In order to check for normality, the data was plotted in a so-called probability plot. Here, the actual quantiles of the data are plotted against the quantiles of a theoretical distribution. The desirable property of this plot is that if the distribution generating the data is normal, the points in the plot will follow a straight line. This behavior can be observed in the plots in Figure 6 for both the high-end and the low-cost LED drivers. The dashed line in the plot was fitted using linear regression and represents the normal distribution estimated from the data. The deviations from this line can be explained by the small sample size. Both samples are therefore regarded to stem from normally distributed populations.

Finally, the homogeneity of variance has to be checked. Instead of the variance, the standard deviation was calculated for both samples. Demanding equal standard deviation for the samples has the same implications as demanding equal variance as the relationship between variance s^2 and standard deviation σ is given as $\sigma = \sqrt{s^2}$. However, the standard deviation is more commonly used in engineering. The calculated values are

$$\sigma_{lc} = 572 \text{ h and } \sigma_{he} = 604 \text{ h.}$$

The standard deviations for both samples are very close to each other, so it can be assumed that the variances of the underlying populations represented by the samples are equal.

t Test for Two Independent Samples

The null hypothesis of the chosen test is given as

$$H_0: \mu_1 \neq \mu_2$$

meaning that the means of the populations from which the samples stem are not equal. The alternative hypothesis states that the mean lifetime of the population of high-end drivers is higher than that of the low-end driver

$$H_1: \mu_1 < \mu_2.$$

As the samples are not of the same size, the generalized form of the t Test for Two Independent Samples is used [15]. Its test statistic is given as

$$t = \frac{\bar{X}_1 - \bar{X}_2}{\sqrt{\left[\frac{(n_1 - 1)\tilde{s}_1^2 + (n_2 - 1)\tilde{s}_2^2}{n_1 + n_2 - 2}\right]\left[\frac{1}{n_1} + \frac{1}{n_2}\right]}}$$

where $\bar{X}_i$ is the mean of sample i, n_i is the size of the sample and $\tilde{s}_i^2$ is the estimated population variance which is computed as

$$\tilde{s}_i^2 = \frac{\Sigma X_i^2 - (\Sigma X_i)^2/n_i}{n_i - 1}$$

where X_i denote the values of sample i. The value of the test statistic evaluates to $t = 0.708$. Before the critical value for the test can be looked up in the relevant literature the degrees of freedom have to be calculated by

$$df = n_1 + n_2 - 2.$$

Both the values for the 0.05 and 0.01 level of significance (95% and 99% confidence levels) were taken from [15] and are presented in Table 1.

	$t_{.05}$	$t_{.01}$
Two-tailed values	2.080	2.831

Tab. 1 t Test critical values for 21 degrees of freedom.

The test statistic is now compared to the critical value, given that the null hypothesis can be rejected if the value of t is positive and equal to or greater than the critical value. It is evident, that in this case the null hypothesis cannot be rejected for neither the 0.05 nor the 0.01 level of significance as the calculated value is smaller than both critical values. The lifetime of high-end LED drivers in this test was therefore not significantly longer than the one of low-cost drivers.

Conclusions

The results of an accelerated lifetime test in a THB condition of two samples of LED drivers have been presented. One sample is considered to represent the high-end segment of the market, the other one the low-cost segment. Two important failure mechanisms were identified: galvanic corrosion and cracking or even melting of metallized film capacitors. High levels of humidity are the trigger for both mechanisms. However, galvanic corrosion can be avoided by cleaning the PCBA after soldering. It is reasoned that PCBA cleaning is the reason why the high-end sample showed much less galvanic corrosion.

Finally, a statistical test approach was devised to decide whether the high-end sample lasts significantly longer in a THB condition compared to the low-cost sample. The test's assumptions were thoroughly checked before it was applied to the data. The test concludes that the high-end sample does not last significantly longer than the low-cost sample.

Acknowledgments

The research project was carried as part of the industrial collective research program (IGF no. 19278 N). It was supported by the Federal Ministry for Economic Affairs and Energy (BMWi) through the AiF (German Federation of Industrial Research Associations eV) based on a decision by the German Bundestag.

References

1. MarketWatch. "Global LED Lighting Market 2019 Trends, Market Share, Industry Size, Opportunities, Analysis and Forecast To 2025." (accessed 2019-10-28).
2. J. L. Davis *et al.*, "System reliability for LED-based products," in *2014 15th International Conference on Thermal, Mechanical and Mulit-Physics Simulation and Experiments in Microelectronics and Microsystems (EuroSimE)*, 2014: IEEE, pp. 1-7.
3. L. Han and N. Narendran, "An accelerated test method for predicting the useful life of an LED driver," *IEEE Transactions on Power Electronics,* Vol. 26, (2011), pp. 2249-2257.
4. S. Tarashioon, W. Van Driel, and G. Zhang, "Multi-physics reliability simulation for solid state lighting drivers," *Microelectronics Reliability,* Vol. 54, no. 6-7 (2014), pp. 1212-1222.
5. R. International, "Hammer Testing Findings for Solid-State Lighting Luminaires," December 2013.
6. S. D. Shepherd, K. C. Mills, R. Yaga, C. Johnson, and J. L. Davis, "New understandings of failure modes in SSL luminaires," in *SPIE Optical Engineering + Applications*, 2014, vol. 9190: SPIE.
7. *Steady-State Temperature-Humidity Bias Life Test*, JESD22-A101D, J. S. S. T. Association, July 2015.
8. *Failure Mechanisms and Models for Semiconductor Devices*, JEP122H, J. S. S. T. Association, September 2016.
9. M. S. Jellesen, V. Verdingovas, H. Conseil, K. Piotrowska, and R. Ambat, "Corrosion in electronics: Overview of failures and countermeasures," in *European Corrosion Congress*, 2014.
10. V. Verdingovas, M. S. Jellesen, and R. Ambat, "Solder flux residues and humidity-related failures in electronics: relative effects of weak organic acids used in no-clean flux systems," *Journal of Electronic Materials,* Vol. 44, no. 4 (2015), pp. 1116-1127.
11. R. Ambat, "Perspectives on climatic reliability of electronic devices and components," in *IMAPS Nordic Annual Conference Proceedings 2012*.
12. S. Zhan, M. H. Azarian, and M. Pecht, "Reliability of Printed Circuit Boards Processed Using No-Clean Flux Technology in Temperature–Humidity–Bias Conditions," *IEEE Transactions on Device and Materials Reliability,* Vol. 8, no. 2 (2008), pp. 426-434.
13. H. Wang and F. Blaabjerg, "Reliability of capacitors for DC-link applications in power electronic converters - An overview," *IEEE Transactions on Industry Applications,* Vol. 50, (2014), pp. 3569-3578.
14. R. Gallay, "Metallized Film Capacitor Lifetime Evaluation and Failure Mode Analysis," *CAS - CERN Accelerator School: Power Converters,* Vol. 003, (2016), pp. 7-14.
15. D. J. Sheskin, <u>Handbook of parametric and nonparametric statistical procedures</u>, Chapman and Hall/CRC (2007).
16. J. M. Chambers, <u>Graphical methods for data analysis</u>, Chapman and Hall/CRC (Belmont, Calif., 1983).

Corrosion Failure Analysis and Coping Strategies of Light Reflecting Devices in Light-Emitting Diode Devices

Lyu tiangang,Wang yuefei,Lyu henan,Wang caixia, Chen lihe,Xu bingjian,Tang leming
Hongli Zhihui Group Co.,Ltd.
Xianke 1st Road, Huadong Town, Huadu District, Guangzhou, China.
ledc51@163.com,15920888599

Abstract

In the research、manufacturing and application process of light-emitting diodes, chemical corrosion is one of the most important causes of blackening of light-emitting diode（hereinafter referred to as LED）light-reflecting devices, which causes light decay and leads to product failure[1].

This paper studies the technical background, formation mechanism, risk traceability, solution, performance evaluation scheme, fault analysis scheme and related technical standards at home and abroad for the corrosion failure of LED reflectors.

1. Advantages of Ag materials

The light path of a LED usually consists of two parts: a lens made of resin or glass, and a reflective device with a silver coating.

Sulfur resistance is a key quality indicator for LED packages because silver is easily attacked by sulfurized gases and causes significant light decay.

Main functions of LED reflector: First, the stray light is reflected to the functional area to improve the utilization rate of light. Second, closely coordinate with the lens to adjust the light output Angle.

Depending on the type of LED package, there are two types of reflective surfaces: horizontal and cup-shaped.

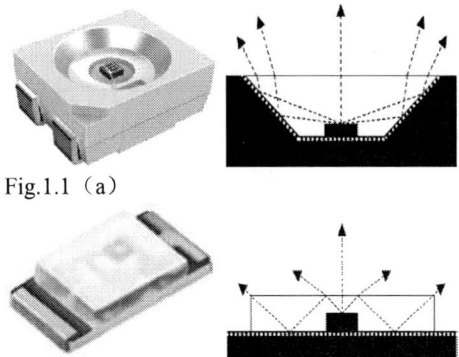

Fig.1.1 （a）

Fig.1.1 （b）

Fig.1.1 Schematic diagram of LED light reflector

Regardless of the type of reflector, its performance has a decisive impact on the performance of LED devices, especially in light efficiency.

The reflector is one of the most important and common ways to improve the efficiency of light extraction.

In addition, to select the appropriate reflective material, it is also necessary to evaluate the stability of the material and the process maturity.

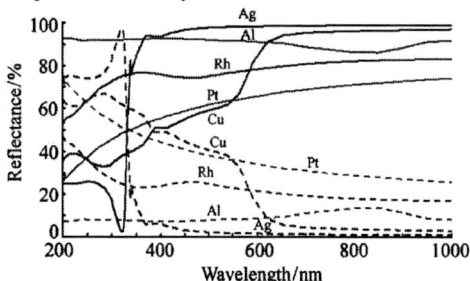

Fig.1.2 Reflectance(solid) and absorptance (dash) of commonly-used metal coatings

Ag silver has a high reflectivity from the visible spectrum (380nm~780nm) to the infrared region, and its optical properties are significantly better than other metal materials. Therefore, Ag is a commonly used metal reflective film material in optical systems[2].

The silver-plated reflective film is a light-reflecting film material comprising a silver alloy film having pure silver or silver as an absolute main component as a protective layer of a light-reflecting film layer, a substrate (film) and a silver reflective film layer. The incident light and the reflected light of the silver-plated reflective film have a one-to-one correspondence, and specular reflection occurs, and the higher the gloss of the surface of the reflective film, the higher the reflectance. The hallmark of silver plating reflectivity is the light reflectivity. The silver-plated reflective film has high reflectivity, high reflectivity in both the visible and infrared regions, and its reflectance is usually above 97%, Moreover, the polarization is small when the light is obliquely incident and has good thermal stability and durability[3].

2. Corrosion failure mechanism

The silver or silver alloy metal film of the LED reflector is affected by the corrosive medium in the environment, and the glossiness, discoloration and coarsening of the reflecting surface will lead to the deviation of spectral response frequency, the decrease of optical reflectivity, and then the decrease of luminous flux/radiation flux, and the deviation of color, finally leading to the complete failure of the LED product.

Silver has excellent optical properties and is suitable for industrial processing. However, since silver is easily oxidized, vulcanized, and chlorinated in the air to form Ag_2O, Ag_2S, $AgCl$, Ag_2SO_4, Ag_2CO_3,etc., the corresponding silver becomes the following color: Ag_2O brown, Ag_2S gray black, $AgCl$ and Ag_2SO_4 white (photolysis black), Ag_2CO_3 light yellow.

978-1-7281-5757-3/19 $31.00 © 2019 IEEE

The reflecting device of the LED is usually packaged inside the organic silica gel mixed with the phosphor, but the organic silica material has a pore of 4 nm ~ 7 nm. The diameter of the gas molecules is substantially less than 1 nm, such as O_2 is only 0.346 nm, so that the air molecules easily permeate through the silicone and contact with the surface of the silver reflective film to react. In addition, the corrosion reaction process of Ag is also affected by factors such as light, temperature, and humidity [4].

The following will introduce the influence factors of Ag reflective film failure from two aspects of corrosive medium and environmental factors.

Part I: corrosive medium

(1) Ag and O_2

Ag is thermodynamically unstable in the presence of oxygen [5] and is easily oxidized by O_2 in the atmosphere to form Ag_2O and AgO.

The reaction is as follows:

$4Ag+O_2 \rightarrow 2Ag_2O$

Fortunately, the internal temperature and oxygen content of the LED are not enough to support the rapid oxidation reaction of Ag, so the early failure problems caused by this generally do not occur.

(2) Ag and H_2S

Ag is extremely sensitive to H_2S, and its concentration of $0.3\mu g/m^3$ is sufficient to cause the sulfuric discoloration of Ag to produce black Ag_2S.This is fatal to the Ag reflective film, which is responsible for the LED.

H_2S concentration is less than $10^{-6}V/V$ and the conversion rate of H_2S on Ag surface is about 0.0001% under dry conditions. The conversion rate remains unchanged when H_2 with $10^{-5} \sim 3.8\%V/V$ concentration is added. Therefore, Volpe et al[6] believed that when Ag was vulcanized in H_2S environment, H_2 did not participate in the reaction, but O_2 acted as an oxidant. The reactions of Ag in H_2S environment are as follows:

$4Ag+2H_2S+O_2 \rightarrow 2Ag_2S+2H_2O$

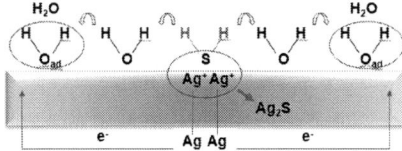

Fig.2.1 Schematic diagram of Ag corrosion discoloration in H_2S environment

(3) Ag and S

When the concentration of sulfur bloom (S8) in the air reaches $3\times10^{-8}V/V$, Ag_2S film can be obviously produced on the surface of Ag[7]. Reagor et al[8] demonstrated that Ag corrosion by sulfur bloom was achieved through the following reactions:

$S_x+2xAg \rightarrow Ag_2S$

Allpress et al[9] showed that there was a linear relationship between sulfur concentration and Ag corrosion rate, that is, diffusion was the control step of vulcanization reaction.

（4）Ag and NO_2+H_2S gas mixture

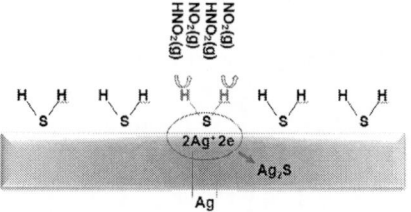

Fig.2.2 Schematic diagram of Ag corrosion discoloration in N_2O and H_2S environments

Kim[10] found that in $0.1\times10^{-9}V/V$ H_2S and $0.1\times10^{-6}V/V$ NO_2, the corrosion rate of Ag showed a linear rule. The corrosion reaction process is as follows:

$H_2S+2 NO_2 \rightarrow S+2HNO_2$

$2Ag+S \rightarrow Ag_2S$

(5) Ag and other substances

In addition to the above O_2, H_2S, elemental S, NO_2+H_2S mixture in vitro,SO_2, NO_2 and Cl compounds also react with Ag.

Part II: environmental factors

(1) Light

Fang Jingli et al[11] studied the effect of irradiation wavelength and irradiation time on Ag discoloration (Table2.1). It can be seen from the table that Ag absorbs ultraviolet light and is prone to discoloration. The order of color wavelength change from strong to weak is:

253.7nm>365.0nm> sunlight

Sinclai [12] experiments confirmed that the corrosion rate of corrosion medium to Ag was significantly accelerated during illumination, and the energy of the illumination was proportional to the corrosion rate.

Table2.1 The effect of wave length and exposure time silver-planting layer

λ_p(nm)	Exposure time （h）				
	6	12	18	24	48
253.7	No change	Local macula	Yellow brown	Brown black	Deep black
365.0	No change	No change	No change	Yellow	——
sunlight	No change	No change	No change	Local macula	——

(2) Temperature

Through the single-element sulfur corrosion test of the LED and the H_2S corrosion experiment, it is found that the corrosion rate is positively correlated with the temperature.

Table2.2 Effect of sulfur corrosion temperature on light decay of LED

T_c(℃)	t(h)	S concentration (g/ml)	Luminous maintenance rate(%)
69.8	4	0.002	97.79
75.3	4	0.002	96.86
80.3	4	0.002	91.25

(3) Humidity

The diameter of water molecules in air is about 0.4nm, which is easy to form water molecules on the surface of Ag reflective film after being inhaled by encapsulating colloid. The thickness of the water molecules is closely related to Ag corrosion rate, as shown in the figure.

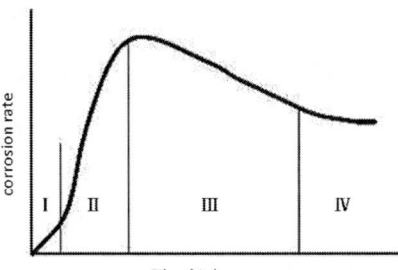

Fig.2.3 relation chart of water film thickness - corrosion rate

Ag has little change in vulcanization rate in an environment where the content of water vapor is $10^{-6} \sim 10^{-3}$ V/V and the concentration of H_2S is 2% ~ 50%. When water vapor reaches saturation, the vulcanization rate increases rapidly [13]. However, in the field exposure experiment of the atmosphere, Rice et al [14] measured that the velocity of Ag forming rust layer was independent of the humidity of the environment.

When H_2S and NO_2 coexist, since H_2O does not participate in the reaction, the corrosion rate is independent of humidity [10], but is affected by corrosive media. It can be seen that the influence of humidity on the corrosion process of Ag is complicated and needs to be considered comprehensively.

3. Risk source analysis

The failure of Ag reflector of LED mainly occurs during the reflow soldering of SMT process and the aging process after the application product assembly. But the sources of risk can be hidden in every corner. Combined with the corrosion denaturation mechanism of Ag mentioned in the previous chapter, all elements of the whole life cycle of LED were sorted out and analyzed, and the source of the corrosion medium and catalytic conditions that LED to the corrosion failure of Ag reflector was found. It is of great significance to improve and optimize the process, solve and prevent the failure.

The inspection covers four processes: bracket process, LED process, PCB process and LED application process, including the main elements involved in these processes, such as process conditions, environment, main raw materials and accessories, etc.

The high risk points are identified through analysis and screening, which can be used as a reference for formulating coping strategies.

a) Fluorescent powder [15][16]

There are three types of commonly used LED phosphors:

Yellow phosphors —— mainly aluminate $Y_3Al_5O_{12}:Ce^{3+}$(YAG:Ce)and silicate $Y_3Al_5O_{12}$（YAG:Ce^{3+}）.

Red phosphors —— commonly used as nitride phosphors, such as $M_2Si_5N_8:Eu^{2+}$ and $MA_1SiN_3:Eu^{2+}$.

In addition, there are sulfide and oxide series phosphors.

Green phosphors —— green phosphors that can be excited by blue light are mainly thiogallium salts, such as $SrGa_2S_4:Eu^{2+}$.

Risk assessment of the above three types of phosphors is as follows:

Yellow phosphor —— the preparation process may use nitric or sulfuric acid.

Red phosphor —— the sulfide phosphor contains the element S.

And this kind of phosphor is very unstable at high temperature.

Green phosphor —— gallium thioate system green phosphor contains S element, and H_2S gas is used in the preparation process.

In conclusion, the risk of fluorescent materials containing or residual corrosive media in the process cannot be ruled out.

b) packaging adhesive

The inspection of the encapsulant is mainly carried out from two aspects, one is whether it contains corrosive medium such as S, and the other is the ability to block the intrusion of corrosive gas.

First, EDS analysis was performed on a common organic silica gel, and no corrosion medium remained.

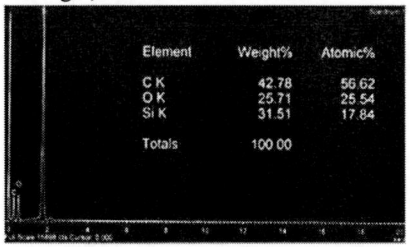

Fig.3.1 EDS element analysis of organic silica gel for packaging

Then, the LED devices made of six different types of package adhesives A, B, C, D, E, and F were subjected to a red ink dye penetration test. According to the test results, the penetration depth is sorted:

B>A>C>D>E>F

B penetration is the most obvious, and F has almost no penetration. Tests have shown that there are differences in the sealing effect of different types of encapsulants, which have a direct impact on the corrosion resistance of the device.

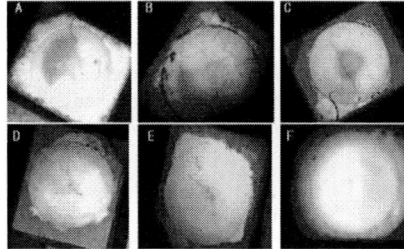

Fig.3.2 Dye penetration test of package adhesive

Finally, new samples of the above six groups of test packaging adhesive were taken for S corrosion test, and the results were ranked according to the degree of corrosion blackening in the area of LED reflector:

A/B/C > D/E/F

Thus, it is proved that the corrosion is correlated with the encapsulation adhesive and positively correlated with the sealing performance.

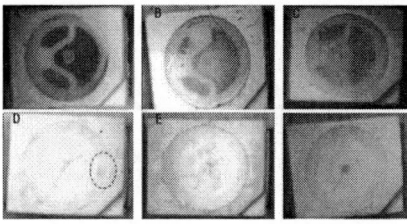

Fig.3.3 S Corrosion test of package adhesive

B. Process analysis

Process analysis, mainly screening LED manufacturing cycle process, equipment, auxiliary materials and temperature/humidity/pressure and other environmental factors may exist in direct or indirect risks.

a) LED manufacturing process

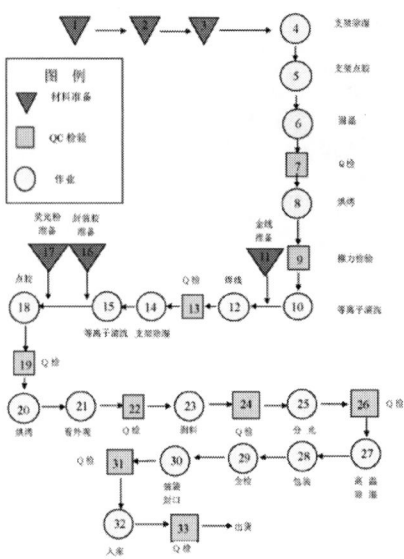

Fig.3.4 LED manufacturing process flow chart

The vulcanization phenomenon in the production process mainly occurs in the solid crystal and dispensing process, and the vulcanization is mainly composed of a silver-containing material and a silicone rubber material. In the production process, we must pay great attention to the two stages of vulcanization prevention.

The main vulcanization pollutions are: sulfur-containing gloves, finger cots, sulfur-containing masks, sulfur-containing PCBs, vulcanized ovens containing sulfur-containing cleaning agents, vulcanized cartridges, fixtures, etc [17].

c) LED application product manufacturing process

Fig.3.5 LED application products flow chart

Application process is the most concentrated link of abnormal LED corrosion, especially in SMD, assembly and aging process. There are mainly two reasons:

First, the type of raw materials is complex and the risk of corrosion source is high.

In the production process, PCB board, solder paste, washing board water, fixing glue, sealant, wire and other materials were found, and the detection record of element S was recorded[17].

Second, the process produces high temperature, accelerating the corrosion of volatile substances and corrosion reaction.

According to the analysis in chapter 2 of this paper, there is a positive correlation between temperature and corrosion rate. Under the same condition, temperature ↗ corrosion rate ↗ reflectivity ↘ luminous effect ↘.

SMT process —— the product enters the high-temperature reflow welding furnace after the automatic SMT placement is completed. In this high-temperature environment, if PCB board, solder paste, washing board water and other materials contain corrosion media similar to S, these corrodes will become extremely active and volatile, which will contact with the coating of LED reflector and cause corrosion reaction.

Assembly and aging process —— after the completion of the assembly, LED components sealed in the application product, when the aging power, LED and drive power to produce high temperature.

If there are corrosive substances in the materials used in the assembly process, such as structural parts, sealant, fixing glue, wire, flux, plate washing water, insulating materials and sealant silicone ring[18], they will quickly evaporate and corrode the LED reflective layer.

4. **Countermeasure**

The main cause of corrosion failure of the LED light reflecting device is the chemical reaction between the Ag reflecting film and the invading corrosive medium. There are mainly the following ways:

First, improve the chemical stability of the reflective layer itself;

Second, blocking the passage between the reflective layer and the corrosive medium;

Third, eliminate the source of environmental corrosive media.

The specific measures are as follows:

(1) Anti-tarnish Ag alloy

Improving and improving the corrosion resistance of Ag reflective layer materials is the most direct means to solve the photocorrosion failure of LEDs.

Britain and Germany studied the vulcanization resistance of silver alloys as early as 1920s.

According to a systematic study conducted by the us national bureau of standards, in order to inhibit the formation of sulfide Ag, alloy must be formed with 40%Pd, 70%Au or 60%Pt.

The common silver alloys, 92.5%Ag and 7.5%Cu, are called standard silver alloys.

Silver alloying is an important way to improve the anti-discoloration ability. Andin et al [19] studied and compared the anti-discoloration performance of au-ag, au-ag-cu and other alloys.

The accelerated vulcanization experiment shows that the Ag content is the main factor that affects the anti-discoloration performance of the alloy, and the lower the content, the less discoloration will occur.

However, the decrease of Ag content has a negative effect on the light reflectivity.

(2) Ag reflective film surface treatment

The Ag_2S produced by vulcanization does not prevent the further occurrence of vulcanization and corrosion like the $Al2O3$ protective film formed by oxidation of the surface of the Al film, thus eventually causing the silver film to be completely corroded. A protective film is coated on the surface of the Ag mirror to effectively block the contact of the corrosive material to the Ag reflective interface.

The protective film not only needs to be easily corroded, but also provides effective protection, and also has a small influence on the reflectance of the working band, and is simple to prepare and dense in film formation.

According to the different forms of silver surface film, the following types are common:

a) Plating other metals on the surface by electroplating or ion sputtering

Ag surface electrodeposition is about 150nm, Au and Pd noble metal layer [20], which can prevent Ag discoloration for a long time, but the cost is high and the surface loses metallic luster.

This method is only used in military products with high reliability and stability.

Ta, Nb, Ti and Al metal coatings can be formed on the surface by ion sputtering technology.

b) Chemical passivation

Chromate passivation is one of the most commonly used chemical passivation methods. Chromate passivation is carried out in an acidic or alkaline solution containing hexavalent chromium compounds to generate Ag_2O and Ag_2CrO_4 films on the surface of Ag, thus playing the role of anti-discoloration. Xiao guizheng et al[21] added NaH_2PO_4 to K_2CrO_7-KCN solution for passivation, which improved the anti-discoloration effect without affecting the metallic color of Ag.

After the formation of chromate layer on the surface of Ag, Ag discoloration can be prevented. However, the mechanical stability of chromate layer is poor, and complex structure and edges and corners are difficult to be coated. Hexavalent chromium is toxic, polluting and requires expensive wastewater treatment facilities.

c) Electrolytic passivation

Electrolytic passivation does not affect the appearance of silver, can maintain the metallic luster of silver. However, the passivation film does not adhere firmly to the silver surface, is not resistant to wiping, is not stable when heated, and is vulnerable to changes in environmental conditions and weakens the anti-ag discoloration effect [22].

Due to the harm of chromate to the environment, researches on chromium-free electrolytic passivation process are gradually carried out, such as chloride solution passivation process proposed by Singh et al [23].

d) Deposited oxide film

On the Ag surface, an oxide film is obtained by a sputtering technique or electrophoretical deposition of the third to fifth periodic metals Al, Be, Zr, Mg, Ti, and Nb in the periodic table in an aqueous solution. Dubkov et al[24] used water-soluble cation electrophoresis to form an oxide or fluoride layer on the surface of Ag, which can effectively prevent Ag discoloration.

e) Organic adsorption passivation layer

With the increasing awareness of surface organics passivation, some new organic passivators have emerged, such as a-mercaptoacetamides such as $Ar-NH-CO-CH_2-SH$ [25]. In addition, organic polymers with heterocycles have also attracted attention. Brusic et al[26] dissolved polyaniline derivatives in an organic solvent and applied a thin layer on the surface of Ag to significantly improve the anti-tarnishing properties of the mask.

Xue et al[27] chemically adsorbed polybenzimidazole after treatment of Ag surface with nitric acid. SERS analysis showed that the polymer reacted with Ag to form a complex membrane, which could prevent high temperature oxidation. However, there are also major problems with organic passivation. Since the coating is almost invisible to the naked eye, if the film is too thin, it does not protect; if the film is too thick, it is neither economical nor easy to cause unevenness of the surface coating.

f) Resin coating

The silver surface is coated with a layer of organic polymer film by spraying or dipping, and the Ag surface is isolated from the surrounding media, so as to prevent discoloration. Nitrocellulose varnish has been widely used in anti - discoloration. At present, the commonly used synthetic resin baking paint includes acrylic acid, polyurethane, organosilicone, amino clear baking paint, polyethylene vinyl acetate and paoic polyether coatings [28].The anti-discoloration effect obtained by this method is enhanced with the increase of film thickness [29].In general, the thickness of the coating should be $10_{\mu m} \sim 15_{\mu m}$. If too thin, anti - discoloration effect is not ideal.

There is a one-to-one correspondence between the incident light and reflected light of Ag reflective film. What happens is specular reflection. The higher the gloss of the reflective film surface, the higher the reflectivity. Ag reflective film surface treatment of the landmark index is the light reflectivity. High reflectivity in the visible region, preferably above 90%, and low polarization, good thermal stability and durability when the light is tilted in.

(3) Other measures

a) Improve the performance of packaging adhesive

Silicone plays an important role in the packaging process of LED devices, including fixing and protecting internal structures such as chip and reflection film, fixing and adjusting the distribution of fluorescent materials, synthesizing white light, and adjusting the beam. With the continuous development of technology, the requirements of organic silicone LED are becoming higher and higher. Indicators include: transmittance, refractive index, hardness, thermal stability, UV resistance, viscosity, hydrophobicity, air tightness, adhesion, solvent resistance, mechanical strength, curing time, curing temperature and so on.

The epoxy modified silicon polymer based on TEOS by shen zheng et al [30] takes TEOS and epoxy resin as the main raw materials and forms polymer with graft and interpenetrating network structure through sol-gel reaction. The general reaction formula is as follows:

The silicon network formed by TEOS in the system through hydrolysis and condensation covered the surface of the substrate, which hindered the penetration of water and electrolyte, reduced the corrosion rate, and played a good protection effect on the substrate.

b) Eliminate environmental corrosion sources

Eliminating the source of corrosive medium in the manufacturing process of LED devices and applied products can effectively prevent corrosion failure of products. Specific measures include: do not use sulfur-containing or sulfur-contaminated raw materials, equipment, clamps, vehicles, labor insurance supplies, consumables and packaging materials. Comprehensive investigation and removal of pollution sources.

As mentioned in the previous chapter, PCB, solder paste and water for washing board in SMT process, as well as plastic parts, paint, fixing glue and sealing glue in the assembly process of finished products, may have sulfur or other ag-sensitive media residues, which will be volatilized at high temperature and then penetrate into LED devices.

Therefore, early failure of Ag reflector of LED device is common in SMT process after high temperature reflow welding and aging process of finished product assembly.

5. Verification scheme

In the development and production of LED products, It is important to objectively evaluate the corrosion resistance of its Ag reflector.

At present, IEC does not have a uniform standard for corrosion test of LED. There are two main methods in the industry, single-level S corrosion test and H_2S gas corrosion test. The H_2S gas corrosion test refers to IEC 60068-2-43 Environmental testing-Part 2-43： Tests-Test Kd: Hydrogen sulphide test for contacts and connections.

(1) Elemental sulfur corrosion test

Test equipment/equipment: oven, beaker, sample placement rack (matched with beaker, placed in the cup), elemental sulfur, light color electricity comprehensive analysis system.

Test conditions: ambient humidity, ambient temperature, duration (there is no uniform standard for this parameter, generally set by the manufacturer or negotiated with the customer).

Test process:

—— check the samples, number and take photos before the test;

—— photoelectric performance test before the test, recording sample information;

—— take sulfur powder and evenly cover the bottom of the beaker;

—— put the tested sample on the sample shelf;

—— placing the sample holder in the beaker;

—— seal the beaker mouth with tin foil;

—— put the beaker in the oven;

—— turn on the oven and set the temperature/time;

—— at the end of baking, take out the beaker;

—— open the beaker mouth tin foil and cool it in a well-ventilated room temperature;

—— taking photos of samples after the test;

—— photoelectric performance test after the test, recording sample information;

—— preparation of test records and test reports.

Judgment criteria:

Compare the appearance, luminous flux, color temperature and color coordinate of the tested samples before and after the test to determine whether the corrosion resistance of the product is qualified.

Different manufacturer, different product model, its judge standard is not identical, do not have unified standard at present.

Notes:

a. Make sure the equipment used is sulfur free and clean before and after the test;

b. Avoid contamination of the test samples against the integrating sphere;

c. Try to reduce the test time of sample integrating sphere after vulcanization, otherwise residual S element will promote further reaction with Ag under the action of spontaneous heat, resulting in inaccurate test results.

Summary: due to simple operation, high efficiency and low cost, the elemental sulfur corrosion test is capable of being completed in general enterprise laboratories and is widely used in large-scale development.

Its disadvantage is that the volatilization and uniformity of sulfur in the test are not easy to control, which leads to a large test error.

(2) H_2S test

Test equipment/equipment: mixed gas test chamber.

Test conditions: ambient humidity, ambient temperature, H_2S concentration and time (there is no unified standard for this parameter, which is generally set by the manufacturer or negotiated with the customer).

Test process:

—— check the samples, number and take photos before the test;

—— photoelectric performance test before the test, recording sample information;

—— the test process shall refer to the national standard IEC 60068-2-43;

—— taking photos of samples after the test;

—— photoelectric performance test after the test, recording sample information;

—— preparation of test records and test reports.

Judgment criteria:

Compare the appearance, luminous flux, color temperature and color coordinate of the tested samples before and after the test to determine whether the corrosion resistance of the product is qualified.

Different manufacturer, different product model, its judge standard is not identical, do not have unified standard at present.

Notes:

a. Test process to ensure that the speed and concentration of H_2S passing through the sample is uniform and controllable;

978-1-7281-5757-3/19 $31.00 © 2019 IEEE

b. Taking measures to ensure test safety.

H_2S is a highly toxic, inflammable and dangerous chemical, which can form explosive mixture when mixed with air and cause combustion and explosion when exposed to open fire and high heat energy.

Conclusion: H_2S test is a mature and standardized accelerated corrosion test method, which was originally used to evaluate the influence of silver and silver alloy discoloration on contact points and connectors, and to predict their corrosion resistance in the industrial atmosphere.

For LED, the test process can refer to the original standard, and the test conditions and judgment can be made according to the actual situation.

(3) Other types of tests

The sealing test for cavity products can assist in verifying the corrosion resistance of the product. The higher the seal rating, the lower the risk of ingress of corrosive gases and the easy protection of the silver reflective layer.

In addition, there is a test method for mixed gas of H_2S and NO_2. Compared with the H_2S gas test alone, the advantage of the mixed gas test is that it does not require an external environment to provide additional O_2 participation, and the reaction rate is high and the corrosion effect is obvious. However, the operation difficulty is higher than that of a single H_2S gas test, so this method is rarely applied in the actual LED industry.

6. Failure analysis scheme

The commonly used LED Ag reflector failure analysis method is to use X-ray energy spectrum analysis (EDS) surface coating structure and composition related parameters, and analyze the residual solid residue composition to reveal the material conditions and failure process.

The principle is that when high-energy electrons enter the sample, they are inelastic scattered by the sample atoms, and their energy is transferred to the atoms so that the electrons in an inner shell are ionized and separated from the atoms, and there is a vacancy in the inner shell, and the atoms are in an unstable high-energy excited state. Within 10 to 12 seconds of excitation, the atom returns to its lowest energy ground state. In this process, a series of outer electrons transition to the inner shell vacancy, releasing excess energy, producing characteristic x-rays and Auger electrons. X-ray radiation is a stream of quantum or photon particles with energy.

The characteristic X-ray energy has a functional relationship with the sample atomic number Z. As early as 1913, Moseley derived the following formula:

$$E = A(Z - C)^2$$

Where, A and C are constants related to the X-ray spectral line system. This relationship, also known as Moseley's law, indicates that if the energy of a characteristic X-ray is detected, the corresponding element must be found. This is the theoretical basis of using characteristic X-ray to analyze the element composition of materials [31].

Analysis steps of abnormal corrosion discoloration of LED light reflection device:

First, collect data. Including time of failure, location, type, quantity, proportion, appearance, etc.

Second, take samples.

Take several failed samples and normal samples respectively.

Third, DE-CAP.

You can choose whether to do this step or not according to the actual situation.

Fourthly, locate the failure point and make physical and chemical analysis.

Finally, the comprehensive analysis, the conclusion.

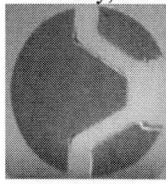 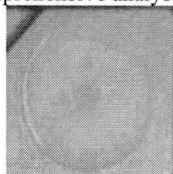

Fig.6.1 （a1）   Fig.6.1 （a2）

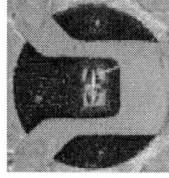 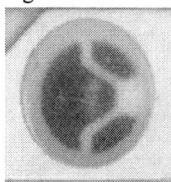

Fig.6.1 （b1）   Fig.6.1 （b2）

a. Control group: a1 normal silver-plated reflective film, a2 normal LED

b. Abnormal group: b1 failure reflective film, b2 failure LED

Fig.6.1 Comparison before and after curing failure of LED reflector

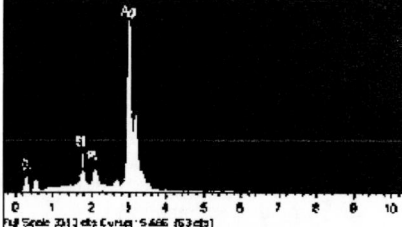

Fig.6.2 （a） Control group

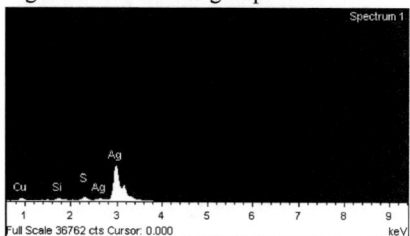

Fig.6.2 （b） Abnormal group

Fig.6.2 EDS element analysis

SEM electron microscope scanning and EDS ray energy spectrum analysis tests were carried out on the fault residues, and the components and phase structures in different color areas were analyzed, so as to determine the main components and structures inside.

By SEM and EDS positioning analysis, the analysis results of the black part in FIG. 5-b1 and the control group before black part in FIG.5-a1 are shown in FIG.6. The main components of the black part in B1 are Ag, Cu and S, and a small amount of Cd and Si. The main components of A1 are Ag and a small amount of S element. The main difference between B1 and control group A1 was 32.42% Cu and 2.13% S.

978-1-7281-5757-3/19 $31.00 © 2019 IEEE   241

Table6.1 EDS element analysis table

Element	abnormal group（b）		control group（a）	
	Weight%	Atomic%	Weight%	Atomic%
O K	0.00	0.00	-	-
Si K	0.73	2.17	0.34	1.28
S K	2.13	5.52	-	-
Cu L	32.42	42.46	-	-
Ag L	62.02	47.85	96.38	95.59
Cd L	2.70	2.00	3.29	3.13
Totals	100.00	-	100.00	-

Combined with the data of the control group and analyzed, two types of suspicious elements were found in the failure area of the abnormal group, namely, 2.13% element S and 32.42% element Cu. All others are normal. It is known that the failure source is the ag-sensitive medium S element, and element Cu is the bottom layer Cu coating exposed after Ag reflection film is damaged by S.

Table6.2 S element path trace analysis table

	Analysis point	S element content %
Group a	Between the reflective film and the colloid	N.D
	Colloid lateral	N.D
	The surface of the pin	N.D
Group b	Between the reflective film and the colloid	2.13
	Colloid lateral	6.22
	The surface of the pin	11.42

Further, the content of S in the internal and external structures of group a&b was compared and analyzed. No S element was detected in group a, while S element was detected in group b, and the content trend was "between reflective film and colloid < colloid outside < pin surface".

Therefore, it can be known that the corroded S element comes from the outside of the LED device and invades the inside of the device through the encapsulated colloid, causing corrosion failure of the Ag reflection device.

7. Relevant technical standards

There is currently no IEC standard for LED corrosion testing. Product development and production processes mainly refer to similar standards, such as: IEC 60068-2-43, IEC 60068-2-60. The company sets the test parameters according to the actual situation.

The international standard for LED corrosion test（IEC 60747-5-13 Semiconductor devices - Part 5-13: Optoelectronic devices – Hydrogen sulphide corrosion test for LED packages）is in the process of being developed. The project is under the responsibility of the IEC TC47/SC47E/WG9 working group and has entered the CD stage.

The details are as follows:

(1) Relevant technical standards

Table7.1 Relevant technical standards

No.	Item	International standard	National Standard of China	Scope	Evaluation purpose
I	Standard Number	IEC 60068-2-43：2003	GB/T 2423.20-2014	H2S test for contacts and connections	Evaluation purpose, Electrical connection performance
	Standard Name	Environmental testing-Part 2-43：Tests-Test Kd: Hydrogen sulphide test for contacts and connections	Environmental testing-Part 2: Tests-Test Kd: Hydrogen sulphide test for contacts and connections		
II	Standard Number	IEC 60068-2-60：1995; IEC 60068-2-60：2015（Effective）	GB/T 2423.51-2012(Effective）; Standard is being developing (Projectnumber: 20171714-T-469)	Flowing mixed gas corrosion test for contacts and connections	Evaluation purpose, Electrical connection performance
	Standard Name	Environmental testing-Part 2-60：Tests-Test Ke: Flowing mixed gas corrosion test	Environmental testing-Part 2: Tests-Test Ke: Flowing mixed gas corrosion test		
III	Standard Number	IEC 60747-5-13（Unpublished）	NULL	H2S & NO2 test for LED	Optical performance
	Standard Name	Semiconductor devices - Part 5-13: Optoelectronic devices - Hydrogen sulphide corrosion test for LED packages	NULL		

(2) Progress of new standard

IEC 60747-5-13 Semiconductor devices - Part 5-13: Optoelectronic devices - Hydrogen sulphide corrosion test for LED packages.

Table7.2 IEC 60747-5-13 standard progress

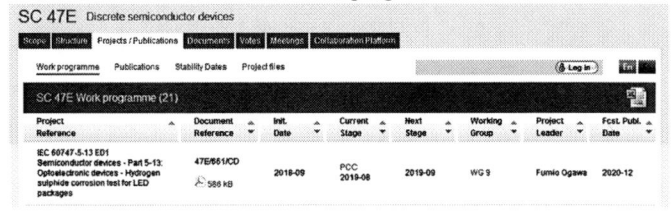

——https://www.iec.ch

Table7.3 TC47/SC47E/WG9 group

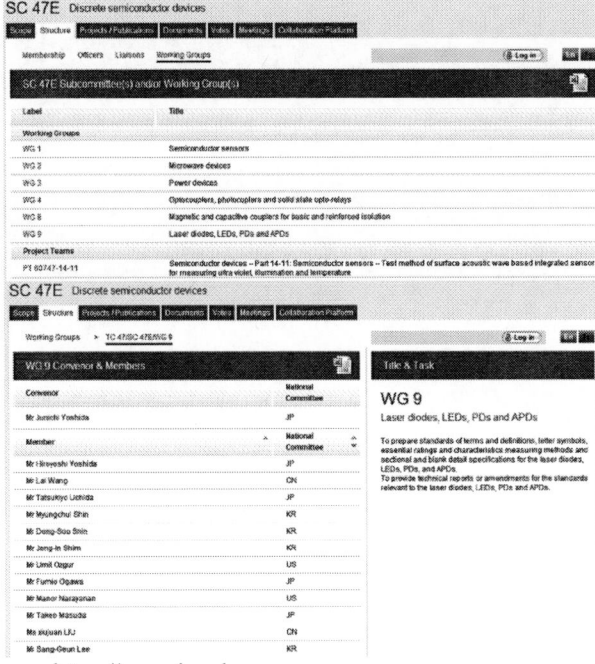

——https://www.iec.ch

Conclusions

This paper based on the composition principle, function and characteristics of LED reflector, studies the cause of corrosion failure of LED reflector, and analyzes the reaction process between Ag reflection film and different corrosion media. The source of corrosion risk in the whole life cycle of LED is analyzed. The prediction and evaluation scheme of anti - vulcanization ability of LED is introduced. And EDS material characterization method for analyzing the cause of corrosion failure. Finally, the status quo of relevant technical standards is introduced.

References

1. Wang huanliang，Yu bin，Qiu hongping，LED Package Light Degradation and Failure Test and Cause Analysis，CHINA LIGHT & LIGHTING，No.9（Sep.2018），pp.14-17.

2. Sun mengzhi, Wang tongtong, Wang Yanchao, *et al*, R esearch development of high reflecting coating for large-diameter mirror, Chinese Optics, Vol.9, No.2（Apr.2016）, pp.203-212.

3. Feng mingzhu, Wang Qun, Ji Xuemei, *et al*, The Study on Preparation of Silver Coated Reflective Film, Materials • engineering, Vol.19, No.11（2018）, pp.46-48.

4. Yang changjiang, Tarnishing Mechanism and Anti-tarnishing Techniques of Gold and Silver Coins, Dalian University of Technology, 2008, pp.1-161.

5. Lee dongliang. Properties and applications of AgAuPt.Peking：Higher education press, 1998.

6. Volpe L, Peterson P J.The atmospheric sulfidation of silver in a tubular corrosion. Corrosion Science, Vol. 29, No. 10 (1989), pp. 1179-1196.

7. Zhu hongfan, Zhou hao, Cai lankun. The causes of discoloration of silverware and the selection of anti-discoloration and corrosion inhibitor. Sciences of Conservation and Archaeology, Vol. 13, No. 1 (2001), pp. 15-21.

8. Reagor B T, Sinclair J D. Tarnishing of silver by sulfur vapor: film characteristics and humidity effects. Journal of the Electrochemical Society, Vol. 128, No. 3 (1981), pp. 701-705.

9. Allpress J G, Sanders J V. Influence of surface structure on a tarnishing reaction. Philosophical Magazine, Vol. 10, No.6 (1964), pp. 827-834.

10. Kim H.Corrosion process of silver in environments containing 0.1 ppm H2S and 1.2 ppm NO2.Materials and Corrosion, Vol. 54, No. 4 (2003), pp. 243-251.

11. Fang jingli, Cai zi. Discoloration and protection of silver coating. ScienceinChinaSeriesB:Chemistry, Vol. 18, No. 5 (1988), pp. 473-451.

12. Sinclair J D.Tarnishing of silver by organic sulfur vapors:Rates and film characteristics.Journal of the Electrochemical Society, Vol. 129, No. 1 (1982), pp. 33-39.

13. Pope D, Gibbens H R, Moss R L.The tarnishing of Ag naturally-occurring H2S and SO2 levels.Corrosion Science, Vol.8, No. 12 (1968), pp. 883-887.

14. Rice D W, Peterson P, Rigby E B, *et al*.Atmospheric corrosion of copper and silver. Journal of the Electrochemical Society, Vol. 128, No. 2 (1981), pp. 275-284.

15. Liu Ronghui, He Huaqiang, Huang Xiaowei, *et al*, New Progress in Study and Application on Phosphors for White LED, Semiconductor Technology, Vol.37, No.3（2012）, pp.221-227.

16. Lee Guanghuan, Preparation and Properties of the Phosphors for LED, Dalian University of Technology, 2012, pp.1-151.

17. Liu Zhanqing, Prevention of vulcanization in the production and use of LED products, No.11（2012）, pp. 61+58.

18. Fu haiyong, Effect of finished LED luminaire material on LED light decay, Vol.35, No.3（2017）, pp.9-12.

19. Randin J P, Ramoni P, Renaud J P.Tarnishing of AuAgCu alloys-Effect of the composition, Materials and Corrosion, Vol.43, No.3(1992), pp.115-123.

20. Mao xiuying, Tian kaishang. Corrosion discoloration mechanism and anti - discoloration process of silver plating [J]. Electroplating & Pollution Control | Electroplat Poll Contrl, Vol.15, No.1（1995）, pp.8-11.

21. Xiao zhenggui, Lin yichao, Leng guicheng, A new way to prevent silver handicraft from changing color, Materials Protection, Vol.18, No.1（1985）, pp.20-24.

22. Randin J P, Chromated layer as an anti-tarnish protection of AuAgCu alloys, Materials and Corrosion, Vol.43, No.4（2004）, pp.172-176.

23. Singh I, Sabita M P, Altekar V A, Silver tarnishing and its prevention.Anti-corrosion Methods and Materials, Vol.30, No.7（1983）, pp.4-8.

24. Dubkov V M, V. N. Rozhdestvenski. Protecting reflecting layers of copper and silver against tarnishing.Journal of Optical Technology. Vol.50, No.1（1983）, pp.35-36.

25. Lee aijie, Xu chun. Study on new inhibitors of silver discoloration. Chemical Engineer, Vol. 89, No. 2 (2002), pp. 54-55.

26. Vlasta Brusic, Marie Angelopoulos.Teresita Graham.Use of polyaniline and its derivatives in corrosion protection of copper and silver, Journal of the Electrochemical Society, Vol. 144, No. 2(1997), pp. 436-442.

27. Gi Xue, Jian Dong, Peiyi Wu.Surface-enhanced Raman scattering study of polymer on metals(Ⅲ):Chemisorbed polybenzimidazole and its corrosion-inhibiting properties at high temperature.Journal of Polymer Science Part B:Polymer Physics, Vol. 30, No. 10(1992), pp. 1097-1102.

28. Xiao xin, Long youqian, Zou xiaoyan. Comprehensive control of discoloration of silver plating. Materials Protection | Mater Protec, Vol. 31, No. 2(1998), pp.27-29.

29. Lin yichao. Discoloration and protection of silver handicraft. Materials Protection | Mater Protec, Vol. 2631, No. 3(1993), pp.28-31.

30. Shen zheng, Xia libin, Zhang li, *et al*, Research Progress of Organosilicon Polymer Modified By Epoxy, Coatings Technology & Abstracts, Vol.37, No.5（2016）, pp.51-54.

31. Ren zhigang, Wang yaqun, Xu xingquan, *et al*, Fault analysis of GIS equipment based on solid residue X-ray energy spectrum, Electrotechnical Application, Vol.37, No.5（2015）, pp.639-641.

Failure Analysis of Glass Transition Temperature of LED Insulation Layer

Yibin Wang[1,2], Fang Fang[3], Jing Wu[3], Kaixuan Lin[3], Tingting Xu[3], Weiqing Liang[3], Luqiao Yin[2,4,*]

[1] School of Materials Science and Engineering, Shanghai University, Shanghai 200072, China
[2] Key Laboratory of Advanced Display and System Applications, Shanghai University, Ministry of Education, Shanghai, 200072, China
[3] Gold Medal Analytical & Testing Group
[4] School of Mechatronics and Automation, Shanghai University, Shanghai, 200072, China

Abstract

With the development of light-emitting diodes (LEDs) technology, the number of LEDs chips in high-power LEDs devices is increasing, and the use of LEDs continues to expand, the reliability has become the key to its widespread use. When the LEDs device fails, the failure analysis of the device can effectively derive the mechanism and condition of the device failure, which has a significant effect on enhancing the reliability of the device. In this paper, by analyzing the instantaneous voltage curve and thermal of the failed device, it can be concluded that when the operating current of the LEDs device is large, due to the poor thermal conductivity of the insulating layer, the insulating layer of the device is in the vitrification transition temperature for a long time. The shape of the insulating layer is easily changed or even burned to cause delamination, resulting in LEDs failure.

1.Introduction

As a new type of lighting source, LEDs is used in various illumination fields, such as LCD display back lighting, automotive lighting and so on.[1] Compared with other traditional light sources, LEDs has many advantages such as high light efficiency, low power consumption, durability, reliability and environmental friendliness.[2, 3] However, LED will fail due to its own defects, electrical, thermal, chemical and mechanical stress during actual use.[4]

The failure of LEDs are complicated into three failure site: the chip, interconnected material layer, and the packages.[5] In the chips ,the failure mechanisms can be divided into defect and dislocation generation and movement, die cracking, dopant diffusion and electromigration, [5-9] this part failure is due to the unfavorable factors in the production process of LED chips.In the interconnected material layer, the failure mechanisms can be divided into electrical overstress–induced bond wire fracture and wire ball bond fatigue, electrical contact metallurgical interdiffusion, and electrostatic discharge.[5, 10-12] In the packages level, the failure mechanisms can be divided into carbonization of the encapsulant, delamination, encapsulant yellowing, lens cracking, phosphor thermal quenching, and solder joint fatigue,[5, 13-16] this part failure is due to the aging of the packaging material and the interface stress.However, LEDs burned due to the external circuit factors, chip defects and packaging problem will happen in all three types of the LEDs failure mechanisms. In order to ensure the reliability of the LEDs, it is important to analyze the specific causes of the burned phenomenon and solve the problem in a targeted manner.

In this paper, the LEDs module were inspected in the failure analysis test by the systematic methods, SEM, power test, temperature test, FT-IR, TMA, DSC

2.Fault analysis and Experiments

2.1 samples describe

The LEDs module used in the failure analysis was packaged with flip chip. Circuit structure was two group of chips in parallel, each group of chips consists of 35 chips in series. There are 5 samples in this failure analysis test, including three good light sources and two failed light sources. Fig.1a show the good light source, Fig.1b show the failed light source, the copper foil in the negative electrode of LEDs tilted after LEDs burned.

2.2 power test

In order to test the power output voltage and current stability under AC 220 V, randomly select a burned light source have a repair experiments .Fig.2 show the stability test results of power output voltage and current. The power supply output voltage is 104 V (Fig.2a) and the output current is 203 mA (Fig.2b). Fig.2c show the current ripple waveform is a triangular wave. After filtering out the high frequency components, the output voltage amplitude is 5.8% (Fig.2b) and the output current amplitude is 9.9% (Fig.2d). The power output has small fluctuation without spike voltage and spike current. Circuit structure was two group of chips in parallel and each group of chips consists of 35 chips in series. The voltage of a single chip is about 2.97 V, the current of a single chip is about 100 mA, but the maximum drive current of this chip is 150 mA. The current of a single chip indicating no over-use of the light source chip case. This means that the LED module's circuit structure is not the cause of the LED burnout

2.3 temperature test

The light source direct contact with the reflective cup

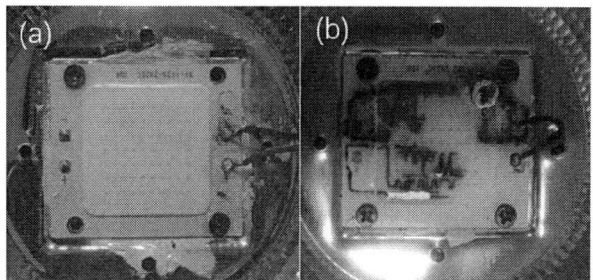

Fig.1 The LEDs light sources

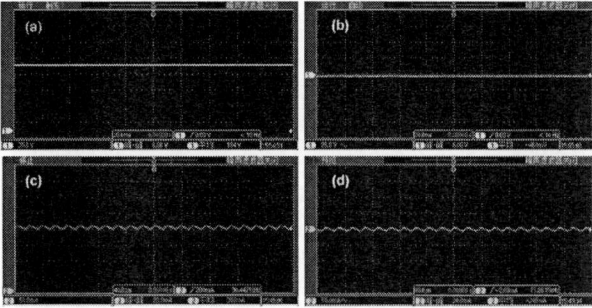

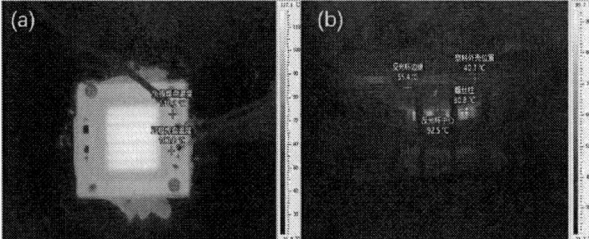

Fig.2 stability test results of power output voltage and current

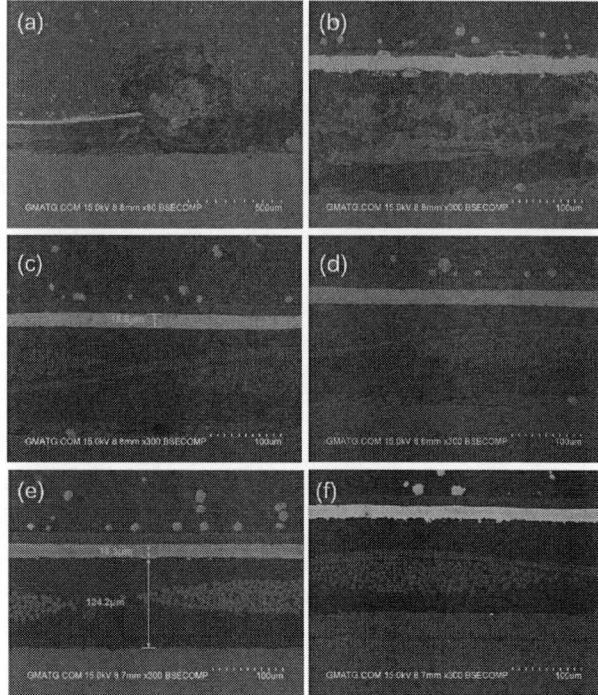

Fig.3 The temperature of the LEDs

through the silicone grease, and use a plastic case as external support. The heat generated by the light source conduct to the reflective cup by the way of heat conduction. Then the heat transfer to the surrounding by the way of heat radiation.

AC 220 V was applied for 5 hours to reach a steady state. Use an infrared thermal camera to test the LEDs. The temperature of the LEDs is show in Fig.3, the temperature of LED negative electrode was 110.2 ℃, but the temperature of the LED positive electrode was 100.8 ℃.The temperature difference between the positive and negative electrode pads was about 9.4 ℃.This phenomenon is due to the current density in negative electrode is bigger than positive electrode. The center temperature of the reflector cup was 92.5 ℃, the edge temperature of the reflector cup was 55.4 ℃, The temperature of the plastic case temperature is 40.1℃. The temperature difference between the center temperature of the heat source and the edge temperature is more than 10 ℃. The plastic housing is not a good lamp heat conductor and this will cause a lot of heat accumulation in the light source part.

3.Results and discussion

The SEM images of the light sources are shown in Fig.4. Fig.4a show that the copper foil and the insulation layer were burned and the encapsulating glue are carbonized at the serious burned-out position. Fig.4b show that insulating layer is melted and deformed due to the high temperature at the slight burned-out position.

The SEM images of the didn't burn location is shown in Fig.4c,Fig.4d. Fig.4c shows that the insulating layer and the copper foil did not occur any significant burnout, but at the interface between the insulating layer and the aluminum substrate, there was an abnormal phenomenon of delamination, which delaminated the heat transfer structure of the aluminum substrate and resulted in heat builds up in the insulation and creates a high temperature at the insulation.

The SEM images of a good product light source is shown in Fig.4.The SEM images shows that the copper foil with the

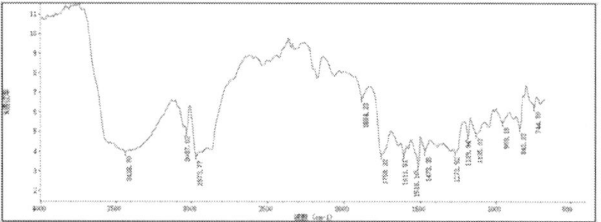

Fig.4 The SEM images of the lighting sources (a), (b): copper foil and the insulation layer were burned and the encapsulating glue are carbonized at the serious burned-out position;(c),(d): insulating layer and the copper foil did not occur any significant burnout;(e),(f): copper foil with the insulation layer and the insulation layer and the aluminum substrate have a complete structural connection;

insulation layer and the insulation layer and the aluminum substrate have a complete structural connection. The thickness of the copper foil is 18.3 μm and the thickness of the insulation layer is 124.2 μm.

The properties of the colloid of insulating layer were characterized by FT-IR, the test results are shown in the Fig.5. It can be seen from the peak of the image at 3438.70 that the colloid composition contains hydroxyl groups. The elemental test result shown that the composition of the filler is oxygen (O), aluminum (Al) and silicon (Si) calcium (Ca). Therefore, it can be judged that the insulating layer colloid is epoxy resin.

Remove the aluminum substrate copper foil by mechanical methods. The properties of insulation material were characterized by thermally mechanically analyze (TMA test). The insulation material obtained by the physical method, use 400 sandpaper polish the upper and lower surface to be

Fig.5 The FT-IR of insulating layer

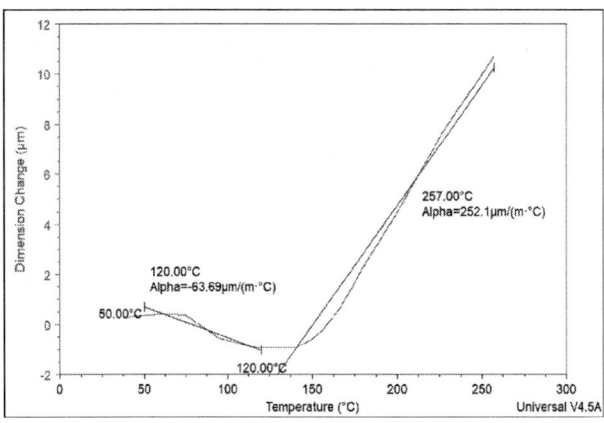

Fig.6 The TMA of insulating layer

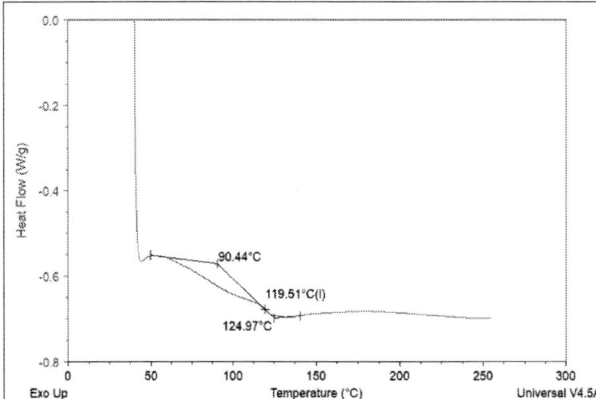

Fig.7 The DSC of insulating layer

smooth and then placed in a nitrogen atmosphere. The nitrogen gas flow rate was 50 ml/min. After eliminating the thermal history and stabilizing the temperature, the insulating layer material is subjected to a temperature raising process from 40 ℃ to 260 ℃, the rate was 10 ℃/ min.

The results show that the material of insulating layer expands with increasing temperature, and the relative thermal expansion coefficient of insulating material is -63.69μm / (m · ℃) at 50-120 ℃ . The relative thermal expansion coefficient of the encapsulant at 120-257 ℃ is 252.1 μm/ (m · ℃), and the material is in an expanded state. And copper foil and aluminum substrate thermal expansion coefficient does not match (complement the thermal expansion coefficient of the two materials), resulting stratified between the layers in high temperature, and LEDSS Thermal conductive structure damage. A large amount of heat accumulated, resulting in the burning of LEDs.

Remove the copper foil from the aluminum substrate by Mechanical methods and take DSC test of the insulation material. The result is shown in the Fig.7.The lamp bead is physically removed and shredded with a scalpel to the bottom of the AL crucible. The test sample mass is 5.7 mg. The crucible was placed in a nitrogen atmosphere with a nitrogen gas flow rate of 50 ml/min. After eliminating the thermal history of the sample and stabilizing the temperature, the temperature was raised from 40 ℃ to 260 ℃ at 20 ℃/min. As

shown in the Fig.7, the test results show that the insulation material glass transition temperature of 119.51 ℃.

Place the sample lamp in a plastic case at 30 ℃ and apply the same power(AC 220V) for 5 hours. Use an infrared thermal camera to test the lamp for temperature, the temperature of negative electrode is 120.8 ℃, greater than the Tg temperature of the insulating layer, thereby affecting the performance of the insulating layer material properties and process performance, resulting in insulation colloid Burn carbonization and LEDs light source failure.

4.Conclusions

The method described in this paper can effectively detect the causes of LEDs light source burnout. Through the power test, SEM, temperature test, the analysis shows that the cooling design of the lamp is poor. When the temperature of the aluminum substrate is close to the Tg temperature of the insulation layer, the interface between the insulation layer and aluminum substrate will delamination. This delamination disrupts the heat transfer structure of the aluminum substrate, causing heat to accumulate at the insulating layer, causing high temperatures at the insulating layer, thus burning the light source.

Acknowledgments

This work was supported by the National Nature Science Foundation of China (NSFC) under the Grant Number (51605272) and Science and Technology Commission of Shanghai Municipality Program (19DZ2281000; 17DZ2281700).

References

[1]   M. R. Krames *et al.*, "Status and Future of High-Power Light-Emitting Diodes for Solid-State Lighting," *Journal of Display Technology,* vol. 3, no. 2, pp. 160-175, 2007.

[2]   C. C. Lin and R. S. Liu, "Advances in Phosphors for Light-emitting Diodes," *J Phys Chem Lett,* vol. 2, no. 11, pp. 1268-77, Jun 2 2011.

[3]   I. T. Ferguson *et al.*, "Progress of solid state lighting technology in Taiwan," presented at the Third International Conference on Solid State Lighting, 2004.

[4]   Z. Chen, Q. Zhang, K. Wang, X. Luo, and S. Liu, "Reliability test and failure analysis of high power LED packages," *Journal of Semiconductors,* vol. 32, no. 1, 2011.

[5]   M.-H. Chang, D. Das, P. V. Varde, and M. Pecht, "Light emitting diodes reliability review," *Microelectronics Reliability,* vol. 52, no. 5, pp. 762-782, 2012.

[6]   D. L. Barton and M. Osinski, "Life tests and failure mechanisms of GaN-AlGaN-InGaN light emitting diodes," in *IEEE International Reliability Physics Symposium*, 1998.

[7]   D. L. Barton *et al.*, "Degradation of blue AlGaN/InGaN/GaN LEDs subjected to high current pulses," in *IEEE International Reliability Physics Symposium*, 2002.

[8] G. Meneghesso *et al.*, "Failure Modes and Mechanisms of DC- Aged GaN LEDs," vol. 194, no. 2, pp. 389-392, 2015.

[9] A. Uddin, A. C. Wei, and T. G. J. T. S. F. Andersson, "Study of degradation mechanism of blue light emitting diodes," vol. 483, no. 1, pp. 378-381, 2005.

[10] M. Dammann, A. Leuther, F. Benkhelifa, T. Feltgen, and W. J. P. S. S. A. Jantz, "Reliability and degradation mechanism of AlGaAs/InGaAs and InAlAs/InGaAs HEMTs," vol. 195, no. 1, pp. 81-86, 2003.

[11] G. Lu, S. Yang, and H. Yun, "Analysis on Failure Modes and Mechanisms of LED," in *International Conference on Reliability*, 2009.

[12] F. Wu, Z. Wei, S. Yang, and C. Zhang, "Failure modes and failure analysis of white LEDs," in *International Conference on Electronic Measurement & Instruments*, 2009.

[13] H. T. Li, C. W. Hsu, and K. C. Chen, "The study of thermal properties and thermal resistant behaviors of siloxane-modified LED transparent encapsulant," in *Microsystems, Packaging, Assembly & Circuits Technology, Impact International*, 2007.

[14] X. Luo, B. Wu, L. J. I. T. o. D. Sheng, and M. Reliability, "Effects of Moist Environments on LED Module Reliability," vol. 10, no. 2, pp. 182-186, 2010.

[15] P. Mccluskey, K. Mensah, C. O'Connor, and A. J. M. R. Gallo, "Reliable use of commercial technology in high temperature environments," vol. 40, no. 8-10, pp. 1671-1678, 2000.

[16] E. F. Schubert, "Light-Emitting Diodes - 2nd Edition," 2006.

Intelligent Control Semiconductor Laser Reliability Test System

Xiaoling HU[1], Wensha LAN[2]

1. Faculty of Information Technology, College of Microelectronics, Beijing University of Technology
2. Tianjin Zhengfang Science and Technology Development Co., Ltd
Pingleyuan 100[#] Chaoyang Distric Beijing, 100124, People's Republic of China
huxiaoling@bjut.edu.cn

Abstract

With the development of electronic technology, semiconductor lasers have been widely used in the field of optoelectronics due to their own distinctive features. In order to improve the performance of semiconductor lasers, reliability testing research has become a hot issue at present. In this paper, the intelligent control semiconductor laser reliability test system is described. Its aging and screening process must be controlled by a highly automated, intelligent control system. There are five parts that make up this system: the master control system, the slave control system, the RS-485 serial data interface standard used to communicate between the master unit and the slave unit, the hot stage used to control the test temperature and the power supply includes +5V, +12V, ±15V, +24V. The test system is mainly controlled by AT89C51 single-chip microcomputer, which can not only realize the continuous adjustment of the drive current (0~3A), but also realize the continuous adjustment of the aging temperature (room temperature~150°C). Moreover, the system's current control accuracy is up to milliamp, and temperature control accuracy is ±0.5°C. The system can also preset the aging current value and the aging temperature value, and display the current working current and working temperature in real time.

1 Introduction

Comparing with the ordinary light，semiconductor lasers has the advantages of high electro-optical conversion efficiency, good monochromaticity, directivity, small size, long life, maintenance-free and pollution-free, etc. So it is widely used in many fields such as communication, information recording, printing and display, material processing, medical, and pumped solid-state lasers.[1-3] In recent years, the launch of semiconductor lighting projects has attracted widespread attention from all over the world. However, for the application of semiconductor lasers in any field, the research of semiconductor laser has been concentrated on how to make the lifetime longer, how to improve reliability and how to increase output light power. Therefore, research on reliability and lifetime testing of semiconductor lasers has become a hot topic. Moreover, the industrialization of semiconductor lasers requires that their

The intelligent test system is mainly controlled by AT89C51 single-chip microcomputer, which can not only realize the continuous adjustment of the drive current (0~3A), but also realize the continuous adjustment of the aging temperature (room temperature~150°C). Taking into account the stability of the system and the design cost, the temperature controller adopts the existing commercial function module. This article mainly describes the master-slave control system.

aging and screening process be controlled by highly automated, intelligent control systems.

2 Intelligent control semiconductor laser reliability test system

According to the burn-in testing theory of semiconductor laser, the intelligent control semiconductor laser reliability test system is designed.[4] And it is highly automated and easy to use on a large scale. There are five parts that make up this system: the master control system, the slave control system, the RS-485 serial data interface standard used to communicate between the master unit and the slave unit, the hot stage used to control the test temperature and the power supply includes +5V, +12V, ±15V, +24V. The master control part is mainly used to set the working current, display the working current, the working condition of each lower unit and the working condition of the power supply. The five-channel slave machine has the same function, which provides reliable and stable current for the normal operation of the laser. MAXIM's MAX485 is used to communicate between the master unit and the slave unit. The power supply part mainly supplies the required power for the master unit and the slave unit. The system can perform aging test on 50 lasers at the same time. And the maximum aging current required by the system is up to 3A. If power is supplied to each laser separately, the system needs to provide a large current of $50 \times 3 = 150A$, which is bound to require a large power supply and an increase in design cost. Based on the above considerations, we use a five-channel parallel design method, in which 10 lasers in each way are connected in series. The intelligent control semiconductor laser reliability test system block diagram is shown in Fig. 2.1.

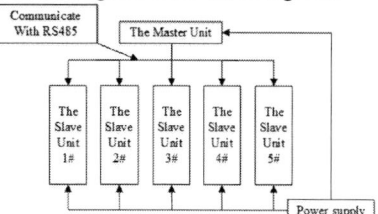

Fig. 2.1 The intelligent control test
system block diagram

3 Master-Slave Control System

3.1 The Master Control System

The master unit is mainly composed of a single chip microcomputer, a digital tube display, a keyboard control circuit, and a communication circuit is shown in Fig. 3.1. It mainly controls the work of the slave unit through the communication circuit, and displays the drive current per channel of the lower unit in real time, and the working of each power supply. In addition, the function keys in the master unit is mainly used to set the operating current of the slave unit per channel, and realize the control of the lower unit through the communication circuit. There are two main types of 7-Seg LED Nixietube. One of the digital tubes is mainly used to display the number of channels which the slave unit currently setting or working. The other is a common cathode four-digit digital tube, one of which is used to display the set current value, and the other is used to display the testing current value from the slave unit.

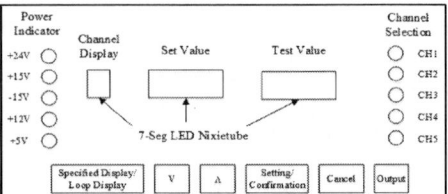

Fig. 3.1 The panel of the master unit

3.2 The Slaver Control system

Since the five-channel slave control unit is exactly the same, we only introduce the specific circuit composition by one of them. In order to effectively predict the lifetime of a semiconductor laser, it is first necessary to provide a reliable and stable drive current for the semiconductor laser. We know that the illuminating power of a semiconductor laser is almost proportional to the driving current within a certain range. Therefore, to stabilize the optical power of the semiconductor laser, a constant current source circuit should be used.

The design scheme refers to the basic structure of the integrated operational amplifier circuit with negative feedback to form the constant current source.[5,6] The constant current driving circuit in the system is realized by a software and hardware closed-loop system composed of single chip microcomputer, D/A converter, adjustment circuit, power transistor drive circuit, sampling circuit, and A/D converter. The constant current source circuit which is the main component of the slave control system is shown in Fig. 3.2.

In order to meet the test requirements of high-power semiconductor lasers, the design specification requires the system to output a maximum current of 3A. So the constant current source uses a high-power Darlington transistor 2SD1559 as the adjustment circuit, its maximum collector current up to 20A, collector dissipation power 100W. The laser diode acts as a load in series with the collector of the Darlington transistor and controls the laser current by controlling the base of 2SD1559. The sampling resistor (R29) is connected to the emitter of 2SD1559. The sampling circuit sends the collected signal to the single-chip microcomputer and compares it with the set current, calculates the offset by the PID algorithm and sends the

offset value to the adjustment circuit to realize the constant current output. In addition, the current output from the single chip microcomputer is digital signal that needs to be converted by the D/A converter before inputting the constant current source circuit. Therefore, the accuracy of this digitally controlled constant current source ultimately depends on the conversion accuracy of the A/D and D/A converters in the circuit.

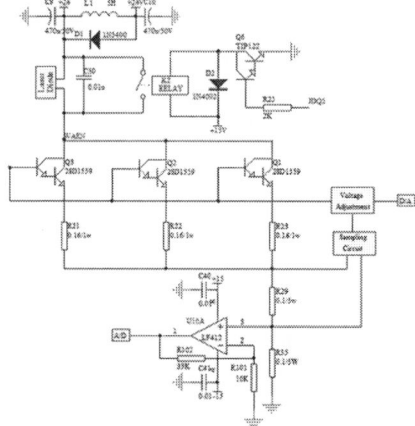

Fig. 3.2 Constant current source circuit

Since the semiconductor laser is easily damaged by surge voltage, current or static charge during operation, the slave control system set the power supply filter current circuit, the delay and soft start circuit to eliminate the impact of the power supply interference and the surge current caused by the power supply on and off. [7,8] The π-type low-pass filter network is added between the DC power supply and the laser diode to further filter the Surge voltage. As shown in Fig. 3.2, the π-type filtering network consisting of L1, D1, C9 and C10 can effectively overcome the transient peak surge voltage.

3 Experimental result

Under the room temperature, a high-power 980 nm semiconductor laser was selected, and the optical power at a driving current of 500 mA was measured by a semiconductor laser integrated tester to be 392 mW. Then, the constant current mode test experiment was performed at 60℃ with the designed intelligent control semiconductor laser reliability test system. During the test, the system sets the current value to 500 mA, the laser diode drive current is measured at equal intervals, so we can obtain an average current value.

Current stability is an important parameter of the constant current circuit. Stability can be expressed as the ratio of the standard deviation to the average to determine the probability of the data deviating from the mean: p = (average value – calibration value) / average value.

The experimental results show that the stability of the driving current in the intelligent control semiconductor laser reliability test system is 0.31%, and the driving current has high precision.

Conclusions

The designed semiconductor laser reliability test system based on single chip microcomputer control can carry out reliability experiments on a variety of semiconductor lasers. To improve the accuracy and stability of the system, it adopts closed-loop feedback control of single-chip and output dual-loop, and is equipped with various software and hardware protection and anti-interference measures. In addition, the accuracy of the current in the system is closely related to the bits of A/D and D/A convertor. If a 16-bit A/D, D/A converter is used for the corresponding closed-loop adjustment, the current accuracy will be further improved. The reliability evaluation system designed in this paper has been well applied in the laboratory.

References

1. Wang Lijun., *et al*, "Development of High Power Diode Laser," *Chinese Journal of Luminescence*, Vol. 36, No. 1 (2015), pp. 1-19.

2. Ma Xiaoyu., *et al*, "Present Situation of Investigations and Applications in High Power Semiconductor Lasers," *Infrared and Laser Engineering,* Vol. 37, No. 2 (2008), pp. 189-194.

3. Chen Heming, Zhao Xinyan, Laser Principles and Application, Publishing House of Electronics industry (Beijing, 2009).

4. Han L, Narendran N., "An Accelerated Test Method for Predicting the Useful Life of an LED Driver," *Power Electronics*, Vol. 26, No. 8 (2011), pp. 2249-2257.

5. Zhang Guoxiong, Huang Chunhui, "Continuously Adjustable Semiconductor Laser Diode Driver with High Stability," *Laser & Infrared*, Vol. 41, No. 2 (2011), pp. 160-163.

6. Shi Quanlin, Xin Desheng, *et al*, "The Driver of Continuous Semiconductor Laser," *Journal of Changchun Institute of Optics and Fine Mechanics*, Vol. 24, No. 1 (2001), pp. 12-15.

7. Chen Wei, Cai Yinbo, *et al*, "Design on Protect Circuit in High Power Laser Diode Driver Power Supple," *Optics & optoelectronic Technology*, Vol. 6, No. 6 (2008), pp. 68-74.

8. Yang Chunli, Jia Hongzhi, *et al*, "Design for the Circuit to Eliminate Surge of the Laser Diode Power Supply," *Applied Laser*, Vol. 28, No. 4 (2008), pp. 310-313.

Junction Temperature Prediction of the Multi-LED Module with the Modified Thermal Resistance Matrix

Fanny Zhao[1, *], Brian Shieh[1], Fangyun Zeng[1], Guoming Yang[1], S. W. Ricky Lee[1, 2]

1. HKUST LED-FPD Technology R&D Center at Foshan

7-304, Block A, Hantian Industrial Park, Foshan, Guangdong, China 528200

2. Department of Mechanical & Aerospace Engineering

Hong Kong University of Science & Technology

Clear Water Bay, Kowloon, Hong Kong, China

*zhaohs@fsldctr.org

Abstract

The junction temperature of LEDs is important for its life, reliability and efficacy. The existing transient measurement method of the junction temperature of using transient thermal tester (T3ster) based on the time-resolved measuring with a constant bias current according to the JESD51-14 is able to obtain the LED junction temperature rise based on the known temperature of thermostat module as the heatsink. In practice, without T3ster and the thermostat module, for multi-LED module with multiple heat sources in the luminaries, the thermal resistance matrix needs to be built to realize the thermal transfer between one another such that their junctions temperature raises can be calculated from their thermal power consumption. It is demonstrated that without measurement of the temperature heatsink temperature, the individual junction temperature of LEDs (groups) in multi-LED module with a thermistor (NTC or PTC) on its PCB measuring the local reference temperature can be obtained from the modified N x (N+1) thermal resistance matrix having N LED heat sources and N+1 junction temperature rise. A multi-LED module sample with 3 groups of LEDs has been studied at 5 different ambient temperatures and 2 additional different power combinations using 3 thermal resistance matrices built at 3 different ambient temperatures for comparison. The result showed that the modified method is effective and convenient for measurement of the junction temperature of the multi-LED modules with a similar accuracy comparing to the conventional one in the local reference temperature range of 20~100°C.

Keywords - *thermal resistance; thermal resistance matrix; multi-LED module; junction temperature;T3ster*

1. Introduction

Thermal coupling effects and thermal resistance matrix for multi-heat sources had been studied where the thermal resistance matrix was established by finite Element Analysis simulation.[1] Their result verified with a good accuracy using IR tests showed the thermal resistance matrix method is suitable to evaluate the thermal performance of chips in MCM. Thermal transient measurements of multi-chip LED using T3ster was implemented by Lee[2]. For application in smart LED light, due to the requirements of precise dimming and chromaticity controlling, the effect of junction temperature shift cannot be ignored. For reliability testing applications, it is necessary to monitor the junction temperature of LED luminaires to ensure a good LED life prediction. The testing method of junction temperature widely used is electrical test according to JEDEC JESD 51-1 using T3ster.

Lee[4] proposed the LED junction temperature measurement technique using T3ster with the IR thermometry to measure the LED surface temperature.

2. Thermal resistance matrix modelling

The thermal model of the multi-LED module composed of N LED groups can be simplified by equation (1). For the LED group i, its junction temperature rise ΔT_i due to all the LED groups, where R is an N x N matrix, and $\mathbf{\Delta T}$ and $\mathbf{P}$ are the N rank array.

$$\mathbf{\Delta T} = \mathbf{R}\mathbf{P} \text{ or } \Delta T_i = \sum_{j=1}^{n} R_{ij}P_j \, , \tag{1}$$

or

$$\mathbf{\Delta T} = \begin{matrix} \Delta T_1 \\ \Delta T_2 \\ \vdots \\ \Delta T_i \\ \vdots \\ \Delta T_n \end{matrix} = \begin{bmatrix} R_{11} & R_{12} & \cdots & R_{1j} & \cdots & R_{1n} \\ R_{21} & R_{22} & \cdots & R_{2n} & \cdots & R_{2n} \\ \vdots & \vdots & \vdots & \vdots & \vdots & \vdots \\ R_{i1} & R_{i2} & \cdots & R_{ij} & \cdots & R_{in} \\ \vdots & \vdots & \vdots & \vdots & \vdots & \vdots \\ R_{n1} & R_{n2} & \cdots & R_{nj} & \cdots & R_{nn} \end{bmatrix} \begin{bmatrix} P_1 \\ P_2 \\ \vdots \\ P_j \\ \vdots \\ P_n \end{bmatrix}$$

As depicted in Fig. 2.1, the P_1, P_2 and P_n represent the dissipated power of 1st to nth LED group of the multi-LED module, respectively; and the R_{ij} represents the thermal resistance between the LED group i and the heatsink due to the dissipated power P_j by LED group j;

$K_{ij} = R_{ij}/R_{ii}$, where $i = j$, $K_{ij} = 1$; $i{\neq}j, 0 < K_{ij} < 1$.

The K_{ij}, the thermal coupling factor, can be regarded as the effective dissipated power P_j of LED group j transferred to the LED group i. [4]

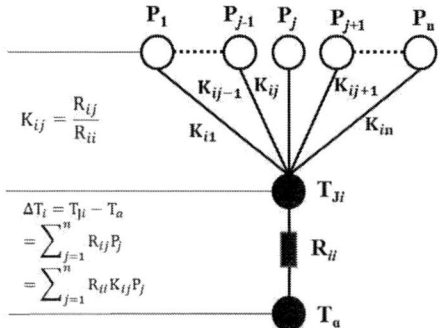

Fig. 2.1 Thermal configuration of the multi-LED module.

And

$$\Delta T_i = T_{Ji} - T_a. \tag{2}$$

With the measurement of the heatsink temperature T_a, the junction temperature T_{Ji} of group i in the multi-LED module can be obtained. However, in practice, for luminaries without direct measurement of the heatsink temperature T_a, it is suggested that the heatsink temperature T_a can be obtained through a design-in thermistor (NTC or PTC) on its PCB measuring the local reference temperature T_s for convenience. The local reference temperature T_s is considered as a virtual thermal source without power, such that $\mathbf{R}$ is modified to be N x (N+1) modified thermal resistance matrix $\underline{R}$ and ΔT became the N+1 junction temperature rise array shown as the following:

$$\Delta T = \begin{array}{c} \Delta T_1 \\ \Delta T_2 \\ \vdots \\ \Delta T_i \\ \vdots \\ \Delta T_n \\ \Delta T_s \end{array} = \underline{R}P = \begin{bmatrix} R_{11}R_{12}\cdots R_{1j}\cdots R_{1n} \\ R_{21}R_{22}\cdots R_{2n}\cdots R_{2n} \\ \vdots\ \vdots\ \vdots\ \vdots\ \vdots\ \vdots \\ R_{i1}R_{i2}\cdots R_{ij}\cdots R_{in} \\ \vdots\ \vdots\ \vdots\ \vdots\ \vdots\ \vdots \\ R_{n1}R_{n2}\cdots R_{nj}\cdots R_{nn} \\ R_{s1}R_{s2}\cdots R_{sj}\cdots R_{sn} \end{bmatrix} \begin{array}{c} P_1 \\ P_2 \\ \vdots \\ P_j \\ \vdots \\ P_n \end{array} \quad (3)$$

The elements of N x (N+1) modified thermal resistance matrix $\mathbf{R}$ above can be measured by the T3ster. If all the element of the modified thermal resistance matrix were measured, the (average) junction temperature rise of each LED group in the multi-LED module can be calculated basing on its dissipated power and the local reference temperature T_s.

According to Equation (3),

$$\Delta T_s = \sum_{j=1}^{n} R_{sj} P_j, \text{ and} \quad (4)$$

$$T_s = T_a + \Delta T_s \quad (5)$$

While the ΔT_s can be calculated by Equation (4). So, the environment temperature T_a can be calculated by equation (5) and the (average) junction temperature T_{Ji} of LED group i also can be calculated by Equation (6).

$$T_{Ji} = T_a + \Delta T_i \quad (6)$$

3. Measurement of multi-LED module sample

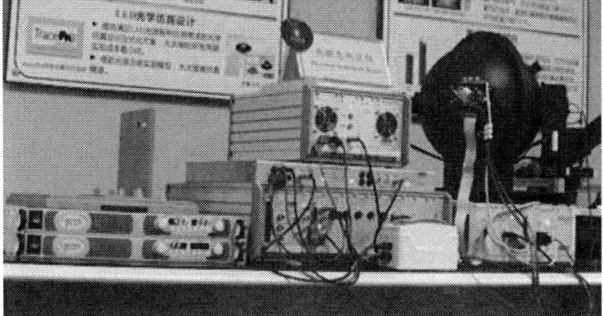

Fig. 3.1 The T3ster for measurement of junction temperature of LEDs.

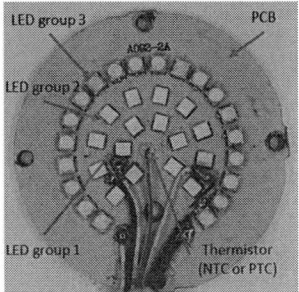

Fig. 3.2 The sample of the LED module.

The T3ster for measurement of junction temperature of LEDs is shown in Fig. 3.1. As shown in Fig. 3.2, the LED module sample for the case study with LED group 1 having 6 LEDs connected in series, LED group 2 having 10 LEDs connected in series and LED group 3 having 20 LEDs connected in series.

Table 1 Conditions of testing for comparison.

Cases		T_a(°C) of thermostat	Driving Current for LED Group 1, 2, and 3 (mA)	T_a of R and $\underline{R}$ built (°C)
1		25, 40, 55, 70, and 85	170,170, and 350	25, 55 and 85
2	A	55	250,250, and 500	
	B	55	100,100, and 200	

The three 3x3 thermal resistance matrices $\mathbf{R}^{25}$, $\mathbf{R}^{55}$, and $\mathbf{R}^{85}$ for three thermostat temperature T_a=25℃, 55℃, and 85℃ respectively can be built with conditions of case 1 in Table 1.

$$\mathbf{R}^{25} = \begin{bmatrix} 3.641 & 0.170 & 0.203 \\ 0.171 & 2.492 & 0.211 \\ 0.144 & 0.174 & 0.470 \end{bmatrix}, \quad (7\text{-}1)$$

$$\mathbf{R}^{55} = \begin{bmatrix} 4.298 & 0.256 & 0.265 \\ 0.244 & 2.831 & 0.266 \\ 0.268 & 0.336 & 0.480 \end{bmatrix}, \quad (7\text{-}2)$$

$$\mathbf{R}^{85} = \begin{bmatrix} 4.723 & 0.321 & 0.278 \\ 0.265 & 3.085 & 0.263 \\ 0.252 & 0.760 & 0.510 \end{bmatrix}. \quad (7\text{-}3)$$

The calculated junction temperature using the thermal resistance matrices $\mathbf{R}^{25}$, $\mathbf{R}^{55}$, and $\mathbf{R}^{85}$ in Equation 7 and the directly measured by T3ster for Case 1 is depicted in Fig.3.3.

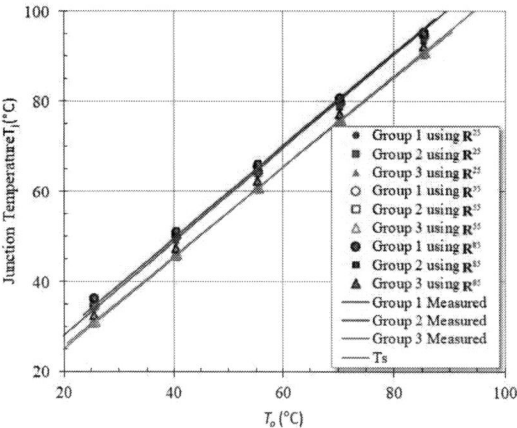

Fig. 3.3 The calculated junction temperature using the thermal resistance matrices $\mathbf{R}^{25}$, $\mathbf{R}^{55}$, and $\mathbf{R}^{85}$ and the directly measured by T3ster for Case 1.

For case 1 and 2, the discrepancy between the calculated junction temperature using the thermal resistance matrices $\mathbf{R}^{25}$, $\mathbf{R}^{55}$, and $\mathbf{R}^{85}$ and the directly measured junction temperature by T3ster for Case 1 and 2 as shown in Fig 3.4 and 3.5.

978-1-7281-5757-3/19 $31.00 © 2019 IEEE   253

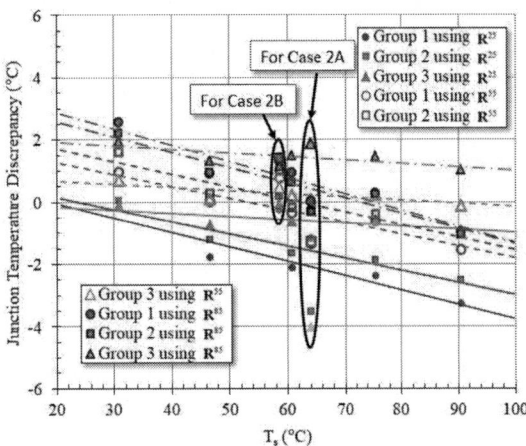

Fig 3.4 The discrepancy (°C) between the calculated junction temperature using the thermal resistance matrices $\mathbf{R}^{25}$, $\mathbf{R}^{55}$, and $\mathbf{R}^{85}$ and the directly measured junction temperature by T3ster for Case 1 and 2 as shown in Fig 3.4.

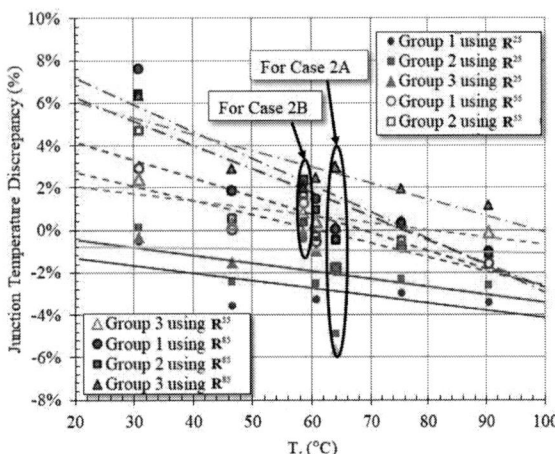

Fig 3.5 The discrepancy (%) between the calculated junction temperature using the thermal resistance matrices $\mathbf{R}^{25}$, $\mathbf{R}^{55}$, and $\mathbf{R}^{85}$ and the directly measured junction temperature by T3ster for Case 1 and 2.

By introducing the local reference temperature T_s, following the equation 3, the 3x4 modified thermal resistance matrices $\underline{\mathbf{R}}^{25}$, $\underline{\mathbf{R}}^{55}$, and $\underline{\mathbf{R}}^{85}$ for three thermostat temperature T_a=25℃, 55℃, and 85℃ respectively, can be built with conditions of case 1 shown in Table 1.

$$\underline{\mathbf{R}}^{25} = \begin{bmatrix} 3.641 & 0.170 & 0.203 \\ 0.171 & 2.492 & 0.211 \\ 0.144 & 0.174 & 0.470 \\ 0.408 & 0.277 & 0.297 \end{bmatrix} \quad (8\text{-}1)$$

is for thermostat of T3ster T_a=25℃,

$$\underline{\mathbf{R}}^{55} = \begin{bmatrix} 4.298 & 0.256 & 0.265 \\ 0.244 & 2.831 & 0.266 \\ 0.268 & 0.336 & 0.480 \\ 0.173 & 0.209 & 0.355 \end{bmatrix}, \quad (8\text{-}2)$$

is for thermostat of T3ster T_a=55℃, and

$$\underline{\mathbf{R}}^{85} = \begin{bmatrix} 4.723 & 0.321 & 0.278 \\ 0.265 & 3.085 & 0.263 \\ 0.252 & 0.760 & 0.510 \\ 0.000 & 0.049 & 0.340 \end{bmatrix} \quad (8\text{-}3)$$

is for thermostat of T3ster T_a=85℃.

The calculated junction temperature using the thermal resistance matrices $\underline{\mathbf{R}}^{25}$, $\underline{\mathbf{R}}^{55}$, and $\underline{\mathbf{R}}^{85}$ in Equation 8 and the direct measured by T3ster for Case 1 is depicted in Fig.3.6.

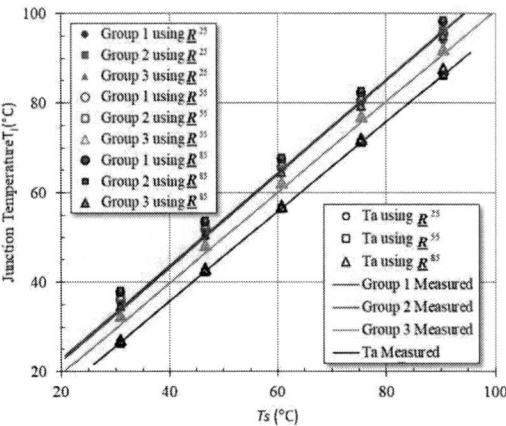

Fig. 3.6 The calculated junction temperature using the thermal resistance matrices $\underline{\mathbf{R}}^{25}$, $\underline{\mathbf{R}}^{55}$, and $\underline{\mathbf{R}}^{85}$ and the directly measured by T3ster for Case 1.

The discrepancy between the calculated junction temperature using the thermal resistance matrices $\underline{\mathbf{R}}^{25}$, $\underline{\mathbf{R}}^{55}$, and $\underline{\mathbf{R}}^{85}$ and the directly measured by T3ster for Case 1 and 2 is shown in Fig 3.7 and 3.8.

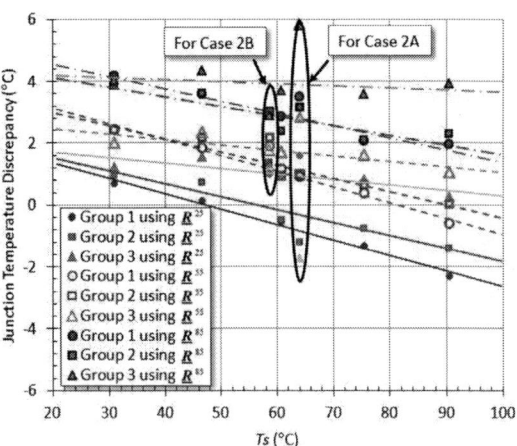

Fig.3.7 The discrepancy (°C) between the calculated and the directly measured junction temperature for Case 1 and 2.

As depicted in the Fig3.3, 3.4, 3.7 and 3.8, the range of discrepancy between the calculated by the three modified thermal resistance matrices $\underline{\mathbf{R}}$ and the directly measured junction temperature is similar but shifted higher about 2°C (2~4%) than by the three thermal resistance matrices $\mathbf{R}$ in the T_s (or T_a) ranging from 20 to 100℃. It also can be found the

978-1-7281-5757-3/19 $31.00 © 2019 IEEE   254

zero discrepancy shifts right when the higher T_a of the thermal resistance matrix is used. This suggests the accuracy of calculation of junction temperature is dependent on the T_a of the thermal resistance matrix.

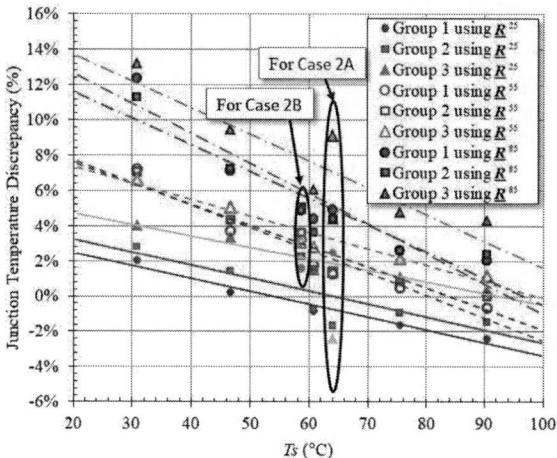

Fig.3.8 The discrepancy (%) between the calculated and the directly measured junction temperature for Case 1 and 2.

To study the effect due to power, two different current combinations were applied in Case 2A and 2B to find the change of junction temperature discrepancy as shown in the circled area of Fig. 3.4, 3.5, 3.7 and 3.8 for comparison with Case 1, where the thermostat temperature of T3ster T_a is set to be 55°C.

It can be found that the discrepancy for case 2A consuming higher power (20.18W) with the higher T_s=64.1°C, is more distribution-divergent than those for the case 1 consuming total power (13.70W) with the T_s=60.7°C. And the discrepancy for case 2B consuming less total power (7.67W) with the lower T_s=58.7°C, is less more distribution-divergent than those for Case 1.

4. Conclusions

In this study, the proposed new junction temperature prediction of the multi-LED module by the modified thermal resistance matrix with a local reference temperature T_s instead of conventional ambient temperature T_a is successfully demonstrated. The result shows accuracy for the modified thermal resistance matrix is similar to the conventional one calculated at local reference temperatures T_s ranging from 20 to 100℃. And the T_s-dependent discrepancy also suggests the calculation of junction temperature by selecting an adequate T_a of the thermal resistance matrix to obtain a better accuracy. It also can be found that more (or less) discrepancy distribution scattering is introduced due to more (or less) current/power applied.

References

1. Ji Cheng, Xiaoqi He, Xunping Li, Bin Zhou, Hengwei Bao, Lianrong Zhou, Zhangchao Wang, "Thermal Coupling Effects and Thermal Resistance Matrix Research of Multi-Heat Sources MCM," 2014 International Conference on Reliability, Maintainability and Safety (ICRMS), P715~719.
2. Sze-Yen Lee，Mutharasu Devarajan, "Thermal analysis of multi-chip LED package with different position and ambient temperatures," 2011 2nd International Conference on Photonics.
3. Dongjun Lee and *et al*, "A Study on the Measurement and Prediction of LED Junction Temperature," International Journal of Heat and Mass Transfer Volume 127, Part B, December 2018, Pages 1243-1252.
4. Huayong Zou, Lingyan Lu, Jiaqi Wang, Brian Shieh, S. W. Ricky Lee, "Thermal Characterization of Multi-Chip Light Emitting Diodes with Thermal Resistance Matrix," 2017 14th China International Forum on Solid State (IFWS).

Smart Lighting with Autonomous Color Tunability

Tianhang Zheng, Wujun, Zhixian Zhou, Wanghui Yan

Front-end innovation department, Opple Lighting Co., LTD., Shanghai, 200233, China

Contact: zhengtianhang@opple.com, henryzheng@outlook.com

Abstract

Smart lighting technologies are very important for high illumination quality and energy-efficiency. In this article, lighting performance controlled by color sensor module has been studied. With simplified system architecture and effective algorithm, the module can be used to obtain precise color and light characteristics for a target area, which can help to tune the specific color rendering property and the intensity of illuminance to an appropriate level automatically for different lighting scenarios. The balance between color preference and color fidelity can be achieved easily with suitable transformation program, and the illuminance information offers a route for governing lighting in high energy-efficiency.

Introduction

Artificial lighting is of great importance for people's everyday life and our society, which is not only capable of offering us general illumination environment, but also has functions from creating various lighting effects, adjusting people mood, influencing industrial output, to curing creatures lifecycles, etc. However, according to statistics from US Energy Department, around 20% of electricity will be consumed for artificial lighting. Therefore, efficient lighting technologies are quite significant for us to improve energy efficiency [1-3].

With the development of solid state lighting technologies, e.g. light emitting diodes (LED), currently, light conversion efficiency has been improved obviously compared with some of conventional lighting sources, e.g. incandescent bulb. Various LED lighting products, systems and applications have been proposed either to improve light qualities for certain purposes, or to enhance lighting efficiency for saving energy. In particular, under the trends of Internet of Things and the concept of smart lighting, more and more lighting products have the capabilities of internet connectivity and environment sensing, which cannot only help people control their lighting remotely, but also the lighting can adjust their state spontaneously. For instance, in the corridor area, bulb with infrared and visible light sensors can turn the lamp on automatically when there is people appearing. More functional sensors will be embedded into lighting for such purposes, lighting will become smarter enough compared with the traditional lighting can only change its brightness. On the other hand, LED offer people more space to tune other lighting parameters including color, correlated color temperature (*CCT*) and color rendering of white light, and even light spectrum in visible range.

Here, a low cost Red-Green-Blue (RGB) color sensor module is developed by combining simplified device architecture and effective algorithm, which has been able to obtain precise illuminance level and color coordinates for a target area, which can help to tune the illuminance and specific color rendering properties to an appropriate level for different application scenarios. The balance between color

preference and color fidelity can be achieved easily with suitable transformation program. Meanwhile, the illuminance information can be used for controlling lighting in high energy-efficiency.

Color sensor and Signal processing

In this work, a RGB color sensor with four channels is used, and three channels correspond to the color component of Red, Green and Blue part (called RGB channel) in visible range, respectively, and the forth channel is called Clear channel used for infrared filtering [4]. The RGB channels can detect the intensity of light at the wavelength of 465nm, 525nm and 615 nm with full width at half maximum (FWHM) around 100nm, and the Clear channel has a wide range of light responsivity from 380 nm to 1100nm. The spectral responsivity curves of the four channels is shown in the Figure 1(a), where the curves of RGB channels are quite similar with the *XYZ* (*XYZ* is the tristimulus value) sensitivity curve in the CIE 1931 color matching functions, as shown in Figure 1 (b). With effective signal processing, it could get enough color and light information, e.g. illuminance, *XYZ* value, *xy* color coordinates, etc., transformed from the output of the RGB color sensor.

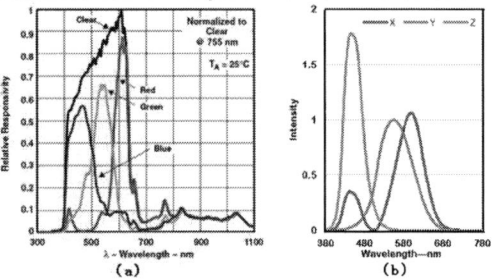

Figure 1. (a) The spectral responsivity of the RGB color sensor [4]. (b) CIE 1931 standard colorimetric observer sensitivity curves for red (*x*), green (*y*) and blue (*z*) cones.

Firstly, it can see the intensity of the light in infrared range from 780nm to 1100 nm is still significant in the spectral responsibity of the color sensor, which will influence the result of Red, Green and Blue color component in visible range. According to the relationship between the RGB and the Clear channels, in visible range, the sum of RGB component is about equal to the value of the corresponding range of Clear channel, and in the infrared range, the intensity is same for the four channels. Thus, the contributing value from the infrared part can be estimated through Function (1).

$$Value_{IR} = \frac{((Value_R + Value_G + Value_B) - Value_{Clear})}{2}$$

(1)

$$TrueValue_R = Value_R - a \times Value_{IR}$$

$$TrueValue_G = Value_G - b \times Value_{IR}$$

$$TrueValue_B = Value_B - c \times Value_{IR}$$

(2)

After eliminating the influences of infrared part on RGB channels, the deviation is still existing between the RGB spectral responsivity and the *XYZ* sensitivity of the color matching functions. Then, several empirical coefficients, *a-c* in function (2), are introduced for further calibration. Once *a-c* value is obtained, it can further help get the true integral value of Red, Green and Blue color component, which have a corresponding relationship with the *XYZ* value of the target lighting area. Meanwhile, the value of *X, Y* and *Z* for the testing sample can be achieved using calibrated light spectrometry meter. With the known *X, Y, Z* value and the real value of Red, Green and Blue color component, the relationship can be built through a matrix transformation program as shown in Function (3). In the program, the correction coefficient matrix (called *CC Matrix*) can be calculated with the known RGB value and *X, Y, Z* results. Generally, in order to get a highly trusted *CC Matrix*, it needs to employ large amount of testing samples that with different spectrum, after several round of calibration and data processing, the satisfied *CC Matrix* can be achieved.

$$\begin{pmatrix} X \\ Y \\ Z \end{pmatrix} = \begin{pmatrix} k_{11} & k_{12} & k_{13} \\ k_{21} & k_{22} & k_{23} \\ k_{31} & k_{32} & k_{33} \end{pmatrix} \begin{pmatrix} TrueValue_R \\ TrueValue_G \\ TrueValue_B \end{pmatrix}$$

(3)

Once the *CC Matrix* is determined, for a given light source, it easily calculates its *X, Y,* and *Z* value, which can be used for computing other relevant parameters, e.g. chromaticity coordinate *x* and *y*, illuminance, d_{uv}, color purity, *CCT* for white light, standard color deviation, color mixing proportion, driver output parameters, etc.

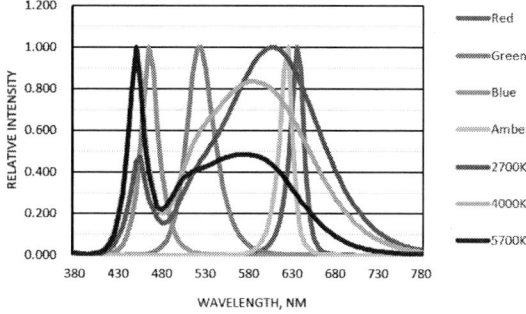

Figure 2. The normalized spectrum of the 7 LEDs used for color sensor calibration.

Here, 7 color LEDs (4 monochromatic light and 3 white light) with different light spectrum, as shown in Figure 2, driven at 3 different current levels respectively have been used for computing the empirical coefficients *a-c* and the *CC Matrix*. Totally, 21 different light spectrums will be produced, and the experimental results are shown in the Table 1.

Table 1. The tested and calculated chromaticity coordinate results of the 21 light spectrums.

LED Samples	Tested true u, v		Sensor signal value				CC Matrix transformation			
	u	v	C	R	G	B	u_1	v_1	duv	SDCM
Amber-1	0.5092	0.3490	3228	2980	176	268	0.5221	0.3478	0.013	
Amber-2	0.5075	0.3492	2925	2696	160	241	0.5212	0.3479	0.0137	
Amber-3	0.5060	0.3493	2583	2379	141	211	0.5206	0.3479	0.0146	
Red-1	0.5385	0.3461	3085	2915	161	294	0.5385	0.3461	0	
Red-2	0.5378	0.3461	2762	2610	143	261	0.5384	0.3462	0.0006	
Red-3	0.5371	0.3462	2417	2281	125	227	0.5376	0.3462	0.0005	
Green-1	0.0780	0.3851	3225	343	2128	632	0.078	0.3851	0	
Green-2	0.0784	0.3853	2949	316	1953	569	0.0779	0.3859	0.0008	
Green-3	0.0791	0.3855	2651	287	1763	502	0.0778	0.3869	0.0018	
Blue-1	0.1399	0.1224	6488	92	1611	4795	0.1399	0.1224	0	
Blue-2	0.1401	0.1221	5859	82	1454	4334	0.1399	0.122	0.0002	
Blue-3	0.1398	0.1221	5208	73	1301	3849	0.1391	0.1229	0.0011	
2700K-1	0.2573	0.3491	2412	1108	778	471	0.2692	0.349	0.012	11.35
2700K-2	0.2572	0.3492	2167	996	698	423	0.2695	0.349	0.0122	11.58
2700K-3	0.2572	0.3493	1916	881	618	373	0.2694	0.3492	0.0121	11.52
4000K-1	0.2241	0.3331	2813	999	991	716	0.2291	0.337	0.0063	6.135
4000K-2	0.2240	0.3332	2526	897	890	643	0.2291	0.337	0.0063	6.137
4000K-3	0.2240	0.3333	2231	793	787	567	0.229	0.3371	0.0063	6.147
5700K-1	0.2052	0.3148	3320	926	1179	1057	0.2045	0.3188	0.0041	5.321
5700K-2	0.2051	0.3148	2975	830	1058	947	0.2044	0.3189	0.0042	5.45
5700K-3	0.2051	0.3150	2626	732	934	835	0.2043	0.319	0.0041	5.382

At first, the samples are tested in calibrated integration sphere and the information including light flux and chromaticity coordinates is obtained, and the true value of coordinate *u* and *v* is listed in Table 1. Then, color sensor was equipped on the surface of the sphere to collect the signal of Red, Green, Blue and Clear channels, as shown in Table 1. According to the introduced method, the data is processed and transferred into *X, Y,* and *Z* results using the adjusted *CC Matrix*. The *CC Matrix* is firstly obtained based on the computation from the data of Red-1, Green-1 and Blue-1 samples. In order to get a reliable and ubiquitous *CC Matrix*, it is better to employ multiple groups of LEDs to achieve a series of *CC Matrix*, and then, use appropriate numerical method to process the data and figure out the best *CC Matrix* according to the lighting sources and the application purposes. With the *CC Matrix*, it can transform R, G, and B signal into *X, Y* and *Z*, and further get the chromaticity coordinates u_1 and v_1. Then, it can compare the true *u* and *v* with the calculated u_1 and v_1 to get *duv*. It can see that the biggest *duv* for color and white light is about 0.0146 and 0.0122, respectively. This error is in an acceptable range, especially the color sensor is used for color tuning or lighting mode controlling. Furthermore, if some filters are used in front of the color sensor to filter the incoming light for making the responsivity curve more similar to the standard sensitivity curves, a smaller *duv* can be achieved [5-6].

Modulization of color sensor

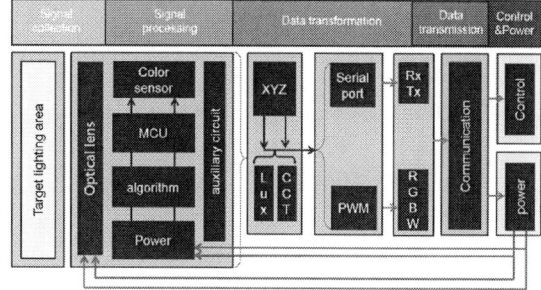

Figure 3. The functional block diagram of the color sensor module.

Now, it is known that the color sensor could detect the color coordinates precisely, in most cases, the tolerance can be controlled within *duv*<0.01 for white and color LED light with proper processing methods. In order to utilize the color

978-1-7281-5757-3/19 $31.00 © 2019 IEEE

sensor more effectively and conveniently, color sensor module with standard data output format and simplified architecture has been developed. Figure 3 is a functional block diagram of the module, which includes several key components to finish the functions of signal collection in target area, signal filtering and processing, data transformation, data communication and control and power management. The data of XYZ, xy, uv, CCT, illuminance and PWM (Pulse width modulation) signal have been formatted through embedding the relevant algorithms and computing in processor unit (called MCU). There is also a serial port on the color module and any data can be communicated through the port with external devices, e.g. LED drivers.

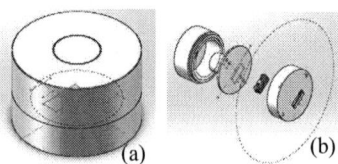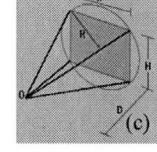

Figure 4. (a) The overall structure of the color sensor module; (b) the structural explosion diagram of the color sensor module; (c) The optical model of the detection area for the color sensor module.

The overall structure of the color sensor module is shown in Figure 4 (a) and (b), which is compact and in "plug and play" type. On top of the sensor PCB board, there is an optical lens to collect light from the target area other than any angle of light can reach to sensor surface to lower the color resolution. The interface between sensor board and the lens is designed in M12 lens type, and the focal length of lens can be chosen accordingly to the optical model in Figure 4 (c). When we use color sensor to detect certain lighting environment, it is needed to know the sensitive area of the used sensor, the size of the area (the width (W) and height (H)), the distance (D) between target area and the sensor module. Once these data is known, the focal length (F) and field view angle (A) can be derived. For instance, for sensors having 4:3 size ratio, the formula (4)-(6) can be used for calculating F and A for the size of 1/4, 1/3 and 1/2 type sensors, respectively. Compare these value with the standard M12 type lens, as shown in Table 2, it can easily select the suitable lens type for the used color sensor.

For 1/4 type lens, $F = \dfrac{3.6D}{W}$ or $F = \dfrac{2.7D}{W}$

$$(4)$$

For 1/3 type lens, $F = \dfrac{4.8D}{W}$ or $F = \dfrac{3.6D}{W}$

$$(5)$$

For 1/2 type lens, $F = \dfrac{6.4D}{W}$ or $F = \dfrac{4.8D}{W}$

$$(6)$$

Table 2. Summary of the parameters for standard optical lens.

Parameters of Standard Optical Lens									
Focal Length, mm	3.6	4	6	8	12	16	25	35	50
Field view angle, degrees	75.7	63.8	44.5	34.7	23.8	17.1	13.7	7.48	5.30

Light sources enhanced by color sensor module

One track spot light used for commercial lighting purpose, e.g. shop lighting, enhanced by the developed color sensor module has been fabricated, as shown in Figure 5. There, color sensor module with F=8mm lens (A=34.7 degrees) is installed near to the light output window, and its power comes from the driver. In the inset of Figure 5, the used LED board is given where color LEDs including Red, Green, Blue and White are used in order that the light from the sources has a high color rendering property and wide range of tunability in CIE color space by mixing them in different ratio. The ratio will be decided by the transformed data from the color sensor. As it has known, the color sensor can detect the overall color properties of the target area, especially the chromaticity coordinates x, y and the illuminance. Thus, the lighting characteristics of the target area is known. Meanwhile, combing the objects in the target area, such as different color clothes, different vegetables, art photos, etc., ideal lighting environment corresponding to the type of objects in the target area., e.g. target chromaticity coordinates x, y and the illuminance, can be scoped based on recognized lighting principles. With these parameters and the known LEDs properties, the output characteristics from the driver could be calculated, such as PWM value for Red, Green, Blue and White channels. By mixing the light from these four channels to the target area, the scoped ideal lighting environment is built. The correlation algorithm and program is burned into the MCU, and now color sensor enhanced program are created. When the module detect the change of the objects in the target area, the program will adjust its active state to the corresponding state, and finally the lighting have the ability to adjust its output automatically [7].

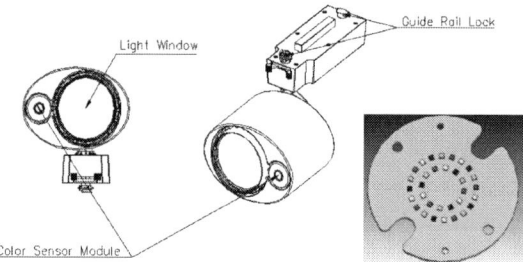

Figure 5. The schematic structure of the color sensor based track spot light. The inset image is the LED pattern on PCB board.

Four color (light blue, green, yellow and pink) simulated objects are chosen to verify the lighting performance of this color sensor enhanced light (CS Light). A correlation program is been built between the main color in the target area and specific color rendering properties for the objects, and the balance between color preference and color fidelity have been optimized, as shown in Figure 6 (a). There, the four objects (Figure 6 (a_2, a_4, a_6, a_8)) are looking more vividly while their color fidelity is well kept under this CS Light, compared with that of the control white COB light. Figure 6 (b) is the spectrum of the CS Light and the control light. The CS Light is assembled by individual RGBW LEDs, which have more flexibility to tune its spectrum according to the signal from the color sensor module. The control device is general COB type LED, which have a fixed spectrum with blue and yellow peak usually. Basically, the two light sources have similar color rendering index value, i.e. R_a, as shown in Figure 6 (c). The reason for the viewing difference is mainly coming from the subtle distinction of the specific color rendering value, e.g. R_9, which demonstrates that the final viewing experience is totally different if only a little variation of certain color characteristics, e.g. spectrum, R_a, etc. Color sensor module is able to detect such small changes and offer great freedom for the adjustment of light output according to the feature of target area and the defined correlation program.

Moreover, these objects have different reflection coefficients and need different illuminance in order to make it have a nice and real appearance. In previous transformation, it is known that the Y value from the sensor gives illuminance value for the testing area. Thus it can be used for modulating the intensity of illuminance simultaneously, which is helpful for the area that need to control the illuminance in order to improve the energy efficiency while keep the flexibility of better illumination conditions.

achieved. Light sources enhanced by this color sensing technology could have great potential applications for both high performance illumination and outstanding energy-efficiency purposes, which offer people a smart way for lighting utilization.

References

[1] S. Muthu, F. Schuurmans, M. Pashley, IEEE J Sel Top Quantum Electron, 8(2002)333.

[2] S. Muthu and J. Gaines, Proc. Industry Applications Conf., (2003)515.

[3] G. F. Dalla Betta, N. Zorzi, P. Bellutti, M. Boscardin, G. Soncini, Microelectronic Test Structures, 15 (2002)217.

[4] https://ams.com/eng/content/.../file/ CS3472_Datasheet_EN_v2

[5] P. Deurenberg, C. Hoelen, J. van Meurs, J. Ansems, Proc. SPIE, 2005.

[6] R. Zach, Osram Opto Semiconductors, Regensburg, 2004.

[7] F. Hailer, LED professional Review, (2007) 45.

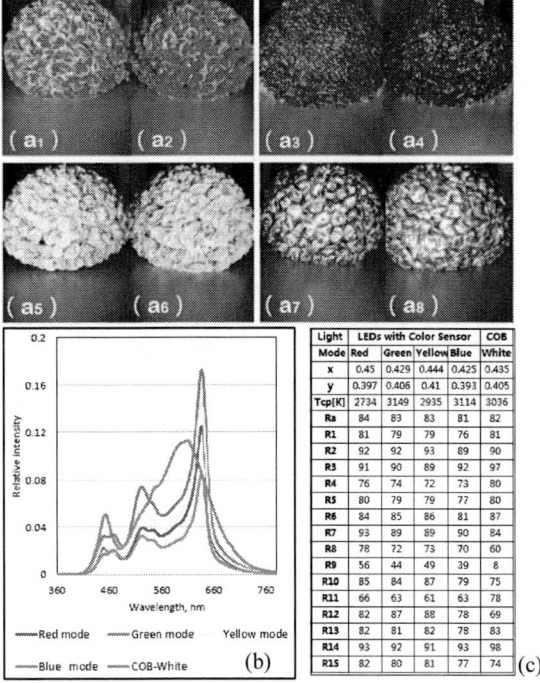

Light Mode	LEDs with Color Sensor				COB
	Red	Green	Yellow	Blue	White
x	0.45	0.429	0.444	0.425	0.435
y	0.397	0.406	0.41	0.393	0.405
Tcp[K]	2734	3149	2935	3114	3036
Ra	84	83	83	81	82
R1	81	79	79	76	81
R2	92	92	93	89	90
R3	91	90	89	92	97
R4	76	74	72	73	80
R5	80	79	79	77	80
R6	84	85	86	81	87
R7	93	89	89	90	84
R8	78	72	73	70	60
R9	56	44	49	39	8
R10	85	84	87	79	75
R11	66	63	61	63	78
R12	82	87	88	78	69
R13	82	81	82	78	83
R14	93	92	91	93	98
R15	82	80	81	77	74

Figure 6. (a1-a8) Objects under white COB light and the CS Light, respectively. (b) The spectrum of the CS Light and the white COB light. (c) The color characteristics of the CS Light and the white COB light.

Summary and prospect

In summary, a color sensor module with simplified structure and effective color sensing capabilities has been developed with a highly versatile data processing algorithm and filtering strategies. The responsivity of the sensor module can adapt to different lighting environment actively, which can be used for wide range of light tunability. Light sources enhanced by the color sensor module could adjust its light output in high performance automatically according to the objects in the target area and the given transformation program. The balance between color preference and color fidelity, and high energy utilization efficiency can be easily

Design of intelligent temperature control driving circuit for high power LED array

Fei Wang[1,2], Houda Zhou[1,2], Jingjing Liu[4], Luqiao Yin[1,3], Jianhua Zhang[1,2*]

[1]Key Laboratory of Advanced Display and System Applications, Shanghai University, Ministry of Education, Shanghai, 200072, China, 1324725476@qq.com
[2]School of Mechanical Engineering and Automation, Shanghai University, Shanghai 200072, China
[3]School of Materials Science and Engineering, Shanghai University, Shanghai, 200072, China
[4]Shanghai Electric Power University, Shanghai, 200090, China

Abstract

A new type of high-power underwater lighting LED driver circuit is proposed. This design studies and designs the electrical performance stability, temperature regulation and LED array arrangement of LED array circuits. The temperature control mainly uses the temperature sensor module for detection, and the temperature feedback is timely performed by the MCU. When the temperature reaches the set temperature gradient threshold, the PWM is adjusted in time to adjust the current to achieve the brightness reduction. The purpose of heat generation is to adjust the temperature; the other point is mainly to design the voltage regulator circuit part. The stability of the circuit is very important for the LED array circuit and the peripheral circuit in this design, which also becomes the judgment circuit. One of the key indicators. This paper mainly designs the circuit stability for input voltage regulation and temperature control adjustment. Underwater illumination flux will also have certain requirements, LED array arrangement is also very important. Through the simulation experiment, the hexagonal array arrangement is adopted. For the above three points, a high-power LED driver circuit is developed. Finally, through the specific experiment to verify the operation of the circuit, the overall operation of the circuit is good, the dimming, voltage regulation is good and the efficiency is high, and the design goal is achieved.

Keywords: LED array; voltage regulation; temperature regulation; efficiency

1.Introduction

LED is a new type of semiconductor lighting source. All kinds of high-power LED devices are developing very rapidly. Nowadays, they are used in all aspects of life. Due to their unique energy-saving characteristics, the market demand is getting larger and larger, and array LED combination circuits are used. High efficiency, low power consumption, low voltage drive, long service life, and energy-saving and environmental protection have attracted worldwide attention[1]. And in recent years, with the advent of the 5G era, the theme of intelligence is often mentioned by us, and more and more LED related products appear on the market, but the advent of these products is based on some key Above the sex indicator, stability is one of them[2, 3].

This paper designed a new high-power LED driver circuit[4, 5]. A key indicator of this design is circuit stability, which is also one of the key indicators for evaluating the quality of a product. Therefore, the stability of the LED peripheral drive circuit is particularly important. The circuit is used in underwater lighting equipment, and a large-area LED array is used, and the LED array and the circuit are placed in a small sealed cavity, and the heat dissipation problem is very severe. It is difficult to ensure the timely release of heat by simply relying on the structure for heat dissipation, and the heat cannot be The timely export will be very harmful to the equipment, so the self-regulating system is added to the drive circuit[6-9]. The voltage used in this design is 40-50V wide input voltage, using step-down circuit, the voltage is supplied to the LED array through the voltage regulator part at 40V, and the voltage is supplied to the MCU. About 5V[10, 11], After experimental verification, the luminous flux of the circuit can reach more than 10000lm, and the circuit efficiency can reach over 92%[12, 13].

2.overall design

The whole circuit is divided into the following parts. The first part is the voltage regulation part, which mainly prevents the circuit from being damaged due to the fluctuation of the peripheral circuit. This part mainly adopts the step-down voltage control circuit for the step-down voltage control. The main output is 40V voltage to supply power to the LED array... The other part is the MCU control part. This part uses the second part of the step-down circuit to stabilize the voltage at the 5V output. The other part is the temperature acquisition module, which collects the temperature in real-time and feeds the information back to the MCU for temperature adjustment. As shown in Figure 1.

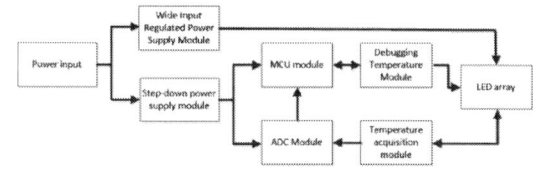

Figure1 overall design

3.Hardware circuit design

3.1Voltage regulator circuit and LED array circuit module

The hardware part of the circuit is mainly divided into three parts: 40V voltage regulator module, 5V voltage regulator module and LED array module. The 40V regulator

978-1-7281-5757-3/19 $31.00 © 2019 IEEE

circuit module mainly adopts the buck regulator module, and the input peak voltage can reach 50V. The step-down module can drive the circuit parameters of 40V3A required by the LED array circuit and has a good linear region, and the voltage regulation performance is superior. The LED array circuit uses two inputs. Each LED is 12 lamps with a single power of about 5W. The overall power is about 120W. The LED substrate adopts a hexagonal array. The substrate is set to two channels to increase the diversity of the lamp beads. When there are other requirements, the lamp bead combination can be changed as required, and the circuit can be finely adjusted to achieve separate control of each channel, and the structural change is very simple. The circuit structure is shown in Figure 2.

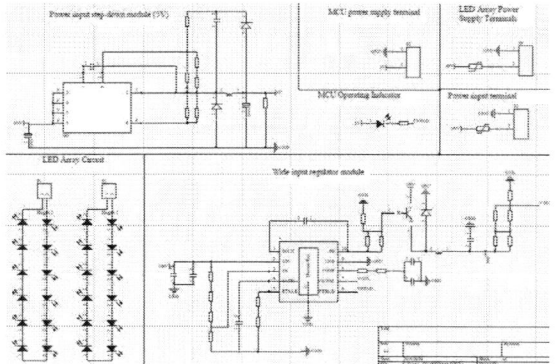

Figure 2 voltage regulator circuit
and LED array circuit module

3.2Temperature control circuit module

The temperature regulation circuit module is mainly divided into four parts. The first part is the sensor part. The accuracy of the temperature sensor used in this design can reach 0.1 °C, which meets the collection accuracy requirements of this experiment. The analog-to-digital conversion part adopts the ADS module, and the acquired temperature is fed back to the MCU module in real-time according to the temperature feedback and the temperature threshold value set in advance in the program. When the temperature exceeds the critical value, the MCU gives The pin sends a signal to adjust the status of the 74 series pins to adjust the PWM. The current is controlled to adjust the brightness of the LED to adjust the temperature. And the power consumption problem is considered in design, and the peripheral circuit is reduced as much as possible to reduce power consumption without affecting stable operation. The circuit structure is shown in Figure 3.

4.System software design

The program part is mainly for real-time detection of temperature and input voltage signal. On the one hand, when the external voltage changes the stone to adjust the output

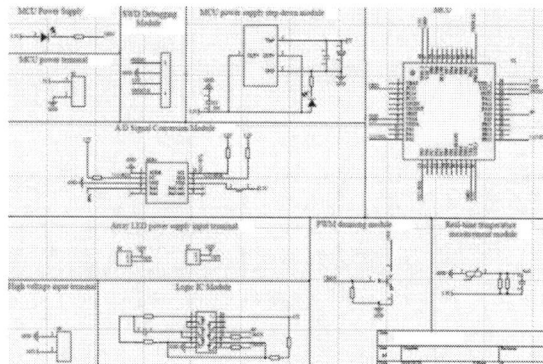

Figure 3 thermostat circuit module

voltage in time, on the other hand, the feedback is adjusted to the temperature to adjust the PWM value to achieve the purpose of cooling. When the system is powered on, the system will work normally. And the work indicator is soldered on the circuit board, and it can also be confirmed according to whether the light is on or off. The temperature zero boundary is set to two parts, and different temperature zero points can be adjusted to adjust the different values of PWM. According to the ratio, it can be divided into 0-100%, but the PWM should not be set too small. Too small may cause the circuit to work normally. The program function flow chart is shown in Figure 4.

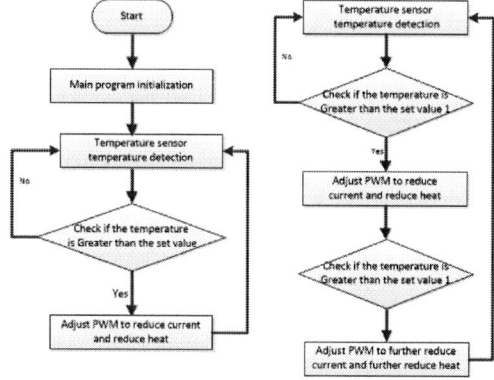

Figure 4 main program flow chart

5.Testing and analysis

The test is mainly divided into the following parts. The first part of the test is to test the two LED connection drive circuit, but the circuit in the air to test whether the circuit can work normally, the display circuit works normally, and the luminous flux is under normal working condition. Up to 10000 lm, the circuit and the designed lamp housing are assembled and placed in the integrating sphere for testing. The luminous flux of the test environment is 0 lm, and the measured luminous flux reaches about 10000 lm. The actual test results are shown in Figure 5.

Table 1 parameter test results

Serial number	Voltage (V)	Current (A)	Luminous flux (lm)
1	39.452	2.997	10521
2	39.027	3.992	12457

Figure 5 actual lighting test

This part mainly tests the stability of the 40V step-down circuit. The test results are shown in the figure. The input range can be changed to about 10V. The output of the voltage signal is adjusted and stabilized at about 40V. The experimental results are compared with the expected targets. The goal was set at the beginning of the design. As shown in Figure 6.

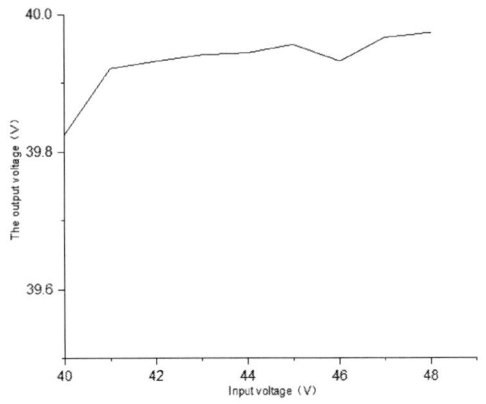

Figure 6 40V output stability test results

This part of the main MCU power supply step-down circuit stability test, this part of the main power supply for the MCU module and ADS analog-to-digital conversion module, through the test input for 40V output is stable at around 5V, can meet the drive MCU and ADS analog-to-digital conversion module, And working in good condition. The experimental results are shown in the figure7.

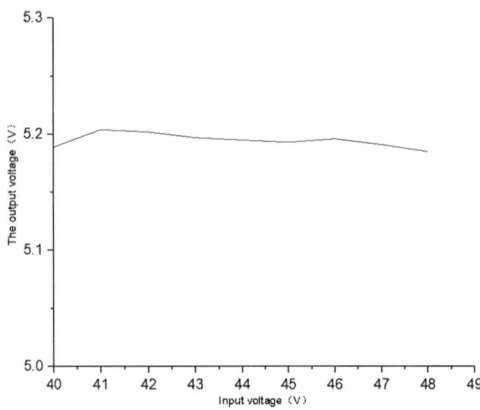

Figure 7 5 V output stability test results

The LED array circuit is loaded into the designed sealed cavity, and placed in a water environment for testing. As the temperature rises continuously when it reaches 50 ° C, the previously set temperature node is reached, and the PWM value is adjusted. Adjusting the current, there are experimental data, the temperature rise slope v is decreasing, and the temperature rising speed is gradually slowing down. The experimental results are shown in Figure 8.

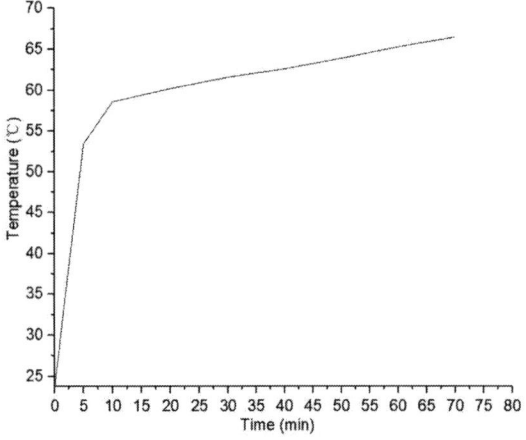

Figure 8 thermostat test

6.Conclusion

The design of the underwater high-power LED array intelligent temperature-regulating driving circuit design mainly realizes three functions. The first realize stable driving of the LED array, and the driving effect is good, which plays a good role in prolonging the service life of the LED lamp bead. The realization of the second over-temperature protection function is beneficial to further protect the normal use of the circuit, prolong the service life of the circuit and the service life of the LED array. The third part adopts a hexagonal LED array, and the luminous flux can reach more than 10000 lm, and the overall design of the circuit is as simple as possible. The peripheral circuit and the circuit can work stably, all of them are universal devices, and the stability and cost performance is very good, which is very popular.

7.Acknowledgments

This work was supported by the National Nature Science Foundation of China (NSFC) under the Grant Number (51605272) and Science and Technology Commission of Shanghai Municipality Program (19DZ2281000; 17DZ2281700). I would like to thank Professor Zhang Jianhua, Professor Yin Luqiao, and Guo Aiying for their support and help in this work.

8.References

[1]   R. Lin, J. Y. Tsai, S. Y. Liu, H. W. J. I. J. o. E. Chiang, and S. T. i. P. Electronics, "Optimal Design of LED Array Combinations for CCM Single-Loop Control LED Drivers," vol. 3, no. 3, pp. 609-616, 2015.

[2]   Y. Yang, Z. Song, and Y. Gao, "Design of High-Power White LED Drive Chip with Fully Integrated PWM Dimming Function," in symposium on photonics and optoelectronics, 2010, pp. 1-4.

[3]   Y. Huang, Y. Lin, W. Wang, and G. Dong, "Design of High-Stability Driver for White LED," in symposium on photonics and optoelectronics, 2009, pp. 1-3.

[4]   X. Hailiu, and W. Yanping, "Design of a New LED Drive Circuit," presented at the 2018 International Conference on Computational Science and Engineering(ICCSE 2018), Qingdao, Shandong, China 2018-11-03.

[5]   Getao, L. Yu, F. Xiansong, N. Pingjuan, and C. Yuhua, "110W high power LED Street lamp power driver," in international conference on educational and information technology, 2010, vol. 3.

[6]   Z. Tong, Y. Yuan, S. Zhenghua, and F. Yongbo, "A dual mode dimming white LED driver based on Buck DC-DC converter," in international conference on computer science and information processing, 2012, pp. 1129-1131.

[7]   M. Tahan and T. Hu, "High performance multiple string LED driver with flexible and wide range PWM dimming capability," in applied power electronics conference, 2017, pp. 1570-1577.

[8]   L. Lohaus, A. Rossius, S. Dietrich, R. Wunderlich, and S. J. I. T. o. I. A. Heinen, "A Dimmable LED Driver With Resistive DAC Feedback Control for Adaptive Voltage Regulation," vol. 51, no. 4, pp. 3254-3262, 2015.

[9]   L. Liu, Y. Niu, J. Zou, Z. Zhu, Y. J. A. I. C. Yang, and S. Processing, "A novel monolithic white LED driver with dual dimming mode," vol. 79, no. 1, pp. 37-44, 2014.

[10]   V. Yousefzadeh, A. Babazadeh, B. Ramachandran, E. Alarcon, L. Y. Pao, and D. J. I. T. o. P. E. Maksimovic, "Proximate Time-Optimal Digital Control for Synchronous Buck DC–DC Converters," vol. 23, no. 4, pp. 2018-2026, 2008.

[11]   A. Prodic, D. Maksimovic, and R. W. Erickson, "Design and implementation of a digital PWM controller for a high-frequency switching DC-DC power converter," in conference of the industrial electronics society, 2001, vol. 2, pp. 893-898.

[12]   T. Liu, S. Wang, S. Song, and Y. Ai, "Research on High-Efficiency Driving Technology for High Power LED Lighting," in ieee pes asia-pacific power and energy engineering conference, 2010, pp. 1-4.

[13]   G. Jane, C. Su, H. Chiu, and Y. Lo, "High-efficiency LED driver for street light applications," in ieee international conference on renewable energy research and applications, 2012, pp. 1-5.

Research on a Smart LED Lighting Based on Improved Flyback Driver

Wenran.Liu, Weiming.Lin

College of Electrical Engineering and Automation Fuzhou university, Fuzhou, China, 350116

649215253@qq.com, weming@fzu.edu.cn

Abstract

Aiming at various lighting energy supply circuits currently used, a improved Flyback circuit is proposed applied in the smart LED lighting in this paper, which can realize photovoltaic power generation and energy storage, and stored energy is inverter to the grid or AC load. The proposed driver can realize multi-form and multi-function utilization of energy. The circuit operating process is analyzed and the key parameters relation equations are set up in detail. The design circuit uses 18V photovoltaic panel as input, 24V 6AH lead-acid battery as storage part, three 1W 350mA LED lamps in series as LED lamp group, and 50 ohms AC load as DC-AC inverter load. The simulation software PSIM is carried out to verify the feasibility of the improved circuit design and analysis.

Key words—smart LED lighting, PV combined with other energy，improved Flyback converter, bidirectional inverter, battery

1.Introduction

In recent years, LED lamps have gradually replaced traditional lighting fixtures and been applied in various lighting occasions. LED lamps have the advantages of high luminous efficiency, long life, convenient control and maintenance free. At the same time, the development and utilization of clean energy, such as solar energy, wind energy, tidal energy, geothermal energy and other forms of renewable energy, has attracted more and more attention and been widely used. Among them, solar energy is widely used because of its universality, harmfulness, large reserves and long-term use. Moreover, photovoltaic power generation is of great significance as the renewable energy storage of smart LED lighting. The energy storage and the battery play a important role in the smart LED lighting, it can ensure the normal operation of LED when the light is insufficient, and can inverter the excess energy back to the AC load to improve the energy utilization rate.

An improved Flyback circuit for smart LED lighting is proposed in this paper. The circuit can realize the bidirectional flow of energy. During normal operation, the photovoltaic panel will convert the energy from the solar energy into a DC voltage input to the circuit, which will supply power to the LED and charge the battery at the same time. At the same time, Flyback converter with center tapped secondary winding is used to convert the excess energy stored in the battery into power frequency sinusoidal voltage and feed back to the power grid or AC load. Finally, the design circuit is established in PSIM simulation software, and the simulation results verify the feasibility of the design circuit.

2.The Proposed Circuit and its Operating Principle

A. Circuit topology

An improved Fyback converter based on a single magnetic element and multiple windings is proposed in this paper, which consists of four parts: photovoltaic input circuit, battery charge-discharge circuit, LED lighting circuit and Flyback inverter circuit. The proposed circuit topology is shown as Fig.1.

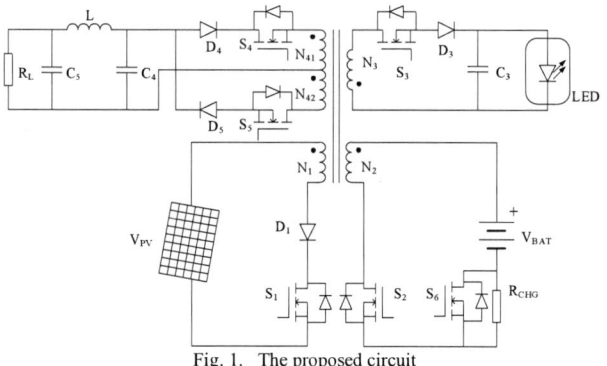

Fig. 1. The proposed circuit

B. Operating principle

There are two operating modes analyzed in detail in the proposed circuit and every mode has two stages. One is to charge and store energy for battery and supply power for LED respectively with photovoltaic panel as input. The other is to DC-AC inverter from battery.

Mode I: Photovoltaic power generation, battery charging and LED on lighting: it is a DC-DC converter mode, MOS_1 is fixed duty cycle high-frequency switch, MOS_2 is turn-off, MOS_3 is turn-on, MOS_4 and MOS_5 signals are blocked, and the inverter circuit does not work. The equivalent circuit is shown as Fig.2.

Stage I: when MOS_1 is on, the N_2 winding flows through the induction current to charge the battery, and the R_{CHG} is connected in series to limit the charging current. And the working current of LED is supplied by capacitor C_3 as shown in Fig.2 (a);

Stage II: when MOS_1 is turn-off, the N_3 winding flows through the induced current to power the LED, at the same time in the photovoltaic input circuit reverse series diode D_1, ensure that there is no reverse current in the circuit as shown in Fig.2 (b).

978-1-7281-5757-3/19 $31.00 © 2019 IEEE

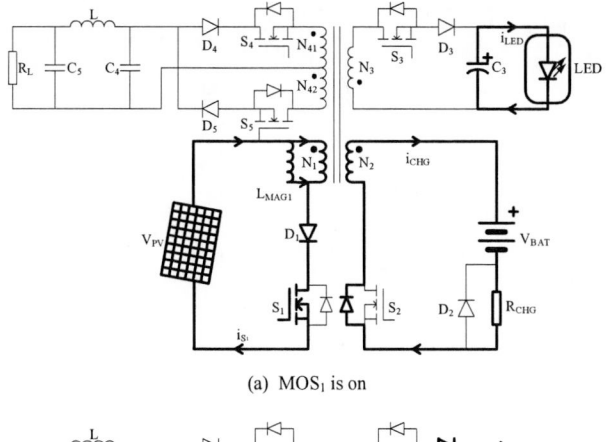

(a) MOS$_1$ is on

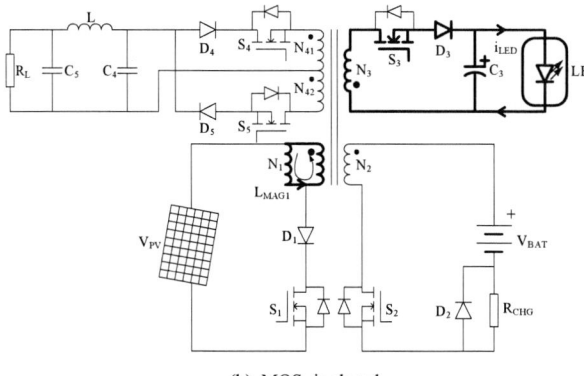

(b) MOS$_1$ is closed

Fig.2. The operating equivalent circuit at Mode I

Mode II: DC-AC inverter from battery to grid or AC load: block MOS$_1$ and MOS$_3$ signals, photovoltaic input circuit, and LED lighting circuit don't work; MOS$_2$ drive with SPWM wave, MOS$_4$ and MOS$_5$ drive with half line period at line frequency complementary.

Stage I: when MOS$_4$ is on and MOS$_5$ is in off state, the output is positive half cycle as shown in Fig.3 (a), current flows out from the central tap of transformer and back to transformer through diode D$_4$ and MOS$_4$ after load;

Stage II: when MOS$_5$ is on and MOS$_4$ in off state, the output is negative half cycle as shown in Fig.3 (b), current flows out of transformer from MOS$_5$ and diode D$_5$ and back to transformer from central tap after load.

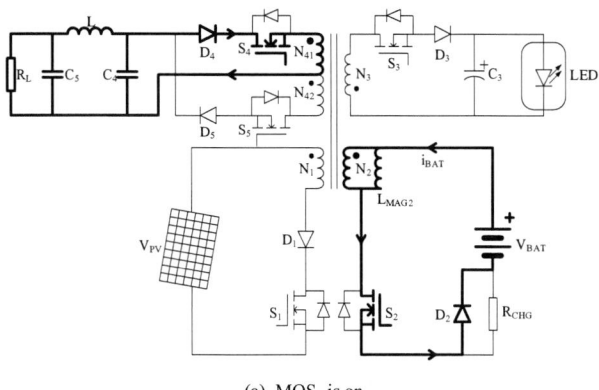

(a) MOS$_4$ is on

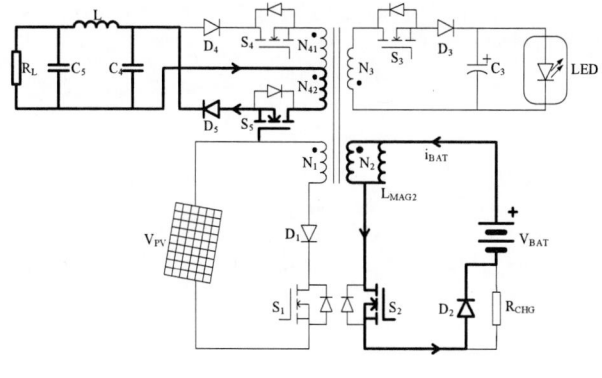

(b) MOS$_5$ is on

Fig.3.   The operating equivalent circuit at Mode II

3.Design Consideration

The design circuit uses 18V photovoltaic power board as input, 24V 6AH lead-acid battery as storage battery, four 1W 350mA LED lamps in series as LED lamp group, and 50 ohms resistance as load on inverter output side.

A. Key parameter design

At mode I, when MOS$_1$ is on, the battery charges; when MOS$_1$ in off state, power supply of LED lamp group works, this gives the relationship between input and output:

$$\frac{V_{O2}}{V_{in}} = D \cdot \frac{N_2}{N_1} \tag{1}$$

$$\frac{V_{O3}}{V_{in}} = \frac{D}{1-D} \cdot \frac{N_3}{N_1} \tag{2}$$

Taking duty cycle $D = 0.5$, the ratio relationship among N_1, N_2, N_3 windings can be deduced, and the turns of N_4 windings can be obtained according to the required output voltage range of the inverter.

At mode II, MOS$_4$, MOS$_5$ alternating switch with line frequency, so that the inverter bridge to get line frequency alternating current, a low-pass filter is added after the inverter bridge is added to filter other harmonic signals except line frequency. The capacitance and inductance of the low-pass filter is obtained by the following formula.

$$f = \frac{1}{2\pi\sqrt{LC}} \tag{3}$$

where f is the cut-off frequency, and the value of inductance L is determined by capacitance C.

According to the operating state of the mode II, the t_{on} and t_{off} of MOS$_2$ can be expressed as,

$$t_{on} = \frac{L_m I_P}{V_{bat}} \tag{4}$$

$$t_{off} = \frac{L_m I_P}{\sqrt{2}V_{ac}} \tag{5}$$

The excitation inductance of transformer can be obtained from eq.(4) and eq.(5).

$$L_m = \frac{\sqrt{2}V_{ac}V_{bat}}{2NfI_P\left(V_{bat} + \sqrt{2}V_{ac}\right)} \qquad (6)$$

Where V_{ac} is the root-mean-square value of ac utility grid line-voltage and V_{bat} is the voltage of battery. And f is the ac utility grid line frequency and N is the total number of the switching periods during the half cycle. I_P is defined as the crest value of the enveloped peak current.

The detail main parameters of the designed circuit are shown in the following table.

Table 1.　Circuit main parameters

	Parameter	Value	Parameter	Value
Parameter setting	V_{PV}/V	18	C_3/mF	10
	V_{BAT}/V	24	N_1	2
	V_{LED}/V	3×3	N_2	1
	R_{CHG}/Ω	24	N_3	6
	R_L/Ω	50	N_{41} N_{42}	55

B. Control system

At mode I, a closed loop is designed to ensure stable output of LED lamp group. The output current of LED lamp group is collected, and the difference is compared with the rated reference value of 350mA. After the difference is adjusted by PI regulator and compared with the triangular wave, and the drive signal control MOS$_1$ is obtained as shown in Fig.4.

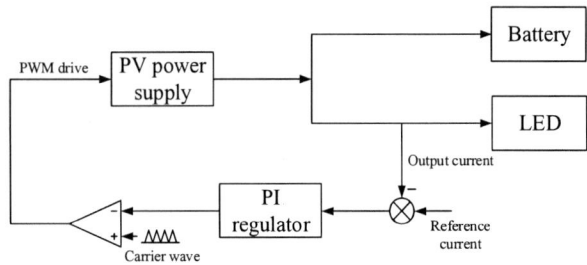

Fig.4.　The control system at Mode I

At mode II, PWM tracking control technology based on carrier wave comparison is adopted to build the closed loop. The actual output current of the inverter circuit is compared with the instructional current, the deviation current is calculated, amplified by amplifier A, and then compared with the triangle wave to generate PWM waveform. The amplifier A usually has the proportional integral characteristic and the proportional characteristic, and its coefficient directly affects the current tracking characteristic of the inverter circuit, as shown in Fig.5.

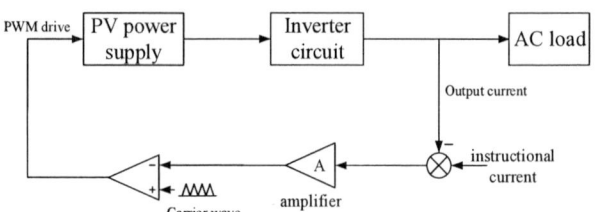

Fig.5.　The control system of Mode II

4.Simulation Results

In order to verify the effectiveness of the proposed circuit analysis, PSIM simulation software is used to build simulation models for two modes. Specific parameters and operating conditions of the circuit are described in detail in the previous section.

At the mode I, MOS$_1$ is fixed duty cycle high-frequency switch, MOS$_2$ is turn-off, MOS$_3$ is turn-on, MOS$_4$ and MOS$_5$ signals are blocked.

The simulation results of mode I are shown in the following figures. Fig.6 shows the operating situation of LED lamp group. V_{LED} is the voltage of LED and I_{LED} is the current of LED, showing the dynamic response process and current ripple. $U_{MOS1.ds}$ is the switching voltage of MOS$_1$ and u_1 is the voltage at both ends of transformer N$_1$ winding.

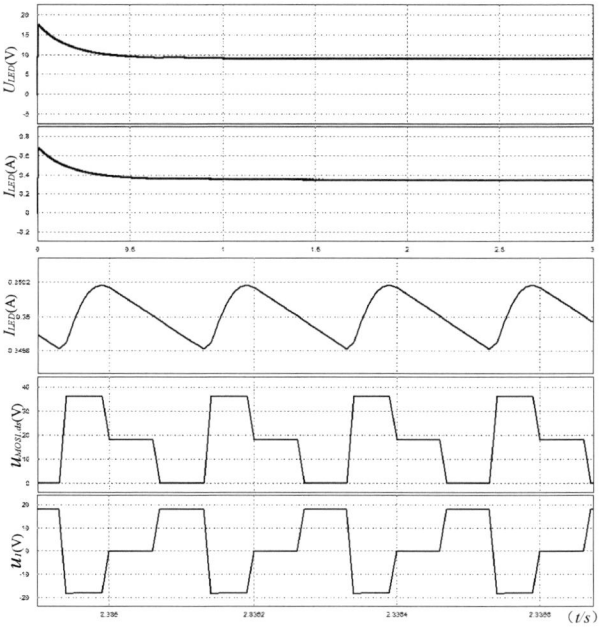

Fig.6.　The LED simulation waveforms at Mode I

As can be seen from the Fig.6, the output current of LED is stable at the rated current of 350mA, and the voltage of three groups of LED beads in series is stable at about 9V. The ripple of output current is analyzed, and its ripple coefficient is 0.011%. When MOS$_1$ is turn-off, LED lighting circuit works. Voltage at both ends of MOS$_1$ is clamped in the sum of voltage at both ends of N$_1$ winding and photovoltaic input V_{PV}, and LED working current increases. When the energy is released in the N$_3$ winding, the voltage at both ends of MOS$_1$

is clamped in the photovoltaic input V_{PV}, and the working current of LED starts to decrease.

Fig.7 shows the operating situation of battery changing circuit. U_{BAT} is the voltage at both ends of battery; i_2 is the charging current of battery charging circuit; i_{MOS1} is the switching current of MOS$_1$. The operating state of the circuit can be clearly distinguished by comparing the switching state of MOS$_1$. When MOS$_1$ is turn-on, the current flowing through the winding MOS$_1$ and N$_1$ winding gradually increases, the charging current i_2 increases, and the voltage at both ends of the battery rises. When MOS$_1$ is turn-off, the current flowing through the winding of MOS$_1$ and N$_1$ winding turns to 0, and the charging current i_2 also turns to 0, and the voltage at both ends of the battery starts to decrease.

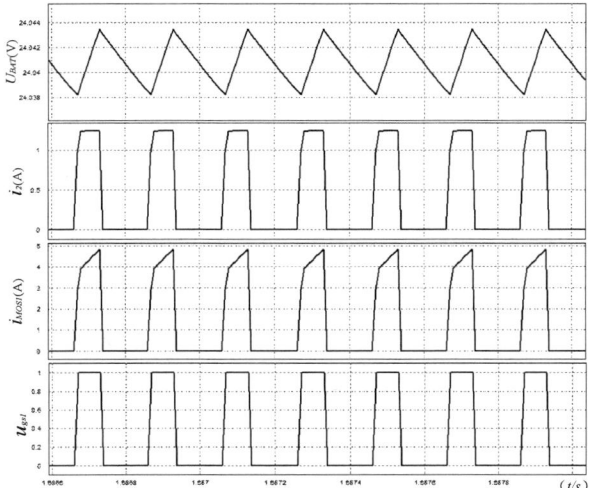

Fig.7. The battery changing circuit simulation waveforms at Mode I

In mode II, MOS$_2$ drive with SPWM wave, MOS$_4$ and MOS$_5$ drive with half line period at line frequency complementary and block MOS$_1$ and MOS$_3$ signals.

The simulation results of mode II are shown in Fig.8. $U_{MOS2.ds}$ and i_{MOS2} are the switching voltage and current of MOS$_2$ respectively. U_{ac} and i_{ac} are the current and voltage of the inverter side. Where the thick line represents the voltage of the inverter side, and the thin line represents the current of the inverter side, which is amplified by 40 times.

The inverter AC load is 50 ohms, and the output reference current of the inverter is the power frequency sinusoidal current with an amplitude of 4A. The switching state of MOS$_2$ shown in the upper part is the corresponding period of one of the output sinusoidal peaks, and the driving signal of MOS$_2$ is compared intuitively.

According to the simulation results, the voltage and current of the inverter side are calculated and analyzed. The *THD* (Total Harmonic Distortion Factor) can be calculated from the following formula:

$$THD = \frac{1}{V_1} \sqrt{\sum_{n=2}^{\infty} V_n^2} \qquad (6)$$

Where V_1 is the effective value of the fundamental wave component, V_n is the effective value of the nth harmonic component.

The *PF* (Power Factor) can be calculated from the following formula according to *THD*:

$$PF = \frac{1}{\sqrt{1 + THD^2}} \qquad (7)$$

Finally, the result of the calculation is the *THD* is 4.963% and the *PF* is 0.9988.

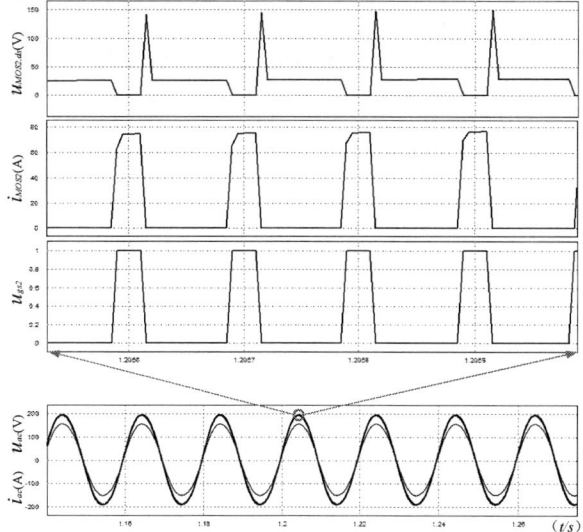

Fig.8. The simulation results at Mode II

5.Conclusion

A improved bidirectional Flyback circuit integrated with a magnetic element is proposed in this paper, which can realize photovoltaic power supply and energy storage, energy storage and power supply and inverter to the grid or AC load, realizing multi-form and multi-function utilization of energy.

The simulation results show that under the operating working conditions, the photovoltaic panel can provide a stable charging voltage of 24V for battery and a operating current of 0.5A for the LED, and the *THD* is small. The AC load of the inverter can also output stably according to the reference current of 4A and the *THD* is about 2.95% and the *PF* is 0.9988.

References

1. Nobuyuki Kasa et al, "Flyback Inverter Controlled by Sensorless Current MPPT for Photovoltaic Power System," *IEEE* Transactions on Industrial Eectronics, Vol. 52, No. 4, August, 2005.

2. Jesus Cardesin *et al*, "LED Permanent Emergency Lighting System Based on a Single Component,"*IEEE* Transactions on Power Eectronics, Vol. 24, No. 5, May, 2009.

3. WANG Yun-fan, XU Zheng-xi, YAO Chuan, ,"WU Da-li, Comparative analysis and simulation on inverter control strategy," *Ship Science and Technology*, Vol. 39, No. 7, July, 2017.

978-1-7281-5757-3/19 $31.00 © 2019 IEEE

4. LI Jun-lin. "Research on repetitive control and dual-loop controltechnology for single-phase inverter,"[D]. Wuhan Huazhong University of Science and Technology, 2004.

5. M. Nagao, H. Horikawa, and K. Harada, "Photovoltaic system using buck-boost PWM inverter," *Trans. Inst. Elect. Eng. Jpn.*, vol. 114-D, pp. 885–892, 1994.

6. M. Nagao and K. Harada, "Power flow of photovoltaic system suing buck-boost PWM power inverters," in *Proc. PEDS'96*, May, 1996, pp. 114-149.

7. T. Shimizu, N. Nakamura, and K. Wada, "A novel flyback-type utility interactive inverter for AC module systems," in *Proc. ICPE'01*, Seoul, Korea, Oct. 2001, pp. 518-522.

8. M. Rico-Secades, E. L. Corominas, J. Garcia-Garcia, J. Ribas, A. J. Calleja, J. M. Alonso, and J. Caresin, "Low cost electronic ballast for a 36-W fluorescent lamp based on a current-mode-controlled boost inverter for a 120V DC bus power distribution," *IEEE Trans. Power Electron.*, Vol. 21, No. 4, pp. 1099-1106, Jul. 2006.

9. M. A. Co, D. S. L. Simonetti, and J. L. F. Vieira, "High power factor eletronic ballast operating at critical conduction mode," in *Proc. IEEE Power Election. Spec. Conf.*, 1996, pp. 962-968.

10. LI Feng, LUO An, TANG Ci, LUO Zhuo-wei, "Study on SVPWM Control Strategy for Current-tracking PWM Inverter," *Power Electronics*, Vol. 42, No. 4, April, 2008, pp. 77-79.

Analysis of Smart Wall Switch without Neutral Wire Compatibility Issue

Yang Hu, Wei Wen, Wanghui Yan and Ran Ding
Research and Innovation Laboratory
Opple Lighting Co., Ltd
Shanghai, China
Huyang1@opple.com , huyangsmail@163.com

Abstract

The smart wall switch without neutral wire is low cost solution to convert ordinary residential lighting into smart lighting. However, for the smart wall switch, the actual circuit is not totally off. There is a tiny electrical current flowing from Line through the switch, through the LED luminaire and to Neutral for keeping the controller work. Although this current is tiny enough, it cannot be compatible with some LED luminaire. This paper analyzes incompatible reason and condition for occurrence, and provide suggestion to avoid incompatibility issue.

Introduction

Over the years, as the number of buildings and the number of rooms within building increases dramatically, it was difficult to manage the waste of energy due to the inefficient light control and illumination distribution. Saving energy and increasing efficiency has become one of the most challenging issues in recent years because of environmental issues such as climate change and global warming [1]-[4].

In order to save energy, it is necessary and effective to convert ordinary residential lighting into smart lighting. Because of the lighting accounts for around approximately 20% of the world's total electricity consumption [5]. A smart lighting retrofits solution is use "smart LED luminaire" to replace the LED luminaire, the "smart LED luminaire" is based on wireless technology that make it communicate with smart phone and can be controlled by it[6]. The smart phone can turn on, off or dim the "smart LED luminaire" via wireless signal. The wireless technology such as Zigbee, WIFI, BLE and so on is suitable for retrofits, since no need to lay out control lines. However, it is not cheaper solution because of the cost of "smart LED luminaries" currently is more expensive than LED luminaries.

A low cost retrofits solution is use smart wall switch to replace traditional wall switch. It also is based on wireless technology that make it communicate with smart phone. The smart phone can control smart wall switch to turn on or off the LED luminaire via wireless signal, although the LED luminaire cannot be dimed. Since this solution do not need "smart LED luminaries", its cost is lower. A typical hardware block diagram of smart wall switch is shown in fig. 1. Nevertheless, the smart wall switch has a drawback for retrofits solution since most of existing traditional wall switches don't include a neutral wire as the fig. 2 shown. It is barrier to apply smart wall switch. The smart wall switch without Neutral Wire (SWSNW) is designed for solving this drawback. However, for the smart wall switch, the actual circuit is not totally off. There is a tiny electrical current flowing from Line through the switch, through the LED luminaire and to Neutral for keeping the controller work.

Although this current is tiny enough, it cannot be compatible with some LED luminaire.

This paper analyzes incompatible reason and condition for occurrence of SWSNW, and provide suggestion to avoid compatibility issue.

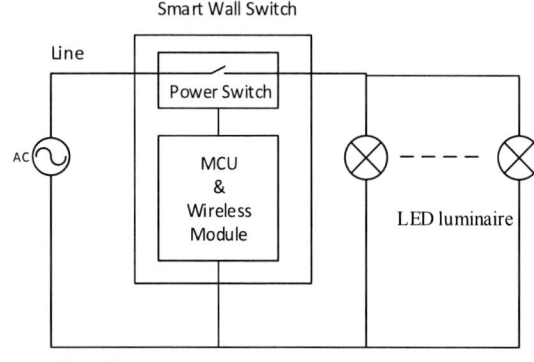

Figure 1, a typical hardware block diagram of smart wall switch.

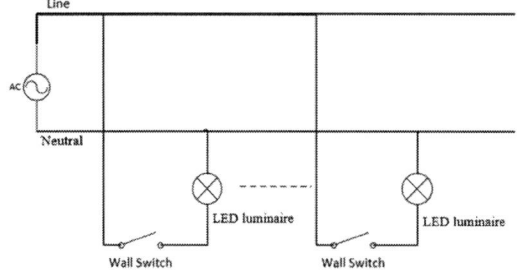

Figure 2, a typical traditional wall switch layout.

Operating Principle of Smart Wall Switch without Neutral Wire

The typical hardware block diagram of SWSNW is shown fig.3. It consists of a power switch unit, an on state power supply unit, an AC-DC converter unit, an energy storage unit and controller & wireless module unit. The SWSNW has only two states: turn off states and turn on states.

Turn off state: Once the controller & wireless module unit receive turn off signal from smart phone, it will turn of switch S1 of power switch and S2 of on state power supply unit. The AC-DC converter unit convert voltage across the SWSNW to a suitable voltage to charge the energy storage unit and supply power for the controller & wireless module unit. The fig. 4 is the diagram of current flow. In this state, the SWSNW and LED luminaire is equal to two resisters Rs and RL in series as fig. 5 shown. The voltage Vs and VL is across the resisters. They are given by:

$$Vs = VAC (Zs / (ZL+Zs))$$

VL = VAC (ZL/ (ZL+Zs))

Where VAC is line voltage, Zs is impedance of Rs and ZL is impedance of RL. The Zs should be much high than ZL, The VL will be lower enough not to turn on LED luminaire and the Vs is almost equal to Vac.

Turn on state: Once the controller & wireless module unit receive turn on signal from smart phone, it will turn on switch S2 of on state power supply. The current I1 will flow through the on state power supply unit to charge the energy storage unit and supply power for the controller & wireless module unit as fig. 6 shown. Once the voltage of energy storage unit rises to setting value, the controller & wireless module will turn on the S1 of power switch unit. The current stop flowing through the on state power supply unit since the impedance of SWSNW and voltage across the SWSNW is very low. The VAC will provide power to LED luminaire, the current I1 will flow through LED luminaire. The controller & wireless module unit will get power from the energy storage unit to keep working, the current I2 will flow through the energy storage unit as fig. 7 shown. The voltage of energy storage unit will be reduced as time goes by. Once the voltage of energy storage unit reduces to setting value, the controller & wireless module will turn on S1 to recharge the energy storage unit. Since the LED luminaire usually has energy storage component, it will continue working. In this state, the SWSNW chop the sine wave of VAC and share the power with LED luminaires. The operating waveform is shown in fig. 8.

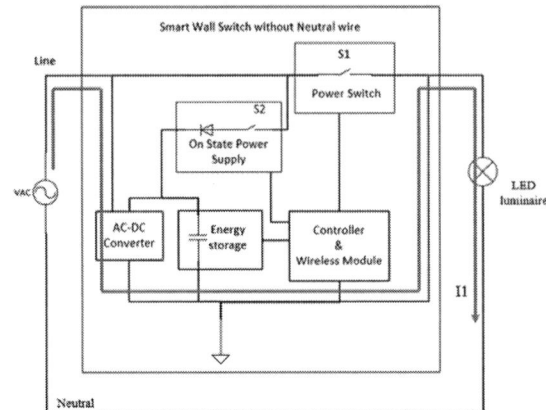

Figure 4, the diagram of current flow in turn off state.

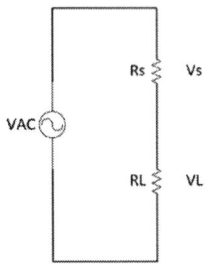

Figure 5, the equivalent circuit in turn off state.

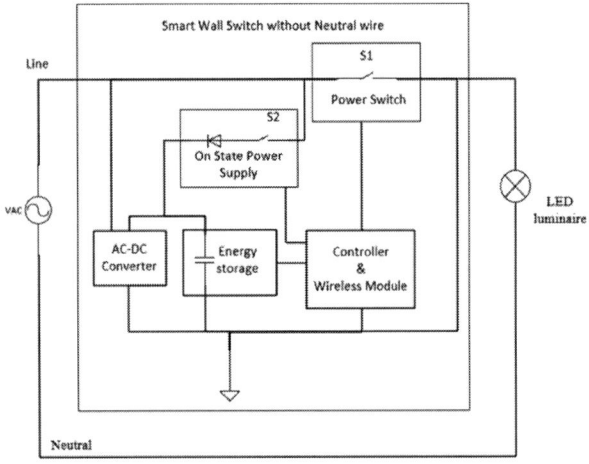

Figure 3, the typical hardware block diagram of smart wall switch without neutral wire.

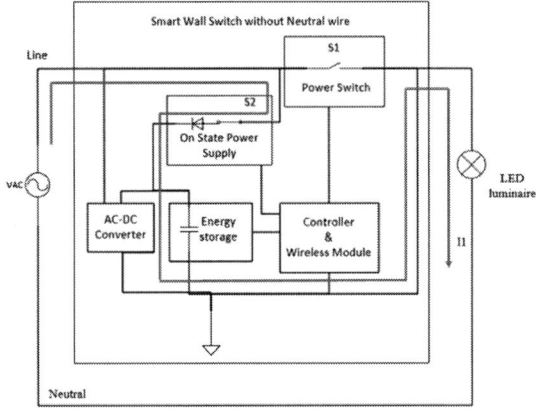

Figure 6, the diagram of current flow in turn on state.

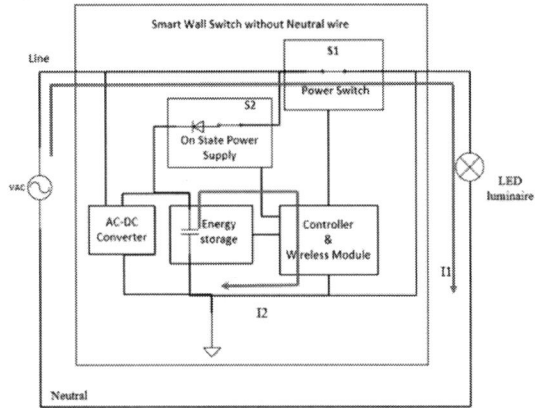

Figure 7, the diagram of current flow in turn on state.

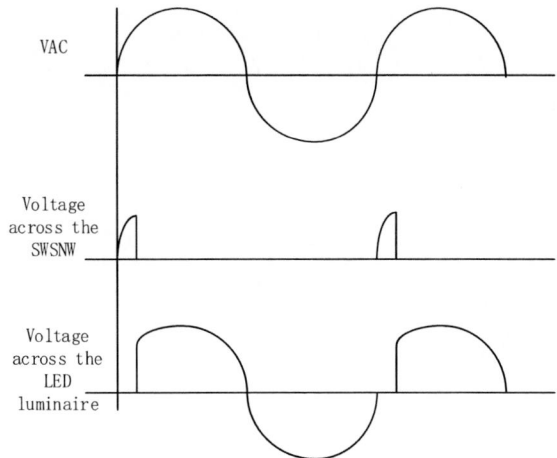

Figure 8, the operating waveform in turn on state.

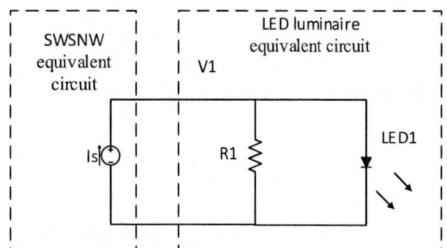

Figure 9, the equivalent circuit of turn off state of SWSNS.

Compatibility Analysis

As previous chapter mentioned, the actual circuit is not totally off for the SWSNW. There is a tiny electrical current flowing from Line through the switch as fig. 6 shown. As long as the Zs is much higher than ZL, the LEDs of luminaire will not be bright. However, the LEDs of luminaire will be bright or flicker on the contrary. This is main incompatibility issue of SWSNW. There are so many LED luminaires, lamps and bulbs on market to sale, so it is difficult to avoid appearing incompatibility issue.

Some manufactory of SWSNW provides minimum LED luminaire wattage suggestion to avoid incompatibility issue. However, the compatibility issue of SWSNS is not related to wattage of luminaire. The fig. 9 is the equivalent circuit of turn off state of SWSNS. In the turn off state, the SWSNW is equal to a current source Is, the value of Is is depended on the efficiency of AC-DC converter and controller & wireless module unit. The LED luminaire is equal to a resister R1 and a LED string LED1 in parallel. The R1 is equivalent impedance, the most of value of R1 is depended on quiescent current of LED driver IC. The condition of compatibility is given by:

V1 < Vled_on;

Where $V1 = Is \times R1$, Vled_on is minimum LED string on voltage. If the voltage of V1 is lower than voltage of Vled_on, the luminaire cannot be bright. Otherwise, the luminaire can be bright or flicker. The compatibility condition for occurrence is related to the value of Is, R1 and minimum LED string on voltage. It is not related to the wattage of LED luminaire power. If the LED string is connected in series, the compatibility condition for occurrence is related to the wattage of LED luminaire power. Because of the Vled_on is increased with wattage of LED luminaire power increasing. However, the LED string of luminaire can be connected in parallel or the quiescent current of LED driver of luminaire is very low. So the wattage of LED luminaire power is not completely related to compatibility, but the wattage of LED luminaire power is good for compatibility.

To verify conclusion as above, a SWSNW product which it is on sale in the market from S company (the real name of company is hided), and select 12 LED bulbs from four different manufacturers. The LED bulbs list is shown in table 1. The S company claims that this SWSNW product can be compatible with LED luminaire which total power is not low than five wattage. The test result is shown in table 2. As the test result shown, even if the power of LED bulb is higher than five wattage, there will be incompatibility.

Table 1.

No.	Luminaire Type	Manufactory	Rated power(W)
1	LED Bulb	Opple	3
2	LED Bulb		6
3	LED Bulb		8
4	LED Bulb	Panasonic	3.5
5	LED Bulb		6
6	LED Bulb		9
7	LED Bulb	Yaming	3
8	LED Bulb		5
9	LED Bulb		7
10	LED Bulb	Philips	3
11	LED Bulb		5
12	LED Bulb		6.5

Table 2

No.	Manufactory	Rated power (W)	Test Result		
			Line voltage 220V	Line voltage 253V	Line voltage 187V
1	Opple	3	ok	ok	ok
2		6	ok	ok	ok
3		8	ok	ok	ok
4	Panasonic	3.5	Fault2	Fault2	Fault2
5		6	ok	ok	ok
6		9	ok	ok	ok
7	Yaming	3	Fault1	ok	Fault1
8		5	Fault1	ok	Fault1
9		7	Fault1	ok	Fault1
10	Philips	3	ok	ok	ok
11		5	ok	ok	ok
12		6.5	ok	ok	ok

Fault1: bulb cannot be turned off;
Falut2: bulb is flicker;

Suggestion for Compatibility

Although the compatibility performance of SWSNW is depended on the value of Is, R1 and Vled_on, The Is is only can be improved by manufactory. Some suggestion of compatibility is shown in below:

1) The controller & wireless module unit of SWSNW is more efficient that is better to compatible with LED luminaire.
2) The AC-DC converter unit is more efficient that is better to compatible with LED luminaire.
3) The 220V line voltage is better than 110V line voltage for compatibility.

Conclusions

Although most of products use luminaire power to indicate compatibility of SWSNW, The compatibility of SWSNW is not related to luminaire power. The test result of SWSNW compatibility also proves this conclusion.

Acknowledgments

The authors gratefully acknowledge support from the research and innovation laboratory of Opple Lighting Co., Ltd.

References

1. S. Tompros, N. Mouratidis, M. Draaijer, A. Foglar, and H. Hrasnica, "Enabling applicability of energy saving applications on the appliances of the home environment," IEEE Network, vol. 23, no. 6, pp. 8-16, Nov.-Dec. 2009.

2. Tao Chen, Yang Yang, Honggang Zhang, Haesik Kim, and K. Horneman, "Network energy saving technologies for green wireless access networks," IEEE Wireless Communications, vol. 18, no. 5, pp. 30-38, Oct. 2011.

3. J. Byun and S. Park, "Development of a self-adapting intelligent system for building energy saving and context-aware smart services," IEEE Trans. on Consumer Electron., vol. 57, no. 1, pp. 90-98, Feb. 2011.

4. J. Han, C.-S. Choi, and I. Lee, "More efficient home energy management system based on ZigBee communication and infrared remote controls," IEEE Trans. on Consumer Electron., vol. 57, no. 1, pp. 85-89, Feb. 2011.

5. Çağdaş Atıcı, Member, IEEE, Tanır Özçelebi, Member, IEEE and Johan J. Lukkien, Member, IEEE, "Exploring user-centered intelligent road lighting design: a road map and future research directions," IEEE Trans. Consum. Electron., vol. 57, no. 2, pp. 788-793, May. 2011.

6. Tse-Hsu Wu, Che-Min Kung, Mei-Tan Wang, Ke-Fang Hsu, Shun-Yi Yang, Mei-Wen Chen, and Jung-Min Hwang, "Design and Implement an AC LED Light Engine with Smart Phone Control" 2015 International Conference on Consumer Electronics-Taiwan (ICCE-TW)

A Wavelength Stabilized GaN based Laser Utilizing Distributed Bragg Reflector

Mingle Liao [1,2], Wuze Xie [1,2], Zejia Deng [1,2] and Junze Li [1,2,*]

[1] Microsystems and Terahertz Research Center, China Academy of Engineering Physics, Chengdu, Sichuan 610200, China;

[2] Institute of Electronic Engineering, China Academy of Engineering Physics, Mianyang, Sichuan 621999, China;

* lijunze@mtrc.ac.cn

Abstract

GaN-based short wavelength laser diode is a promising lasing source for a variety of applications in the visible light spectrum. In this work, we report on a wavelength stabilized blue laser utilizing distributed bragg reflector (DBR) based on InGaN/GaN. The passive DBR is located on the rear side of the ridge waveguide, acting as a high-reflective reflector to the cavity. The uniform grating structure has a period of 1.55μm, duty ratio of 75% and etched depth of 500nm, and is defined by electron-beam lithography (EBL) and etched by Inductively Coupled Plasma (ICP). The electrical and optical characterizes of 19th order DBR laser emitting at a wavelength of 403nm are measured under the pulse drive condition to avoid thermal accumulation in the diode. A minimum emission linewidth of 0.45nm is observed, indicating the mode selection function is realized by the DBR. High wavelength stabilization with driving currents can be obtained due to the separation of active area and grating region in DBR laser. Experiment results also show that temperature stable emission with a wavelength shift of 0.013 nm/K is obtained within the DBR laser.

1. Introduction

Narrow-linewidth and temperature-stabilized lasers sources are key components in emerging laser technologies in visible spectrum region such as visible light communication (VLC) [1, 2], laser radar [3], and others. Although Fabry–Pérot (FP) lasers are attractive sources for VLC system [4-6], the emission spectrum is broad and a sufficiently stable lasing wavelength is not assured. Typically the spectral widths of FP lasers are bigger than 1 nm and the center wavelength is sensitive to temperature.

Distributed bragg reflector (DBR) laser based on GaN is a promising candidate for VLC system due to its incomparable advantages, such as the small size, excellent beam quality, long life time, and good stability. Compared with buried DBR structure, surface-etched DBR requires much simpler fabrication process, since it can simultaneously fabricated with the ridge waveguide with a single epitaxial growth step. Moreover, the DBR grating selects the longitudinal modes found in the F-P cavity, and is separated from the active region, which is capable of obtaining a more stable and narrower laser emission. In recent years, some studies have been carried out for the GaN-based DBR lasers. Cho et al. demonstrated the room temperature operation of an electrically injected InGaN/GaN-based DBR laser and reduced the threshold current density [7]. The influence of the surface-etched DBR lasers had been studied, for instance, the optimization of refractive index contrast [8, 9].

In this work, we fabricated the 19th order DBR laser utilizing electron-beam lithography (EBL) and etched by Inductively Coupled Plasma (ICP). The power-current-voltage (P-I-V) and spectral characteristics for the GaN based DBR lasers would be presented. Besides, lasing wavelength stability of the laser as a function of operating currents and temperature will be measured.

2. Device Simulation and Fabrication

A schematic view of the DBR laser on the GaN based epitaxial wafer is shown in Figure 1(a). The active gain region and the passive DBR region is spatially separated, however, both share the same epitaxial layer to simplify the fabrication process. The vertical laser structure underlying the simulation and used for fabrication of the laser is based on InGaN triple quantum well active layer, surrounded by GaN confinement layers and AlGaN cladding layers.

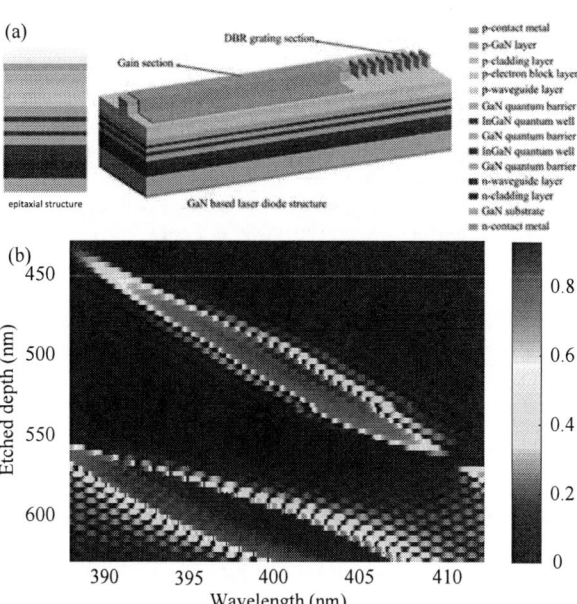

Figure 1. (a) Schematic cross-section of the GaN based epitaxial wafer. (b) simulation results of reflection spectra of varied etched depth.

The center wavelength λ_0 can be calculated by the Bragg condition, $\lambda_0 = 2n_{eff}\Lambda/m$, where n_{eff} is the longitudinally averaged effective refractive index of the grating, Λ is the grating period, and m is the order of the Bragg grating. For a given width of etched grooves o, the average effective refractive index n_{eff} is determined by the duty ratio $D = (\Lambda-o)/\Lambda$, and is greatly influenced by the etched depth. The etched grooves is covered by air and the grating pair number is fixed at 15, which is sufficient to provide high reflectivity with moderate computing resource. Figure 1(b) shows the dependence of reflectivity R on etched depth, where etched depth varied from 0.43 μm to 0.64 μm. To mimic the

978-1-7281-5757-3/19 $31.00 © 2019 IEEE

experimental condition, the grating is assumed to consist of rectangular shaped grooves. Generally speaking, the increase in etched depth leads to an increase in (averaged) reflective index, resulting in the red shift of center wavelength. Further increasing the depth will result in an increase in the relative overlap of the optical mode with the grating and thus increase reflectivity, despite the fact that the fabrication process difficulty will greatly increase. Maximum reflectivity of >80% can be obtained with appropriate etched depth around 500 nm. By comparison, the calculated reflectivity at the GaN-air interface is less than 20% due to the low refractive index contrast between the GaN-based materials and air, yielding suppressing the F-P mode by DBR. Finally, 19th order brag grating with a period of 1.55μm, duty ratio of 75% and etched depth of 500nm is designed. The large period grating is easier to achieve a high quality compare with small period grating, due to the structure fluctuation of large scale period grating can be easier confined on the grating manufacture procedure. What's more, the higher order grating has a narrower reflection bandwidth [10], which is favorable in reducing linewidth of the laser.

The sample was firstly processed into a 10um wide ridge structure including an 800um length gain section and distributed Bragg Reflector (DBR) section, with the ridge etch stopping at p-electron block layer by ion beam etching. Before the deposition of p-contact metal, a 200nm thick SiO_2 layer was deposited on the side of ridge in order to confine the injection section on the top of ridge and prevent additional optical loss from metal contact. Aiming to an emission wavelength at 400nm, a period of 1550nm was chosen for19th order Bragg grating contains 100 pairs gratings. This period grating is large enough, that it can be very precisely defined over the pre-etched DBR section by electron beam lithography, and followed by inductively coupled plasma(ICP) etch to realize the grating structure transferred into the epitaxial layer simultaneously. A 10/20/30 nm Pd/Ni/Au p ohm contact metal was deposited on the top of the ridge, and then, a 15/300 nm Ti/Au contact pad was deposited on the top of p ohm contact metal. The GaN substrate was thinned to a thick 100um by chemical mechanical polish method. A 50/50/200 nm Ti/Pt/Au n contact metal was deposited on the back of GaN substrate. Finally, the sample was cleaved to form the laser facet with a perpendicular direction to ridge structure.

Figure 2(a) and (b) show the scanning electron microscope (SEM) picture before and after the deposition of SiO_2 layer of DBR grating. The duty ratio looks smaller after the deposition of SiO_2 layer as shown in Figure 2. Obviously, the grating was precisely fabricated over the desired DBR section at the end of ridge waveguide of laser diode. The sidewall of DBR grating structure looks smooth at the glance of SEM. Previous works show that the influence of relative large notch angle of 70° could be ignored to offer high enough reflectivity for the gratings [11], which is the same in our case. The shape of fabricated DBR grating may have a little deviation compare with the designed structure, due to the slight change in dose of e-beam writing process would change the line/space width of grating. However, the influence of this deviation on the laser performance can be ignored.

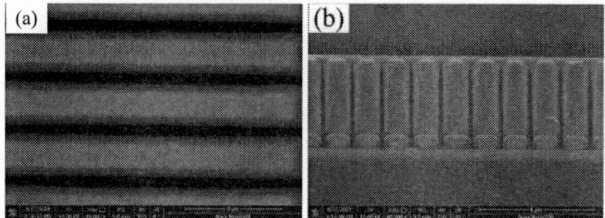

Figure 2. The SEM inagine of (a) top and (b) side view of the fabricated DBR grating

The power-current-voltage (P-I-V) measurement was characterized under pulse drive condition, 1us pulse width and 10 kHz repetition rate. The spectral characteristics was measured by an Ocean Optics spectrometer with a resolution 0.16nm under pulse drive condition, 500ns pulse width and 1 kHz repetition rate. The DBR laser is mounted on a temperature-controlled aluminum plate with temperature instability less than 0.1 °C.

3. Results

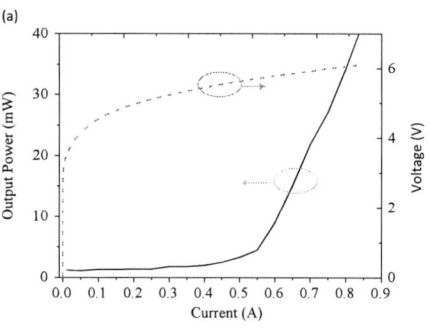

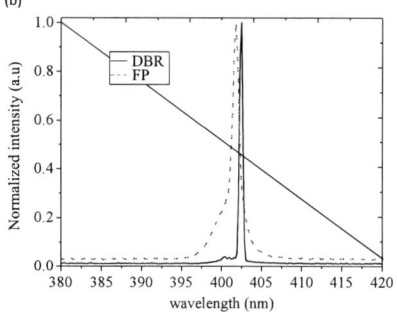

Figure 3. Eelectrical and optical characterizes of DBR laser. (a) the PIV curve; (b) lasing spectrum of DBR laser, the dot line shows the F-P spectrum for comparison.

Figure 3(a) shows the power-current-voltage (P-I-V) characteristics of DBR laser at room temperature. The current through ridge waveguide varied from 10 mA to 850mA with a step of 50mA. The measured threshold current (I_{th}) is ~500mA, the measured average slope efficiency is ~0.15W/A, and the series resistance is ~1.8 Ω. The relative high threshold current and low slop efficiency may be attributed to the power loss of the guided mode caused by the scattering and diffraction in the grating area. Besides, the imperfect quality of the epitaxial material can also degrade the optical performance. The emission spectrum of the DBR laser diodes under the injection current of $1.2I_{th}$ is shown in Figure 3(b). The emission wavelengths and the minimum full width at half maxima (FWHM) are 403nm and 0.45nm, respectively. The FWHM is evaluated by fitting in the least-squares sense a Lorentzian lineshape to the measured spectra. For comparison, the F-P laser fabricated on the same wafer has a FWHM of 1.29nm, which clearly declare that the fabricated DBR grating meets the design expectation and mode selection is realized by it.

shift is about 0.013 nm/K in the DBR laser, which is almost half lower than that of the conventional F-P laser fabricated on the same wafer. For the DBR laser, the wavelength shift with temperature is mainly determined by the temperature dependence of the effective refractive index of the optical mode. The calculated thermo-optic coefficient of the waveguide is approximate 0.0076/K. By comparison, the shift in emission wavelength for F-P laser is caused by the modal gain spectrum varied with temperature. It should be pointed out that, despite the fact that measurement is limited by the resolution of our spectrometer, the wavelength shift and thermo-optic coefficient is very close to the result of the similar work [12-14].

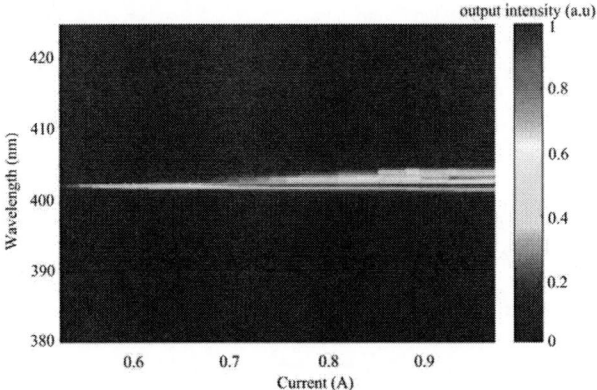

Figure 4. Mapping of the optical spectra versus driving current of the DBR laser, the output power is nomorized to the maximum output power of all.

Mapping of measured the optical spectra versus current is shown in Figure 4. The lasing wavelength keeps at 403 nm and shows no significant shift with current, revealing high wavelength stabilization characteristic of the DBR laser. Although efficiency is not high, the laser is operated in pulsed condition and thermal accumulation is minimized, so that no substantial heat flow to the DBR section occurs. However, an obvious spectral broadening with increasing current is observed. Typically, three peaks may be found at 0.95A. The spectral broadening may be attributed to several factors, including the stimulation of higher order lateral modes or temperature gradient along the DBR section. Several ways could be considered to overcome the problem. First, narrowing the ridge width to achieve single mode operation may reduce the generation of high order modes at high current. Second, a thermal isolation structure may be helpful to improve the heat diffusion to the DBR section in some degrees. Moreover, the quality improvement of the epitaxial material to maximize the efficiency may also be helpful to improve the situation.

The laser wavelength variation with temperature is shown in Figure 5. Typical emission spectra for temperatures of 15, 30 and 45 °C are shown in Figure 5(a). As can be seen, the spectra shapes keep almost unchanged at different temperature. Moderate mode hoping of the peak emission wavelength is observed. Assuming a linear dependence on the temperature range from 10 to 45°C, the observed wavelength

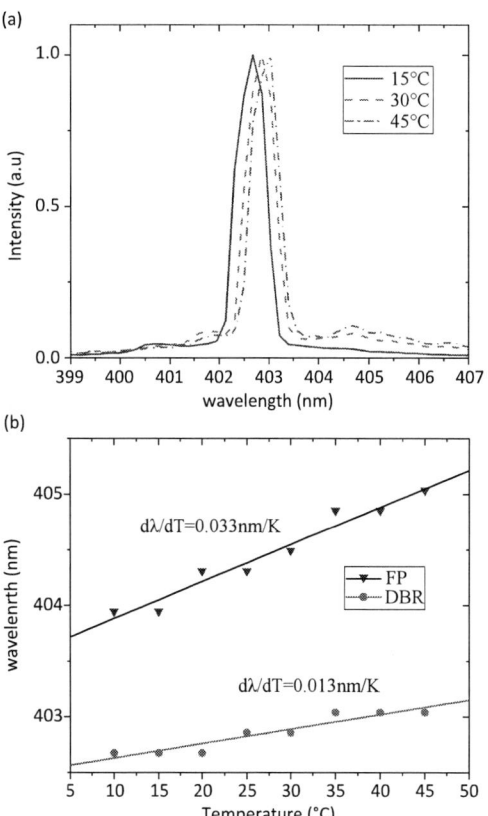

Figure 5. (a) normalized spectra at different temperature; (b) wavelength against temperature

4. Conclusions

In summary, we have demonstrated electrically pumped DBR laser based on GaN with surface etched grating fabricated by EBL and ICP. A minimum linewidth of 0.45nm is obtained at a wavelength of 403nm on the 19th order DBR laser diode. High wavelength stabilization with varied currents and temperatures is obtained utilizing the grating. Research on active tuning of bragg grating by carrier injection or thermal effect is on the way.

Acknowledgments

This work was funded by Science Challenge Project (No. TZ2016003-2), National Natural Science Foundation of China (No. 61804140), and National Key R&D Program of China (No. 2017YFB0403103).

References

1. Haas, H, "Visible light communication," *2015 Optical Fiber Communications Conference and Exhibition*, Los Angeles, CA, USA, Mar. 2015, pp. 1-72.
2. Pathak, P. H. *et al*, "Visible light communication, networking, and sensing: A survey, potential and challenges," *IEEE Communi. Surv. & Tut.* Vol. 17, No. 4 (2015), pp. 2047-2077.
3. Molebny, V. *et al*, "Laser radar: historical prospective—from the East to the West," *Opt. Eng.*, Vol. 56, No. 3 (2016), pp. 031220.
4. Retamal, J. R. D. *et al*, "4-Gbit/s visible light communication link based on 16-QAM OFDM transmission over remote phosphor-film converted white light by using blue laser diode," *Opt. Express*, Vol. 23, No. 26 (2015), pp. 33656-33666.
5. Wu, T. C. *et al*, "Blue laser diode enables underwater communication at 12.4 Gbps," *Sci. Rep.* 7, (2017), pp. 40480.
6. Zafar, F. *et al*, "Laser-diode-based visible light communication: Toward gigabit class communication," *IEEE Commun. Mag.*, Vol. 55, No. 2 (2017), pp. 144-151.
7. Cho, J. *et al*, "InGaN/GaN multi-quantum well distributed Bragg reflector laser diode," *Appl. Phys. Lett.* , Vol. 76, No. 12 (2000), pp. 1489–1491.
8. Ren, Q. *et al*, "Micro-zone optical measurements on GaN based nitride/air distributed Bragg reflector (DBR) mirrors made by focused ion beam milling," *Phys. Status. Solidi. (c)*, Vol. 1, No. 10 (2004), pp. 2450–2453.
9. Pengchong, L, *et al*, "Optimization design and preparation of near ultraviolet AlGaN/GaN distributed Bragg reflectors," *Superlattice. & Microst.* Vol. 122, (2018), pp. 661-666.
10. Fricke, J, *et al*, "Properties and fabrication of high-order bragg gratings for wavelength stabilization of diode lasers," *Semicond. Sci. & Tech.* Vol. 27, No. 5, (2012), pp. 055009.
11. Bao, S, *et al*, "The influence of grating shape formation fluctuation on DFB laser diode threshold condition," *Opt. Rev.* Vol. 25, No. 3 (2018), pp. 330-335.
12. Dumitru, V, *et al.* "InGaN/GaN multi-quantum well distributed Bragg reflector laser diode with second-order gratings," *Electron. Lett.* Vol. 39, No. 4 (2003), pp. 372-373.
13. Slight, T. J, *et al*, "InGaN/GaN laser diodes with high order notched gratings," *IEEE Photonics Technol. Lett.*, Vol. 29, No. 23 (2017), pp. 2020-2022.
14. Lutgen, S, *et al*, "Progress of blue and green InGaN laser diodes," *Proc. SPIE*, Vol. 7616 (2010), pp. 76160G.

Perovskite liquid quantum dots as a color converter for LD-based white lighting system for visible light communication

Shunming Liang[1], Zhou Lu[1], Xinrui Ding[1*], Jiexin Li[1], Yong Tang[1], Zongtao Li[1] Binhai Yu[1]

[1]*National & local joint engineering research center of semiconductor display and optical communication devices, South China University of Technology*
Guangzhou 510641, China.
*Corresponding Author: Xinrui Ding (e-mail: dingxr@scut.edu.cn).

Abstract

Perovskite material has been widely researched for illumination. Since its shorter PL lifetime (nanosecond), compared with the traditional illumination phosphor (microsecond), perovskite has been considered as a promising material for communication. In this work, by ultrasonication, we synthesized the CsPbBr3 perovskite quantum dots and found that liquid perovskite quantum dots (LPQDs) presents shorter luminescence lifetime of 24ns than solid state perovskite (SPQDs). We demonstrated a white visible light communication (VLC) system by using the LPQDs as a color converter. The LD-based white lighting VLC system can reach a 1Gbps data rate at a receivable BER. Our study indicates that LPQDs has great potential in communication performance.

Keywords— *VLC, Liquid PQDs, LD-based white lighting system*

Introduction

The advanced 5G wireless communication technologies has recently gained plenty of attention for its unprecedented information transfer rate. The existing wireless communications technologies such as Wi-Fi and Bluetooth, can't afford to satisfy the next generation indoor communication. Visible light communication (VLC) has been regarded as a potential resolution for the next generation communication technology[1]. VLC technology has attracted considerable attention for its plentiful merit such as unlimited license for frequency spectra, no electromagnetic interference, high data rate and security, and good infrastructure compatibility. In order to improve the performance of the VLC, a series of researches have been done including modulation scheme (such as OOK, OFDM), different light sources (such as LED, SLD, LD) and novel material (such as YAG, quantum dots).

Considering the short PL lifetime of the QDs, Xiao et al. used the Cd/Se QDs solution for VLC. However, their modulation performance was limited by LED light source which is much lower than LD[2]. Chi et al. demonstrated a YAG phosphorous LD white light system and indoor communication, their works indicates the low bandwidth of conventional YAG limits VLC system's communication performance[3].

Laser diode (LD) is an ideal device for communication for its high modulation bandwidth. For the shorter PL lifetime than traditional YAG phosphor, quantum dots (QDs) have higher bandwidth and better communication performance. Liu et al. using the RGB LDs for underwater lighting and VLC[4]. Zhou et al. proposed a convenient treatment for blue-emissive carbon dots (CDs) to produce green-emissive CDs[5]. Using this CDs as a color converter, they demonstrated a LD-based white light

system and considering the short PL lifetime of the novel CDs, they investigated communication performance of them and it showed an impressive bandwidth and data rate. Liu et al. also proposed a novel treatment for CDs for LD-based white light system, and gained a considerable communication performance[6]. Wang et al. proposed a white lighting system based on a violet LD through mixing green and red CdSe/ZnS core shell QDs, which supports the maximal transmission data rate of 9.6Gbps[7].

Plenty of new materials are applied in LD-based illumination system as a color converter for their short PL lifetime. Inorganic perovskite has a shorter lifetime than conventional YAG phosphor. Ibrahim et al. firstly used hot junction process perovskite as a color converter for VLC and proposed a perovskite-enhanced white light source with high CRI enabling dual VLC and lighting function[8]. However, perovskite's unstable characteristic, especially in solid state, prevents its application for LD-based VLC and illumination. Considering the unstable property of perovskite is mainly caused by thermal quenching, liquid perovskite quantum dots (LPQDs) is proposed as a color converter for its good thermal diffusivity. Li et al. proved that LPQDs have a great enhancement in lifetime for LD illumination[9].

The PL decay curve is showed in Fig 1. Time correlated single-photon counting (TCSPC) is used to measure the PL lifetime of both LPQDs (dispersed in toluene) and solid perovskite quantum dots (SPQDs, packaged in PMMA film). The PL decay curve was obtained with a testing condition of excitation at 405nm and collection at 515nm and could be well fitted by single exponential function. The PL lifetime of SPQDs is 34ns while LPQDs expresses a shorter PL lifetime of 24ns. The difference of PL lifetime between this two sample could be ascribed to the radiative recombination process of LPQDs and SPQDs, and it is still in the study, and even so, LPQDs shows promising character for VLC.

In order to take full advantage of LPQDs (short PL lifetime and weak thermal quenching), in this work, we proposed a liquid perovskite for both LD-based white light system and VLC system. Compared with solid perovskite quantum dots (SPQDs, packaged in PMMA film), LPQDs express better stability in illumination and communication. Moreover, LPQDs have a better performance in communication than SPQDs.

Experimental Methods

Synthesis of Materials. The PQDs is synthesis by ultrasonication with the following formula, which is modified from [10]: 0.11g lead bromide (PbBr2, Macklin, AR, 99.0%), 0.0326g caesium carbonate (Cs2CO3, Macklin, AR, 99%), 0.1ml Oleic acid (Macklin AR), 2ml Oleylamine (Macklin,

C18: 80-90%) and 10ml 1-Octadecene (Macklin, >90%(GC)). The mixed solution then is treated by 5min ultrasonication (ShangHaiShengXi, FS-300N), and after that the mixed solution is processed an ice-bath. First centrifugation (10000rpm, 10mins) is to take out redundant Oleic acid and Oleyamine. Second centrifugation (2500rpm, 5mins) is to deposit large grain size PQDs.The PL decay curve is obtained by a HAMAMATSU fluorescence lifetime measuring instrument (Quantaurus-Tau C11367-11).

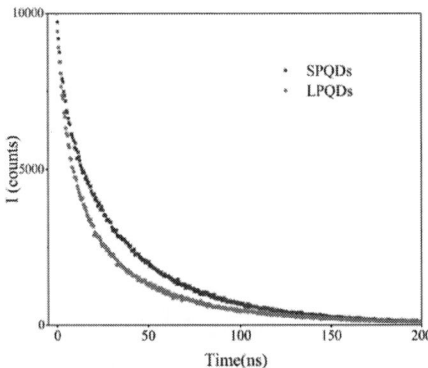

Fig 1 PL decay of LPQDs and SPQDs.

Illumination performance. The optical spectrum is obtained by an Ocean Optics spectrometer (USB+2000). The LPQDs is excited by a 450nm LD from Laserwave (LWBL450-10W-F BL18121434). Absorption spectrum and emission spectrum is measured by fluorophotometer (SHIMADZU, RF-6000) and spectrophotometer (PERSEE, TU-1901) respectively.

Communication performance. The transmitter of the VLC including a 450nm laser diode (Thorlab, LP450-SF15) which is mounted in a power source (Thorlab, CLD1010LP), a bit error rate tester (BERT, Agilent N4903A) and a vector signal generator (VSG, KEYSIGHT MXG N5182B). The receiver of VLC consists of an avalanche photodetector (APD, Meno System APD210), a digital communications analyzer (DCA, Agilent DCA-J 86100C) and a digital phosphor oscilloscope (DPO, Tektronix TDS7104).

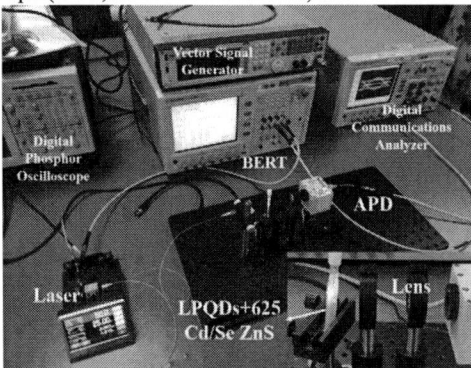

Fig 2. LD-based VLC system.

The VLC system is showed in Fig 2. The bandwidth measurement is conducted by VSG and DPO, and the BER and eye diagram is measured by BERT and DCA.

Results and discussions

The absorption spectrum and emission spectrum of LPQDs is showed in Fig 3. The absorption peak is at 500nm and emission peak is at 525nm, which is the same as SPQDs in the same concentration.

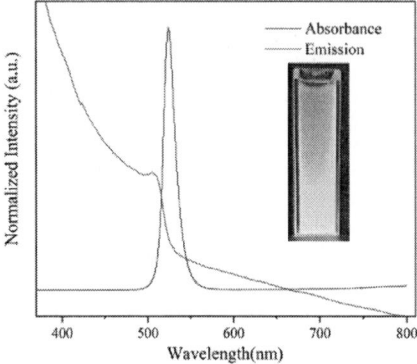

Fig 3 Absorption spectrum and emission spectrum of LPQDs (Inset: photographs of the LPQDs under UV light).

Considering the short PL lifetime of Cd/Se ZnS QDs and in order to investigate both illumination and communication performance of LPQDs, a LD-based white lighting system was proposed, using LPQDs and red Cd/Se ZnS QDs mixed solution as color converter. The emission spectrum of the proposed white lighting system is showed in Fig 4. The proposed white light has a CIE coordinate of (0.43,0.40) and correlated color temperature (CCT) of 3017K.

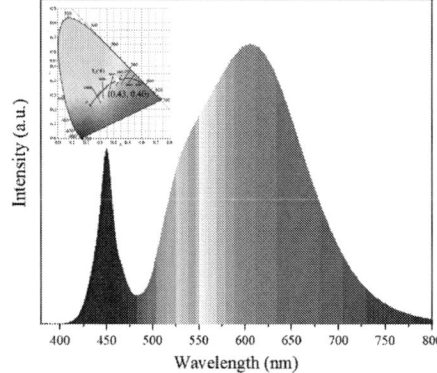

Fig 4. Emission spectrum of LD-based white lighting system (Inset: CIE coordinate).

The bandwidth measurement of the proposed LD-based white lighting system is showed in Fig 5. Bandwidth of LPQDs is up to 39.8MHz, which is almost 30% higher than bandwidth of SPQDs of 31.4MHz. The improvement of LPQDs can be attributed to the faster recombination process of liquid state PQDs which is showed in Fig 1. In addition, the quenching is less serious in LPQDs than SPQDs, which help stabilize the signal transmission process. Using the proposed LD-based white lighting system, a VLC system is demonstrated and a high data rate of 1Gbps can be transmitted with BER of 1.3×10^{-3}, which is below the FEC threshold. The eye diagram is showed in the inset of Fig 5. The open eye diagram is showed in inset of Fig 5, indicating that the proposed white lighting system has a remarkable performance in communication application.

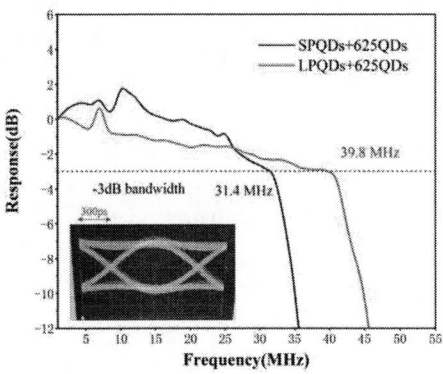

Fig 5. Modulation bandwidth measurements (Inset: eye diagram under 1Gbps data rate).

Conclusions

In this work, we investigated the PL decay of the LPQDs and SPQDs, and LPQDs shows a potential performance in communication. Employing LPQDs as a color converter, LD-based white lighting system is proposed and the illumination performance of the proposed white lighting system were investigated, which has a CIE of (0.43, 0.40), and CCT of 3017K. For the first time, a VLC system is demonstrated based on the proposed white lighting system, using LPQDs as a color converter, to evaluate its modulation and communication performance. The bandwidth of LPQDs is 30% higher than SPQDs and the proposed white lighting system can achieve a 1Gbps data rate. The data rate of our system is limited by the small power of the LD, and higher speed transmission can be achieved with high power laser diode. Our works indicated that LPQDs has a promising potential in communication application.

Acknowledgments

This work was supported in part by the National Natural Science Foundation of China (51805173), the Science and Technology Program of Guangzhou (20190401025), Natural Science Foundation of Guangdong Province (2019A1515011741), the Fundamental Research Funds for the Central Universities(2018MS44)

References

[1] M. B. Rahaim and T. D. C. Little, "TOWARD PRACTICAL INTEGRATION OF DUAL-USE VLC WITHIN 5G NETWORKS," *Ieee Wireless Communications,* vol. 22, no. 4, pp. 97-103, Aug 2015.

[2] X. Xiao *et al.*, "Improving the modulation bandwidth of LED by CdSe/ZnS quantum dots for visible light communication," *Optics Express,* vol. 24, no. 19, pp. 21577-21586, Sep 19 2016.

[3] Y.-C. Chi *et al.*, "Phosphorous Diffuser Diverged Blue Laser Diode for Indoor Lighting and Communication," *Scientific Reports,* vol. 5, Dec 21 2015, Art. no. 18690.

[4] X. Liu *et al.*, "Laser-based white-light source for high-speed underwater wireless optical communication and high-efficiency underwater solid-state lighting," *Optics Express,* vol. 26, no. 15, pp. 19259-19274, Jul 23 2018.

[5] Z. Zhou *et al.*, "Hydrogen Peroxide-Treated Carbon Dot Phosphor with a Bathochromic-Shifted, Aggregation-Enhanced Emission for Light-Emitting Devices and Visible Light Communication," *Advanced Science,* vol. 5, no. 8, Aug 2018, Art. no. 1800369.

[6] E. Liu *et al.*, "Highly Emissive Carbon Dots in Solid State and Their Applications in Light-Emitting Devices and Visible Light Communication," *Acs Sustainable Chemistry & Engineering,* vol. 7, no. 10, pp. 9301-9308, May 20 2019.

[7] W.-C. W. H.-Y. W. T.-Y. C. C.-T. T. Chih-Hsien, "CdSe/ZnS core-shell quantum dot assisted color conversion of violet laser diode for white lighting communication," *Nanophotonics,* vol. 10, p. 1515, 2019.

[8] I. Dursun *et al.*, "Perovskite Nanocrystals as a Color Converter for Visible Light Communication," *Acs Photonics,* vol. 3, no. 7, pp. 1150-1156, Jul 2016.

[9] Z. L. X. T. T. J. Y. Y. T. B. Y. Y. H. B. L. X. Ding, "Lifetime Enhancement of a Circulated Cooling Perovskite Quantum Dots Colloidal Solution System for Laser Illuminations," *IEEE Access,* vol. 7, pp. 136214-136222, 05 September 2019.

[10] Y. Tong *et al.*, "Highly Luminescent Cesium Lead Halide Perovskite Nanocrystals with Tunable Composition and Thickness by Ultrasonication," *Angewandte Chemie-International Edition,* vol. 55, no. 44, pp. 13887-13892, Oct 2016.

IEEE
445 Hoes Lane
Piscataway, NJ 08854-4141

ISBN 978-1-7281-5757-3